Algorithm Design Practice for Collegiate Programming Contests and Education

Algorithm Design Practice for Collegiate Programming Contests and Education

by Yonghui Wu
and Jiande Wang

CRC Press
Taylor & Francis Group
Boca Raton London New York

CRC Press is an imprint of the
Taylor & Francis Group, an **informa** business

CRC Press
Taylor & Francis Group
6000 Broken Sound Parkway NW, Suite 300
Boca Raton, FL 33487-2742

© 2019 by Taylor & Francis Group, LLC

CRC Press is an imprint of Taylor & Francis Group, an Informa business

No claim to original U.S. Government works

Printed on acid-free paper

International Standard Book Number-13: 978-1-4987-7663-9 (Hardback)

Library of Congress Cataloging-in-Publication Data
A catalog record has been requested for this book

Visit the Taylor & Francis Web site at
http://www.taylorandfrancis.com

and the CRC Press Web site at
http://www.crcpress.com

Contents

Preface

Programming contests are contests solving problems by programming. Starting in the 1990s, the ACM International Collegiate Programming Contest (ACM-ICPC) has become a worldwide programming contest. Every year, 6 continents, over 110 countries, 50,000 students, 5,000 coaches, and 3,000 universities participate in ACM-ICPC local contests, preliminary contests, and regional contests all over the world. Alongside, some international programming contests, such as Google Code Jam, TopCoder Open Algorithm, Facebook Hacker Cup, Internet Problem Solving Contest (IPSC), and so on, are held every year. Programmers from all over the world, in addition to students, can participate in these contests through the Internet.

Based on these programming contests, programming contests' problems from all over the world can be obtained, analyzed, and solved by students. These contest problems can be used not only for programming contest training, but also for education.

In our opinion, not only programming contestants' ability to solve problems, but also computer students' programming skills are based on their programming knowledge system and programming strategies for solving problems. The programming knowledge system can be summarized as: "Algorithms + Data Structures = Programs." It is also the foundation for the knowledge system of computer science and engineering. Strategies for solving problems are strategies for data modeling and algorithm design. When data models and algorithms for problems are not standard, we need to take some strategies to solve these problems.

Based on these facts, we published a series of books, not only for systematic programming contest training, but also for polishing computer students' programming skill better, using programming contests problems: *Data Structure Experiment for Collegiate Programming Contest and Education, Algorithm Design Experiment for Collegiate Programming Contest and Education,* and *Programming Strategies Solving Problems* in Mainland China. And the traditional Chinese version for *Data Structure Experiment for Collegiate Programming Contest and Education* and *Programming Strategies Solving Problems* were also published in Taiwan. In 2016, the first book's English version *Data Structure Practice: for Collegiate Programming Contest and Education* was published by CRC Press.

Algorithm Design Practice for Collegiate Programming Contest and Education is the English version for *Algorithm Design Experiment for Collegiate Programming*

Contest and Education. There are 9 chapters and 247 programming contest problems in this book.

Chapter 1, "Practice for Ad Hoc Problems", focuses on solving problems that there are no classical algorithms to solve. There are two methods to solve such problems: the mechanism analysis method and the statistical analysis method. In Chapter 2, "Practice for Simulation Problems", experiments and practices for simulation problems are shown. In problem descriptions, solution procedures or rules are shown. Simulation problems are solved by implementing rules or simulating solution procedures. Chapter 3, "Practice for Number Theory", Chapter 4, "Practice for Combinatorics", and Chapter 8, "Practice for Computational Geometry", introduce the mathematical background for number theory, combinatorics, and computational geometry, respectively, and then show problems solved by mathematical methods. Greedy algorithms and dynamic programming are used to solve optimization problems. Chapter 5, "Practice for Greedy Algorithms", and Chapter 6, "Practice for Dynamic Programming", introduce greedy algorithms and dynamic programming respectively, and show problems solved by greedy algorithms and dynamic programming. Chapter 7, "Practice for Advanced Data Structures", describes using suffix arrays, segment trees, and some graph algorithms to solve problems. Search technologies are fundamental to computer science and technology. Chapter 9, "Practice for State Space Search", describes the implementation of state space search through solving contest problems.

The features of the book are as follows:

1. The book's outlines are based on the outlines of algorithms. Programming contest problems and their analyses and solutions are used as experiments. For each chapter, there is a "Problems" section to let students solve programming contests' problems, and hints for these problems are also included.
2. Problems in the book are all selected from the ACM-ICPC regional and world finals programming contests, universities' local contests, and online contests, from 1990 to now.
3. Not only analyses and solutions, or hints to problems are shown, but also test data for most of the problems are provided. Sources and IDs for Online Judge for these problems are also provided. This can help readers polish their programming skills better and more easily. In addition, there are problems and test data available for download at https://www.crcpress.com/9781498776639.

The book can be used not only as an experiment book, but also for training for systematic programming contests.

We appreciate Professors Steven Skiena and Rezaul Chowdhury, from Stony Brook University; C. Jinshong Hwang, Ziliang Zong, and Hongchi Shi, from Texas State University; Normaziah Abdul Aziz, from International Islamic University Malaysia; Abul L. Haque, from North South University; Jiannong Cao, from

The Hong Kong Polytechnic University; and Rudolf Fleischer, from German University of Technology in Oman. They provided us platforms in which English is the native language that improved our manuscript. We also appreciate Miss Jiaqi Chen, an undergraduate student from the Georgia Institute of Technology, who reviewed and used several chapters in the manuscript, and pointed out some errors.

Online Judge systems for problems in this book are as follows:

Online Judge Systems	Abbreviations	Web Sites
Peking University Online Judge System	POJ	http://poj.org/
Zhejiang University Online Judge System	ZOJ	http://acm.zju.edu.cn/onlinejudge/
UVA Online Judge System	UVA	http://uva.onlinejudge.org/ http://livearchive.onlinejudge.org/
Ural Online Judge System	Ural	http://acm.timus.ru/
SGU Online Judge System	SGU	http://acm.sgu.ru/

If you discover anything you believe to be an error, please contact us through Yonghui Wu's email id: yhwu@fudan.edu.cn. Your help is appreciated.

Yonghui Wu, Jiande Wang
June, 2018

Author Biographical Information

Yonghui Wu, Ph.D., Associate Professor, Fudan University. He acted as the coach of Fudan University Programming Contest teams from 2001 to 2011. Under his guidance, Fudan University qualified for ACM-ICPC World Finals every year and won three medals (bronze medal in 2002, silver medal in 2005, and bronze medal in 2010) in ACM-ICPC World Finals. Since 2012, he has published a series of books for programming contest and education in simplified and traditional Chinese and English. Since 2013, he has given lectures in Oman, Taiwan, HongKong, Macau, Malaysia, Bangladesh, Mainland China, and the United States for programming contest training. He is the chair of ACM-ICPC Asia Programming Contest Training Committee now.

Jiande Wang, High School Senior Teacher. He is a famous coach for Olympiad in Informatics in China. He has published 24 books for programming contests since 1990s. Under his guidance, his students won seven gold medals, three silver medals, and two bronze medals in International Olympiad in Informatics for China.

Chapter 1

Practice for Ad
Hoc Problems

Ad hoc means "for the special purpose or end presently under consideration." There are no classical algorithms that can solve these ad hoc problems. Programmers need to design specific algorithms to solve ad hoc problems. There are two strategies to design algorithms for solving ad hoc problems: mechanism analysis and statistical analysis.

To solve an ad hoc problem, we need to see past its appearance and understand its essence.

In this chapter, two kinds of analyses solving ad hoc problems are shown:

- Mechanism Analysis;
- Statistical Analysis.

1.1 Solving Problems by Mechanism Analysis

Mechanism analysis examines the characteristics and internal mechanisms of an object to find a mathematical representation of the problem. Therefore, the key to mechanism analysis is mathematical modeling. Solving problems by mechanism analysis is a top-down method.

1.1.1 Factstone Benchmark

Amtel has announced that it will release a 128-bit computer chip by 2010, a 256-bit computer by 2020, and so on, continuing its strategy of doubling the word size every ten years. (Amtel released a 64-bit computer in 2000, a 32-bit computer in 1990, a 16-bit computer in 1980, an 8-bit computer in 1970, and a 4-bit computer, its first, in 1960.)

Amtel will use a new benchmark—the *Factstone*—to advertise the vastly improved capacity of its new chips. The *Factstone* rating is defined to be the largest integer n such that $n!$ can be represented as an unsigned integer in a computer word.

Given a year $1960 \leq y \leq 2160$, what will be the *Factstone* rating of Amtel's most recently released chip?

Input

There are several test cases. For each test case, there is one line of input containing y. A line containing 0 follows the last test case.

Output

For each test case, output a line giving the Factstone rating.

Sample Input	Sample Output
1960	3
1981	8
0	

Source: Waterloo local 2005.09.24

IDs for Online Judges: POJ 2661, UVA 10916

 Analysis

For a given year, first the number of bits for the computer in this year is calculated, and then the largest integer n (the *Factstone* rating) that $n!$ can be represented as an unsigned integer in a computer word is calculated.

The computer was a 4-bit computer in 1960. Amtel doubles the word size every ten years. That is, the number of bits for the computer in year Y is $K = 2^{2 + \left\lfloor \frac{Y - 1960}{10} \right\rfloor}$. The largest unsigned integer for K-bit is $2^K - 1$. If $n!$ is the largest unsigned integer not greater than $2^K - 1$, then n is the Factstone rating in year Y. There are two calculation methods.

Method 1: Calculate $n!$ directly. This method is slow and easily leads to overflow.
Method 2: Logarithms are used to calculate $n!$. Based on the following formula:

$$\log_2 n! = \log_2 n + \log_2 (n-1) + \ldots\ldots + \log_2 1 \leq \log_2 (2^K - 1) < K,$$

n can be calculated. Initially i is 1, repeat i++, and $\log_2 i$ is accumulated until the sum is larger than K. Then $i-1$ is the *Factstone* rating.

 Program

```
#include <stdio.h>
#include <math.h>
int y,Y,i,j,m;      // Year y
double f,w;     // f: the sum of accumulation for log₂ i
main(){
    while (1 == scanf("%d",&y) && y){      //Input test cases
        w = log(4);
        for (Y=1960; Y<=y; Y+=10){
            w *= 2;
        }
        i = 1;      //accumulation log₂ i until larger than w
        f = 0;
        while (f < w) {
            f += log((double)++i);
        }
        printf("%d\n",i-1);      //Output the Factstone rating
    }
    if (y) printf("fishy ending %d\n",y);
}
```

1.1.2 Bridge

Consider that n people wish to cross a bridge at night. A group of at most two people may cross at any time, and each group must have a flashlight. Only one flashlight is available among the n people, so some sort of shuttle arrangement must be arranged in order to return the flashlight so that more people may cross.

Each person has a different crossing speed; the speed of a group is determined by the speed of the slower member. Your job is to determine a strategy that gets all n people across the bridge in the minimum time.

Input

The first line of input contains n, followed by n lines giving the crossing times for each of the people. There are not more than 1000 people, and nobody takes more than 100 seconds to cross the bridge.

Output

The first line of output must contain the total number of seconds required for all *n* people to cross the bridge. The following lines give a strategy for achieving this time. Each line contains either one or two integers, indicating which person or people form the next group to cross. (Each person is indicated by the crossing time specified in the input. Although many people may have the same crossing time, the ambiguity is of no consequence.) Note that the crossings alternate directions, as it is necessary to return the flashlight so that more may cross. If more than one strategy yields the minimal time, any one will do.

Sample Input	Sample Output
4	17
1	1 2
2	1
5	5 10
10	2
	1 2

Source: POJ 2573, ZOJ 1877, UVA 10037

IDs for Online Judge: Waterloo local 2000.09.30

 Analysis

The strategy that gets all *n* people across the bridge in the minimum time is: fast people should return the flashlight to help slow people.

Because a group of at most two people may cross the bridge each time, we solve the problem by analyzing members of groups. First, *n* people's crossing times are sorted in descending order. Suppose that in the current sequence, *A* is the current fastest person's crossing time, *B* is the current second fastest person's crossing time, *a* is the current slowest person's crossing time, and *b* is the current second slowest person's crossing time.

There are two methods for making the current slowest person and the current second slowest person to cross the bridge:

Method 1: The fastest person helps the slowest person and the second slowest person to cross the bridge. The steps are as follows:
   Step 1: The fastest person and the slowest person cross the bridge;
   Step 2: The fastest person is back;
   Step 3: The fastest person and the second slowest person cross the bridge;
   Step 4: The fastest person is back.
It takes time $2 \times A + a + b$.

Method 2: The fastest person and the second fastest person help the current slowest person and the current second slowest person to cross the bridge.

Step 1: The fastest person and the second fastest person cross the bridge;

Step 2: The fastest person is back and returns the flashlight to the slowest person and the second slowest person;

Step 3: The slowest person and the second slowest person cross the bridge and give the flashlight to the second fastest person;

Step 4: The second fastest person is back.

It takes time $2 \times B + A + a$.

Each time, we need to compare method 1 and method 2. If ($2 \times A + a + b < 2 \times B + A + a$), then we use method 1, else we use method 2. And each time the current slowest person and the current second slowest person cross the bridge. Finally, there are two cases:

Case 1: If there are only two persons who need to cross the bridge, then the two persons cross the bridge. It takes time B.

Case 2: There are three persons who need to cross the bridge. First, the fastest person and the slowest person cross the bridge. Then, the fastest person is back. Finally, the last two persons cross the bridge. It takes time $a + A + b$.

 Program

```
#include<iostream>
#include<algorithm>
#include<cstdio>
#include<cstring>
#include<cstdlib>
#include<cmath>
#include<string>
using  namespace std;
int n,i,j,k,a[111111];      //n: the number of persons, a[ ]: n
people's crossing times
int ans=0;      // ans: the total number of seconds for all n
people to cross the bridge
int main () {
    scanf("%d",&n);      //Input
    for(i=1;i<=n;i++)scanf("%d",a+i);
    if(n==1){      //only 1 person
        printf("%d\n%d\n",a[1],a[1]);return 0;
    }
```

```
    int nn=n;
    sort(a+1,a+n+1);    //n people's crossing times are sorted
in descending order
    while(n>3){    //calculate the total number of seconds for
all n people to cross the bridge
      if(a[1]+a[n-1]<2*a[2]){    //Method 1
          ans+=a[n]+a[1]*2+a[n-1];
      }else{    //Method 2
            ans+=a[2]+a[1]+a[2]+a[n];
        }
        n-=2;    //the two slowest persons cross the bridge
    }
    if(n==2)ans+=a[2];    //only two persons need to cross the
bridge
    else  ans+=a[1]+a[2]+a[3];    //three persons need to
cross the bridge
    printf("%d\n",ans);    //the total number of seconds for
all n people to cross the bridge
    n=nn;
    while(n>3){    //output the strategy for achieving this
time
        if(a[1]+a[n-1]<2*a[2])    //Method 1
            printf("%d%d\n%d\n%d%d\n%d\n",a[1],a[n],a[1],
a[1],a[n-1],a[1]);
        else    //Method 2
            printf("%d%d\n%d\n%d%d\n%d\n",a[1],a[2],a[1],
a[n-1],a[n],a[2]);
        n-=2;    //the two slowest persons cross the bridge
    }
    if(n==2)printf("%d %d\n",a[1],a[2]);    //only two persons
need to cross the bridge
    else    //three persons need to cross the bridge
        printf("%d %d\n%d\n%d %d\n",a[1],a[3],a[1],a[1],a[2]);
    return 0;
}
```

1.2 Solving Problems by Statistical Analysis

Unlike mechanism analysis, statistical analysis begins with a partial solution to the problem, and the overall global solution is found based on analyzing the partial solution. Solving problems by statistical analysis is a bottom-up method.

1.2.1 Ants

An army of ants walk on a horizontal pole of length l cm, each with a constant speed of 1 cm/s. When a walking ant reaches an end of the pole, it immediately falls off it. When two ants meet, they turn back and start walking in opposite directions.

We know the original positions of ants on the pole; unfortunately, we do not know the directions in which the ants are walking. Your task is to compute the earliest and the latest possible times needed for all ants to fall off the pole.

Input

The first line of input contains one integer giving the number of cases that follow. The data for each case start with two integer numbers: the length of the pole (in cm) and n, the number of ants residing on the pole. These two numbers are followed by n integers giving the position of each ant on the pole as the distance measured from the left end of the pole, in no particular order. All input integers are not bigger than 1000000, and they are separated by whitespace.

Output

For each case of input, output two numbers separated by a single space. The first number is the earliest possible time when all ants fall off the pole (if the directions of their walks are chosen appropriately), and the second number is the latest possible such time.

Sample Input	Sample Output
2	4 8
10 3	38 207
2 6 7	
214 7	
11 12 7 13 176 23 191	

Source: Waterloo local 2004.09.19

IDs for Online judges: POJ 1852, ZOJ 2376, UVA 10714

 Analysis

The upper limit of the number of ants is 1000000. The upper limit of the number of combinations for ants' walking is $2^{1000000}$. Therefore, the problem can't be solved by enumerating ants walking.

First, we analyze the case that a few ants walk on a horizontal pole (Figure 1.1).

In Figure 1.1, when two ants meet, that is, " 🐜🐜 ", they'll turn back and start walking in opposite directions, that is, " 🐜🐜 ". All ants are the same. Therefore, all ants walk in their original directions no matter whether they meet or not. There are two values for the time that an ant falls off the pole: the ant walks to the left, or the ant walks to the right.

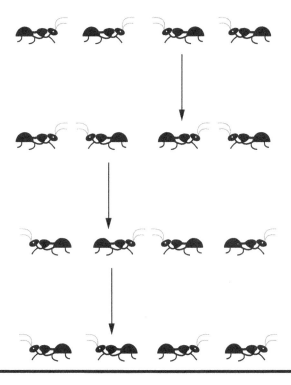

Figure 1.1

Suppose l_i is the position of ant i on the pole, that is, the distance measured from the left end of the pole, $1 \le i \le n$; *little* is the earliest possible time when all ants fall off the pole; and *big* is the latest possible time when all ants fall off the pole. Based on these facts, the algorithm is as follows:

$$little = \min_{1 \le i \le n}\{l_i, L - l_i\}, \ big = \max_{1 \le i \le n}\{l_i, L - l_i\}.$$

 Program

```
#include <stdio.h>
int c,big,little,L,i,j,k,n;    //c: number of test cases; L:
the length of the pole; n: number of ants on the pole
main(){
    scanf("%d",&c);    // input the number of test cases
    while (c-- && (2 == scanf("%d%d",&L,&n))) {    //Input the
length of the pole and the number of ants on the pole
```

```
        big = little = 0;    //Initialization
        for (i=0;i<n;i++) {    //Input original positions for
all ants and adjust times
            scanf("%d",&k);
            if (k > big) big = k;    //adjust the earliest
possible time
            if (L-k > big) big = L-k;
            if (k > L-k) k = L-k;    //adjust the latest possible
time
            if (k > little) little = k;
        }
        printf("%d %d\n",little,big);    //Output the result
    }
    if (c != -1) printf("missing input\n");
}
```

1.2.2 Matches Game

Here is a simple game. In this game, there are several piles of matches and two players. The two players play in turn. In each turn, one can choose a pile and take away an arbitrary number of matches from the pile (of course, the number of matches, which is taken away, cannot be zero and cannot be larger than the number of matches in the chosen pile). If, after a player's turn, there is no match left, the player is the winner. Suppose that the two players are all very clear. Your job is to tell whether the player who plays first can win the game or not.

Input

The input consists of several lines, and in each line there is a test case. At the beginning of a line, there is an integer M ($1 \leq M \leq 20$), which is the number of piles. Then come M positive integers, which are not larger than 10000000. These M integers represent the number of matches in each pile.

Output

For each test case, output "Yes" in a single line, if the player who play first will win; otherwise output "No."

Sample Input	Sample Output
2 45 45	No
3 3 6 9	Yes

Source: POJ Monthly, readchild

ID for Online Judge: POJ 2234

Analysis

The problem is a Nimm's Game problem. Cases for the game are analyzed as follows:

Case 1: There is only one pile of matches. The player who plays first will take away all matches from the pile and win the game.

Case 2: There are two piles of matches. Numbers of matches in the two piles are N_1 and N_2 respectively.

If $N_1 \neq N_2$, the player who plays first will take away some matches from the larger pile to make the two piles have the same number of matches. Then, by mimicking the player who plays second and taking the same number of matches that he takes, just from the opposite pile, the player who plays first will win the game.

If $N_1 = N_2$, the player who plays second will take the same number of matches as the player who plays first takes, just from the opposite pile, and then the player who plays second will win the game.

Case 3: There are more than two piles of matches.

Each natural number can be represented as a binary number. For example, $57_{(10)} = 111001_{(2)}$, that is, $57_{(10)} = 2^5 + 2^4 + 2^3 + 2^0$. A pile with 57 matches can be regarded as 4 little piles, a pile with 2^5 matches, a pile with 2^4 matches, a pile with 2^3 matches, and a pile with 2^0 matches.

Suppose there are k piles of matches, $k > 2$, and the numbers of matches in the k piles are N_1, N_2, ..., and N_k respectively. N_i can be represented as a $(s+1)$-digit binary number, that is, $N_i = n_{is}...n_{i1}n_{i0}$, n_{ij} is a binary digit, $0 \leq j \leq s$, $1 \leq i \leq k$. If the digit of a binary number is less than $s+1$, leading zeros are added.

The game state is balanced if $n_{10} + n_{20} + ... + n_{k0}$ is even, $n_{11} + n_{21} + ... + n_{k1}$ is even, ..., and $n_{1s} + n_{2s} + ... + n_{ks}$ is even, that is, n_{10} XOR n_{20} XOR...XOR n_{k0} is 0, n_{11} XOR n_{21} XOR...XOR n_{k1} is 0, ..., and n_{1s} XOR n_{2s} XOR...XOR n_{ks} is 0; else the game state is unbalanced. If a player faces an unbalanced state, he can take away some matches from a pile to make the state a balanced state. And if a player faces a balanced state, no matter what strategies he takes, the state will become an unbalanced state. The final state for the game is that all binary numbers are zero, that is, the final state is balanced. Therefore, the strategy for winning the game (**Bouton's Theorem**) is as follows:

The player who plays first will win the game if the initial state is unbalanced. And the player who plays second will win the game if the initial state is balanced.

For example, there are four piles of matches. There are 7, 9, 12, and 15 matches in the four piles respectively. 7, 9, 12, and 15 can be represented as binary numbers 0111, 1001, 1100, and 1111. This is shown in the following list.

Size of a Pile	$2^3 = 8$	$2^2 = 4$	$2^1 = 2$	$2^0 = 1$
7	0	1	1	1
9	1	0	0	1
12	1	1	0	0
15	1	1	1	1
	Odd	Odd	Even	Odd

The initial state for the game is unbalanced. The player who plays first takes away some matches from a pile to make the state become a balanced state. There are many choices. For example, the player who plays first takes away 11 matches from a pile with 12 matches to make the state become a balanced state. This is shown in the following list.

Size of a Pile	$2^3 = 8$	$2^2 = 4$	$2^1 = 2$	$2^0 = 1$
7	0	1	1	1
9	1	0	0	1
12⇒1	0	0	0	1
15	1	1	1	1

The method that the player who plays first takes away some matches from a pile to make the state become a balanced state is to select a row (a pile), and to flip values of bits in odd columns in the row. After flipping values of bits in odd columns, the number of matches is less than the original number of matches in the row. The number of matches that the player who plays first takes away from the corresponding pile is the difference between the original number of matches and the new number of matches. Then, the player who plays second takes away matches under a balanced state. The state will become an unbalanced state. And the player who plays first can make the state balance no matter how the player who plays second takes away matches. The process is repeated until the player who plays second takes away some matches under a balanced state last time, and then the player who plays first can take away all remainder matches.

For the same reason, the player who plays second will win the game when the initial state is a balanced game.

Therefore, the algorithm is as follows:

N piles of matches are represented as *N* binary numbers. If the initial state is unbalanced, the player who plays first will win the game, else the player who plays second will win the game.

 Program

```
# include <cstdio>
# include <cstring>
# include <cstdlib>
# include <iostream>
# include <string>
# include <cmath>
# include <algorithm>
using namespace std;
int main(){
  int n;
  while(~scanf("%d",&n)){    //number of piles
    int a=0,b;    //a: result, b: number of matches in the
current pile
    for(int i=0;i<n;i++){    //input numbers of matches in
all piles
      scanf("%d",&b);
      a^=b;    //XOR operations
    }
    printf("%s\n",a?"Yes":"No");    //if a isn't balanced,
output "Yes", else output "No"
  }
  return 0;
}
```

1.3 Problems

1.3.1 Perfection

From the article Number Theory in the 1994 Microsoft Encarta: "If *a*, *b*, *c* are integers such that *a* = *bc*, a is called a multiple of *b* or of *c*, and *b* or *c* is called a divisor or factor of *a*. If *c* is not ±1, *b* is called a proper divisor of *a*. Even integers, which include 0, are multiples of 2, for example, −4, 0, 2, 10; an odd integer is an integer that is not even, for example, −5, 1, 3, 9. A perfect number is a positive integer that is equal to the sum of all its positive, proper divisors; for example, 6, which equals 1 + 2 + 3, and 28, which equals 1 + 2 + 4 + 7 + 14, are perfect numbers. A positive number that is not perfect is imperfect and is deficient or abundant according to whether the sum of its positive, proper divisors is smaller or larger than the number itself. Thus, 9, with proper divisors 1, 3, is deficient; 12, with proper divisors 1, 2, 3, 4, 6, is abundant."

Given a number, determine if it is perfect, abundant, or deficient.

Input

A list of *N* positive integers (none greater than 60,000), with 1<*N*<100. A 0 will mark the end of the list.

Output

The first line of output should read PERFECTION OUTPUT. The next *N* lines of output should list for each input integer whether it is perfect, deficient, or abundant, as shown in the following example. Format counts: the echoed integers should be right-justified within the first five spaces of the output line, followed by two blank spaces, followed by the description of the integer. The final line of output should read END OF OUTPUT.

Sample Input	Sample Output

15 28 6 56 60000 22
496 0
	PERFECTION OUTPUT
	   15 DEFICIENT
	   28 PERFECT
	    6 PERFECT
	   56 ABUNDANT
	60000 ABUNDANT
	   22 DEFICIENT
	  496 PERFECT
	END OF OUTPUT

Source: ACM Mid-Atlantic 1996

IDs for Online Judges: POJ 1528, ZOJ 1284, UVA 382

 Hint

First, all proper divisors of the current integer are calculated. Then the sum of all proper divisors is calculated.

> If the current integer > the sum of all proper divisors, then output "DEFICIENT";
> If the current integer < the sum of all proper divisors, then output "ABUNDANT";
> If the current integer = the sum of all proper divisors, then output "PERFECT".

1.3.2 Uniform Generator

Computer simulations often require random numbers. One way to generate pseudo-random numbers is via a function of the form:

$$seed\,(x+1)=\left[\,seed\,(x)+STEP\,\right]\%MOD$$ where "%" is the modulus operator.

Such a function will generate pseudo-random numbers (*seed*) between 0 and *MOD*-1. One problem with functions of this form is that they will always generate the same pattern over and over. In order to minimize this effect, selecting the *STEP* and *MOD* values carefully can result in a uniform distribution of all values between (and including) 0 and *MOD*-1.

For example, if *STEP*=3 and *MOD*=5, the function will generate the series of pseudo-random numbers 0, 3, 1, 4, 2 in a repeating cycle. In this example, all of the numbers between and including 0 and *MOD*-1 will be generated every *MOD* iterations of the function. Note that by the nature of the function to generate the same *seed*(*x*+1) every time, *seed*(*x*) occurs means that if a function will generate all the numbers between 0 and *MOD*-1, it will generate pseudo-random numbers uniformly with every *MOD* iteration.

If *STEP*=15 and *MOD*=20, the function generates the series 0, 15, 10, 5 (or any other repeating series if the initial seed is other than 0). This is a poor selection of *STEP* and *MOD* because no initial seed will generate all of the numbers from 0 and *MOD*-1.

Your program will determine whether choices of *STEP* and *MOD* will generate a uniform distribution of pseudo-random numbers.

Input

Each line of input will contain a pair of integers for *STEP* and *MOD* in that order (1≤*STEP*,*MOD*≤100000).

Output

For each line of input, your program should print the *STEP* value right-justified in columns 1 through 10, the *MOD* value right-justified in columns 11 through 20, and either "Good Choice" or "Bad Choice" left-justified starting in column 25. The "Good Choice" message should be printed when the selection of *STEP* and *MOD* will generate all the numbers between and including 0 and *MOD*-1 when MOD numbers are generated. Otherwise, your program should print the message "Bad Choice." After each output test set, your program should print exactly one blank line.

Sample Input	Sample Output		
3 5	3	5	Good Choice
15 20	15	20	Bad Choice
63923 99999	63923	99999	Good Choice

Source: ACM South Central USA 1996

IDs for Online Judges: POJ 1597, ZOJ 1314, UVA 408

Hint

Suppose $seed_i$ is the i-th pseudo-random number. Based on the problem description, the next pseudo-random number (the $(i+1)$-th pseudo-random number) is $seed_{i+1}=(seed_i+step)\%MOD$.

From $seed_0$, the function is iterated MOD-1 times. If produced MOD-1 pseudo-random numbers are all the numbers between 1 and MOD-1, it generates a uniform distribution of pseudo-random numbers; else it doesn't generate a uniform distribution of pseudo-random numbers.

1.3.3 WERTYU

A common typing error is to place the hands on the keyboard one row to the right of the correct position (see Figure 1.2). So "Q" is typed as "W" and "J" is typed as "K" and so on. You are to decode a message typed in this manner.

Input

Input consists of several lines of text. Each line may contain digits, spaces, uppercase letters (except Q, A, Z), or punctuation shown above (except back-quote ['']). Keys labelled with words (*Tab, BackSp, Control*, etc.) are not represented in the input.

Figure 1.2

Output

You are to replace each letter or punctuation symbol by the one immediately to its left on the QWERTY keyboard shown above. Spaces in the input should be echoed in the output.

Sample Input	Sample Output
O S, GOMR YPFSU/	I AM FINE TODAY.

Source: Waterloo local 2001.01.27

IDs for Online Judges: POJ 2538, ZOJ 1884, UVA 10082

 Hint

First, the offline method is used to calculate the conversion table based on the keyboard figure. Then, for each letter, the corresponding letter in the conversion table is output.

1.3.4 Soundex

Soundex coding groups together words that appear to sound alike based on their spelling. For example, "can" and "khawn", "con" and "gone" would be equivalent under Soundex coding.

Soundex coding involves translating each word into a series of digits in which each digit represents a letter:

1 represents B, F, P, or V
2 represents C, G, J, K, Q, S, X, or Z
3 represents D or T
4 represents L
5 represents M or N
6 represents R

The letters A, E, I, O, U, H, W, and Y are not represented in Soundex coding, and repeated letters with the same code digit are represented by a single instance of that digit. Words with the same Soundex coding are considered equivalent.

Input

Each line of input contains a single word, all uppercase, less than 20 letters long.

Output

For each line of input, produce a line of output giving the Soundex code.

Sample Input	Sample Output
KHAWN	25
PFISTER	1236
BOBBY	11

Source: Waterloo local 1999.09.25

IDs for Online Judges: POJ 2608, ZOJ 1858, UVA 10260

 Hint

For each word, letters are transferred into corresponding digits from left to right. And based on the problem description, letters A, E, I, O, U, H, W, and Y are not represented in Soundex coding, and repeated letters with the same code digit are represented by a single instance of that digit.

1.3.5 Minesweeper

The game *Minesweeper* is played on an *n* by *n* grid. In this grid are hidden *m* mines, each at a distinct grid location. The player repeatedly touches grid positions. If a position with a mine is touched, the mine explodes and the player loses. If a position not containing a mine is touched, an integer between 0 and 8 appears, denoting the number of adjacent or diagonally adjacent grid positions that contain a mine. A sequence of moves in a partially played game is illustrated below in Figure 1.3.

Here, *n* is 8, *m* is 10, blank squares represent the integer 0, raised squares represent unplayed positions, and the figures resembling asterisks represent mines. The left-most image represents the partially played game. From the first image to the second,

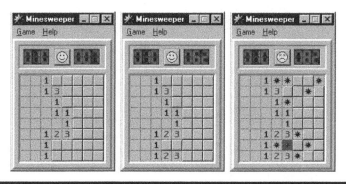

Figure 1.3

the player has played two moves, each time choosing a safe grid position. From the second image to the third, the player is not so lucky; he chooses a position with a mine and therefore loses. The player wins if he continues to make safe moves until only *m* unplayed positions remain; these must necessarily contain the mines.

Your job is to read the information for a partially played game and to print the corresponding board.

Input

The first line of input contains a single positive integer $n \leq 10$. The next *n* lines represent the positions of the mines. Each line represents the contents of a row using *n* characters: a period indicates an unmined positon, while an asterisk indicates a mined position. The next *n* lines are each *n* characters long: touched positions are denoted by an *x*, and untouched positions by a period. The sample input corresponds to the middle section of Figure 1.3.

Output

Your output should represent the board, with each position filled in appropriately. Positions that have been touched and do not contain a mine should contain an integer between 0 and 8. If a mine has been touched, all positions with a mine should contain an asterisk. All other positions should contain a period.

Sample Input	Sample Output
8	001.....
...**..*	0013....
......*.	0001....
....*...	00011...
....	00001...
....	00123...
.....*..	001.....
...**.*.	00123...
.....*..	
xxx.....	
xxxx....	
xxxx....	
xxxxx...	
xxxxx...	
xxxxx...	
xxx.....	
xxxxx...	

Source: Waterloo local 1999.10.02

IDs for Online Judges: POJ 2612, ZOJ 1862, UVA 10279

Hint

Suppose *g*[*i*][*j*] is the matrix for mines, and *try*[*i*][*j*] is the touch matrix, 1≤*i*, *j*≤*n*.

First we need to determine whether a mine is touched or not, that is, whether there exists such a grid that (*try*[*i*][*j*]=='x'&&*g*[*i*][*j*]=='*'). The mark *mc*=

$$\begin{cases} \text{'*'} & \text{There exists a touched mine.} \\ \text{'.'} & \text{There is no touched mine.} \end{cases}$$ shows whether there is a touched mine or not.

Then calculate and output the state for every grid (*i,j*) from left to right, and from top to bottom, 1≤*i*, *j*≤*n*.

If grid (*i,j*) is touched and doesn't contain a mine (*try*[*i*][*j*]== 'x'&&*g*[*i*][*j*]== '.'), then the number of adjacent or diagonally adjacent grid positions that contain a mine *x* is calculated and is filled into (*i,j*); else (i.e., *try*[*i*][*j*]== '.'||*g*[*i*][*j*]== '*'), if grid (*i,j*) contains a mine, *mc* is filled into (*i,j*); else '.' is filled into (*i,j*).

1.3.6 Tic Tac Toe

Tic Tac Toe is a child's game played on a 3 by 3 grid. One player, X, starts by placing an X at an unoccupied grid position. Then the other player, O, places an O at an unoccupied grid position. Play alternates between X and O until the grid is filled or one player's symbols occupy an entire line (vertical, horizontal, or diagonal) in the grid.

We will denote the initial empty Tic Tac Toe grid with nine dots. Whenever X or O plays, we fill in an X or an O in the appropriate position. The example in Figure 1.4 illustrates each grid configuration from the beginning to the end of a game in which X wins.

Your job is to read a grid and to determine whether or not it could possibly be part of a valid Tic Tac Toe game. That is, is there a series of plays that can yield this grid somewhere between the start and end of the game?

Input

The first line of input contains *N*, the number of test cases. 4*N*−1 lines follow, specifying *N* grid configurations separated by empty lines.

...	X..	X.O	X.O	X.O	X.O	X.O	X.O
...	...	...	...	.O.	.O.	OO.	OO.
...	...	...	..X	..X	X.X	X.X	XXX

Figure 1.4

Output

For each case, print "yes" or "no" on a line by itself, indicating whether or not the configuration could be part of a Tic Tac Toe game.

Sample Input	Sample Output
2 X.O OO. XXX O.X XX. OOO	yes no

Source: POJ 2361, ZOJ 1908, UVA 10363

IDs for Online Judges: Waterloo local 2002.09.21

 Hint

Based on the problem description, a configuration for part of a valid Tic Tac Toe game must satisfy the following properties:

1. The number of Os must be one less than or equal to the number of Xs;
2. If the number of Os is one less than the number of Xs, O doesn't win the game;
3. If the number of Os is equal to the number of Xs, X doesn't win the game.

That is to say, if a configuration isn't part of a valid Tic Tac Toe game, it must satisfy the following properties:

1. The number of Os must be larger than the number of Xs; or
2. The number of Os is two less than the number of Xs at least; or
3. Both O and X win the game; or
4. O wins the game; and the number of Os isn't equal to the number of Xs; or
5. X wins the game; and the number of Os is equal to the number of Xs.

Otherwise, the configuration is part of a valid Tic Tac Toe game.

1.3.7 Rock, Scissors, Paper

Bart's sister Lisa has created a new civilization on a two-dimensional grid. At the outset, each grid location may be occupied by one of three life forms: *Rocks, Scissors,* or *Papers*. Each day, differing life forms occupying horizontally or vertically adjacent grid locations wage war. In each war, Rocks always defeat Scissors, Scissors always defeat Papers, and Papers always defeat Rocks. At the end of the day, the victor expands its territory to include the loser's grid position. The loser vacates the position.

Your job is to determine the territory occupied by each life form after *n* days.

Input

The first line of input contains *t*, the number of test cases. Each test case begins with three integers not greater than 100: *r* and *c*, the number of rows and columns in the grid, and *n*. The grid is represented by the *r* lines that follow, each with *c* characters. Each character in the grid is R, S, or P, indicating that it is occupied by Rocks, Scissors, or Papers respectively.

Output

For each test case, print the grid as it appears at the end of the *n*th day. Leave an empty line between the output for successive test cases.

Sample Input	Sample Output
2	RRR
3 3 1	RRR
RRR	RRR
RSR	
RRR	RRRS
3 4 2	RRSP
RSPR	RSPR
SPRS	
PRSP	

Source: POJ 2339, ZOJ 1921, UVA 10443

IDs for Online Judges: Waterloo local 2003.01.25

 Hint

Because the two-dimensional grid is changed at the end of the day, two matrices are used to represent yesterday's two-dimensional grid and today's two-dimensional grid respectively. Today's two-dimensional grid is calculated based on yesterday's two-dimensional grid.

- An 'R' will be changed into a 'P' if and only if the 'R' is adjacent to a 'P' in yesterday's two-dimensional grid. That is, if an 'R' is adjacent to a 'P' in yesterday's two-dimensional grid, then the 'R' is changed into 'P' in today's two-dimensional grid.
- An 'S' will be changed into an 'R' if and only if the 'S' is adjacent to an 'R' in yesterday's two-dimensional grid. That is, if an 'S' is adjacent to an 'R' in yesterday's two-dimensional grid, then the 'S' is changed into 'R' in today's two-dimensional grid.
- A 'P' will be changed into an 'S' if and only if the 'P' is adjacent to an 'S' in yesterday's two-dimensional grid. That is, if a 'P' is adjacent to an 'S' in yesterday's two-dimensional grid, then the 'P' is changed into 'S' in today's two-dimensional grid.

For example,

$$
\begin{vmatrix} R & S & P & R \\ S & P & R & S \\ P & R & S & P \end{vmatrix} \Rightarrow
\begin{vmatrix} R & R & S & P \\ R & S & P & R \\ S & P & R & S \end{vmatrix} \Rightarrow
\begin{vmatrix} R & R & R & S \\ R & R & S & P \\ R & S & P & R \end{vmatrix} \Rightarrow
$$

$$
\begin{vmatrix} R & R & R & R \\ R & R & R & S \\ R & R & S & P \end{vmatrix} \cdots\cdots
$$

The grid as it appears at the end of the nth day is calculated based on the above rules.

1.3.8 Prerequisites?

Freddie the freshman has chosen to take k courses. To meet the degree requirements, he must take courses from each of several categories. Can you assure Freddie that he will graduate, based on his course selection?

Input

Input consists of several test cases. For each case, the first line of input contains $1 \le k \le 100$, the number of courses Freddie has chosen, and $0 \le m \le 100$, the number of categories. One or more lines containing k four-digit integers follow; each is the number of a course selected by Freddie. Each category is represented by a line containing $1 \le c \le 100$, the number of courses in the category; $0 \le r \le c$, the minimum number of courses from the category that must be taken; and the c course numbers in the category. Each course number is a four-digit integer. The same course may fulfil several category requirements. Freddie's selections, and the course numbers in any particular category, are distinct. A line containing 0 follows the last test case.

Output

For each test case, output a line containing "yes" if Freddie's course selection meets the degree requirements; otherwise output "no."

Sample Input	Sample Output
3 2	yes
0123 9876 2222	no
2 1 8888 2222	
3 2 9876 2222 7654	
3 2	
0123 9876 2222	
2 2 8888 2222	
3 2 7654 9876 2222	
0	

Source: Waterloo local 2005.09.24

IDs for Online Judges: POJ 2664, UVA 10919

Hint

Suppose c_i is the number of courses in the i-th category, $done_i$ is the set of courses in the i-th category, and r_i is the minimum number of courses from the i-th category that must be taken, $1 \leq i \leq m$.

First, k courses that Freddie has chosen to take are put into a set *take*[ ].

Then courses that Freddie has chosen to take are analyzed. For courses in the i-th category, if $r_i \leq |take[\ ] \cap done_i|$, the number of courses in the i-th category that Freddie has chosen to take is larger than or equal to the minimum number of courses from the i-th category that must be taken, and set the mark yes_i=true.

Finally, if $\bigcap_{1 \leq i \leq m} \{yes_i\}$ ==true, then Freddie's course selection meets the degree requirements, else Freddie's course selection doesn't meet the degree requirements.

1.3.9 Save Hridoy

It would be great if banners with good words could inspire us all. Then we could make large banners with good words on them to make this world beautiful. With all good wishes, we will make such a banner today—a banner to save a life[1], a banner to save humanity.

In this problem, the program-generated banners will contain the text "SAVE HRIDOY". We will make this banner with different text sizes and two possible types of orientations: horizontal and vertical (see Figure 1.5). As we will make banners of different size in plain monochrome text, we will use two different ASCII characters to denote black-and-white pixels. In this process, the smallest possible banner (font size 1) for us in horizontal orientation is:

```
*****..***..*...*.*****...*...*.*****.*****.***...*****.*...*
*.....*...*.*...*.*.......*...*.*...*...*..*...*..*...*..*.*.
*****.*****.*...*.***.....*****.*****...*...*...*.*...*...*..
....*.*...*..*.*.*...*.....*...*.*.*.....*...*..*..*...*...*..
*****.*...*...*...*****...*...*.*.**.*****.***...*****...*..
```

Figure 1.5

You can see that here black pixels are formed with the "*" character and white pixels are marked with the "." character. In this banner, each character is represented in a (5 × 5) grid, two consecutive characters in a single word are separated by a single vertical dotted line, and the two words are separated by three vertical dotted lines. In the case of vertical banners (of font size 1), two consecutive letters in a single word are separated by a horizontal dotted line, and two words are separated by three horizontal dotted lines. Look at the second output for sample input to know how vertical banners are formed. In the case of a banner of font size 2, each pixel is represented by a (2 × 2) grid of pixels. So actually a banner of font size two has double the width and double the height of a banner of font size 1.

Input

The input file contains at most 30 lines of inputs. Each line contains an integer N $(0 < N < 51)$. This value of N denotes the font size and orientation of a banner. Input is terminated by a line containing a single zero. This line should not be processed.

Output

If N is positive, then you have to draw a banner of horizontal orientation, and if N is negative, then you have to draw a banner of vertical orientation. The detailed description of output for these two types of cases is given below:

1. If N is positive, then produce $5N$ lines of output. These lines actually draw the horizontal banner. Two consecutive letters in a word are separated by N vertical dotted lines. Two words are separated by $3N$ vertical dotted lines.
2. If N is negative, then produce $5L \times 10 + 11L$ lines of output, where L is the absolute value of N. Two consecutive characters in a word are separated by L horizontal dotted lines, and two words are separated by $3L$ horizontal dotted lines.

After the output of each test case, print two blank lines.

Sample Input	Sample Output
| –<br>1<br>2<br>0 | ```
*****
*....
*****
....*
*****

.....
.***.
*...*
*****
*...*
*...*
.....
*...*
*...*
*...*
.*.*.
..*..
.....
*****
*....
***..
*....
*****

.....
.....
.....
*...*
*...*
*****
*...*
*...*
.....
*****
*...*
*****
*.*..
*.**.
.....
*****
..*..
..*..
``` |

(continued)
```

Sample Input	Sample Output
|  | ```
..*..
*****
.....
***..
*.*.
*..*
*.*.
***..
.....
*****
*..*
*..*
*..*
*****
.....
*..*
.*.*.
..*..
..*..
..*..
``` |

Source: UVA Monthly Contest August 2005

ID for Online Judge: UVA 10894
```

Hint

First, the offline method is used to construct a matrix $F[][]$, representing a banner of horizontal orientation, and a matrix $G[][]$, representing a banner of vertical orientation (font size 1).

Then, for each test case N, $F[][]$, or $G[][]$ is magnified. If N is positive, then $F[][]$ is magnified N times. That is, a horizontal banner with $5N \times 61N$ is produced, where (i,j) is $F\left[\left\lfloor \dfrac{i-1}{N} \right\rfloor + 1\right]\left[\left\lfloor \dfrac{j-1}{N} \right\rfloor + 1\right]$. If N is negative, then $G[][]$ is magnified $|N|$ times. That is, a horizontal banner with $61N \times 5N$ is produced, where (i,j) is $G\left[\left\lfloor \dfrac{i-1}{-N} \right\rfloor + 1\right]\left[\left\lfloor \dfrac{j-1}{-N} \right\rfloor + 1\right]$.

1.3.10 Find the Telephone

In some places, it is common to remember a phone number by associating its digits to letters. In this way, the expression **MY LOVE** means **69 5683**. Of course, there are some problems, because some phone numbers cannot form a word or a phrase and the digits **1** and **0** are not associated to any letter.

Your task is to read an expression and find the corresponding phone number based on the table below. An expression is composed by the capital letters (**A-Z**), hyphens (-) and the numbers **1** and **0**.

Letters	Number
ABC	2
DEF	3
GHI	4
JKL	5
MNO	6
PQRS	7
TUV	8
WXYZ	9

Input

The input consists of a set of expressions. Each expression is in a line by itself and has **C** characters, where $1 \leq C \leq 30$. The input is terminated by end of file (EOF).

Output

For each expression, you should print the corresponding phone number.

Sample Input	Sample Output
1-HOME-SWEET-HOME MY-MISERABLE-JOB	1-4663-79338-4663 69-647372253-562

Source: UFRN-2005 Contest 1

ID for Online Judge: UVA 10921

Hint

In an expression, characters are analyzed from left to right. If a character is a hyphen, '1', or '0', the character is output directly; else the number that the character corresponds to is output.

1.3.11 2 the 9s

A well-known trick to know if an integer N is a multiple of nine is to compute the sum S of its digits. If S is a multiple of nine, then so is N. This is a recursive test, and the depth of the recursion needed to obtain the answer on N is called the 9-degree of N.

Your job is, given a positive number N, to determine if it is a multiple of nine and, if it is, its 9-degree.

Input

The input is a file such that each line contains a positive number. A line containing the number 0 is the end of the input. The given numbers can contain up to 1000 digits.

Output

The output of the program shall indicate, for each input number, if it is a multiple of nine, and in case it is, the value of its 9-degree. See the sample output for an example of the expected formatting of the output.

Sample Input	Sample Output
999999999999999999999	999999999999999999999 is a multiple of 9 and has 9-degree 3.
9999999999999999999999999999980	9 is a multiple of 9 and has 9-degree 1. 9999999999999999999999999999998 is not a multiple of 9.

Source: UFRN-2005 Contest 1

ID for Online Judge: UVA 10922

Hint

For this problem, the statistical analysis method is used. First, two sample test cases are analyzed.

1. $N = 999999999999999999999$
   a. The first level for the recursion: There are 21 digits for 999999999999999999999.
   The sum of 21 digits is 9×21=189;
   The second level for the recursion: The sum of three digits for 189 is 1+8+9=18;
   The third level for the recursion: The sum of two digits for 18 is 9. The recursion ends.
   Therefore, 999999999999999999999 is a multiple of nine and has 9-degree 3.
2. $N = 9999999999999999999999999999998$
   b. The first level for the recursion: There are 31 digits for 99999999999999 99999999999999998. The sum of 31 digits is 30×9+8=278;
   The second level for the recursion: The sum of three digits for 278 is 2+7+8=17;
   The third level for the recursion: The sum of two digits for 17 is 8. 8 isn't a multiple of 9. The recursion ends.

Therefore, 9999999999999999999999999999998 is not a multiple of 9.

The method determining whether a positive number N is a multiple of nine or not is a recursive method. And the algorithm can be implemented with the above method to solve the problem.

1.3.12 You Can Say 11

Your job, given a positive number N, is to determine whether it is a multiple of eleven.

Input

The input is a file such that each line contains a positive number. A line containing the number 0 is the end of the input. The given numbers can contain up to 1000 digits.

Output

The output of the program shall indicate, for each input number, if it is a multiple of eleven or not.

Sample Input	Sample Output
112233	112233 is a multiple of 11.
30800	30800 is a multiple of 11.
2937	2937 is a multiple of 11.
323455693	323455693 is a multiple of 11.
5038297	5038297 is a multiple of 11.
112234	112234 is not a multiple of 11.
0	

Source: UFRN-2005 Contest 2

ID for Online Judge: UVA 10929

Hint

Suppose the given large positive number can be represented as a high precision number $A=a_0...a_{l-1}$. From right to left, sums of odd positions and even positions for the number are calculated respectively. Then the difference for the two sums is calculated. If the difference is a multiple of 11 (including 0), that is,

$$\sum_{i=0}^{\lfloor \frac{l}{2} \rfloor} a_{2 \cdot i} - \sum_{i=1}^{\lceil \frac{l}{2} \rceil} a_{2 \cdot i-1} = 11 \cdot k,$$ then A is a multiple of 11. Otherwise, A is not a multiple

of 11.

Another method is to simply shift and mod. A number A is divisible by 11 if A mod 11 is 0. We can shift and mod with primitive types.

1.3.13 Parity

We define the parity of an integer $\underline{n}$ as the sum of the bits in binary representation computed in modulo two. As an example, the number $21=10101_2$ has three 1s in its binary representation, so it has parity 3 (mod 2), or 1.

In this problem, you have to calculate the parity of an integer $1 \leq I \leq 2147483647$.

Input

Each line of the input has an integer I and the end of the input is indicated by a line where $I=0$ that should not be processed.

Output

For each integer I in the input, you should print a line "The parity of B is P (mod 2).", where B is the binary representation of I.

Sample Input	Sample Output
1	The parity of 1 is 1 (mod 2).
2	The parity of 10 is 1 (mod 2).
10	The parity of 1010 is 2 (mod 2).
21	The parity of 10101 is 3 (mod 2).
0	

Source: UFRN-2005 Contest 2

ID for Online Judge: UVA 10931

Hint

The problem requires you to figure out how many 1's are in a binary number for the decimal number they give you. This is most easily done by keeping track of the 1's and continuously bitshifting until the number is 0. The constraints are 31 bits of all 1's 2^31−1, so just an integer will suffice.

1.3.14 Not That Kind of Graph

Your task is to graph the price of a stock over time. In one unit of time, the stock can either Rise, Fall, or stay Constant. The stock's price will be given to you as a string of R's, F's, and C's. You need to graph it using the characters '/' (slash), '\' (backslash) and '_' (underscore).

Input

The first line of input gives the number of cases, N. N test cases follow. Each one contains a string of at least 1 and at most 50 uppercase characters (R, F, or C).

Output

For each test case, output the line "Case #x:", where x is the number of the test case. Then print the graph, as shown in the sample output, including the x- and y-axes. The x-axis should be one character longer than the graph, and there should be one space between the y-axis and the start of the graph. There should be no trailing spaces on any line. Do not print unnecessary lines. The *x*-axis should always appear directly below the graph. Finally, print an empty line after each test case.

Sample Input	Sample Output
1RCRFCRFFCCRRC	Case #1: | _ | _\/\ / |/ __/ +---------------

Source: Abednego's Graph Lovers' Contest, 2005

ID for Online Judge: UVA 10800

 Hint

The problem explanation covers the problem with enough detail to solve it without really needing much insight. We are given a string of characters, each of which is R (rise), C (constant), or F (fall), and we have to draw the corresponding line. Just make a 2D matrix of characters and draw to the matrix, and then output the matrix.

For the problem, the two points should be noted. The stock price does not necessarily start at its minimum. Don't output spaces at the end of the line.

1.3.15 Decode the Tape

Your boss has just unearthed a roll of old computer tapes. The tapes have holes in them and might contain some sort of useful information. It falls to you to figure out what is written on them.

Input

The input will contain one tape.

Output

Output the message that is written on the tape.

Sample Input	Sample Output
_____	A quick brown fox jumps over the lazy dog.
|o . o|	
| o . |	
| ooo . o|	
| ooo .o o|	
| oo o. o|	
| oo . oo|	
| oo o. oo|	
| o . |	
| oo . o |	
| ooo . o |	
| oo o.ooo|	
| ooo .ooo|	
| oo o.oo |	
| o . |	
| oo .oo |	
| oo o.ooo|	
| oooo. |	
| o . |	
| oo o. o |	
| ooo .o o|	
| oo o.o o|	
| ooo . |	
| ooo . oo|	
| o . |	
| oo o.ooo|	
| ooo .oo |	
| oo .o o|	
| ooo . o |	
| o . |	
| ooo .o |	
| oo o. |	
| oo .o o|	
| o . |	
| oo o.o |	
| oo . o|	
| oooo. o |	
| oooo. o|	
| o . |	
| oo .o |	
| oo o.ooo|	
| oo .ooo|	
| o o.oo |	
| o. o|	
_____	

Source: Abednego's Mathy Contest 2005

ID for Online Judge: UVA 10878

Hint

From the sample input, there are 10 characters $a_0 \ldots a_9$ in a line in a tape, where a_0 is leading flag '|', a_6 is a space, spaces in other positions represent 0, and 'o' represents 1.

That is, if the i-th is 'o', the position represents an integer $a_i = \begin{cases} 2^{9-i} & 7 \le i \le 9 \\ 2^{8-i} & 2 \le i \le 5 \end{cases}$. A line

corresponds to an ASCII code representing a character. The character string is the message that is written on the tape.

1.3.16 Fractions Again?!

It is easy to see that for every fraction in the form $\dfrac{1}{k}(k>0)$, we can always find two positive integers x and y, $x \ge y$, such that:

$$\frac{1}{k} = \frac{1}{x} + \frac{1}{y}.$$

Now our question is: can you write a program that counts how many such pairs of x and y there are for any given k?

Input

Input contains no more than 100 lines, each giving a value of k ($0 < k \le 10000$).

Output

For each k, output the number of corresponding (x, y) pairs, followed by a sorted list of the values of x and y, as shown in the sample output.

Sample Input	Sample Output
2	2
12	1/2 = 1/6 + 1/3
	1/2 = 1/4 + 1/4
	8
	1/12 = 1/156 + 1/13
	1/12 = 1/84 + 1/14
	1/12 = 1/60 + 1/15
	1/12 = 1/48 + 1/16
	1/12 = 1/36 + 1/18
	1/12 = 1/30 + 1/20
	1/12 = 1/28 + 1/21
	1/12 = 1/24 + 1/24

Source: Return of the Newbies 2005

ID for Online Judge: UVA 10976

Hint

For a given positive integer k, find all pairs of positive integers x and y, $x \geq y$, such that $\dfrac{1}{k} = \dfrac{1}{x} + \dfrac{1}{y}$. Obviously $k+1 \leq y \leq 2k$. For every possible y, check whether the corresponding x is an integer or not. That is, because $\dfrac{y-k}{k*y} = \dfrac{1}{x}$, $x = \dfrac{k*y}{y-k}$. If $(k \times y)\%(y-k) == 0$, then x is an integer.

1.3.17 Factorial! You Must be Kidding!!!

Arif has bought a supercomputer from Bongobazar. Bongobazar is a place in Dhaka where secondhand goods are available. So the supercomputer he bought is also secondhand and has some bugs. One of the bugs is that the range of unsigned long integers of this computer for a C/C++ compiler has changed. Now its new lower limit is 10000 and the upper limit is 6227020800. Arif writes a program in C/C++ which determines the factorial of an integer. The factorial of an integer is defined recursively as:

$$\text{Factorial}(0) = 1$$

$$\text{Factorial}(n) = n \times \text{Factorial}(n-1).$$

Of course, one can manipulate these expressions. For example, it can be written as:

$$\text{Factorial}(n) = n \times (n-1) \times \text{Factorial}(n-2)$$

This definition can also be converted to an iterative one.

But Arif knows that his program will not behave correctly in the supercomputer. You are to write a program which will simulate that changed behavior in a normal computer.

Input

The input file contains several lines of input. Each line contains a single integer n. No integer has more than 6 digits. Input is terminated by end of file.

Output

For each line of input, you should output a single line. This line will contain a single integer $n!$ if the value of $n!$ fits within the unsigned long integer of Arif's computer. Otherwise, the line will contain one of the following two words:

Overflow! //(When $n! > 6227020800$)
Underflow! //(When $n! < 10000$)

Sample Input	Sample Output
2	Underflow!
10	3628800
100	Overflow!

Source: GWCF Contest 4 - The Decider

ID for Online Judge: UVA 10323

Hint

The concept behind the problem is quite simple: given n, if $n!$ is greater than 6227020800, then print "Overflow!"; if $n!$ is less than 10000, print "Underflow!"; otherwise print $n!$.

Though a negative factorial is normally undefined, this problem stretches the limit of well-known definitions.

For this problem, we have $F(n)=n\times F(n-1)$, and $F(0)=1$. With some manipulations, for negative factorials, we can get: $F(0)=0\times F(-1)$, or $F(-1)=\dfrac{F(0)}{0}=\infty$. Continuing with this logic: $F(-1)=-1\times F(-2)$, or $F(-2)=-F(-1)$. Similarly, $F(-1)=F(-3)=-F(-2)$.

First, the offline method is used to calculate $f[i]=i!$, $8\le i\le 13$. Then, for each n:

> if n is between 8 to 13, then print $f[n]$;
>
> if $(n\ge14||(n<0\&\&(-n)\%2==1))$, then print "Overflow!";
>
> if $(n\le7||(n<0\&\&(-n)\%2==0))$, then print "Underflow!".

1.3.18 Squares

A children's board game consists of a square array of dots that contains lines connecting some of the pairs of adjacent dots. One part of the game requires that the players count the number of squares of certain sizes that are formed by these lines. For example, in Figure 1.6, there are three squares, two of size 1 and one of size 2. (The "size" of a square is the number of line segments required to form a side.)

Your problem is to write a program that automates the process of counting all the possible squares.

Input

The input file represents a series of game boards. Each board consists of a description of a square array of n^2 dots (where $2\le n\le 9$) and some interconnecting horizontal

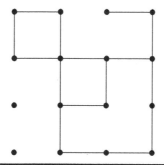

Figure 1.6

and vertical lines. A record for a single board with n^2 dots and m interconnecting lines is formatted as follows:

Line 1: n the number of dots in a single row or column of the array
Line 2: m the number of interconnecting lines

Each of the next m lines are of one of two types: Hij, indicates a horizontal line in row i which connects the dot in column j to the one to its right in column $j+1$; or Vij, indicates a vertical line in column i which connects the dot in row j to the one below in row $j+1$.

Information for each line begins in column 1. The end of input is indicated by end of file. The first record of the sample input below represents the board of the square above.

Output

For each record, label the corresponding output with "Problem #1", "Problem #2", and so forth. Output for a record consists of the number of squares of each size on the board, from the smallest to the largest. If no squares of any size exist, your program should print an appropriate message indicating this. Separate output for successive input records by a line of asterisks between two blank lines, as shown in the sample below.

Sample Input	Sample Output
4	Problem #1
16	
H 1 1	2 square (s) of size 1
H 1 3	1 square (s) of size 2
H 2 1	

(continued)

Sample Input	Sample Output
H 2 2	***********************************
H 2 3	
H 3 2	Problem #2
H 4 2	
H 4 3	No completed squares can be found.
V 1 1	
V 2 1	
V 2 2	
V 2 3	
V 3 2	
V 4 1	
V 4 2	
V 4 3	
2	
3	
H 1 1	
H 2 1	
V 2 1	

Source: ACM World Finals 1989

ID for Online Judge: UVA 201

Hint

Since $N \leq 9$, we can simply iterate all the possible squares.

We can think of vertical or horizontal lines as edges between two adjacent points. After that, we can take a three-dimensional array (say $a[N][N][2]$) to store the count of horizontal ($a[i][j][0]$) edges and vertical ($a[i][j][1]$) edges. $a[i][j][0]$ contains the number of horizontal edges at row i up to column j. And $a[i][j][1]$ contains the number of vertical edges at column j up to row i. Next you use a $O(n^2)$ loop to find a square. A square of size 1 is found if there is an edge from (i, j) to $(i, j+1)$ and $(i, j+1)$ to $(i+1, j+1)$ and (i, j) to $(i+1, j)$ and $(i+1, j)$ to $(i+1, j+1)$. We can get this just by subtracting the values calculated above.

1.3.19 The Cow Doctor

Texas is the state with the largest number of cows in the United States: according to the 2005 report of the National Agricultural Statistics Service, the bovine population of Texas is 13.8 million. This is higher than the population of the two runner-up states combined: there are only 6.65 million cows in Kansas and 6.35 million cows in Nebraska.

There are several diseases that can threaten a herd of cows, the most feared being "Mad Cow Disease" or Bovine Spongiform Encephalopathy (BSE); therefore, it is very important to be able to diagnose certain illnesses. Fortunately, there are many tests available that can be used to detect these diseases.

A test is performed as follows. First, a blood sample is taken from the cow, and then the sample is mixed with a test material. Each test material detects a certain number of diseases. If the test material is mixed with a blood sample having any of these diseases, then a reaction takes place that is easy to observe. However, if a test material can detect several diseases, then we have no way to decide which of these diseases is present in the blood sample, as all of them produce the same reaction. There are materials that detect many diseases (such tests can be used to rule out several diseases at once), and there are tests that detect only a few diseases (they can be used to make an accurate diagnosis of the problem).

The test materials can be mixed to create new tests. If we have a test material that detects diseases A and B, and there is another test material that detects diseases B and C, then they can be mixed to obtain a test that detects diseases A, B, and C. This means that if we have these two test materials, then there is no need for a test material that tests diseases A, B, and C—such a material can be obtained by mixing these two.

Producing, distributing, and storing many different types of test materials is very expensive, and in most cases, unnecessary. Your task is to eliminate as many unnecessary test materials as possible. It has to be done in such a way that if a test material is eliminated, then it should be possible to mix an equivalent test from the remaining materials. ("Equivalent" means that the mix tests exactly the same diseases as the eliminated material, not more, not less.)

Input

The input contains several blocks of test cases. Each case begins with a line containing two integers: the number $1 \leq n \leq 300$ of diseases, and the number $1 \leq m \leq 200$ of test materials. The next m lines correspond to the m test materials. Each line begins with an integer, the number $1 \leq k \leq 300$ of diseases that the material can detect. This is followed by k integers describing the k diseases. These integers are between 1 and n.

The input is terminated by a block with $n=m=0$.

Output

For each test case, you have to output a line containing a single integer: the maximum number of test materials that can be eliminated.

Sample Input	Sample Output
10 5	2
2 1 2	4
2 2 3	
3 1 2 3	
4 1 2 3 4	
1 4	
3 7	
1 1	
1 2	
1 3	
2 1 2	
2 1 3	
2 3 2	
3 1 2 3	
0 0	

Source: ACM Central Europe 2005

IDs for Online Judges: POJ 2943, UVA 3524

 Hint by the Problemsetter

The doctor has some test materials. Each test material can test a set of diseases. A mixture of two test materials gives a new test material that can test diseases for at least one of the mixed materials tested.

Given is a set of test materials. Determine how many of them are redundant, i.e., can be obtained by mixing some other test materials.

This problem is pretty straightforward. How can you check whether a given test material *M* is redundant? Consider the set *S* of all other test materials that test a subset of *M*'s diseases. *M* is redundant if and only if the mixture of all materials in *S* tests exactly the same set of diseases as *M*.

It is convenient to represent the materials as bit vectors.

1.3.20 Wine Trading in Gergovia

As you may know from the comic "Asterix and the Chieftain's Shield", Gergovia consists of one street, and every inhabitant of the city is a wine salesman. How does this economy works? Simple enough: everyone buys wine from other inhabitants of the city. Every day, each inhabitant decides how much wine he wants to buy or sell. Interestingly, demand and supply is always the same, so that each inhabitant gets what he wants.

There is one problem, however: Transporting wine from one house to another results in work. Since all wines are equally good, the inhabitants of Gergovia don't care which persons they are doing trade with; they are only interested in selling or buying a specific amount of wine. They are clever enough to figure out a way of trading so that the overall amount of work needed for transports is minimized.

In this problem, you are asked to reconstruct the trading during one day in Gergovia. For simplicity, we will assume that the houses are built along a straight line with equal distance between adjacent houses. Transporting one bottle of wine from one house to an adjacent house results in one unit of work.

Input

The input consists of several test cases.

Each test case starts with the number of inhabitants n ($2 \leq n \leq 100000$). The following line contains n integers a_i ($-1000 \leq a_i \leq 1000$). If $a_i \geq 0$, it means that the inhabitant living in the i-th house wants to buy a_i bottles of wine, otherwise if $a_i < 0$, he wants to sell $-a_i$ bottles of wine. You may assume that the numbers a_i sum up to 0.

The last test case is followed by a line containing 0.

Output

For each test case, print the minimum number of work units needed so that every inhabitant has his demand fulfilled. You may assume that this number fits into a signed 64-bit integer (in C/C++ you can use the data type "long long", or in JAVA the data type "long").

Sample Input	Sample Output
5 5 –4 1 –3 1 6 –1000 –1000 –1000 1000 1000 1000 0	9 9000

Source: Ulm Local 2006

ID for Online Judge: POJ 2940

Hint by the Problemsetter (http://www.informatik .uni-ulm.de/acm/Locals/2006/)

This problem is based on the so-called "Earth Mover's Distance", which is used to calculate a measure of similarity between two histograms. In the one-dimensional case, the following greedy algorithm gives optimal results:

Go through the values from left to right, and try to reduce them to 0 by using greedily the closest values. To get the required linear time complexity, notice that only values to the right can be used to reduce the current value to 0 (since all values to the left are already 0). Therefore, we can add the current value to the next value and add the absolute value to the number of work units needed.

Judges' test data consists of 25 test cases, and most of them are random-generated.

1.3.21 Power et al.

Finding the exponent of any number can be very troublesome as it grows exponentially. But in this problem you will have to do a very simple task. Given two non-negative numbers m and n, you have to find the last digit of m^n in the decimal number system.

Input

The input file contains less than 100000 lines. Each line contains two integers m and n (less than 10^{101}). Input is terminated by a line containing two zeros. This line should not be processed.

Output

For each set of input, you must produce one line of output, which contains a single digit. This digit is the last digit of m^n.

Sample Input	Sample Output
2 2	4
2 5	2
0 0	

Source: June 2003 Monthly Contest

IDs for Online Judge: UVA 10515

 Hint

First, the regularity of the last digit of 8^n is analyzed. And through it, the regularity of the last digit of m^n is obtained.

The last digit of 8^1 is 8. The last digit of 8^2 is 4. The last digit of 8^3 is 2. The last digit of 8^4 is 6. The last digit of 8^5 is 8. The last digit of 8^6 is 4. .... That is, there are four times for one cycle. For example, for 8^{1998}, because 1998 mod 4=2, the last digit of 8^{1998} is 6.

Likewise, for 2, 3, and 7, there are also four times for one cycle; for 4 and 9, there are also two times for one cycle; and the last digit of any power of 5 and 6 is itself.

Therefore, the algorithm is as follows:

Suppose the last digit of *m* is *k*, and the last two digits of *n* is *d*. The last digit of

$$m^n \ ans = (k^p)\%10, \text{ where } p = \begin{cases} 4 & d\%4 == 0 \\ d\%4 & d\%4 \neq 0 \end{cases}.$$

1.3.22 Connect the Cable Wires

Asif is a student of East West University, and he is currently working for the EWUISP to meet his relatively high tuition fees. One day, as a part of his job, he was instructed to connect cable wires to *N* houses. All the houses lie in a straight line. He wants to use only the minimum number of cable wires required to complete his task, such that all the houses receive the cable service. A house can either get the connection from the main transmission center, or it can get it from a house to its immediate left or right, provided the latter house is already getting the service.

You are to write a program that determines the number of different combinations of the cable wires that is possible so that every house receives the service.

> **Example:** If there are two houses, then three combinations are possible, as shown in Figure 1.7.

Input

Each line of input contains a positive integer *N* ($N \leq 2000$). The meaning of *N* is described in the above paragraph. A value of **0** for *N* indicates the end of input which should not be processed.

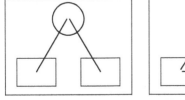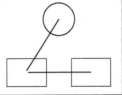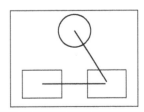

Figure 1.7 **Circles represent the transmission center and the small rectangles represent the houses.**

Output

For each line of input you have to output, on a single line, the number of possible arrangements. You can safely assume that this number will have less than **1000** digits.

Sample Input	Sample Output
1	1
2	3
3	8
0	

Source: The Next Generation - Contest I 2005

ID for Online Judge: UVA 10862

 Hint

Let $f(n)$ be the number of ways to connect the main transmission center and n houses. By removing the main transmission center and its cables to the houses, there will be one or more connected components of houses. Let k be the number of houses of the rightmost connected component. Then, there are k ways to connect one cable from the main transmission center to this component, and there are $f(n-k)$ ways to connect the main transmission center to the rest $n-k$ houses.

So, there are $k \times f(n-k)$ ways to connect them all. Since the range of k is from 1 to n inclusive, by setting $f(0)=1$, we then have $f(n)=1 \times f(n-1)+2 \times f(n-2)+...+ (n-1) \times f(1)+n \times f(0)$. $fib(2 \times n)=fib(n+1) \times fib(n)+fib(n) \times f(n-1)$ (Fibonacci). Therefore, $f(n)=fib(2 \times n)$.

Chapter 2

Practice for Simulation Problems

In the real world, there are many problems that we can solve by simulating their processes. Such problems are called simulation problems. For these problems, solution procedures or rules are shown in problem descriptions. Programs must simulate procedures or implement rules based on descriptions.

In this chapter, three kinds of simulations are introduced:

- Simulation of Direct Statement;
- Simulation by Sieve Method;
- Construction Simulation.

2.1 Simulation of Direct Statement

For problems for simulation of direct statement, programmers are required to solve these problems by strictly implementing rules shown in the descriptions of the problems. Programmers must read such problems carefully, and simulate processes based on descriptions. A problem for simulation of direct statement becomes harder as the number of rules increases. It causes the amount of code to increase and become more illegible.

There are two kinds of simulations of direct statement: simulations based on a sequence of instructions, and simulations based on a sequence of time intervals.

2.1.1 The Hardest Problem Ever

Julius Caesar lived in a time of danger and intrigue. The hardest situation Caesar ever faced was keeping himself alive. In order to survive, he decided to create one of the first ciphers. This cipher was so incredibly sound that no one could figure it out without knowing how it worked.

You are a subcaptain of Caesar's army. It is your job to decipher the messages sent by Caesar and provide the text of the messages to your general. The code is simple. For each letter in a plaintext message, you shift it five places to the right to create the secure message (i.e., if the letter is 'A', the cipher text would be 'F'). Since you are creating plain text out of Caesar's messages, you will do the opposite:

Cipher text: A B C D E F G H I J K L M N O P Q R S T U V W X Y Z
Plain text: V W X Y Z A B C D E F G H I J K L M N O P Q R S T U

Only letters are shifted in this cipher. Any non-alphabetical character should remain the same, and all alphabetical characters will be uppercase.

Input

Input to this problem will consist of a (non-empty) series of up to 100 data sets. Each data set will be formatted according to the following description, and there will be no blank lines separating data sets. All characters will be uppercase.

A single data set has three components:

1. Start line: A single line, "START";
2. Cipher message: A single line containing from 1 to 200 characters, inclusive, comprising a single message from Caesar;
3. End line: A single line, "END".

Following the final data set will be a single line, "ENDOFINPUT".

Output

For each data set, there will be exactly one line of output. This is the original message by Caesar.

Sample Input	Sample Output
START NS BFW, JAJSYX TK NRUTWYFSHJ FWJ YMJ WJXZQY TK YWNANFQ HFZXJX END START	IN WAR, EVENTS OF IMPORTANCE ARE THE RESULT OF TRIVIAL CAUSES

Sample Input	Sample Output
N BTZQI WFYMJW GJ KNWXY NS F QNYYQJ NGJWNFS ANQQFLJ YMFS XJHTSI NS WTRJ END START IFSLJW PSTBX KZQQ BJQQ YMFY HFJXFW NX RTWJ IFSLJWTZX YMFS MJ END ENDOFINPUT	I WOULD RATHER BE FIRST IN A LITTLE IBERIAN VILLAGE THAN SECOND IN ROME DANGER KNOWS FULL WELL THAT CAESAR IS MORE DANGEROUS THAN HE

Source: ACM South Central USA 2002

IDs for Online Judges: POJ 1298, ZOJ 1392, UVA 2540

 Analysis

Obviously, the problem is solved by strictly implementing the rule in the problem description. The rule creating plain text out of Caesar's messages is as follows:

A letter in the plain text = 'A'+(A letter in the cipher text–'A'+21)%26.

 Program

```
#include <iostream>
#include <string>
using namespace std;
int main()
{
   string str;    //Caesar's message
   int i;
   while (cin >> str)    //Input Caesar's message
   {
     cin.ignore(INT_MAX, '\n');
     if (str == "ENDOFINPUT") break;
     getline(cin, str, '\n');
     for (i = 0; i < str.length(); i++)    // The rule
creating plain text
        if (isalpha(str[i]))
          str[i] = 'A' + (str[i] - 'A' + 21) % 26;
     cout << str << endl;    // Output the original message by
Caesar
```

```
   cin >> str;    //Next Caesar's message
  }
  return 0;
}
```

2.1.2 Rock-Paper-Scissors Tournament

Rock-paper-scissors is a game for two players, A and B, who each choose, independently of the other, one of *rock, paper,* or *scissors.* A player choosing *paper* wins over a player choosing *rock*; a player choosing *scissors* wins over a player choosing *paper*; a player choosing *rock* wins over a player choosing *scissors.* A player choosing the same thing as the other player neither wins nor loses.

A tournament has been organized in which each of *n* players plays *k* rock-paper-scissors games with each of the other players—$k\dfrac{n(n-1)}{2}$ games in total. Your job is to compute the *win average* for each player, defined as $\dfrac{w}{w+l}$, where *w* is the number of games won and *l* is the number of games lost by the player.

Input

Input consists of several test cases. The first line of input for each case contains $1 \le n \le 100$, $1 \le k \le 100$ as defined above. For each game, a line follows containing p_1, m_1, p_2, m_2. $1 \le p_1 \le n$ and $1 \le p_2 \le n$ are distinct integers identifying two players; m_1 and m_2 are their respective moves ("rock", "scissors", or "paper"). A line containing 0 follows the last test case.

Output

Output one line each for player 1, player 2, and so on, through player *n*, giving the player's win average rounded to three decimal places. If the win average is undefined, output "-". Output an empty line between cases.

Sample Input	Sample Output
2 4	0.333
1 rock 2 paper	0.667
1 scissors 2 paper	
1 rock 2 rock	0.000
2 rock 1 scissors	1.000
2 1	
1 rock 2 paper	
0	

Source: Waterloo local 2005.09.17

IDs for Online Judges: POJ 2654, UVA 10903

Analysis

This is a problem for simulation of direct statement. In the problem description, a player choosing *paper* wins over a player choosing *rock*; a player choosing *scissors* wins over a player choosing *paper*; and a player choosing *rock* wins over a player choosing *scissors*. A player choosing the same thing as the other player neither wins nor loses. A tournament has been organized in which each of *n* players plays *k* rock-scissors-paper games with each of the other players—$k\dfrac{n(n-1)}{2}$ games in total. For each test case, cases are input one by one; and for each player, the number of games won and the number of games lost are accumulated. Finally, the win average for each player is calculated. For a player, if the number of games won and the number of games lost are all 0, then the win average is undefined; else the win average for the player is $\dfrac{w}{w+l}$.

Program

```c
#include <stdio.h>
#include <string.h>
int w[200], l[200];      //For player i, the number of games won
w[i], and the number of games lost l[i]
int p1,p2,i,j,k,m,n;     // n players, m test cases, k games
for each player, p1 and p2 play a game
char m1[10], m2[10];     //player p1 chooses m1[]; player p2
chooses m2[]
main(){
    for (m=0; 1<=scanf("%d%d",&n,&k)&& n; m++) {//n players
play k rock-scissors-paper games with each of the other
players
        if (m) {
            printf("\n");
            memset(w,0,sizeof(w));     //initialization
            memset(l,0,sizeof(l));
        }
        for(i=0; i<k*n*(n-1)/2;i++){//Input players and moves
            scanf("%d%s%d%s",&p1,m1,&p2,m2);
            if (!strcmp(m1,"rock") && !strcmp(m2,"scissors") ||
                !strcmp(m1,"scissors") && !strcmp(m2,"paper") ||
                !strcmp(m1,"paper") && !strcmp(m2,"rock")) {
                w[p1]++; l[p2]++;     //p1 wins and p2 loses
            }
```

```
            if (!strcmp(m2,"rock") && !strcmp(m1,"scissors") ||
                !strcmp(m2,"scissors") && !strcmp(m1,"paper") ||
                !strcmp(m2,"paper") && !strcmp(m1,"rock")) {
                w[p2]++; l[p1]++;    //p2 wins and p1 loses
            }
        }
// the win average
        for (i=1;i<=n;i++) {
            if (w[i]+l[i]) printf("%0.3lf\n",(double)w[i]/
(w[i]+l[i]));
            else printf("-\n");    //
        }
    }
    if (n) printf("extraneous input! %d\n",n);
}
```

2.1.3 Robocode

Robocode is an educational game designed to help learn Java. The players write programs that control tanks fighting with each other on a battlefield. The idea of this game may seem simple, but it takes a lot of effort to write a winning tank's program. Today we are not going to write an intelligent tank, but design a simplified Robocode game engine.

Assume that the whole battlefield is 120×120 (pixels). Each tank can *only* move in the vertical and horizontal directions on the fixed path. (There are paths every 10 pixels in the battlefield in both vertical and horizontal directions. In all, there are 13 vertical and 13 horizontal paths available for tanks, as shown in Figure 2.1.) The shape and size of the tank are negligible, and one tank has (x, y) $(x, y \in [0, 120])$ representing its coordinate position and α ($\alpha \in \{0, 90, 180, 270\}$) representing its facing direction ($\alpha = 0, 90, 180,$ or 270 means facing right, up, left, or down, respectively). They have a constant speed of 10 pixels/second when they move and they can't move out of the boundary (on touching any boundary of the battlefield,

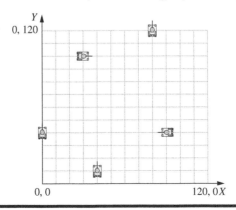

Figure 2.1

the tanks will stop moving, staying in the direction that they are currently facing). The tank can shoot in the direction it's facing whether it's moving or still. The shot moves at the constant speed of 20 pixels/second, and the size of the shot is also negligible. It will explode when it meets a tank on the path. It's possible for more than one shot to explode in the same place if they all reach a tank at the exact same time. The tank being hit by the explosion will be destroyed and removed from the battlefield at once. A shot exploding or flying out of the boundary will also be removed.

When the game begins, all the tanks are stopped at different crosses of the vertical and horizontal paths. Given the initial information of all the tanks and several commands, your job is to find the winner—the last living tank when all the commands are executed (or omitted) and no shot exists in the battlefield (meaning that no tank may die in the future).

Input

There are several test cases. The battlefield and paths are all the same for all test cases as shown in Figure 2.1. Each test case starts with integers N ($1 \leq N \leq 10$) and M ($1 \leq M \leq 1000$), separated by a blank. N represents the number of the tanks playing in the battlefield, and M represents the number of commands to control the movement of the tanks. The following N lines give the initial information (at time 0) of each tank, in the format:

Name x y α

The Name of a tank consists of no more than 10 letters. x, y, α are integers and $x, y \in \{0, 10, 20, ..., 120\}$, $\alpha \in \{0, 90, 180, 270\}$. Each field is separated by a blank.

The following M lines give commands in this format:

Time Name Content

Each field is separated by a blank. All the commands are given in the ascending order of *Time* ($0 \leq Time \leq 30$), which is a positive integer meaning the timestamp when the commands are sent. *Name* points out which tank will receive the command. The *Content* has different types as follows:

MOVE	When receiving the MOVE command, the tank starts to move in its facing direction. If the tank is already moving, the command takes no effect.
STOP	When receiving the STOP command, the tank stops moving. If the tank has already stopped, the command takes no effect.
TURN *angle*	When receiving the TURN command, the tank changes the facing direction α to be ((α + *angle* + 360) mod 360), regardless of whether it is moving or not. You are guaranteed that ((α + *angle* + 360) mod 360) $\in \{0, 90, 180, 270\}$. The TURN command doesn't affect the moving state of the tank.
SHOOT	When receiving the SHOOT command, the tank will shoot one shot in the direction it's facing.

Tanks take the corresponding action as soon as they receive the commands. For example, if the tank at (0, 0), α=90, receives the command MOVE at time 1, it will start moving at once and will reach (0, 1) at time 2. Notice that a tank could receive multiple commands in one second and take the action one by one. For example, if the tank at (0, 0), α=90, receives a command sequence of "TURN 90; SHOOT; TURN −90", it will turn to the direction α=180, shoot, and then turn back. If the tank receives a command sequence of "MOVE; STOP", it will remain in the original position.

Some more notes you need to pay attention to:

If a tank is hit by an explosion, it will take not act on any of the commands received at that moment. Of course, all the commands sent to the already destroyed tank should also be omitted.

Although the commands are sent at discrete seconds, the movement and explosions of tanks and shots happen in the continuous time domain.

No two tanks will meet on the path guaranteed by the input data, so you don't need to consider that situation.

All the input contents will be legal for you.

A test case with *N*=*M*=0 ends the input, and should not be processed.

Output

For each test case, output the winner's name in one line. The winner is defined as the last living tank. If there is no tank, or more than one tank living at the end, output "NO WINNER!" in one line.

Sample Input	Sample Output
2 2	A
A 0 0 90	NO WINNER!
B 0 120 180	B
1 A MOVE	
2 A SHOOT	
2 2	
A 0 0 90	
B 0 120 270	
1 A SHOOT	
2 B SHOOT	
2 6	
A 0 0 90	
B 0 120 0	
1 A MOVE	
2 A SHOOT	
6 B MOVE	

Sample Input	Sample Output
30 B STOP 30 B TURN 180 30 B SHOOT 0 0	

Source: ACM Beijing 2005

IDs for Online Judges: POJ 2729, UVA 3466

 Analysis

The problem is a simulation problem based on a sequence of time intervals. For each command, $0 \leq Time \leq 30$ (seconds), and states may be changed after the last command is sent. Therefore, Robocode must be simulated for 45 seconds at most.

If a tank at (0, 0) shoots at a tank at (0, 1), and the tank at (0, 1) moves to the tank at (0, 0), then the moving tank is shot after it moves 10/3 pixels, and after 0.5 seconds. Therefore, the map should be enlarged six times, and states should be simulated every 1/6 seconds.

Attributes for tanks and shots are as follows: positions, directions, move (or stop), and removed (or unremoved).

Starting at *Time* 0, commands are processed one by one. If the timestamp when the current command is sent is t_2, and the timestamp when the last command is sent is t_1, states from t_1 to t_2 must be simulated. Then attributes for the tank receiving the current command are set as follows:

If the command is the "MOVE" command, the tank receiving the command moves in its facing direction;

If the command is the "STOP" command, the tank receiving the command stops moving;

If the command is the "SHOOT" command and the tank receiving the command isn't removed, then a shot is added, and its attributes are same as the attributes of a tank, except MOVE;

If the command is TURN *angle*, then the tank receiving the command adjusts its facing direction as (the original number of direction $+ \left(\dfrac{angle}{90} \% 4 \right) + 4) \% 4$, where the number of direction is $\dfrac{angle}{90}$.

After all commands are processed, states are simulated for 15 seconds continuously.

Finally, the number of living tanks at the end is calculated. If all tanks are removed, or more than one tank lives, then output "NO WINNER!"; else output the last living tank.

Program

```
#include <iostream>
#include <map>
#include <cstdio>
#include <cstring>
#include <string.h>
#include <string>
using namespace std ;
const int DirX[4] = { 10 , 0 , -10 , 0 } ;      // Horizontal
increment and vertical increment
const int DirY[4] = { 0 , 10 , 0 , -10 } ;
#define mp make_pair
int   N , M , Shoot ;      //N: number of tanks, M: number of
commands, N+1..Shoot: Shots
int   x[1050] , y[1050] , d[1050] ;     // (x[ ], y[ ]) :
positions for tanks and shots; d[ ]: their directions
bool run[1050],die[1050] ;      // run[ ]: flags for tanks'
moving, die[ ]: flags for tanks' or shots' removing
string symbol[1050] ;      // symbol[i]: the i-th tank's name
map<string,int> Name ;      // Name[s]: the sequence number of
the tank whose name is s
void  Init()
{
      Name.clear() ;
      for ( int i = 1 ; i <= N ; i ++ )     //Initialization
      {
          cin >> symbol[i] >> x[i] >> y[i] >> d[i] ;
          x[i] *= 6 ; y[i] *= 6 ;d[i] /= 90 ;     // the map is
enlarged six times, direction numbers are calculated
          run[i] = false ;die[i] = false ;
          Name[symbol[i]] = i ;
      }
      Shoot = N ;
}
bool  In( int x , int y )     //whether (x,y) is in the
boundary or not
{
      if ( x >= 0 && x <= 6*120 && y >= 0 && y <= 6*120 )
return true ;
      return false ;
}
void  RunAll()     // Situation in 1 time unit is simulated
{
      for ( int i = 1 ; i <= N ; i ++ )     // All tanks are
simulated
```

```
        {
            if ( run[i] && !die[i] )
            {
                if ( In( x[i] + DirX[d[i]] , y[i] + DirY[d[i]]
) )
                {
                    x[i] += DirX[d[i]] ;y[i] += DirY[d[i]] ;
                }
                else run[i] = false ;
            }
        }
        for ( int i=N+1 ; i <= Shoot ; i ++ )    //All shots are
simulated
        {
            if ( !die[i] )
            {
                if ( In( x[i] + DirX[d[i]] * 2 , y[i] +
DirY[d[i]] * 2 ) )
                {
                    x[i] += DirX[d[i]] * 2 ;y[i] += DirY[d[i]]
* 2 ;
                }
                else die[i] = true ;
            }
        }
        for ( int i = 1 ; i <= N ; i ++ )     //unremoved tank i
        {
            if ( die[i] ) continue ;
            for ( int j = N+1 ; j <= Shoot ; j ++ ) if ( !die[j] )
//if tank i is shot by shot j
            {
                if ( x[i] == x[j] && y[i] == y[j] )
                {
                    die[j] = true ; die[i] = true ;
                }
            }
        }
    }
}
void  Solve()    //Process commands and output results
{
    int now = 0 ;    // Since Time 0
    for ( int i = 1 ; i <= M ; i ++ )    //Time, Tank,
Content for each command
    {
        int t ; string sym , s ; int th ;
        cin >> t >> sym >> s ;
        t *= 6 ;    // Time *6
        while ( t > now ) { RunAll() ; now ++ ; }
//Simulating situations now ..t
```

```
            int symId = Name[sym] ;      //sequence number for the
tank receiving command
            if ( s == "MOVE" )
                 run[symId] = true ;
            else if ( s == "STOP" )
                 run[symId] = false ;
            else if ( s == "SHOOT" )
            {
              if ( !die[symId] )
                 {
                      Shoot ++ ;
                      run[Shoot] = true ;die[Shoot] = false ;
                      d[Shoot] = d[symId] ; x[Shoot] = x[symId] ;
y[Shoot] = y[symId] ;
                 }
            }
            else    //changing direction
            {
                 cin >> th ; th /= 90 ;
                 d[symId] = (d[symId] + (th % 4) + 4 ) % 4 ;
            }
        }
        for ( int i = 1 ; i <= 15*6 ; i ++ ) RunAll() ;
// simulating 15 seconds
        int cnt = 0 ;     //cnt: the number of last living tanks
        for ( int i = 1 ; i <= N ; i ++ ) if ( !die[i] ) cnt ++ ;
        if ( cnt != 1 ) cout << "NO WINNER!\n" ;
        else
        {
             for ( int i = 1 ; i <= N ; i ++ ) if ( !die[i] )
cout << symbol[i] << "\n" ;
        }
}
int   main()
{
while ( cin>>N>>M && ( N || M ) )
  {
      Init() ;
      Solve() ;
  }
}
```

2.1.4 Eurodiffusion

On January 1, 2002, 12 European countries abandoned their national currency for a new currency, the euro. No more francs, marks, lires, guldens, kroner, ... only euros, all over the eurozone. The same banknotes are used in all countries. And the same coins? Well, not quite. Each country has limited freedom to create its own euro coins.

"Every euro coin carries a common European face. On the obverse, member states decorate the coins with their own motif. No matter which motif is on the coin, it can be used anywhere in the 12 member states. For example, a French citizen is able to buy a hot dog in Berlin using a euro coin with the imprint of the King of Spain." (*Source:* http://europa.eu.int/euro/html/entry.html.)

On January 1, 2002, the only euro coins available in Paris were French coins. Soon the first non-French coins appeared in Paris. Eventually, one may expect all types of coins to be evenly distributed over the 12 participating countries. (Actually this will not be true. All countries continue minting and distributing coins with their own motifs. So even in a stable situation, there should be an excess of German coins in Berlin.) So, how long will it be before the first Finnish or Irish coins are in circulation in the south of Italy? How long will it be before coins of each motif are available everywhere?

You must write a program to simulate the dissemination of euro coins throughout Europe, using a highly simplified model. Restrict your attention to a single euro denomination. Represent European cities as points in a rectangular grid. Each city may have up to four neighbors (one to the north, east, south, and west). Each city belongs to a country, and a country is a rectangular part of the plane. Figure 2.2 shows a map with three countries and 28 cities. The graph of countries is connected, but countries may border holes that represent seas, or non-euro countries, such as Switzerland or Denmark. Initially, each city has one million (1000000) coins in its country's motif. Every day a representative portion of coins, based on the city's beginning day balance, is transported to each neighbor of the city. A representative portion is defined as one coin for every full 1000 coins of a motif.

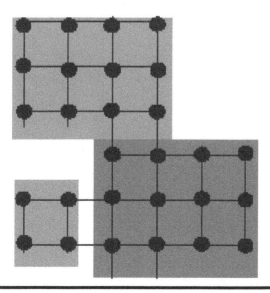

Figure 2.2

A city is complete when at least one coin of each motif is present in that city. A country is complete when all of its cities are complete. Your program must determine the time required for each country to become complete.

Input

The input consists of several test cases. The first line of each test case is the number of countries ($1 \leq c \leq 20$). The next c lines describe each country. The country description has the format: *name* $x_l\, y_l\, x_h\, y_h$, where *name* is a single word with 25 characters at the most; $x_l\, y_l$ are the lower-left city coordinates of that country (most southwestward city) and $x_h\, y_h$ are the upper-right city coordinates of that country (most northeastward city): $1 \leq x_l \leq x_h \leq 10$ and $1 \leq y_l \leq y_h \leq 10$.

The last case in the input is followed by a single zero.

Output

For each test case, print a line indicating the case number, followed by a line for each country with the country's name and the number of days for that country to become complete. Order the countries by days to completion. If two countries have identical days to completion, order them alphabetically by name.

Use the output format shown in the example.

Sample Input	Sample Output
3	Case Number 1
France 1 4 4 6	Spain 382
Spain 3 1 6 3	Portugal 416
Portugal 1 1 2 2	France 1325
1	Case Number 2
Luxembourg 1 1 1 1	Luxembourg 0
2	Case Number 3
Netherlands 1 3 2 4	Belgium 2
Belgium 1 1 2 2	Netherlands 2
0	

Source: ACM World Finals 2003 - Beverly Hills

IDs for Online Judges: UVA 2724

 Analysis

In Europe there are n countries ($1 \leq n \leq 20$). Each country is a rectangular part of the plane. Each city belongs to a country and is a point in the corresponding rectangular grid. Initially, each city has one million (1000000) coins in its country's motif. Every day a representative portion of coins, based on the city's beginning day balance, is transported to each neighbor of the city. In a day, if a city has $x(x>10^3)$ coins of a motif, $d\left(d=\left\lfloor\dfrac{x}{10^3}\right\rfloor\right)$ coins of the motif can be transported to each neighbor. The problem requires you to output the number of days that each country has to become complete, that is, at least one coin of each motif is present in every city.

The problem is a simulation problem based on a sequence of time intervals. Because the data range is small, we can simulate the dissemination of coins every day, and use arrays to store all information.

1. **Constructing a graph for the dissemination of coins.**

    Cities are represented as vertices. Neighboring relations between cities are represented as edges. The information for a vertex includes:

    i. The country which the city belongs to;

    ii. The state for the city, including

    - Marks for all motifs, represented as a binary number with n digits. Initially, the digit corresponding to its country's motif is 1, and other digits are 0. Obviously, when the city is complete, n digits are all 1, that is, the value of the mark is 2^n-1. When values of marks for all vertices are 2^n-1, the algorithm ends.

    - Numbers of all motifs. Initially, each city has one million (1000000) coins in its country's motif, and numbers of other motifs are 0.

    Each city is numbered according to the sequence of input. If there are n countries and m cities ($n\leq m\leq10^2$), in the first country's rectangular grid, the city in its lower-left corner is numbered 1; and in the last country's rectangular grid, the city in its upper-right corner is numbered m. Based on vertices' information and relations, degrees for vertices and the adjacency list for the graph are calculated. Suppose $g[i]$ is the degree of vertex i ($1\leq i\leq m$, $0\leq g[i]\leq4$); and $edge[i][l]$ is the number of the l-th neighboring vertex for vertex i ($0\leq i\leq m-1, 0\leq l\leq4, 0\leq edge[i][l]\leq m-1$).

2. **Simulating the dissemination of coins every day based on a sequence of time intervals.**

    Today's dissemination of coins is only based on yesterday's dissemination of coins. Therefore, in the simulation there are two states: precursor state $o1$ and current state $o2$. And calculating $o2$ is only based on $o1$. The simulation process of each day is as follows: Initially, $f[o2]\leftarrow f[o1]$, and $st[o2]\leftarrow st[o1]$. It flips after simulating the dissemination of coins, that is $o1\leftrightarrow o2$. Suppose $f[o1][i][j]$ is the number of coins in motif j in city i yesterday; $st[o1][i]$ are marks for all motifs in city i yesterday; $f[o2][i][j]$ is the number of coins in motif j in city i today, $st[o2][i]$ are marks for all motifs in city i today; $a[k].ans$ is the number of days for country k to become complete; and $a[k].name$ is the name of country k.

Initially, $o1=0$, $o2=1$. For vertex j in country k ($0 \leq k \leq n-1$, the number of the first vertex in country $k \leq j \leq$ (the number of the last vertex in country k), $f[o1][j][i]=10^6$, $st[o1][j]=2^k$, and other values for $f[o1]$ and $st[o1]$ are all 0.

The goal for the simulation is to calculate two variables.

a. *cnt*: The number of current cities that become complete. Obviously, initially *cnt* is 0. And the simulation ends when $cnt==m$.

b. *day[y]*: The number of days for city y to become complete. The number of days for country k to become complete is the maximal value for numbers of days for its cities to become complete, that is, $a[k].ans=\max_{y \in \text{country } k} day[y]$.

From day 0 ($ans \leftarrow 0$), the dissemination of coins is simulated day by day until $cnt==m$:

```
++ans;
The current state o2 is calculated based on the precursor
state o1(f[o2]=f[o1], st[o2]=st[o1]);
Each city i(0≤i≤m-1) is enumerated:
{ The binary digit k whose value is 1 in st[o1][i] is
enumerated:
```

The number of motif k transported to each neighborhood of city i d is calculated $\left(d = \left\lfloor \dfrac{f[o1][i][k]}{10^3} \right\rfloor \right)$;

```
    if (d≠0)
        { f[o2][i][k]-=g[i]*d;
```

Each neighboring city y for city i is enumerated ($y = edge[i][l], 0 \leq l \leq g[i]-1$): if city y has no motif k, and city y will have n motifs after it has motif k ($f[o2][y][k]==0$ && $(st[o2][y] | = 2^k)==2^n - 1$), then the number of days for city y to become complete is *ans* ($day[y]=ans$), and the number of complete cities increases 1 ($++cnt$);

The number of motif k in city y increases d ($f[o2][y][k]+=d$);

```
        }
    }

o1 ↔ o2;
```

After the above simulation, number of days for m cities to become complete is *day[ ]*. Based on that, $a[k].ans=\max_{y \in \text{country } k} day[y]$; $0 \leq k \leq n-1$.

Finally, *a[ ]* is sorted: *a[ ].ans* is as the first key, and *a[ ].name* is as the second key. And *a[i].name* and *a[i].ans* ($0 \leq i \leq n-1$) are output line by line.

Program

```cpp
#include <cstdio>
#include <cstring>
#include <algorithm>
using namespace std;
#define ms(x, y) memset(x, y, sizeof(x))
#define mc(x, y) memcpy(x, y, sizeof(x))
const int dir[4][2] = {{1, 0}, {-1, 0}, {0, 1}, {0, -1}};
// shift for 4 directions
struct city {      // city
  char name[30];     // city name
  int ans;     // the number of days for the city to become
complete
};
int cs(0);
int log2[1 << 21];     // log2[2^i]=i
int n, tot, full;     //n: the number of countries, tot: the
number of cities, full: the mark that n countries become
complete
city a[22];     //the sequence of countries
int bl[22], br[22];     // for country i, bl[i]: the first
city, br[i]: the last city
int num[11][11], belong[122];     //num[x][y]: the number of
the city at (x, y), belong[t]: the country that city t belongs
to
int g[122];     //g[i]: the number of neighboring cities for
city i
int edge[122][4];     // edge[i][l] is the number of lth
neighboring vertex for vertex i
int o1, o2, f[2][122][22];     // precursor state o1 and
current state o2; f[o][i][j] is the number of coins in motif j
in city i in state o
int day[122], st[2][122];     // day[y]: Number of days for
city y to become complete, st[o][i] are marks for all motifs
in city i, represented as a binary number with n digits: if
the k-th digit is 1, city i has the coin in motif k;
otherwise, city i does not have the coin in motif k; 0≤k≤n-1.
bool cmp(const city &a, const city &b) {     // Compare country
a and b (the first key is the number of days for a country to
become complete, and the second key is names of countries)
  return a.ans < b.ans || a.ans == b.ans && strcmp(a.name,
b.name) < 0;
}
void print() {     //Output the solution to the current
test case
```

```
   sort(a, a + n, cmp);      //Sorting countries a[ ]
   printf("Case Number %d\n", ++cs);      //Output the number of
test cases
   for (int i = 0; i < n; ++i)      // Output countries' names
and the number of days for countries to become complete in
a[ ]
     printf("   %s    %d\n", a[i].name, a[i].ans);
}
int main() {
   for (int i = 0; i < 21; ++i) log2[1 << i] = i;
// log₂[2ⁱ]=i
   while (scanf("%d", &n), n) {      //repeatedly input
countries' names until 0
     tot=0; full=(1 << n)-1;      //tot: number of cities, the
mark for a city to become complete 2ⁿ-1
     ms(num, 0xFF);      //num[][] is initialized 255
     for (int i=0, x1, y1, x2, y2; i<n; ++i) {      // Input
countries' names and coordinates
       scanf("%s%d%d%d%d", a[i].name, &x1, &y1, &x2, &y2);
       --x1, --y1, --x2, --y2;
       bl[i] = tot;      //start city for country i
         for (int x=x1; x<=x2; ++x)      //every city in the
rectangles belongs to country i
           for (int y = y1; y <= y2; ++y) {
             num[x][y] = tot; belong[tot++] = i;
           }
       br[i] = tot;      //end city for country i
     }
     if (n == 1) {      //If there is only one country
       a[0].ans = 0;
       print();
       continue;
     }
// Initialization: calculate the number of neighbors for each
city, and construct edge[ ][ ]
     ms(g, 0);
     for (int i=0; i<10;++i)      //Enumeration
       for (int j = 0; j < 10; ++j)
       if (num[i][j]!= -1)      //If (i, j) is a city
         for (int k = 0, nx, ny; k < 4; ++k) {
             nx = i + dir[k][0], ny = j + dir[k][1];
             if(nx>=0 && nx<10 && ny>=0 && ny<10 && num[nx]
[ny]!=-1)
               edge[num[i][j]][g[num[i][j]]++] = num[nx][ny];
         }
     o1 = 0, o2 = 1;      //Initialize states
     ms(f[o1], 0); ms(st[o1], 0);
     for (int i = 0; i < n; ++i)      //Enumerate each country
     for (int j = bl[i]; j < br[i]; ++j) {      // Enumerate
city j in country i. Initially city j has 10⁶ coins in motif i
```

```
        f[o1][j][i] = 1000000; st[o1][j] = 1 << i;
      }
  ms(day, 0xFF);
  int ans = 0, cnt = 0;
  do {
     ++ans;
     mc(f[o2], f[o1]); mc(st[o2], st[o1]);   //calculate
the current state based on the precursor state
     for (int i = 0; i < tot; ++i)   // Enumerate city i
        for(int j=st[o1][i], k, d; j; j-=1<<k){
// Enumerate motif k in city i
        k = log2[j - (j & (j - 1))];
        d=f[o1][i][k] / 1000;     // The number of motif k
transported to each neighbor of city i d is calculated
           if(d){     //if motif k can be transported
              f[o2][i][k] -= g[i] * d;
              for (int l=0, y; l<g[i]; ++l){
                 y = edge[i][l];
                 if (f[o2][y][k]==0 && (st[o2][y] |= 1 << k) ==
full) {
                    day[y]=ans;
                    ++cnt;
                 }
                 f[o2][y][k] += d; d
              }
            }
        }
     swap(o1, o2);    // o1↔o2
  } while (cnt < tot);   // until tot cities become
complete
// numbers of days for all countries to become complete
    for (int i = 0; i < n; ++i) {    // Enumerate every
country
       a[i].ans = 0;
       for (int j = bl[i]; j < br[i]; ++j) a[i].ans =
max(a[i].ans, day[j]);
     }
     print();    //Output the solution to the current test
case
   }
   return 0;
}
```

2.2 Simulation by Sieve Method

In some problems, constraints are given in descriptions. And these constraints constitute a sieve. All possible solutions are put on the sieve to filter out solutions that do not meet constraints. Finally, solutions settling on the sieve are solutions to the problem. The method for solving such problems is called the simulation by sieve method. The structure and idea for the simulation by sieve method is concise and clear, but it is also blind. Therefore, its time efficiency may not be good. The key to the simulation by sieve method is to find the constraints. Any errors and omissions will lead to failure. Because filtering rules do not need complex algorithm design, such problems are usually simple simulation problems.

2.2.1 The Game

A game of Renju is played on a 19×19 board by two players. One player uses black stones and the other uses white stones. The game begins on an empty board and two players alternate in placing black stones and white stones. Black always goes first. There are 19 horizontal and 19 vertical lines on the board, and the stones are placed on the intersections of the lines.

Horizontal lines are marked 1, 2, …, 19 from up to down, and vertical lines are marked 1, 2, …, 19 from left to right, as shown in Figure 2.3.

The objective of this game is to put five stones of the same color consecutively along a horizontal, vertical, or diagonal line. So, black wins in Figure 2.3. But, a player does not win the game if more than five stones of the same color were put consecutively.

Given a configuration of the game, write a program to determine whether white has won, or black has won, or nobody has won yet. There will be no input data

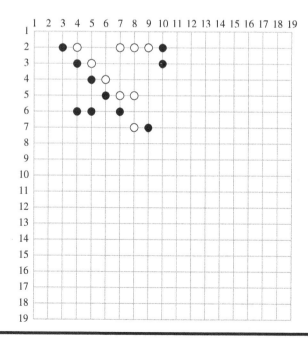

Figure 2.3

where the black and the white both win at the same time. Also, there will be no input data where the white or the black wins in more than one place.

Input

The first line of the input file contains a single integer t ($1 \leq t \leq 11$), the number of test cases, followed by the input data for each test case. Each test case consists of 19 lines, each having 19 numbers. A black stone is denoted by 1, a white stone is denoted by 2, and 0 denotes no stone.

Output

There should be one or two line(s) per test case. In the first line of the test case output, you should print 1 if black wins, 2 if white wins, and 0 if nobody wins yet. If black or white won, in the second line print the horizontal line number and the vertical line number of the leftmost stone among the five consecutive stones. (Select the uppermost stone if the five consecutive stones are located vertically.)

Sample Input	Sample Output
1	1
0 0 0 0 0 0 0 0 0 0 0 0 0 0 0 0 0 0 0	3 2
0 0 0 0 0 0 0 0 0 0 0 0 0 0 0 0 0 0 0	
0 1 2 0 0 2 2 2 1 0 0 0 0 0 0 0 0 0 0	
0 0 1 2 0 0 0 0 1 0 0 0 0 0 0 0 0 0 0	
0 0 0 1 2 0 0 0 0 0 0 0 0 0 0 0 0 0 0	
0 0 0 0 1 2 2 0 0 0 0 0 0 0 0 0 0 0 0	
0 0 1 1 0 1 0 0 0 0 0 0 0 0 0 0 0 0 0	
0 0 0 0 0 0 2 1 0 0 0 0 0 0 0 0 0 0 0	
0 0 0 0 0 0 0 0 0 0 0 0 0 0 0 0 0 0 0	
0 0 0 0 0 0 0 0 0 0 0 0 0 0 0 0 0 0 0	
0 0 0 0 0 0 0 0 0 0 0 0 0 0 0 0 0 0 0	
0 0 0 0 0 0 0 0 0 0 0 0 0 0 0 0 0 0 0	
0 0 0 0 0 0 0 0 0 0 0 0 0 0 0 0 0 0 0	
0 0 0 0 0 0 0 0 0 0 0 0 0 0 0 0 0 0 0	
0 0 0 0 0 0 0 0 0 0 0 0 0 0 0 0 0 0 0	
0 0 0 0 0 0 0 0 0 0 0 0 0 0 0 0 0 0 0	
0 0 0 0 0 0 0 0 0 0 0 0 0 0 0 0 0 0 0	
0 0 0 0 0 0 0 0 0 0 0 0 0 0 0 0 0 0 0	
0 0 0 0 0 0 0 0 0 0 0 0 0 0 0 0 0 0 0	

Source: ACM Tehran Sharif 2004 Preliminary

IDs for Online Judge: POJ 1970, ZOJ 2495

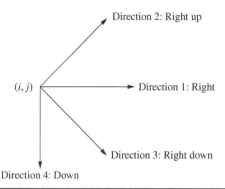

Figure 2.4

 Analysis

Initially all stones on the 19×19 board constitute a sieve. Every stone is scanned from top to down and from left to right. If there is a stone at position (i, j), its adjacent stones in direction k are analyzed ($0 \le k \le 3$, $0 \le i, j \le 18$), as shown in Figure 2.4.

The objective of this game is to put five stones of the same color consecutively along a horizontal, vertical, or diagonal line. Therefore, the constraint conditions for winning a game are as follows:

1. The number at position (i, j) is different from the number at the adjacent position in the opposite direction for direction k;
2. From (i, j) and along direction k, five positions are in the board;
3. From (i, j) and along direction k, numbers at five continuous positions are the same, and the number at the sixth position is different, or the sixth position is out of the board.

If the above constraint conditions hold, the stone at position (i, j) wins the game. If four directions are examined and the above constraint conditions don't hold, the stone at position (i, j) is filtered out.

If all stones are filtered out, nobody wins the game.

 Program

```
#include <iostream>
using namespace std;
const int d[4][2] = {{0, 1}, {1, 0}, {1, 1}, {-1, 1}};
//displacements for 4 directions
```

```
inline bool valid(int x, int y)      //(x, y) in the board or
not
{
   return x >= 0 && x < 19 && y >= 0 && y < 19;
}
int a[20][20];      //board
int main()
{
   int i, j, k, t, x, y, u;
   scanf("%d", &t);      //the number of test cases
   while (t--)      //Input test cases
   {
      for (i = 0; i < 19; ++i)      //Input a board
         for (j = 0; j < 19; ++j) scanf("%d", &a[i][j]);
      for (j = 0; j < 19; ++j)      // Every stone (i, j) is
scanned from top to down and from left to right.
      {
         for (i = 0; i < 19; ++i)
         {
            if (a[i][j] == 0) continue;
            for (k = 0; k < 4; ++k)      //4 directions are
enumerated
            {
               x = i - d[k][0];y = j - d[k][1];
               if (valid(x, y) && a[x][y] == a[i][j]) continue;
               x = i + d[k][0] * 4;y=j + d[k][1] * 4;
               if (!valid(x, y)) continue;
               for (u = 1; u < 5; ++u)
               {
                  x = i + d[k][0] * u;y = j + d[k][1] * u;
                  if (a[x][y] != a[i][j]) break;
               }
               x = i+d[k][0]*5;y = j+d[k][1]*5;
               if (valid(x, y) && a[x][y] == a[i][j]) continue;
               if (u == 5) break;
            }
            if (k < 4) break;
         }
         if (i < 19) break;
      }
      if (j < 19)      // five stones of the same color
consecutively along a direction from (i,j), the color wins the
game;
      {
         printf("%d\n", a[i][j]);
         printf("%d %d\n", i + 1, j + 1);
      }
      else puts("0");      // nobody wins the game
   }
   return 0;
}
```

2.2.2 Game Schedule Required

Sheikh Abdul really loves football. So you better not ask how much money he has spent to make famous teams join the annual tournament. Of course, having spent so much money, he would like to see certain teams play each other. He has worked out a complete list of games that he would like to see. Now it is your task to distribute these games into rounds according to the following rules:

1. In each round, each remaining team plays at most one game;
2. If there is an even number of remaining teams, every team plays exactly one game;
3. If there is an odd number of remaining teams, there is exactly one team which plays no game (it advances with a wildcard to the next round);
4. The winner of each game advances to the next round, and the loser is eliminated from the tournament;
5. If there is only one team left, this team is declared the winner of the tournament.

As can be proved by induction, in such a tournament with n teams, there are exactly n–1 games required until a winner is determined.

Obviously, after round 1, teams may already have been eliminated which should take part in another game. To prevent this, for each game you also have to tell which team should win.

Input

The input contains several test cases. Each test case starts with an integer n ($2{\le}n{\le}1000$), the number of teams participating in the tournament. The following n lines contain the names of the teams participating in the tournament. You can assume that each team name consists of up to 25 letters of the English alphabet ('a' to 'z' or 'A' to 'Z').

Then follow n–1 lines, describing the games that Sheikh Abdul would like to see (in any order). Each line consists of the two names of the teams which take part in that game. You can assume that it is always possible to find a tournament schedule consisting of the given games.

The last test case is followed by a zero.

Output

For each test case, write the game schedule, distributed in rounds.

For each round, first write "Round #X" (where X is the round number) in a line by itself. Then write the games scheduled in this round in the form: "A defeats B", where A is the name of the advancing team and B is the name of the team being eliminated. You may write the games of a round in any order. If a wildcard is needed for the round, write "*A* advances with wildcard" after the last game of the round, where *A* is the name of the team which gets the wildcard. After the last round, write the winner in the format shown below. Print a blank line after each test case.

Sample Input	Sample Output
3	Round #1
A	B defeats A
B	C advances with wildcard
C	Round #2
A B	C defeats B
B C	Winner: C
5	
A	Round #1
B	A defeats B
C	C defeats D
D	E advances with wildcard
E	Round #2
A B	E defeats A
C D	C advances with wildcard
A E	Round #3
C E	E defeats C
0	Winner: E

Source: Ulm Local 2005

IDs for Online Judges: POJ 2476, ZOJ 2801

 Analysis

There are n teams and $n-1$ games. For $n-1$ games that Sheikh Abdul would like to see, the two names of the teams which take part in the game are stored in $a[i]$ and $b[i]$ respectively, $1 \leq i \leq n-1$. Numbers of games that teams take part in are stored in $cnt[i]$, $1 \leq i \leq n$.

Constraints in the problem description constitute a sieve. Initially all teams are put on the sieve.

Sheikh Abdul would like to see every game in each round. In a round, a team which will take part in other games will win the game. Constraints constituting a sieve are as follows:

> In each round, the number of games is the number of teams in the current round divided by 2.
> In each round, $n-1$ games that Sheikh Abdul would like to see are searched sequentially. For game i, $1 \leq i \leq n-1$, if $a[i]$ and $b[i]$ are in the sieve, and one

team can only take part in one game, then the game that $a[i]$ and $b[i]$ take part in is in the round, and the team that has only one game is defeated and filtered out. After $n-1$ games are searched, teams in the sieve enter the next round.

The above process is repeated until only one team is left.

 Program

```
#include<iostream>
#include<cstdlib>
#include<cstdio>
#include<cstring>
#include<cmath>
#include<algorithm>
#include<map>
using namespace std;
const int maxN=1010;
int n,a[maxN],b[maxN],cnt[maxN];     //n: the number of teams;
teams taking part in game i are a[i] and b[i], 1≤i≤n-1; the
number of games that team k is taking part in is cnt[k], 1≤k≤n
char name[maxN][30];      //teams' names
bool flag[maxN];      //the flag indicates whether a team is in
the sieve or not
map<string,int> que;
bool cmp(int a,string s)     //determine whether the name for
team a is s or not
{
   for (int i=0;i<s.size();i++)
     if (name[a][i]!=s[i]) return false;
   return true;
}
void init()     //Input a test case: n teams and n-1 games
{
   que.clear();
   for (int i=1;i<=n;i++)     //team's name
   {
   scanf("%s",name[i]);
   que.insert(map<string,int>::value_type(name[i],i));
//teams' numbers
   }
   string s;
   int p;
   char ch;scanf("%c",&ch);
   for (int i=1;i<n;i++)     // n-1 games
```

```
    {
      scanf("%c",&ch);s="";
      while (ch!=' ') { s+=ch;scanf("%c",&ch);}
      p=que[s];
      cnt[p]++;a[i]=p;
      scanf("%c",&ch);s="";
      while (ch!='\n') { s+=ch;scanf("%c",&ch);}
      p=que[s];
      cnt[p]++;b[i]=p;
    }
}
void work()     // calculate and output the game schedule,
distributed in rounds
{
  int rnd=1,tm=n,s=n/2,now=0;    //rnd: the number of the
current round, tm: the number of teams in the sieve, s: the
number of games in a round, now: the number of hold games in a
round
  memset(flag,1,sizeof(flag));    // Initially all teams are
put on the sieve
  while (tm!=1)                   // only one team left
    for (int i=1;i<n;i++)         // n-1 games are searched
sequentially
      if (flag[a[i]]&&flag[b[i]]&&((cnt[a[i]]==1)||(cnt
[b[i]]==1)))//two teams are on the sieve, at least one team
can only take part in a game
      {
        if (now==0)printf("Round #%d\n",rnd);// the round
number
        now++;tm--;      //number of hold games in the current
round +1,
        cnt[a[i]]--;cnt[b[i]]--;
// if only b[i] take part in one game, b[i] is defeated; if
only a[i] take part in one game, a[i] is defeated; and if a[i]
and b[i] take part in one game, b[i] wins
        if (cnt[a[i]]) printf("%s defeats %s\n",name[a[i]],
name[b[i]]);
          else if (cnt[b[i]]) printf("%s defeats %s\n",
name[b[i]],name[a[i]]);
            else{
                  printf("%s defeats %s\n",name[b[i]],
                  name[a[i]]);
                  printf("Winner: %s\n",name[b[i]]);}
        flag[a[i]]=false;flag[b[i]]=false;

        if (now==s)
        {
          now=0;rnd++;s=tm/2;
          for (int i=1;i<=n;i++)    // wildcard for the team
that doesn't take part in a game in the round
          {
```

```
                if (flag[i] && cnt[i])
                    printf("%s advances with wildcard\n",name[i]);
                if (cnt[i]) flag[i]=true;else flag[i]=false;
            }
        }
    }
  printf("\n");
}
int main()
{
  while (scanf("%d",&n),n)      //number of teams
  {
    init();       // Input a test case: n teams and n-1 games
    work();       // calculate and output the game schedule,
distributed in rounds
  }
  return 0;
}
```

2.3 Construction Simulation

Construction simulation is a kind of relatively complex simulation method. It requires a mathematical model to represent and solve a problem. We need to design parameters of the model, and calculate a simulation result. Because such mathematical models represent objects and their relationships accurately, the efficiencies are relatively high.

2.3.1 Packets

A factory produces products packed in square packets of the same height h and of the sizes 1×1, 2×2, 3×3, 4×4, 5×5, and 6×6. These products are always delivered to customers in the square parcels of the same height h as the products have and of the size 6×6. Because of the expenses, it is in the interest of the factory as well as of the customer to minimize the number of parcels necessary to deliver the ordered products from the factory to the customer. A good program, solving the problem of finding the minimum number of parcels necessary to deliver the given products according to an order, would save a lot of money. You are asked to create such a program.

Input

The input file consists of several lines specifying orders. Each line specifies one order. Orders are described by six integers separated by one space, representing successively the number of packets of individual size from the smallest size 1×1 to the biggest size 6×6. The end of the input file is indicated by the line containing six zeros.

Output

The output file contains one line for each line in the input file. This line contains the minimal number of parcels into which the order from the corresponding line of the input file can be packed. There is no line in the output file corresponding to the last "null" line of the input file.

Sample Input	Sample Output
0 0 4 0 0 1	2
7 5 1 0 0 0	1
0 0 0 0 0 0	

Source: ACM Central Europe 1996

IDs for Online Judges: POJ 1017, ZOJ 1307, UVA 311

 Analysis

The simulation problem is solved by the construction method. The greedy method is also used. Packets are packed in parcels in descending order by size. Because the parcels' size is 6×6, each packet sized 4×4, 5×5, or 6×6 is packed in a parcel. The strategy is as follows:

A packet sized 6×6 is packed in a parcel.
A packet sized 5×5 is packed in a parcel. Packets sized 1×1 are packed into the remaining space of the parcel.
A packet sized 4×4 is packed in a parcel. Packets sized 2×2 are packed into the remaining space of the parcel. If there is no packet sized 2×2, packets sized 1×1 are packed into the remaining space of the parcel.
Four packets sized 3×3 are packed in a parcel.

The algorithm is as follows:
Suppose the number of packets sized $i×i$ is a_i ($1≤i≤6$).

The number of parcels in which packets sized 6×6, 5×5, 4×4, and 3×3 are packed is $M = a_6 + a_5 + a_4 + \left\lceil \dfrac{a_3}{4} \right\rceil$.

The number of packets sized 2×2 and which can be packed in above M parcels is $L_2 = a_4 \times 5 + u[a_3 \bmod 4]$, where $u[0]=0$, $u[1]=5$, $u[2]=3$, and $u[3]=1$. If there are any remaining packets sized 2×2 ($a_2 > L_2$), they are packed in new $\left\lceil \dfrac{a_2 - L_2}{9} \right\rceil$ parcels. And $M += \left\lceil \dfrac{a_2 - L_2}{9} \right\rceil$.

The number of packets sized 1×1 and which can be packed in above M parcels is $L_1 = M \times 36 - a_6 \times 36 - a_5 \times 25 - a_4 \times 16 - a_3 \times 9 - a_2 \times 4$. If there are remaining packets sized 1×1 ($a_1 > L_1$), they are packed in new $\left\lceil \dfrac{a_1 - L_1}{36} \right\rceil$ parcels. And $M += \left\lceil \dfrac{a_1 - L_1}{36} \right\rceil$.

Obviously, M is the minimum number of parcels.

Program

```cpp
#include <iostream>
using namespace std;
int main()
{
    int a[10],i,j,sum,m,left1,left2;     // the number of packets
whose size are i*i is a[i], the number of packets is sum, the
minimal number of parcels is m; the number of parcels in which
2*2 can be packed is left2, the number of parcels in which 1*1
can be packed is left1
    int u[4]={0,5,3,1};     // u[a[3]% 4]
    while (1)
    {
        sum=0;
        for(i=1;i<=6;i++)     //Input the number of packets
        {
            cin>>a[i];
            sum+=a[i];
        }
        if(sum==0) break;
        m=a[6]+a[5]+a[4]+(3+a[3])/4;     // The number of parcels
in which packets whose size are 6*6, 5*5, 4*4, and 3*3 are
packed
        left2=a[4]*5+u[a[3]%4];     // the number of parcels in
which 2*2 can be packed is left2
        if(a[2]>left2)     //If there are remaining 2*2 packets,
new parcels are needed
            m+=(a[2]-left2+8)/9;
        left1=m*36-a[6]*36-a[5]*25-a[4]*16-a[3]*9-a[2]*4;     //
the number of parcels in which 1*1 can be packed is left1
        if(a[1]>left1)     // If there are remaining 1*1 packets,
new parcels are needed
            m+=(a[1]-left1+35)/36;
        cout<<m<<endl;     // the minimal number of parcels
    }
    return 0;
}
```

2.3.2 Paper Cutting

ACM managers need business cards to present themselves to their customers and partners. After the cards are printed on a large sheet of paper, they are cut with a special cutting machine. Since the machine operation is very expensive, it is necessary to minimize the number of cuts made. Your task is to find the optimal solution to produce the business cards.

There are several limitations you have to comply with. The cards are always printed in a grid structure of exactly *a*×*b* cards. The structure size (number of business cards in a single row and column) is fixed and cannot be changed due to printing software restrictions. The sheet is always rectangular and its size is fixed. The grid must be perpendicular to the sheet edges, that is, it can be rotated by 90° only. However, you can exchange the meaning of rows and columns and place the cards into any position on the sheet; they can even touch the paper edges.

For instance, assume the card size is 3×4 cm, and the grid size 1×2 cards. The four possible orientations of the grid are depicted in Figure 2.5. The minimum paper size needed for each of them is stated.

The cutting machine used to cut the cards is able to make an arbitrary long continuous cut. The cut must run through the whole piece of the paper; it cannot stop in the middle. Only one free piece of paper can be cut at once—you cannot stack pieces of paper onto each other, nor place them beside each other to save cuts.

Input

The input consists of several test cases. Each of them is specified by six positive integer numbers, *A*, *B*, *C*, *D*, *E*, and *F*, on one line separated by a space. The numbers are:

> *A* and *B* are the size of a rectangular grid, $1 \leq A, B \leq 1000$; *C* and *D* are the dimensions of a card in cms, $1 \leq C, D \leq 1000$; and *E* and *F* are the dimensions of a paper sheet in cms, $1 \leq E, F \leq 1000000$.

The input is terminated by a line containing six zeros.

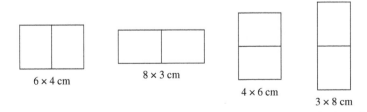

6 × 4 cm   8 × 3 cm   4 × 6 cm   3 × 8 cm

Figure 2.5

Output

For each of the test cases, output a single line. The line should contain the text: "The minimum number of cuts is X.", where X is the minimum number of cuts required. If it is not possible to fit the card grid onto the sheet, output the sentence "The paper is too small." instead.

Sample Input	Sample Output
1 2 3 4 9 4	The minimum number of cuts is 2.
1 2 3 4 8 3	The minimum number of cuts is 1.
1 2 3 4 5 5	The paper is too small.
3 3 3 3 10 10	The minimum number of cuts is 10.
0 0 0 0 0 0	

Source: CTU Open 2003

IDs for Online Judges: POJ 1791, ZOJ 2160

 Analysis

First, the cutting machine cuts the paper to produce grids. Then it cuts grids to produce cards. Suppose $A \times B$ is the size of a rectangular grid; $C \times D$ is the size of a card; and $E \times F$ is the size of a paper sheet. In the longitudinal direction, there are A cards whose length is C. That is, the length is $A \times C$ in the longitudinal direction. In the horizontal direction, there are B cards whose length is D. That is, the length is $B \times D$ in the horizontal direction. The constraint condition is $(A \times C \le E) \&\& (B \times D \le F)$.

In order to produce $A \times B$ rectangular grids, at least $A \times B - 1$ cuts are needed. If $A \times C < E$ in the longitudinal direction, a cut is needed. And if $B \times D < F$ in the horizontal direction, a cut is added. Therefore, the minimal number of cuts $C_0 = A \times B - 1 + (A \times C < E) + (B \times D < F)$.

Grids can be turned 90°, 180°, and 270°. Cases are as follows:

1. $B \times A$ is the size of a rectangular grid; $C \times D$ is the size of a card; and $E \times F$ is the size of a paper sheet.
2. $A \times B$ is the size of a rectangular grid; $D \times C$ is the size of a card; and $E \times F$ is the size of a paper sheet.
3. $B \times A$ is the size of a rectangular grid; $D \times C$ is the size of a card; and $E \times F$ is the size of a paper sheet.

Based on the above method, the minimum numbers of cuts are C_1, C_2, and C_3, respectively. If the constraint condition doesn't hold, the minimum number of cut is ∞. Obviously the minimum number of cuts is $Ans=\min\{C_0, C_1, C_2, C_3\}$.

If $Ans=\infty$, it is not possible to fit the card grid onto the sheet.

 Program

```c
#include <stdio.h>
#include <stdlib.h>
#include <limits.h>
#define TOOBIG INT_MAX
int ncuts(int a,int b,int c,int d,int e,int f) ;
void do_solve(int a,int b,int c,int d,int e,int f)
//enumerate four cases and calculate the minimal numbers of
cuts
{
  int x,m ;
  m=ncuts(a,b,c,d,e,f) ;     //Case 0: C₀
  if ((x=ncuts(b,a,c,d,e,f))<m) m=x ;     // Case 1: C₁
  if ((x=ncuts(a,b,d,c,e,f))<m) m=x ;     // Case 2: C₂
  if ((x=ncuts(b,a,d,c,e,f))<m) m=x;     // Case 3: C₃
  if (m==TOOBIG)
    puts("The paper is too small.") ;
  else
    printf("The minimum number of cuts is " "%d.\n",m) ;
}
int ncuts(int a,int b,int c,int d,int e,int f)
{
  if (a*c>e || b*d>f) return TOOBIG ;     // constraint
condition
 return a*b-1+(a*c<e)+(b*d<f) ;     // the minimal number of
cuts
}
int main()
{ int a,b,c,d,e,f ;
  for(;;) {
    a=0 ; b=0 ; c=0 ; d=0 ; e=0 ; f=0 ;
    scanf("%d %d %d %d %d %d",&a,&b,&c,&d,&e,&f) ;     //a test
case
    if (!a && !b && !c && !d && !e && !f) break ;
    do_solve(a,b,c,d,e,f) ;
  }
  return 0 ;
}
```

2.4 Problems

2.4.1 Mileage Bank

The Mileage program of ACM (Airline of Charming Merlion) is good for travelers who fly frequently. Once you complete a flight with ACM, you can earn ACMPerk miles in your ACM Mileage Bank, depending on the mileage you actually fly. In addition, you can use the ACMPerk mileage in your Mileage Bank to exchange for a free flight ticket from ACM in the future.

The following table helps you calculate how many ACMPerk miles you can earn when you fly on ACM.

When you fly ACM	Class Code	You'll Earn
First Class	F	Actual mileage + 100% mileage bonus
Business Class	B	Actual mileage + 50% mileage bonus
Economy Class 1–500 miles 500+ miles	Y	500 miles Actual mileage

The ACMPerk mileage consists of two parts. One is the actual flight mileage (the minimum ACMPerk mileage for the economy class for one flight is 500 miles), and the other is the mileage bonus (its accuracy is up to one mile) when a traveller flies in business class and first class. For example, one can earn 1329 ACMPerk miles, 1994 ACMPerk miles, and 2658 ACMPerk miles for Y, B, or F class, respectively, for the flight from Beijing to Tokyo (the actual mileage between Beijing and Tokyo is 1329 miles). When one flies from Shanghai to Wuhan, one can earn ACMPerk 500 miles for economy class and ACMPerk 650 miles for business class (the actual mileage between Shanghai and Wuhan is 433 miles).

Your task is to help ACM build a program for automatic calculation of ACMPerk mileage.

Input

The input file contains several data cases. Each case has many flight records, each per line. The flight record is in the following format:

OriginalCity   DistanceCity   ActualMiles   ClassCode

Each case ends with a line of one zero.
A line of one # presents the end of the input file.

Output

Output the summary of ACMPerk mileages for each test case, one per line.

Sample Input	Sample Output
Beijing Tokyo 1329 F Shanghai Wuhan 433 Y 0 #	3158

Source: ACM Beijing 2002

IDs for Online Judges: POJ 1326, ZOJ 1365, UVA 2524

Hint

The problem is a simple, straightforward simulation problem. First, flight records are input one by one. Then, based on the rule in the problem description, the summary of ACMPerk mileages is calculated.

2.4.2 Cola

You see the following special offer by a convenience store:

"A bottle of Choco Cola for every 3 empty bottles returned"

Now you decide to buy some (say *N*) bottles of cola from the store. You would like to know how you can get the most cola from them.

Figure 2.6 shows the case where *N*=8. Method 1 is the standard way: after finishing your eight bottles of cola, you have eight empty bottles. Take six of them and you get two new bottles of cola. Now after drinking them, you have four empty

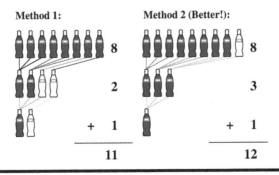

Figure 2.6

bottles, so you take three of them to get yet another new cola. Finally, you have only two bottles in hand, so you cannot get a new cola any more. Hence, you have enjoyed $8 + 2 + 1 = 11$ bottles of cola.

You can actually do better! In method 2, you first borrow an empty bottle from your friend (or the storekeeper??), and then you can enjoy $8 + 3 + 1 = 12$ bottles of cola! Of course, you will have to return your remaining empty bottle back to your friend.

Input

Input consists of several lines, each containing an integer N ($1 \leq N \leq 200$).

Output

For each case, your program should output the maximum number of bottles of cola you can enjoy. You may borrow empty bottles from others, but if you do that, make sure that you have enough bottles afterward to return to them.

Sample Input	Sample Output
8	12

Source: Contest of Newbies 2006

ID for Online Judge: UVA 11150

Hint

Suppose n is the number of bottles of cola you buy from the store initially; i is the number of empty bottles you borrow; *cnt* is the total number of bottles, initially $cnt = n+i$; *tot* is the number of bottles of cola you can enjoy, initially $tot = n$; and *ans* is the maximum number of bottles of cola you can enjoy, initially $ans = 0$.

The "trick" is that borrowing more than two bottles does not help—you would have to return the extra bottles without trading them in, and the borrowing should be done in the beginning, since it would cascade down otherwise. Therefore, the program only needs to simulate borrowing either 0, 1, or 2 bottles. For each case, we simulate the process as follows:

Repeat the process until *cnt*<3:

```
    The number of produced empty bottles tmp=cnt%3;
    The number of increased bottles of cola cnt/=3;
    The number of bottles of cola you can enjoy tot+=cnt;
    The number of increased empty bottles cnt+=tmp;
if (cnt≥i && tot>ans) ans=tot;     // you can return remaining
empty bottles back to your friend, and drink more.
```

We can easily take the best out of the three simulations.

2.4.3 The Collatz Sequence

An algorithm given by Lothar Collatz produces sequences of integers, and is described as follows:

Step 1: Choose an arbitrary positive integer A as the first item in the sequence.
Step 2: If $A = 1$ then stop.
Step 3: If A is even, then replace A by $A/2$ and go to Step 2.
Step 4: If A is odd, then replace A by $3 \times A + 1$ and go to Step 2.

It has been shown that this algorithm will always stop (in Step 2) for initial values of A as large as 10^9, but some values of A encountered in the sequence may exceed the size of an integer on many computers. In this problem, we want to determine the length of the sequence that includes all values produced until either the algorithm stops (in Step 2), or a value larger than some specified limit would be produced (in Step 4).

Input

The input for this problem consists of multiple test cases. For each case, the input contains a single line with two positive integers, the first giving the initial value of A (for Step 1) and the second giving L, the limiting value for terms in the sequence. Neither of these, A or L, is larger than 2147483647 (the largest value that can be stored in a 32-bit signed integer). The initial value of A is always less than L. A line that contains two negative integers follows the last case.

Output

For each input case, display the case number (sequentially numbered starting with 1), a colon, the initial value for A, the limiting value L, and the number of terms computed.

Sample Input	Sample Output
3 100	Case 1: A = 3, limit = 100, number of terms = 8
34 100	Case 2: A = 34, limit = 100, number of terms = 14
75 250	Case 3: A = 75, limit = 250, number of terms = 3
27 2147483647	Case 4: A = 27, limit = 2147483647, number of terms = 112
101 304	Case 5: A = 101, limit = 304, number of terms = 26
101 303	Case 6: A = 101, limit = 303, number of terms = 1
−1 −1	

Source: ACM North Central Regionals 1998

ID for Online Judge: UVA 694

Hint

This is a "follow the instructions" problem. Given the initial value of *a* (for Step 1) and the limiting value for terms in the sequence *l*, the number of terms *ans* is calculated as follows:

```
ans=0;
   while(a<=l&&a!=1){
      ans++;
      a=a&1?3*a+1:a/2;
      }
if(a==1) ans++;
```

2.4.4 Let's Play Magic!

You have seen a card magic trick named "Spelling Bee." The process goes as follows:

1. The magician first arranges 13 cards in a circle, as shown in Figure 2.7.
2. Starting from the marked position, he counts the cards clockwise, saying "*A—C—E.*"
3. He turns the card at the "*E*" position, and... it is an Ace!
4. Next, he takes away the Ace and continues to count the cards, saying "*T—W—O.*"
5. He turns over the card at position "*O*"... it is a Two!!
6. He continues to do this with the rest of the cards from Three to King. :-)

Now, how does the magician arrange the cards?

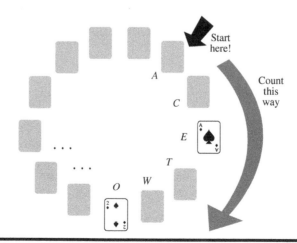

Figure 2.7

Input

Input consists of several test cases. Each case begins with an integer N ($1 \leq N \leq 52$), the number of cards to be used in the magic trick. The following N lines show the order of the turning over of the cards and the words to be spelled. None of the words will have more than 20 characters. The format for each card is a string with two characters: first the value, and second the suit.

Input ends with a test case where $N=0$. This test case should not be processed.

Output

For each case, your program should output the initial arrangement of the cards.

Sample Input	Sample Output
13	QH 4C AS 8D KH 2S 7D 5C TH JH 3S 6C 9D
AS ACE	
2S TWO	
3S THREE	
4C FOUR	
5C FIVE	
6C SIX	
7D SEVEN	
8D EIGHT	
9D NINE	
TH TEN	
JH JACK	
QH QUEEN	
KH KING	
0	

Source: Return of the Newbies 2005

ID for Online Judge: UVA 10978

 Hint

N cards are arranged in a circle. Starting from a certain card, the magician counts cards in a clockwise direction and spells N words. When the last letter of a word is pronounced, he turns over the card and removes it from the circle.

Given the sequence of words, and the order in which the cards are removed, find the initial arrangement of cards in the circle.

The algorithm simulates the magician's actions and recovers the arrangement of cards.

2.4.5 Throwing Cards Away

Given is an ordered deck of n cards numbered 1 to n with card 1 at the top and card n at the bottom. The following operation is performed as long as there are at least two cards in the deck:

> Throw away the top card and move the card that is now on the top of the deck to the bottom of the deck.

Your task is to find the sequence of discarded cards and the last remaining card.

Input

Each line of input (except the last) contains a number $n \leq 50$. The last line contains 0, and this line should not be processed.

Output

For each number from the input, produce two lines of output. The first line presents the sequence of discarded cards, and the second line reports the last remaining card. No line will have leading or trailing spaces. See the sample for the expected format.

Sample Input	Sample Output
7	Discarded cards: 1, 3, 5, 7, 4, 2
19	Remaining card: 6
10	Discarded cards: 1, 3, 5, 7, 9, 11, 13, 15, 17, 19, 4, 8, 12, 16, 2, 10, 18, 14
6	Remaining card: 6
0	Discarded cards: 1, 3, 5, 7, 9, 2, 6, 10, 8
	Remaining card: 4
	Discarded cards: 1, 3, 5, 2, 6
	Remaining card: 4

Source: A Special Contest 2005

ID for Online Judge: UVA 10935

 Hint

Simulate the problem as described. A queue is used to simulate efficiently.

2.4.6 Gift?!

There is a beautiful river in a small village. There are n rocks arranged in a straight line numbered 1 to n from the left bank to the right bank, as shown below:

[Left Bank] – [Rock1] – [Rock2] – [Rock3] – [Rock4] ... [Rock n] – [Right Bank]

The distance between two adjacent rocks is exactly 1 meter, while the distance between the left bank and rock 1, and between rock n and the right bank, is also 1 meter.

Frog Frank was about to cross the river. His neighbor Frog Funny came to him and said,

"Hello, Frank. Happy Children's Day! I have a gift for you. See it? A little parcel on Rock 5."

"Oh, that's great! Thank you! I'll get it."

"Wait...This present is for smart frogs only. You can't get it by jumping to it directly."

"Oh? Then what should I do?"

"Jump more times. Your first jump must be from the left bank to Rock 1, then, jump as many times as you like—no matter forward or backward, but your i-th jump must cover $2 \times i - 1$ meters. What's more, once you return to the left bank or reach the right bank, the game ends, and no more jumps are allowed."

"Hmmm, not easy... let me think!" answered Frog Frank. "Should I give it a try?"

Input

The input will contain no more than 2000 test cases. Each test case contains a single line. It contains two positive integers n ($2 \leq n \leq 10^6$), and m ($2 \leq m \leq n$), m indicates the number of the rock on which the gift is located. A test case in which $n=0$, $m=0$ will terminate the input and should not be regarded as a test case.

Output

For each test case, output a single line containing "Let me try!" if it's possible to get to rock m; otherwise, output a single line containing "Don't make fun of me!".

Sample Input	Sample Output
9 5	Don't make fun of me!
12 2	Let me try!
0 0	

Note: In test case 2, Frank can reach the gift in this way:

Forward (to rock 4), Forward (to rock 9), Backward (to rock 2, got the gift!)

Note that if Frank jumps forward in his last jump, he will land on the right bank (assume that banks are large enough) and thus, he would lose the game.

Source: OIBH Online Programming Contest 1

IDs for Online Judge: ZOJ 1229, UVA 10120

Hint

Suppose n rocks are arranged in a straight line numbered 1 to n from the left bank to the right bank, and m indicates the number of the rock on which the gift is located. It can be proved, if $n>50$, Frog Frank can reach each rock. If $n \leq 50$, we need to determine whether Frog Frank can reach a rock or not. First, the offline method is to determine whether Frog Frank can reach a rock or not.

$$ans[n][m] = \begin{cases} true & \text{Frank Frog can reach Rock } m \\ false & \text{Frank Frog can't reach Rock } m \end{cases},$$

$$\left(1 \leq n \leq 50, 1 \leq m \leq n\right).$$

Then, for each test case n and m, if $n \leq 50$, output the result based on $ans[n][m]$.

2.4.7 A-Sequence

For this problem an A-sequence is a sequence of positive integers a_i satisfying $1 \leq a_1 < a_2 < a_3 < \dots$ and every a_k of the sequence is not the sum of two or more distinct earlier terms of the sequence.

You should write a program to determine if a given sequence is or is not an A-sequence.

Input

The input consists of a set of lines; each line starts with an integer $2 \leq D \leq 30$ that indicates the number of integers that the current sequence has. Following this number there is the sequence itself. The sequence is composed by integers; each integer is greater than or equal to **1** and less than or equal to **1000**. The input is terminated by end of file (EOF).

Output

For each test case in the input you should print two lines: the first line should indicate the number of the test case and the test case itself; in the second line you should print "**This is an A-sequence.**", if the corresponding test case is an A-sequence, or "**This is not an A-sequence.**", if the corresponding test case is not an A-sequence.

Sample Input	Sample Output
2 1 2 3 1 2 3 10 1 3 16 19 25 70 100 243 245 306	Case #1: 1 2 This is an A-sequence. Case #2: 1 2 3 This is not an A-sequence. Case #3: 1 3 16 19 25 70 100 243 245 306 This is not an A-sequence.

Source: UFRN-2005 Contest 2

ID for Online Judge: UVA 10930

 Hint

The problem requires you to determine whether a sequence of positive integers is an *A*-sequence or not. If a sequence of positive integers is an *A*-sequence, then the sequence of positive integers is in the ascending order, and every element of the sequence is not the sum of two or more distinct earlier terms of the sequence.

For a test case, positive integers are input one by one. Suppose sums of two or more distinct earlier terms of the sequence are stored in $g[]$; z is the current input integer; and la is the previous integer.

For the current integer z, if z is the sum of two or more distinct earlier terms of the sequence, or $z \leq la$, then the sequence isn't an *A*-sequence, and exit the process. Else, first, elements in $g[]$ and z are analyzed: if for all $g[i]$, $g[i]+z$ isn't in $g[]$, a new element $g[i]+z$ is added into $g[]$; then $la=z$, and the next integer z is input. After all elements in the sequence are dealt with, the sequence is an *A*-sequence.

2.4.8 Building Design

An architect wants to design a very high building. The building will consist of some floors, and each floor will have a certain size. The size of a floor must be greater than the size of the floor immediately above it. In addition, the designer (who is a fan of a famous Spanish football team) wants to paint the building in blue and red, each floor a color, and in such a way that the colors of two consecutive floors are different.

To design the building, the architect has *n* available floors, with their associated sizes and colors. All the available floors are of different sizes. The architect wants to design the highest possible building with these restrictions, using the available floors.

Input

The input file consists of a first line with the number p of cases to solve. The first line of each case contains the number of available floors. Then, the size and color of each floor appear in one line. Each floor is represented with an integer between -999999 and 999999. There is no floor with size 0. Negative numbers represent red floors and positive numbers represent blue floors. The size of the floor is the absolute value of the number. There are no two floors with the same size. The maximum number of floors for a problem is 500000.

Output

For each case the output will consist of a line with the number of floors of the highest building with the mentioned conditions.

Sample Input	Sample Output
2	2
5	5
7	
−2	
6	
9	
−3	
8	
11	
−9	
2	
5	
18	
17	
−15	
4	

Source: IV Local Contest in Murcia 2006

ID for Online Judge: UVA 11039

 Hint

First, the sizes of the floors are sorted in descending order. Second, we set the lowest floor for the building as blue, and design the highest possible building with the restrictions in the problem description. The number of floors of the highest building is l_1. Third, we set the lowest floor for the building as red, and design the highest possible building with the restrictions in the problem description. The number of floors of the highest building is l_2.

Obviously, the number of floors of the highest building is $max\{l_1, l_2\}$.

2.4.9 *Light Bulbs*

Hollywood's newest theater, the Atheneum of Culture and Movies, has a huge computer-operated marquee composed of thousands of light bulbs. Each row of bulbs is operated by a set of switches that are electronically controlled by a computer program. Unfortunately, the electrician installed the wrong kind of switches, and tonight is the ACM's opening night. You must write a program to make the switches perform correctly.

A row of the marquee contains n light bulbs controlled by n switches. Bulbs and switches are numbered from 1 to n, left to right. Each bulb can be either ON or OFF. Each input case will contain the initial state and the desired final state for a single row of bulbs.

The original lighting plan was to have each switch control a single bulb. However, the electrician's error caused each switch to control two or three consecutive bulbs, as shown in Figure 2.8. The leftmost switch ($i=1$) toggles the states of the two leftmost bulbs (1 and 2); the rightmost switch ($i=n$) toggles the states of the two rightmost bulbs ($n-1$ and n). Each remaining switch ($1<i<n$) toggles the states of the three bulbs with indices $i-1$, i, and $i+1$. (In the special case where there is a single bulb and a single switch, the switch simply toggles the state of that bulb.) Thus, if bulb 1 is ON and bulb 2 is OFF, flipping switch 1 will turn bulb 1 OFF and bulb 2 ON. The minimum cost of changing a row of bulbs from an initial configuration to a final configuration is the minimum number of switches that must be flipped to achieve the change.

You can represent the state of a row of bulbs in binary, where 0 means the bulb is OFF and 1 means the bulb is ON. For instance, 01100 represents a row of five bulbs in which the second and third bulbs are both ON. You could transform this state into 10000 by flipping switches 1, 4, and 5, but it would be less costly to simply flip switch 2.

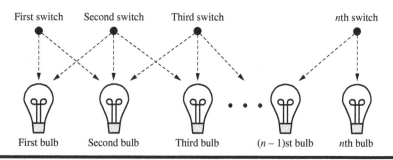

Figure 2.8

You must write a program that determines the switches that must be flipped to change a row of light bulbs from its initial state to its desired final state with minimal cost. Some combinations of initial and final states may not be feasible. For compactness of representation, decimal integers are used instead of binary for the bulb configurations. Thus, 01100 and 10000 are represented by the decimal integers 12 and 16.

Input

The input file contains several test cases. Each test case consists of one line. The line contains two non-negative decimal integers, at least one of which is positive and each of which contains at most 100 digits. The first integer represents the initial state of the row of bulbs, and the second integer represents the final state of the row. The binary equivalent of these integers represents the initial and final states of the bulbs, where 1 means ON and 0 means OFF.

To avoid problems with leading zeros, assume that the first bulb in either the initial or the final configuration (or both) is ON. There are no leading or trailing blanks in the input lines, no leading zeros in the two decimal integers, and the initial and final states are separated by a single blank.

The last test case is followed by a line containing two zeros.

Output

For each test case, print a line containing the case number and a decimal integer representing a minimum-cost set of switches that need to be flipped to convert the row of bulbs from initial state to final state. In the binary equivalent of this integer, the rightmost (least significant) bit represents the nth switch, 1 indicates that a switch has been flipped, and 0 indicates that a switch has not been flipped. If there is no solution, print "impossible". If there is more than one solution, print the one with the smallest decimal equivalent.

Print a blank line between cases. Use the output format shown in the example.

Sample Input	Sample Output
12 16	Case Number 1: 8
1 1	Case Number 2: 0
3 0	Case Number 3: 1
30 5	Case Number 4: 10
7038312 7427958190	Case Number 5: 2805591535
4253404109 657546225	Case Number 6: impossible
0 0	

Source: ACM World Finals - Beverly Hills - 2003

ID for Online Judge: UVA 2722

Hint

Every switch is either flipped or not, and can't be flipped more. After the first switch's operation is determined, only the second switch's operation can control the first bulb. The second switch's operation is determined. Therefore, all switches' operation can be determined, and so on. The simulation algorithm is as follows. First, determine whether the first switch should be flipped or not, and then every switch is enumerated one by one to determine all operations. High precision numbers are used to represent the states of the row of bulbs.

2.4.10 Link and Pop—the Block Game

Recently, Robert found a new game on the Internet that is the newest version of "Link and Pop." The game rule is very simple. Initially, a board of size $n\times m$ is filled with $n\times m$ blocks. Each of these blocks has a symbol on it. All you need to do is to find a pair of blocks with the same symbol on them, which can be linked with a line that consists of at most three straight horizontal or vertical line segments. Note that the line segments cannot cross the other blocks on the board (see Figure 2.9 for some examples of possible links; note that some blocks have already been removed from the board).

If you successfully find such a pair of blocks, the two blocks can be popped (i.e., removed) together. After this, some of the blocks may be moved to new positions on the board following the rules described later. Then, you can start to find the next pair. The game continues until there are no blocks left on the board or you cannot find such a pair.

The blocks are moved according to the following rules. First, each block has a static moving attribute, which is one of 'up', 'down', 'left', 'right', and 'stand still'. After a pair of blocks is removed, the blocks are checked one by one to see whether they can be moved towards the direction of its moving attribute. The blocks in the top row are checked first. Inside the same row, the blocks on the left are checked first. If the adjacent position at the direction of the block's moving attribute is not occupied, the block will be moved to that position immediately. No block can be moved beyond the boundary of the game board. Of course, a block with attribute 'stand still' will always stay at its original position. After all the blocks are checked, which is called a turn of checking, another turn of checking is started.

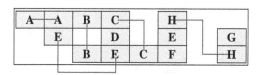

Figure 2.9

This continues until no more blocks can be moved to a new position following the moving rules. Note that inside each turn of checking, each of the blocks is checked and possibly moved only once. Blocks must not be checked and moved on its new position in one turn of checking.

Robert felt that the game was very interesting. However, after some time of playing, he found that when the size of the board is rather large, finding a pair of blocks becomes a very tough job. Furthermore, he often gets a 'Game Over' because no more blocks can be popped. Robert felt that it is not his fault that not all the blocks are being popped. It is only that there is a great chance that the game cannot be finished if the blocks are placed randomly at first. However, it will be very time-consuming to prove this by playing the game many times. So, Robert asks you to write a program for him that will simulate his behavior in the game and see if the game can be finished.

In order to make such a program possible, Robert summarizes his rules of selecting block pairs as follows. First, the pair of blocks that can be linked with one straight line segment must be found and popped because such pairs are easy to find. Next, if such a pair does not exist, the pairs that can be linked by two straight line segments must be found and popped. Finally, if both the above described pairs do not exist, the pairs that can be linked by three straight line segments must be found and popped. If more than one pair that can be linked with the same number of straight line segments exists, the pair that contains a block, which is positioned at the topmost row (or leftmost if two more blocks are positioned in the same row), will be selected first. If this rule still cannot break the tie (more than one pair may share one block that is positioned at the most top, left position), the other blocks in these pairs are compared according to the same rules. Figure 2.10 shows a trace of a mini game of "Link and Pop" that follows the above rules.

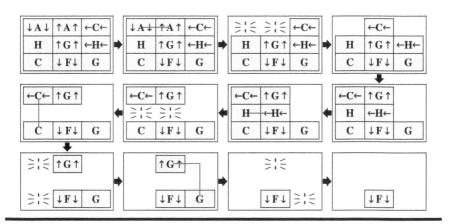

Figure 2.10

Input

The input contains no more than 30 test cases. The first line of each test case contains two integers n, $m(1 \le n$, $m \le 30)$, which is the size of the board. After this line, there will be n more lines. Each of these lines contains m strings, separated by single spaces. Each of these strings represents one block in the initial configuration. Each string always consists of two capital letters. The first letter is the symbol of the block. The second letter is always one of the letters 'U', 'D', 'L', 'R', and 'S', which shows the block's moving attribute, that is, up, down, left, right, and stand still, respectively. There are no blank lines between test cases. The input ends with a line of two 0's: '0 0'.

Output

For each test case, first output the test case number. After this line, you must output the final configuration of the board with n lines, each containing m characters. If there is a block on the position, output the symbol of the block. If there is no block on the position, output a period instead. Do not output blank lines between test cases.

Sample Input	Sample Output
3 3	Case 1
AD AU CL	...
HS GU HL	...
CS FD GS	.F.
1 2	Case 2
BS BL	..
0 0	

Source: ACM Shanghai 2004

IDs for Online Judges: POJ 2281, ZOJ 2391, UVA 3260

Hint

This is a simulation problem. The time limit for the problem is ample. Therefore, based on Robert's rules of selecting block pairs, the solution can be obtained. The simulation method is as follows.

First, we try to find whether there is a pair of blocks that can be linked with one straight line segment. Second, if there is no such pair of blocks, we try to find whether there is a pair of blocks that can be linked by two straight line segments. Finally, if both of the above pairs do not exist, we try to find a pair that can be linked by three straight line segments. If more than one pair that can be linked with the same number of straight line segments exists, the pair that contains a block, which is positioned at the topmost row (or leftmost if two more blocks are positioned in the same row), will be selected first. If this rule still cannot break the tie (more than one pair may share one block that is positioned at the most top, left position), the other block in these pairs are compared according to the same rules.

Each block has a static moving attribute, which is one of 'up', 'down', 'left', 'right', and 'stand still'. After a pair of blocks is removed, the blocks are checked one by one to see whether they can be moved toward the direction of its moving attribute. The blocks in the top row are checked first. Inside the same row, the blocks on the left are checked first. If the adjacent position at the direction of the block's moving attribute is not occupied, the block will be moved to that position. No block can be moved beyond the boundary of the game board. Of course, a block with attribute 'stand still' will always stay at its original position. After all the blocks are checked, which is called a turn of checking, another turn of checking is started. This continues until no more blocks can be moved to a new position following the moving rules.

BFS is used to find the popped pair every time. A dequeue is used in BFS: if a pair of blocks is linked with one straight line segment, it is added at the front of the queue; and if a pair of blocks is linked with more than one straight line segment, it is added at the rear of the queue. A pair of blocks is popped based on rules in the problem description.

2.4.11 Packing Rectangles

Four rectangles are given. Find the smallest enclosing (new) rectangle into which these four may be fitted without overlapping. By smallest rectangle, we mean the one with the smallest area.

All four rectangles should have their sides parallel to the corresponding sides of the enclosing rectangle. Figure 2.11 shows six ways to fit four rectangles together. These six are the only possible basic layouts, since any other layout can be obtained from a basic layout by rotation or reflection. There may exist several different enclosing rectangles fulfilling the requirements, all with the same area. You have to produce all such enclosing rectangles.

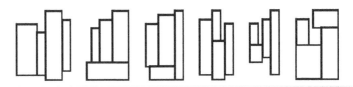

Figure 2.11

Input

Your program is to read from standard input. The input consists of four lines. Each line describes one given rectangle by two positive integers: the lengths of the sides of the rectangle. Each side of a rectangle is at least 1 and at most 50.

Output

Your program is to write to standard output. The output should contain one line more than the number of solutions. The first line contains a single integer: the minimum area of the enclosing rectangles. Each of the following lines contains one solution described by two numbers p and q, with $p \leq q$. These lines must be sorted in ascending order of p, and must all be different.

Sample Input	Sample Output
1 2	40
2 3	4 10
3 4	5 8
4 5	

Source: IOI 1995

ID for Online Judges: POJ 1169

 Hint

1. **Calculating the length and width of the enclosing rectangle.**
   There are six ways to fit four rectangles together. These six ways are the only possible basic layouts, since any other layout can be obtained from a basic layout by rotation or reflection. Therefore, the key to the problem is to calculate areas of rectangles for the six ways. Suppose the four rectangles are as follows:

   Rectangle w whose length and width are w_1 and w_2, respectively;
   Rectangle x whose length and width are x_1 and x_2, respectively;
   Rectangle y whose length and width are y_1 and y_2, respectively;
   Rectangle z whose length and width are z_1 and z_2, respectively.

   The first layout is as shown in Figure 2.12. The length of the enclosing rectangle is $MAX(w_1, x_1, y_1, z_1)$, and the width is $w_2 + x_2 + y_2 + z_2$.

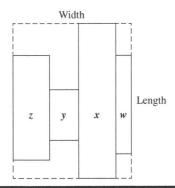

Figure 2.12

The second layout is as shown in Figure 2.13. The length of the enclosing rectangle is $MAX(w_1, x_1, y_1)+z_1$, and the width is $MAX(z_2, w_2+x_2+y_2)$.

The third layout is as shown in Figure 2.14. The length of the enclosing rectangle is $MAX(w_1, x_1+MAX(z_1, y_1))$, and the width is $w_2+MAX(x_2, z_2+y_2)$.

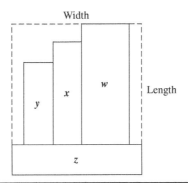

Figure 2.13

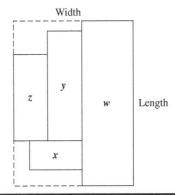

Figure 2.14

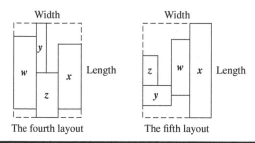

The fourth layout   The fifth layout

Figure 2.15

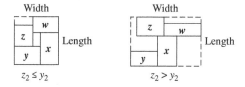

$z_2 \leq y_2$ $z_2 > y_2$

Figure 2.16

The fourth and the fifth layout are as shown in Figure 2.15. The common character for the two layouts is that two rectangles are stacked, and the other two rectangles aren't. The length of both the enclosing rectangles is $MAX(MAX(w_1, x_1), y_1+z_1)$, and their width is $w_2+x_2+MAX(y_2, z_2)$.

The sixth layout is as shown in Figure 2.16. Every two rectangles are stacked, where $z_1 \geq w_1$, $x_1 \geq y_1$, and there are two different ways. There are two cases for the sixth layout.

Case 1: The length of both the enclosing rectangles is $MAX(w_1+x_1, z_1+y_1)$, and their width is $MAX(w_2+z_2, x_2+y_2)$.

Case 2: The length of both the enclosing rectangles is $MAX(w_1+x_1, z_1+x_1)$, and their width is $MAX(w_2+z_2, x_2+y_2)$.

Any other case for the sixth layout that every two rectangles are stacked can be obtained from the above two cases by rotation or reflection.

2. **The minimum area of the enclosing rectangle is calculated by enumeration.**

All cases of enclosing rectangles of the above six layouts are enumerated. For the six layouts, there are seven enclosing rectangles. And for an enclosing rectangle, another enclosing rectangle will be generated if a rectangle in the enclosing rectangle is rotated 90°. Therefore, there are $4! \times 7 \times 2^4$ enclosing rectangles. All of these enclosing rectangles are enumerated to calculate the minimum area of the enclosing rectangle.

Chapter 3

Practice for
Number Theory

Number theory is a branch of pure mathematics that studies the properties of integers. In this chapter, experiments are organized in three parts:

1. Prime Numbers;
2. Indeterminate Equations and Congruence;
3. Multiplicative Functions.

3.1 Practice for Prime Numbers

Prime numbers are natural numbers greater than 1 that have no positive divisors other than 1 and the number itself. Natural numbers greater than 1 that are not prime numbers are called composite numbers.

Two kinds of experiments for prime numbers are discussed next:

1. Calculating all prime numbers in an integer interval [2, n] by a sieve.
2. Testing big prime numbers.

3.1.1 Calculating Prime Numbers by a Sieve

First, the sieve of Eratosthenes is introduced. The sieve of Eratosthenes is used to calculate all prime numbers in an integer interval [2, n].

Suppose $u[]$ is a sieve. Initially all numbers in the interval are in the sieve. In the sieve, the smallest number is found in ascending order, multiples of the number are composite numbers, and the sieve will filter out these numbers. Finally, only prime numbers are in the sieve. The algorithm for the sieve of Eratosthenes is as follows:

```
int i, j, k;
for (i=2; i<=n; i++) u[i]=true;      // all numbers in the
interval are in the sieve
for (i=2; i<=n; i++)      // find the smallest number in the
sieve
if (u[i]){
     for (j=2; j*i<=n; j++)      // the sieve filters out
multiples of i
         u[j*i]=false;
}
for (i=2; i<=n; i++) if (u[i]) {      //prime numbers in the
sieve are put into su[]
     su[++num]=i;
}
```

The sieve of Eratosthenes is a simple algorithm to find prime numbers. Its time complexity is $O(n \times \log \log n)$. There are other more efficient algorithms for finding prime numbers. For example, the algorithm for Euler's sieve is as follows:

```
int i, j, num=1;
memset(u, true, sizeof(u));
for (i=2; i<=n; i++){      //for each number i in the integer
interval
if (u[i]) su[num++]=i;      // the smallest number in the
sieve is put into the prime list
for (j=1; j<num; j++) {      //for each number in the prime
list
if (i*su[j]>n) break;      //if the product of i and the
current prime is greater than n, the next integer i is
analyzed
    u[i*su[j]]=false;      // the sieve filters out the product
of i and the current prime
if (i% su[j]==0) break;      // if the current prime is the
divisor for i, the next integer i is analyzed
  }
}
```

The time complexity for Euler's sieve is $O(n)$.

3.1.1.1 Goldbach's Conjecture

In 1742, Christian Goldbach, an amateur German mathematician, sent a letter to Leonhard Euler, in which he made the following conjecture:

> Every even number greater than four can be written as the sum of two odd prime numbers. For example: 8=3+5. Both 3 and 5 are odd prime numbers. 20=3+17=7+13; 42=5+37=11+31=13+29=19+23.

Today it is still unproven whether the conjecture is right. (I have the proof, of course, but it is too long to write it on the margin of this page.)

Anyway, your task now is to verify Goldbach's conjecture for all even numbers less than a million.

Input

The input file will contain one or more test cases. Each test case consists of one even integer n with $6 \leq n < 1000000$. Input will be terminated by a value of 0 for n.

Output

For each test case, print one line of the form $n=a+b$, where a and b are odd primes. Numbers and operators should be separated by exactly one blank line as shown in the sample output below. If there is more than one pair of odd primes adding up to n, choose the pair where the difference $b-a$ is maximized. If there is no such pair, print a line saying "Goldbach's conjecture is wrong."

Sample Input	Sample Output
8	8 = 3 + 5
20	20 = 3 + 17
42	42 = 5 + 37
0	

Source: Ulm Local 1998

IDs for Online Judges: POJ 2262, ZOJ 1951, UVA 543

 Analysis

First, the offline method is used to calculate the prime list $su[]$ and prime sieve $u[]$ in the interval [2, 1000000]. Then, for each test case (one even integer n), for each prime number in $su[]$ ($2 \times su[i] \leq n$), if $n - su[i]$ is also a prime number (i.e., $u[n - su[i]]$==true), then $su[i]$ and $n - su[i]$ is the solution to the problem.

Program

```cpp
#include<cmath>
#include<cstring>
#include<cstdlib>
#include<cstdio>
using namespace std;
bool u[1111111];     //sieve
int su[1111111],num;     // prime list su[], num: the length of
the su[]
void prepare(){     //Construct su[], sieve of Eratosthenes
  int i,j,k;
    for(i=2;i<=1000000;i++)u[i]=true;
    for(i=2;i<=1000000;i++)
    if(u[i]){
        for(j=2;j*i<=1000000;j++)
            u[j*i]=false;
    }
    for(i=2;i<=1000000;i++)if(u[i]){
        su[++num]=i;
    }
}
int main () {
    prepare();     // Construct su[]
    int i,j,k,n;
    while(scanf("%d",&n)>0&&n)     //Input test cases
    {
        bool ok=false;
        for(i=2;i<=num;i++)     //search each prime number in
the prime list in ascending order
        {
            if(su[i]*2>n)break;     //search ends
            if(u[n-su[i]]){     // the even number can be
written as the sum of two odd prime numbers
                ok=true;
                break;
            }
        }
        if(!ok)puts("Goldbach's conjecture is wrong.");
//Output result
        else printf("%d = %d + %d\n",n,su[i],n-su[i]);
    }
    return 0;
}
```

3.1.1.2 Summation of Four Primes

Euler proved in one of his classic theorems that prime numbers are infinite in number. But can every number be expressed as a summation of four positive primes? I don't know the answer. Perhaps you can help! I want your solution to be very efficient as I have a 386 machine at home. But the time limit specified is for a Pentium III 800 machine. The definition of prime number for this problem is: "A prime number is a positive number which has exactly two distinct integer factors." For example, 37 is prime as it has exactly two distinct integer factors, 37 and 1.

Input

The input contains one integer number N ($N \le 10000000$) in every line. This is the number you will have to express as a summation of four primes. Input is terminated by end of file.

Output

For each line of input, there is one line of output, which contains four prime numbers according to the given condition. If the number cannot be expressed as a summation of four prime numbers, print **"Impossible."** in a single line. There can be multiple solutions. Any good solution will be accepted.

Sample Input	Sample Output
24	3 11 3 7
36	3 7 13 13
46	11 11 17 7

Source: Regionals 2001 Warmup Contest

ID for Online Judge: UVA 10168

 Analysis

The problem is solved based on Goldbach's conjecture. The algorithm is as follows:
First, the prime list $su[]$ and its length *num* in the integer interval [2, 9999999] are calculated. Then, for each test case N,

1. **if $N \le 12$:**
   $N<8$, N can't be expressed as a summation of four prime numbers;
   $N==8$, N can be expressed as a summation of four prime numbers: 2 2 2 2;
   $N==9$, N can be expressed as a summation of four prime numbers: 2 2 2 3;

$N==10$, N can be expressed as a summation of four prime numbers: 2 2 3 3;
$N==11$, N can be expressed as a summation of four prime numbers: 2 3 3 3;
$N==12$, N can be expressed as a summation of four prime numbers: 3 3 3 3;

2. **if $N>12$:**

First, two prime numbers are subtracted from N. If N is an even number ($N\%2==0$), the two prime numbers, 2 and 2, are subtracted from N, that is, $N-=4$; else the two prime numbers, 2 and 3, are subtracted from N, that is, $N-=5$. Obviously, N is an even number greater than four. Based on Goldbach's conjecture, every even number greater than four can be written as the sum of two odd prime numbers. Search the prime list $su[]$($1\leq i\leq num$, $2\times su[i]\leq n$). If $su[n-su[i]]==$true, N can be expressed as a summation of two prime numbers: $su[i]$ and $n-su[i]$.

Finally, output the result.

 Program

```cpp
#include<iostream>
#include<cstdio>
#include<cstring>
#include<cmath>
#include<algorithm>
#include<cstdio>
#include<cstdlib>
using namespace std;
bool u[10000001];      //sieve
int su[5000000],num;      // the prime list su[] and its length
num
void prepare(){      //construct the prime list su[] in the
interval [2, 9999999]
   int i,j,num;
memset(u,true,sizeof(u));      //initially all numbers in the
sieve
for (i=2; i<=9999999; i++){      //analyze all numbers in the
interval one by one
   if (u[i]) su[++num]=i;      //the least number is put into
the prime list
   for (j=1; j<=num; j++) {      //analyze every number in the
prime list
      if (i*su[j]>n) break;
      u[i*su[j]]=false;
      if (i% su[j]==0) break;
   }
```

```
   }
}
int main ()
{
    prepare();      // construct the prime list su[] in the
interval [2, 9999999]
    int n,i,j,k;
    while(scanf("%d",&n)>0){      // Input integer n
        if(n==8){puts("2 2 2 2");continue;}
        if(n==9){puts("2 2 2 3");continue;}
        if(n==10){puts("2 2 3 3");continue;}
        if(n==11){puts("2 3 3 3");continue;}
        if(n==12){puts("3 3 3 3");continue;}
        if(n<8){puts("Impossible.");continue;}
        if(n%2==0){printf("2 2 ");n-=4;}
        else{printf("2 3 ");n-=5;}
        for(i=1;i<=num;i++)      // based on Goldbach's
conjecture
        {
            if(su[i]*2>n)break;
            if(u[n-su[i]]){      //if su[i] and n-su[i] are two
prime numbers
                printf("%d %d\n",su[i],n-su[i]);
                break;
            }
        }
    }
}
```

3.1.1.3 Digit Primes

A prime number is a positive number, which is divisible by exactly two different integers. A digit prime is a prime number whose sum of digits is also prime. For example, the prime number 41 is a digit prime because 4+1=5 and 5 is a prime number. The number 17 is not a digit prime because 1+7=8, and 8 is not a prime number. In this problem, your job is to find out the number of digit primes within a certain range less than 1000000.

Input

The first line of the input file contains a single integer N ($0<N\leq500000$) that indicates the total number of inputs. Each of the next N lines contains two integers t_1 and t_2 ($0<t_1\leq t_2<1000000$).

Output

For each line of input except the first line, produce one line of output containing a single integer that indicates the number of digit primes between t_1 and t_1 (inclusive).

Sample Input	Sample Output
3	1
10 20	10
10 100	576
100 10000	

Note: You should at least use scanf() and printf() to take input and produce output for this problem. cin and cout are too slow for this problem to get it within the time limit.

Source: The Diamond Wedding Contest: Elite Panel's 1st Contest 2003

ID for Online Judge: UVA 10533

 Analysis

Suppose $u[]$ is the prime sieve for the interval [2, 1100001]; $u2[]$ are numbers of digit primes, where $u2[i]$ is the number of digit primes in the interval [2, i], $2 \le i \le 1100001$.

First, the offline method is used to calculate $u2[]$. The prime sieve $u[]$ for the interval [2, 1100001] is calculated. For each number i in [2, 1100001], if i is a digit prime, that is, $u[i] \&\& u[$the sum of digits for $i] ==$ true, then $u2[i]=1$. Then calculate $u2[i]$: $u2[i] += u2[i-1]$ ($2 \le i \le 1100001$). Finally, based on $u2[]$, calculate the number of digit primes within a certain range $[i,j]$: $u2[j]-u2[i-1]$.

 Program

```
#include<iostream>
#include<cstdio>
#include<cstring>
#include<cmath>
#include<algorithm>
#include<cstdio>
#include<cstdlib>
using namespace std;
bool u[1100001];      // prime sieve
int u2[1100001];      // numbers of digit primes
```

```
void prepare(){      // Calculate the prime sieve u[] in
[2, 1100001]
    int i,j,k;
    for(i=2;i<1100001;i++)u[i]=1;       // Initially all numbers
in the sieve
    for(i=2;i<1100001;i++)      // the least is a prime, and its
multiples are taken out
        if(u[i])
          for(j=i+i;j<1100001;j+=i)
            u[j]=false;
}
bool ok(int x){      //Determine whether the sum of digits for x
is a prime
    int i,j,k=0;
    while(x){      // the sum of digits for x
        k+=x%10;x/=10;
    }
    return u[k];
}
int main (){
    int i,j,k;
    prepare();      // Calculate the prime sieve u[] in [2,
1100001]
    for(i=2;i<1100001;i++)      // Calculate digit primes  in
[2, 1100001]
        if(u[i])&&(ok(i))  u2[i]=1;
    for(i=2;i<1100001;i++)u2[i]+=u2[i-1];      // u2[i] is the
number of digit primes in [2, i]
    scanf("%d",&k);      // the number of test cases
    while(k--){
        scanf("%d %d",&i,&j);      //input a test case, an
interval [i, j]
        printf("%d\n",u2[j]-u2[i-1]);      // the number of
digit primes within [i, j]
    }
}
```

3.1.1.4 Prime Gap

The sequence of $n-1$ consecutive composite numbers (positive integers that are not prime and not equal to 1) lying between two successive prime numbers p and $p+n$ is called a prime gap of length n. For example, <24, 25, 26, 27, 28> between 23 and 29 is a prime gap of length 6.

Your mission is to write a program to calculate, for a given positive integer k, the length of the prime gap that contains k. For convenience, the length is considered 0 in case no prime gap contains k.

Input

The input is a sequence of lines each of which contains a single positive integer. Each positive integer is greater than 1 and less than or equal to the 100000th prime number, which is 1299709. The end of the input is indicated by a line containing a single zero.

Output

The output should be composed of lines each of which contains a single non-negative integer. It is the length of the prime gap that contains the corresponding positive integer in the input if it is a composite number, or 0 otherwise. No other characters should occur in the output.

Sample Input	Sample Output
10	4
11	0
27	6
2	0
492170	114
0	

Source: ACM Japan 2007

IDs for Online Judges: POJ 3518, UVA 3883

 Analysis

Suppose $ans[k]$ is the length of the prime gap that contains k. If k is a prime number, then $ans[k]=0$. For any two successive prime numbers p_1 and p_2, $ans[p_1+1]=ans[p_1+2]=...=ans[p_2-1]=p_2-p_1$. The algorithm is as follows:

1. Calculating $ans[]$:
   Calculating the prime sieve $u[]$ in the interval $[2, 1299709]$;
   Enumerate every number i in the interval $[2, 1299709]$. If i is a prime number ($u[i]==$true), then $ans[i]=0$. If i is a composite number, then find the next prime number $j(u[i]==u[i+1]==...==u[j-1]==$false, $u[j]==$true), $ans[i]=ans[i+1]=ans[j-1]=j-i+1$, and $i=j$.
2. For each test case k, output $ans[k]$.

Program

```
#include<iostream>
#include<cstdio>
#include<cstring>
#include<cmath>
#include<algorithm>
#include<cstdio>
#include<cstdlib>
using namespace std;
const int maxn=1299710;
bool u[maxn];      //prime sieve
int ans[maxn];      // the length of the prime gap
void prepare(){
    int i,j,k;
    for(i=2;i<maxn;i++)u[i]=1;      //Calculate prime sieve u[]
in [2, 1299710]
    for(i=2;i<maxn;i++)
       if(u[i])     // i is a prime number
            for(j=2;j*i<maxn;j++) u[i*j]=0;
    for(i=2;i<maxn;i++)     // Enumerate every number i in the
interval
    if(!u[i]){     // i is a composite number
        j=i;
        while(j<maxn&&!u[j]) j++;
        j--;
        for(k=i;k<=j;k++) ans[k]=j-i+2;     //calculate the
length of the prime gap
        i=j;
    }else ans[i]=0;     // i is a prime number
}
int main ()
{
    int i,j,k;
    prepare();
    while(scanf("%d",&k)>0&&k>0){
        printf("%d\n",ans[k]);
    }
}
```

3.1.2 Testing the Primality of Large Numbers

Trial division is the simplest method to test whether a given number n is a prime number or not. n is a prime number if and only if n is not a multiple of any integer between 2 and $\sqrt{n}$. But trial division is also slow for testing the primality of large numbers. There are two optimization methods for trial division: "Sieve + Trial Division", and the Miller–Rabin primality test.

"Sieve + Trial Division" is as follows. First, the prime sieve $u[]$ and prime list $su[]$ for the interval $[2, \sqrt{n}]$ are calculated. The length of $su[]$ is num. x is a prime number if and only if x is a prime number in the interval $[2, \sqrt{n}]$ ($u[x]==1$), or x is not a multiple of any integer between 2 and $\sqrt{n}$ ($x\%su[0]\neq0$, ..., $x\%su[num-1]\neq0$). The time complexity is $O(\sqrt{n})$.

3.1.2.1 Primed Subsequence

Given a sequence of positive integers of length n, we define a primed subsequence as a consecutive subsequence of length at least two that sums to a prime number greater than or equal to two. For example, given the sequence: 3 5 6 3 8, there are two primed subsequences of length 2 (5+6=11 and 3+8=11), one primed subsequence of length 3 (6+3+8=17), and one primed subsequence of length 4 (3+5+6+3=17).

Input

Input consists of a series of test cases. The first line consists of an integer t ($1<t<21$), the number of test cases. Each test case consists of one line. The line begins with the integer n, $0<n<10001$, followed by n non-negative numbers less than 10000 comprising the sequence. You should note that 80 percent of the test cases will have at most 1000 numbers in the sequence.

Output

For each sequence, print the "Shortest primed subsequence is length x:", where x is the length of the shortest primed subsequence, followed by the shortest primed subsequence, separated by spaces. If there are multiple such sequences, print the one that occurs first. If there are no such sequences, print "This sequence is anti-primed.".

Sample Input	Sample Output
3	Shortest primed subsequence is length 2: 5 6
5 3 5 6 3 8	Shortest primed subsequence is length 3: 4 5 4
5 6 4 5 4 12	This sequence is anti-primed.
21 15 17 16 32 28 22 26 30 34 29	
31 20 24 18 33 35 25 27 23 19 21	

Source: June 2005 Monthly Contest

ID for Online Judge: UVA 10871

Analysis

There are *n* non-negative numbers less than 10000 comprising the sequence, $0 < n < 10001$.

First, the prime sieve *u*[] and prime list *su*[] for the interval [2, 10010] are calculated. The length of *su*[] is *num*. If *x* is a prime number in the interval [2, 10010] ($u[x]==1$), or *x* isn't a multiple of any integer in $su[](x\%su[0]\neq0,\ldots,x\%su[num-1]\neq0)$, then *x* is a prime number.

Then, based on the above, the shortest primed subsequence is calculated.

Input a sequence whose length is *n*, and calculate the sum of the first *i* integers $s[i](1\leq i\leq n, s[i]+=s[i-1])$:

Dynamic Programming is used to calculate the shortest primed subsequence:

```
Enumerate the length i(2≤i≤n):
Enumerate the front pointer j(1≤j≤n-i+1):
        If (s[i+j-1]-s[j-1] is a prime number)
              Output the subsequence from the jth integer to
the (j+i-1)th integer, and exit;
        Output "This sequence is anti-primed.";
```

Program

```cpp
#include<iostream>
#include<algorithm>
#include<cmath>
#include<cstdio>
#include<cstring>
#include<cstdlib>
using namespace std;
bool u[10010];      // prime sieve
int su[10010],num;     //prime list and its length
void prepare(){    //construct the prime list su[] in the
interval [2, 10010]
  int i,j,num;
memset(u,true,sizeof(u));
for(i=2;i<=10010;i++){
   if(u[i]) su[++num]=i;
   for(j=1;j<=num;j++) {
     if (i*su[j]>n)break;
     u[i*su[j]]=false;
     if (i% su[j]==0) break;
   }
```

```
  }
}
bool pri(int x){      // If x is a prime number in the interval
[2, 10010] or (u[x]==1), or x isn't a multiple of any integer
in su[], return true; else return false
    int i,j,k;
    if(x<10010)return u[x];
    for(i=1;i<=num;i++)
    if(x%su[i]==0)return false;
    return true;
}
int n,s[10010];      //the sum of the first i integers is s[i]
int main()
{
    int i,j,k;
    prepare();      // calculate the prime list su[]
    int te;
    scanf("%d",&te);      //number of test cases
    while(te--){
        scanf("%d",&n);      //the length of sequence
        s[0]=0;
        for(i=1;i<=n;i++)      //calculate s[]
        {
            scanf("%d",&s[i]);
            s[i]+=s[i-1];
        }
        bool ok=false;
        for(i=2;i<=n;i++){      //enumerate lengths of
subsequence
        for(j=1;j+i-1<=n;j++)      //enumerate front pointers
        {
            k=s[i+j-1]-s[j-1];      //calculate the sum of
subsequences
            if(pri(k)){      // if k is a prime number
                ok=true;
                printf("Shortest primed subsequence is length
%d:",i);
                for(k=1;k<=i;k++)printf("
%d",s[j+k-1]-s[j+k-2]);
                puts("");
                break;
            }
        }
        if(ok)break;
        }
        if(!ok)puts("This sequence is anti-primed.");
// there are no primed sequences
    }
    // system("pause");
}
```

3.2 Practice for Indeterminate Equations and Congruence

Experiments in this section are for the following problems: Greatest Common Divisor (GCD), Indeterminate Equations, Congruence, and Congruence Equations.

3.2.1 Greatest Common Divisors and Indeterminate Equations

The GCD for integers a and b can be found by repeated application of the division algorithm, known as the Euclidean algorithm. The Euclidean algorithm is as follows:

$$GCD(a,b) = \begin{cases} b & a = 0 \\ GCD(b \bmod a, a) & \text{Otherwise} \end{cases} = \begin{cases} a & b = 0 \\ GCD(b, a \bmod b) & \text{Otherwise} \end{cases}.$$

Proof. The key to the proof is $GCD(a, b)$ and $GCD(b \bmod a, a)$ can be divided by each other. $b \bmod a$ can be represented as an integer linear combination of a and b: $b \bmod a = b - \left\lfloor \dfrac{b}{a} \right\rfloor \times a$. Because a and b can be divided by $GCD(a, b)$, $b - \left\lfloor \dfrac{b}{a} \right\rfloor \times a$ can also be divided by $GCD(a, b)$. Therefore $GCD(b \bmod a, a)$ can be divided by $GCD(a, b)$. Similarly, $GCD(a, b)$ can also be divided by $GCD(b \bmod a, a)$. Therefore, $GCD(a, b) = \pm GCD(b \bmod a, a)$.

Similarly, $GCD(a, b)$ and $GCD(b, a \bmod b)$ can be divided by each other.

For example, $GCD(319, 377) = GCD(58, 319) = GCD(29, 58) = GCD(0, 29) = 29$.

Theorem 3.2.1.1 (Bezout's Theorem). If a and b are integers, then there are integers x and y, such that $ax + by = GCD(a, b)$.

Corollary 3.2.1.1 Integers a and b are relatively prime integers if and only if there are integers x and y such that $ax + by = 1$.

Given an indeterminate equation $ax + by = GCD(a, b)$, where a and b are integers, the Extended Euclidean algorithm can be used to calculate integer roots (x, y) of the equation.

Suppose $ax_1 + by_1 = GCD(a, b)$, $bx_2 + (a \bmod b)y_2 = GCD(b, a \bmod b)$. Because $GCD(a, b) = GCD(b, a \bmod b)$, $ax_1 + by_1 = bx_2 + (a \bmod b)y_2$. Because $a \bmod b = a - \left\lfloor \dfrac{a}{b} \right\rfloor \times b$, $ax_1 + by_1 = bx_2 + \left(a - \left\lfloor \dfrac{a}{b} \right\rfloor \times b \right)y_2 = ay_2 + b\left(x_2 - \left\lfloor \dfrac{a}{b} \right\rfloor \right)y_2$. Therefore $x_1 = y_2$, and $y_1 = x_2 - \left\lfloor \dfrac{a}{b} \right\rfloor \times y_2$. Therefore (x_1, y_1) is based on (x_2, y_2). Repeat the recursive process to calculate (x_3, y_3), (x_4, y_4), ..., until $b==0$. At that time $x=1$, $y=0$. Therefore, the Extended Euclidean algorithm is as follows:

```
int exgcd(int a, int b, int &x, int &y)
{
    if (b==0) {x=1; y=0; return a;}
    int t=exgcd(b, a%b, x, y);
    int x0=x, y0=y;
    x=y0; y=x0-(a/b)*y0;
    return t;
}
```

Given an indeterminate equation $ax+by=c$, where a, b, and c are integer constants, x and y are integer variables, $x \in [x_l, x_r]$, and $y \in [y_l, y_r]$, integer roots (x, y) of the equation are required to calculate.

Method 1: Enumeration

Enumerate each pair of (x, y) and find integer roots. That is, the indeterminate equation should be calculated $(x_r - x_l + 1) \times (y_r - y_l + 1)$ times.

Method 2: Extended Euclidean Algorithm

For an indeterminate equation $ax+by=c$, c must be a multiple of $GCD(a, b)$. If c isn't a multiple of $GCD(a, b)$, the indeterminate equation is unsolvable; else the Extended Euclidean algorithm is used to solve the problem.

Suppose $d=GCD(a, b)$, $a'=a$ DIV d, $b'=b$ DIV d, and $c'=c$ DIV d. Then the indeterminate equation $ax+by=c$ can be written as $a'x+b'y=c'$ and $GCD(a', b')==1$. The Extended Euclidean algorithm is used to solve $a'x+b'y=1$ and (x', y') is the integer root. Suppose $x_0=x' \times c'$, $y_0=y' \times c'$. Then (x_0, y_0) is a solution to $ax+by=c$, that is, $ax_0+by_0=c$. Based on that, $a(x_0+b)+b(y_0-a)=c$, $a(x_0+2 \times b)+b(y_0-2 \times a)=c$, ..., $a(x_0+k \times b)+b(y_0-k \times a)=c$, $k \in Z$. Therefore, general solutions to an indeterminate equation $ax+by=c$ are $x= x_0+k \times b$, $y= y_0-k \times a$, $k \in Z$.

3.2.1.1 The Equation

There is an equation $ax+by+c=0$. Given a, b, c, $x1$, $x2$, $y1$, $y2$, you must determine how many integer roots of this equation will satisfy the following conditions: $x1 \le x \le x2$, $y1 \le y \le y2$. The integer root of this equation is a pair of integer numbers (x, y).

Input

Input contains integer numbers a, b, c, $x1$, $x2$, $y1$, $y2$ delimited by spaces and line breaks. All numbers are not greater than 10^8 by absolute value.

Output

Write the answer to the output.

Sample Input	Sample Output
1 1 −3 0 4 0 4	4

ID for Online Judge: SGU 106

 Analysis

First, for the equation $ax+by+c=0$, several special cases for the problem are considered.

1. If $a==0$, $b==0$, and $c\neq0$, then there is no solution. If $a==0$, $b==0$, and $c==0$, then the number of integer roots of an equation is $((x2-x1+1)\times(y2-y1+1))$.
2. If $a==0$, and $b\neq0$, then $by=c$. If c isn't a multiple of b, or c/b isn't an element in $[y1,y2]$, then there is no solution; else for each number x in $[x1,x2]$, $(x,c/b)$ is an integer root.
3. If $b==0$, and $a\neq0$, it is the same as 2.
4. If c isn't a multiple of $GCD(a, b)$, there is no solution.

Then, the solution process is as follows:

1. The equation $ax+by+c=0$ is written as $ax+by=-c$.
2. If a is negative, the value of a needs to be flipped. And we must flip the value of x, if we flip the value of a. That is, the interval $[x1, x2]$ is changed into $[-x2, -x1]$. It is the same for b and y.
3. The Extended Euclidean algorithm is used to calculate the initial solution x_0 and y_0.
4. The integer roots of this equation (x, y) are calculated: $x=x_0+k\times b$, $y=y_0-k\times a$, $k\in Z$. If $x\in[x1,x2]$ and $y\in[y1,y2]$, (x, y) is an integer root.

There is a problem in division: how to transfer reals into integers? For the upper bound, *floor*( ) is used for round down; and for the lower bound, *ceil*( ) is used for round up. For example, if $2.5\leq k\leq5.5$, k can be 3, 4, and 5; and if $-5.5\leq k\leq-2.5$, k can be −3, −4, and −5.

Program

```
#include<cstdio>
#include<cmath>
long long a,b,c,x1,x2,yy1,y2,x0,yy0;     // an equation
ax+by+c=0, the interval for x is[x1, x2], the interval for y
is [yy1, y2], initial solution (x0, yy0)
inline long long cmin(const long long &x,const long long &y)
{return x<y?x:y;}
inline long long cmax(const long long &x,const long long &y)
{return x>y?x:y;}
long long gcd(long long a,long long b)     //GCD(a, b)
{
    if (b==0) return a;
    return gcd(b, a % b);
}
void exgcd(long long a,long long b)     // Extended Euclidean
algorithm is used to calculate the initial solution (x0, yy0)
for ax+by=1
{
    if (b==0){x0=1;yy0=0;return;}
    exgcd(b, a%b);
    long long t=x0; x0=yy0; yy0=t-a/b*yy0;
    return;
}
int main()
{
    scanf("%I64d%I64d%I64d%I64d%I64d%I64d%I64d",&a,&b,&c,&x1,&
x2,&yy1,&y2);
// indeterminate equation: ax+by+c=0, x1≤x≤x2, yy1≤y≤y2
    c=-c;     // ax+by+c=0 is changed to ax+by=-c
    if (c<0) {a=-a; b=-b; c=-c;}
    if (a<0) {a=-a; long long t=x1; x1=-x2; x2=-t;}     //adjust
intervals for x and y
    if (b<0) {b=-b; long long t=yy1; yy1=-y2; y2=-t;}
    if (a==0 && b==0)     // special case: a==0 && b==0
    {
        if (c==0)
        {
            printf("%I64d",(x2-x1+1)*(y2-yy1+1));
            return 0;
        }
        printf("0");return 0;
    }
```

```
    else if (a==0)      // special case: a==0 && b≠ 0
    {
        if(c%b==0)if(c/b<=y2 && c/b>=yy1){ printf("%I64d",x2-
x1+1);return 0;}
        printf("0");
return 0;
    }
    else if (b==0)      // special case: a≠0 && b==0
    {
        if(c%a==0) if(c/a<=x2 && c/a>=x1){ printf("%I64d",y2-
yy1+1);return 0;}
        printf("0");return 0;
    }
  long long d=gcd(a,b);      // d=GCD(a, b). If (c%d!=0), there
is no solution; else Extended Euclidean algorithm is used to
calculate the initial solution x₀ and yy₀
    if (c%d!=0){printf("0");return 0;}
    a=a/d;b=b/d;c=c/d;
    exgcd(a,b);
    x0=x0*c;yy0=yy0*c;
//the upper bound r and the lower bound l
    double tx2=x2,tx1=x1,tx0=x0,ta=a,tb=b,tc=c,ty1=yy1,ty2=y2,
ty0=yy0;
    long long down1=floor(((tx2-tx0)/
tb)),down2=floor(((ty0-ty1)/ta));
    long long r=cmin(down1,down2);
    long long up1=ceil(((tx1-tx0)/tb)),up2=ceil(((ty0-ty2)/
ta));
    long long l=cmax(up1,up2);
    if (r<l) printf("0");      // number of solutions
        else printf("%I64d",r-l+1);
    return 0;
}
```

3.2.2 *Congruences and Congruence Equations*

Given a positive integer m and two integers a and b, if $((a-b) \bmod m)=0$, we say a is congruent to b modulo m, written as $a \equiv b \pmod{m}$. For example, $-7 \equiv -3 \equiv 1 \equiv 5 \equiv 9 \pmod 4$, $-5 \equiv -1 \equiv 3 \equiv 7 \equiv 11 \pmod 4$. On the other hand, if $((a-b) \bmod m) \neq 0$, we say a and b are incongruent modulo m.

Given a set of integers Z and a positive integer m, congruences modulo m satisfy reflexive property, symmetric property, and transitive property. Therefore Z can be divided into m disjoint subsets, called congruence classes modulo m, containing integers that are mutually congruent modulo m.

1. **Congruence Equation**

   A congruence of the form $ax \equiv b \pmod{m}$, where a and b are integers, m is a positive integer, and x is an unknown integer, is called a linear congruence in one variable. The method for calculating x is as follows:

   Step 1: The Euclidean algorithm and Extended Euclidean algorithm are used to calculate $d = GCD(a, m)$ and (x', y') where $d = ax' + my'$, and x' is a solution to $ax' \equiv d \pmod{m}$.

   Step 2: If $b \bmod d \neq 0$, there is no solution for $ax \equiv b \pmod{m}$; else there are d incongruent solutions modulo m, where the first solution is $x_0 = x' \times \left\lfloor \dfrac{b}{d} \right\rfloor \bmod m$, and the other $d-1$ solutions are

   $$x_i = \left(x_0 + i \times \left\lfloor \frac{m}{d} \right\rfloor \right) \bmod m, \ 1 \leq i \leq d-1.$$

   In order to prove the correctness of the two steps, the following three theorems are used.

   1. **Theorem 3.2.2.1** If $ac \equiv bc \pmod{m}$ and $GCD(c, m) = d$, then $a \equiv b \left(\bmod \left\lfloor \dfrac{m}{d} \right\rfloor \right)$.
   2. **Theorem 3.2.2.2** If $d \neq 0$ and $ad \equiv bd \pmod{md}$, then $a \equiv b \pmod{m}$.
   3. **Theorem 3.2.2.3** If $GCD(a, m) = 1$, there are solutions for $ax + b \equiv 0 \pmod{m}$.
      Step 1: Suppose $d = GCD(a, m)$. If $b \bmod d = 0$, solutions to

      $$\left\lfloor \frac{a}{d} \right\rfloor x \equiv \left\lfloor \frac{b}{d} \right\rfloor \left(\bmod \left\lfloor \frac{m}{d} \right\rfloor \right)$$ and $ax \equiv b \pmod{m}$ are the same.

      Proof. Based on **Theorem 3.2.2.1** and **Theorem 3.2.2.2**, solutions to $ax \equiv b \pmod{m}$ and $\left\lfloor \dfrac{a}{d} \right\rfloor x \equiv \left\lfloor \dfrac{b}{d} \right\rfloor \left(\bmod \left\lfloor \dfrac{m}{d} \right\rfloor \right)$ are the same. Because $GCD(a, m) = d > 1$, $\left\lfloor \dfrac{a}{d} \right\rfloor$ and $\left\lfloor \dfrac{m}{d} \right\rfloor$ are relatively prime integers. Based on **Theorem 3.2.2.3**, there are solutions for $\left\lfloor \dfrac{a}{d} \right\rfloor x \equiv \left\lfloor \dfrac{b}{d} \right\rfloor \left(\bmod \left\lfloor \dfrac{m}{d} \right\rfloor \right)$, that is, there is a congruence class $[x]$, where $[x] = \left\{ x + k \times \left\lfloor \dfrac{m}{d} \right\rfloor \middle| k = 0, \pm 1, \pm 2, \ldots \right\}$, and $[x]$ are solutions to $ax \equiv b \pmod{m}$, $0 \leq x \leq \left\lfloor \dfrac{m}{d} \right\rfloor$.

Because x, $x + \left\lfloor \dfrac{m}{d} \right\rfloor$, $x + 2 \times \left\lfloor \dfrac{m}{d} \right\rfloor$, ..., and $x + (d-1) \times \left\lfloor \dfrac{m}{d} \right\rfloor$ are all in $[x]$,

$0 \le x + i \times \left\lfloor \dfrac{m}{d} \right\rfloor < m$, $1 \le i \le d-1$; and they are incongruent modulo m; x, $x + \left\lfloor \dfrac{m}{d} \right\rfloor$,

$x + 2 \times \left\lfloor \dfrac{m}{d} \right\rfloor$, ..., and $x + (d-1) \times \left\lfloor \dfrac{m}{d} \right\rfloor$ are d incongruent solutions modulo m

to $ax \equiv b \pmod{m}$.

Step 2: For $ax \equiv b \pmod{m}$, there are exactly d incongruent solutions modulo

m: x modulo m, $\left(x + \left\lfloor \dfrac{m}{d} \right\rfloor \right)$ modulo m, $\left(x + 2 \times \left\lfloor \dfrac{m}{d} \right\rfloor \right)$ modulo m, ..., and

$\left(x + (d-1) \times \left\lfloor \dfrac{m}{d} \right\rfloor \right)$ modulo m.

Proof. Suppose $x + t \times \left\lfloor \dfrac{m}{d} \right\rfloor$ is a solution to $ax \equiv b \pmod{m}$. Because $t \equiv i \pmod{d}$, $i \in \{0, 1, 2, ..., d-1\}$, based on **Theorem 3.2.2.2**, $t \times \left\lfloor \dfrac{m}{d} \right\rfloor \equiv i \times \left\lfloor \dfrac{m}{d} \right\rfloor \pmod{m}$,

that is, $x + t \times \left\lfloor \dfrac{m}{d} \right\rfloor$ is one of x, $x + \left\lfloor \dfrac{m}{d} \right\rfloor$, $x + 2 \times \left\lfloor \dfrac{m}{d} \right\rfloor$,..., $x + (d-1) \times \left\lfloor \dfrac{m}{d} \right\rfloor$.

Therefore, there are d incongruent solutions modulo m, $\left(x + \left\lfloor \dfrac{m}{d} \right\rfloor \right)$ modulo m,

$\left(x + 2 \times \left\lfloor \dfrac{m}{d} \right\rfloor \right)$ modulo m, ..., and $\left(x + (d-1) \times \left\lfloor \dfrac{m}{d} \right\rfloor \right)$ modulo m.

Based on the above discussions, **Theorem 3.2.2.4** holds.

Theorem 3.2.2.4 Given a positive integer m and two integers a and b, suppose $GCD(a, \ m) = d$. If b mod $d \ne 0$, then $ax \equiv b \pmod{m}$ has no solutions. And if b mod $d = 0$, then $ax \equiv b \pmod{m}$ has exactly d incongruent solutions modulo m.

For example, given a congruence equation $9x \equiv 8 \pmod{3}$, $GCD(9,3) = 3$. Because 8 mod $3 \ne 0$, there is no solution for $9x \equiv 8 \pmod{3}$.

Given a congruence equation $9x \equiv 12 \pmod{15}$, $GCD(9,15) = 3$. Because 12 mod $3 = 0$, $9x \equiv 12 \pmod{15}$ has exactly three incongruent solutions modulo 15. The Extended Euclidean algorithm is used to calculate (x', y') where $3 = 9x' + 15y'$, $x' = 2$, $y' = -1$, 2 is a solution to $9x' \equiv 3 \pmod{15}$. Therefore, $x_0 = 8$ mod $15 = 8$, $x_1 = (x_0 + 5)$ mod $15 = 13$, and $x_2 = (x_0 + 10)$ mod $15 = 18$ mod $15 = 3$.

2. Congruence Equations

Definition 3.2.2.1. Given an integer a with $GCD(a, m)=1$, an integer solution x to $ax \equiv 1 \pmod{m}$ is called an inverse of a modulo m.

By **Theorem 3.2.2.4**, a Congruence Equation $ax \equiv 1 \pmod{m}$ has solutions if and only if $GCD(a, m)=1$ and all solutions are congruent modulo m.

For example, solutions to $6x \equiv 1 \pmod{41}$ satisfy $x \equiv 7 \pmod{41}$. Therefore, 7 is an inverse of 6 modulo 41, and all integers congruent to 7 modulo 41 are inverses of 6 modulo 41. Because $7 \times 6 \equiv 1 \pmod{41}$, 6 and all integers congruent to 6 modulo 41 are inverses of 7 modulo 41.

Theorem 3.2.2.5 (The Chinese Remainder Theorem). Let n_1, n_2, ..., n_k be pairwise relatively prime positive integers. Then the system of congruences

$$a \equiv a_1 \pmod{n_1}$$
$$a \equiv a_2 \pmod{n_2}$$
$$\cdots \cdots \cdots$$
$$\cdots \cdots \cdots$$
$$\cdots \cdots \cdots$$
$$a \equiv a_k \pmod{n_k}$$

has a unique solution modulo $n=n_1 n_2 \ldots n_k$.

The system of congruences can be transformed as a polynomial $a=(a_1 \times c_1 + \ldots a_i \times c_i + \ldots + a_k \times c_k) \bmod (n_1 \times n_2 \times \ldots \times n_k)$. Based on the polynomial, a can be calculated. Now we prove that the system of congruences can be transformed as a polynomial $a=(a_1 \times c_1 + \ldots a_i \times c_i + \ldots + a_k \times c_k) \bmod (n_1 \times n_2 \times \ldots \times n_k)$, and show the method for calculating c_i $(1 \leq i \leq k)$.

Proof. Because n_1, n_2, ..., n_k are pairwise relatively prime positive integers, $GCD(n_i, n_j)=1$, $i \neq j$. Suppose $m_i = \dfrac{n}{n_i}$, $1 \leq i \leq k$. $GCD(n_i, m_i)=1$, $1 \leq i \leq k$. There exist integers n_i' and m_i', such that $m_i m_i' + n_i n_i' = 1$. That is, there exists an integer m_i' such that

$$m_i m_i' \equiv 1 \pmod{n_i} \qquad i = 1, 2, \ldots, k \qquad (1)$$

On the other hand, because $GCD(n_i, n_j)=1$ and $m_i = \dfrac{n}{n_i}$, $n_i | m_j$, $i \neq j$. Therefore

$$a_j m_j m_j' \equiv 0 \pmod{n_i} \qquad i, j = 1, 2, \ldots, k \qquad (2)$$

Based on (1) and (2),

$$a_1 m_1 m_1' + a_2 m_2 m_2' + \ldots + a_k m_k m_k' \equiv a_i m_i m_i' \pmod{n_i},$$
$$a_i m_i m_i' \equiv a_i \pmod{n_i}, \quad i = 1, 2, \ldots, k.$$

Therefore, $a = a_1 m_1 m_1' + a_2 m_2 m_2' + \ldots + a_k m_k m_k' \pmod{n}$ is the unique solution modulo n to the system of congruences.

For example, $a\equiv2(\bmod\ 3)$, $a\equiv4(\bmod\ 7)$, and $a\equiv5(\bmod\ 8)$. 3, 7, and 8 are pairwise relatively prime positive integers. $m_1=n_2\times n_3=56$, $m_2=n_1\times n_3=24$, and $m_3=n_1\times n_2=21$. $n=3\times7\times8=168$. $56\times2=112\equiv1(\bmod\ 3)$, $24\times5=120\equiv1(\bmod\ 7)$, and $21\times5=105\equiv1(\bmod\ 8)$. $2\times112+4\times120+5\times105=1229$. $a=1229$ mod $n=53$.

Steps for calculating the system of congruences are as follows:

Step 1: Calculate m_i, $i=1$, 2, ..., k. Suppose $n=n_1\times n_2\times...\times n_k$; $m_1=\dfrac{n}{n_1}=n_2\times n_3\times...\times n_k$; $m_2=\dfrac{n}{n_2}=n_1\times n_3\times n_4\times...\times n_k$; ......; $m_i=\dfrac{n}{n_i}=n_1\times...\times n_{i-1}\times n_{i+1}...\times n_k$; ......; $m_k=\dfrac{n}{n_k}=n_1\times...\times n_{k-2}\times n_{k-1}$.

Step 2: Calculate an inverse m_i^{-1} of m_i modulo n_i, that is, $m_i\times m_i^{-1}\equiv1(\bmod\ n_i)$, $i=1$, 2, ..., k. There are two methods for calculating m_i^{-1}:

1. Congruence Equation

   Because m_i and n_i are relatively prime integers, that is, $GCD(m_i,\ n_i)=1$, by $m_1\times m_1^{-1}\equiv1(\bmod\ n_1)$; ...; $m_i\times m_i^{-1}\equiv1(\bmod\ n_i)$; ...; $m_k\times m_k^{-1}\equiv1(\bmod\ n_k)$, m_1^{-1}, ..., m_i^{-1}, ..., m_k^{-1} are calculated. There is exactly one solution m_i^{-1} to $m_i\times m_i^{-1}\equiv1(\bmod\ n_i)$, $1\le i\le k$.

2. Extended Euclidean algorithm

   The Extended Euclidean algorithm is used to calculate x and y for $GCD(n_i,\ m_i)=n_i\times x+m_i\times y=1$, and y is m_i^{-1} $(1\le i\le k)$.

Step 3: Calculate $c_i=m_i\times(m_i^{-1}\bmod\ n_i)$, $1\le i\le k$.

Step 4: Calculate $a=(a_1\times c_1+...+a_i\times c_i+...+a_k\times c_k)\bmod\ n$.

3.2.2.1 C Looooops

A compiler mystery: We are given a C-language style for a loop of type

```
for (variable=A; variable!=B; variable+=C)
    statement;
```

that is, a loop which starts by setting a variable to value A, and while *variable* is not equal to B, repeats the statement, followed by increasing the *variable* by C. We want to know how many times does the statement get executed for particular values of A, B, and C, assuming that all arithmetic is calculated in a k-bit unsigned integer type (with values $0\le x<2^k$) modulo 2^k.

Input

The input consists of several instances. Each instance is described by a single line with four integers A, B, C, k separated by a single space. The integer k ($1\le k\le32$) is the number of bits of the control variable of the loop and A, B, C ($0\le A,B,C<2^k$) are the parameters of the loop.

The input is finished by a line containing four zeros.

Output

The output consists of several lines corresponding to the instances on the input. The *i*-th line contains either the number of executions of the statement in the *i*-th instance (a single integer number) or the word FOREVER if the loop does not terminate.

Sample Input	Sample Output
3 3 2 16	0
3 7 2 16	2
7 3 2 16	32766
3 4 2 16	FOREVER
0 0 0 0	

Source: CTU Open 2004

IDs for Online Judges: POJ 2115, ZOJ 2305

 Analysis

Based on the problem description, a loop which starts by setting *variable* to value A and while *variable* is not equal to B, repeats the statement, followed by increasing the *variable* by C. All arithmetic is calculated in a k-bit unsigned integer type (with values $0 \leq x < 2^k$) modulo 2^k. Therefore $D=(B-A) \bmod 2^k$ is equivalent to $x \times C \equiv D (\bmod\ 2^k)$. Obviously, the number of the loop is 0 if and only if $D=(B-A) \bmod 2^k = 0$.

There are solutions to $x \times C \equiv D (\bmod 2^k)$ if and only if $D \bmod GCD(C, 2^k) == 0$. The Extended Euclidean algorithm is used to calculate the minimal non-negative integer solution x to $x \times C + y \times 2^k = GCD(C, 2^k)$. That is, x is a solution to $Cx \equiv GCD(C, 2^k)$ $(\bmod\ 2^k)$. If $((D \bmod GCD(C, 2^k)) \neq 0)$, there is no solution to $x \times C \equiv D (\bmod 2^k)$, and the program enters an endless loop; else $(x \times D) \bmod 2^k$ is the solution to $x \times C \equiv D (\bmod 2^k)$; that is, the number of executions of the statement.

 Program

```
#include<cmath>
#include<cstring>
#include<cstdlib>
#include<cstdio>
#define ll long long
#include<iostream>
using namespace std;
```

```
ll exgcd(ll a,ll b,ll &x,ll &y){     //Extended Euclidean
algorithm: calculate x and y for d=GCD(a,b)=ax+by (x and y can
be 0 or negative)
   if(b==0){
      x=1;y=0;return a;
   }
   ll t=exgcd(b,a%b,y,x);
   y-=a/b*x;
   return t;
}
ll gcd(ll a,ll b){     //Euclidean algorithm returns GCD(a,b)
    if(b==0)return a;
    return gcd(b,a%b);
}
int main () {
    int A,B,C,K;
    ll i,j,ans;
    while(1){
        scanf("%d%d%d%d",&A,&B,&C,&K);     // a test case
        if(!A&&!B&&!C&&!K)break;     // four zeros, break
        ll a,b,c,k;
        a=A,b=B,c=C,k=K;
        ll d=b-a;     //d=(b-a)%2^k. If d=0, the number of loops
is 0; If d%GCD(c, 2^k) ≠0, endless loop
        k=(1ll)<<k;
        d%=k;
        if(d<0)d+=k;
        if(d==0){
            puts("0");continue;
        }
        ll tem=gcd(c,k);
        if(d%tem){
            puts("FOREVER");continue;
        }
        c/=tem,k/=tem,d/=tem;
        exgcd(c,k,ans,j);     //solution ans to GCD(c,
k)=c*ans+k*j
        ans*=(d);
        ans%=k;
        if(ans<0)ans+=k;
        cout<<ans<<endl;
    }
    return 0;
}
```

3.2.2.2 Biorhythms

Some people believe that there are three cycles in a person's life that start the day he or she is born. These three cycles are the physical, emotional, and intellectual cycles, and they have periods of lengths 23, 28, and 33 days, respectively. There is one peak in each period of a cycle. At the peak of a cycle, a person performs at his or her best in the corresponding field (physical, emotional, or mental). For example, if it is the mental curve, thought processes will be sharper and concentration will be easier.

Since the three cycles have different periods, the peaks of the three cycles generally occur at different times. We would like to determine when a triple peak occurs (the peaks of all three cycles occur in the same day) for any person. For each cycle, you will be given the number of days from the beginning of the current year at which one of its peaks (not necessarily the first) occurs. You will also be given a date expressed as the number of days from the beginning of the current year. Your task is to determine the number of days from the given date to the next triple peak. The given date is not counted. For example, if the given date is 10 and the next triple peak occurs on day 12, the answer is 2, not 3. If a triple peak occurs on the given date, you should give the number of days to the next occurrence of a triple peak.

Input

You will be given a number of cases. The input for each case consists of one line of four integers p, e, i, and d. The values p, e, and i are the number of days from the beginning of the current year at which the physical, emotional, and intellectual cycles peak, respectively. The value d is the given date and may be smaller than any of p, e, or i. All values are non-negative and at most 365, and you may assume that a triple peak will occur within 21252 days of the given date. The end of input is indicated by a line in which $p=e=i=d=-1$.

Output

For each test case, print the case number followed by a message indicating the number of days to the next triple peak, in the following form:

Case 1: the next triple peak occurs in 1234 days.

Use the plural form "days" even if the answer is 1.

Sample Input	Sample Output
0 0 0 0	Case 1: the next triple peak occurs in 21252 days.
0 0 0 100	Case 2: the next triple peak occurs in 21152 days.
5 20 34 325	Case 3: the next triple peak occurs in 19575 days.
4 5 6 7	Case 4: the next triple peak occurs in 16994 days.
283 102 23 320	Case 5: the next triple peak occurs in 8910 days.
203 301 203 40	Case 6: the next triple peak occurs in 10789 days.
−1 −1 −1 −1	

Source: ACM East Central North America 1999

IDs for Online Judges: POJ 1006, ZOJ 1160, UVA 756

Analysis

These three cycles are the physical, emotional, and intellectual cycles, and they have periods of lengths 23, 28, and 33 days, respectively. These three integers are pairwise relatively prime positive integers. Suppose x is the number of days to the next triple peak. The system of congruences is as follows:

$$\begin{cases} x \equiv p \ (mod \ 23) \\ x \equiv e \ (mod \ 28) \\ x \equiv i \ (mod \ 33) \end{cases}$$

Based on **The Chinese Remainder Theorem**, x is the only solution in the interval [1, 23×28×33=21253]. Suppose a_i and n_i are the number of days to the next triple peak and the period of length respectively, that is, $a_1=p$, $a_2=e$, $a_3=i$, $n_1=23$, $n_2=28$, and $n_3=33$. The system of congruences is as follows:

$$x \equiv a_i \ (\text{mod} \ n_i), \ (1 \leq i \leq 3).$$

The above four steps are used to calculate $s = \sum_{i=1}^{3} m_i * a_i * (m_i^{-1} \text{ mod } n_i)$, where $m_1=28×33$, $m_2=23×33$, and $m_3=23×28$. The Extended Euclidean algorithm is used to calculate the inverse m_i^{-1} for m_i modulo n_i, that is, $m_i × m_i^{-1} \equiv 1 (\text{mod} \ n_i)$.

Suppose d is the given date. The number of days to the next triple peak is the minimum positive integer for $(s-d) \text{mod} \ n$, where $n=23×28×33$.

Program

```
#include<iostream>
#include<algorithm>
#include<cmath>
#include<cstdio>
#include<cstring>
#include<cstdlib>
#include<string>
using namespace std;
typedef long long ll;
ll power(ll a,ll p,ll mo){      //Calculate a^p%(mo)
    ll ans=1;
    for(;p;p>>=1){
        if(p&1){
            ans*=a;
```

```
        if(mo>0)ans%=mo;
    }
    a*=a;
    if(mo>0)a%=mo;
  }
  return ans;
}
ll exgcd(ll a,ll b,ll &x,ll &y){      // Extended Euclidean
algorithm: calculate x for GCD(a, b)=ax+by
  if(b==0){
     x=1;y=0;return a;
  }
  ll t=exgcd(b,a%b,y,x);
  y-=a/b*x;
  return t;
}
ll niyuan(ll a,ll p){      //calculate a⁻¹%p
  ll x,y;
  exgcd(a,p,x,y);      //calculate x for ax≡GCD(a, p)(%p)
  return (x%p+p)%p;
}
int main(){
    int   a,b,c,d,i,j,k,u,v,te=0;
    while(1){
       scanf("%d%d%d%d",&a,&b,&c,&d);      //test case
       if(a==b&&b==c&&c==d&&a==-1)break;      //end case
```

//calculate $an = \left(\sum_{i=1}^{3} (m_i * a_i * (m_i^{-1} \bmod n_i)) - d \right) \% (23*28*33)$, that

```
is, the number of days to the next triple peak
       ll an=0;
       an=28*33*a*niyuan(28*33,23)+23*33*b*niyuan(23*33,28)+23*
28*c*niyuan(28*23,33);
       an-=d;
       an%=(28*33*23);
       if(an<=0)an+=28*33*23;
       printf("Case %d: the next triple peak occurs in %d
days.\n",++te,(int)an);
    }
}
```

3.3 Multiplicative Functions

Definition 3.3.1 (Multiplicative Function). An arithmetic function *f* is a multiplicative function if *f*(*ab*)=*f*(*a*)*f*(*b*), where *a* and *b* are relatively prime positive integers. An arithmetic function *f* is a completely multiplicative function if *f*(*ab*)=*f*(*a*)*f*(*b*), where *a* and *b* are positive integers.

Definition 3.3.2 (Euler Phi-Function $\varphi(n)$). Suppose n is a positive integer. The Euler phi-function $\varphi(n)$ is defined to be the number of positive integers not exceeding n that are relatively prime to n.

For example, $\varphi(1)=\varphi(2)=1$, $\varphi(3)=\varphi(4)=2$.

Theorem 3.3.1 If n is a prime number, $\varphi(n)=n-1$. And if n is a composite number, $\varphi(n)<n-1$.

For example, $\varphi(7)=6$.

Theorem 3.3.2 (Phi-Function Formula).

1. If p is a prime and $k\geq1$, then $\varphi(p^k)=p^k-p^{k-1}$.
2. If m and n are relatively prime numbers, $\varphi(mn)=\varphi(m)\varphi(n)$.

Therefore, $\varphi(n)$ is a multiplicative function, but $\varphi(n)$ isn't a completely multiplicative function.

Theorem 3.3.3 A number m can be written as a product of primes: $m=p_1^{k_1}\times p_2^{k_2}\times...\times p_r^{k_r}$, where p_1, p_2, ..., p_r are all different primes. $\varphi(m)=\varphi(p_1^{k_1})\times\varphi(p_2^{k_2})\times...\times\varphi(p_r^{k_r})$.

For example, $\varphi(18)=\varphi(2\times3^2)=\varphi(2)\times\varphi(3^2)=3^2-3=6$.

Definition 3.3.3 (Reduced Residue System Modulo n). A reduced residue system modulo n is a set of $\varphi(n)$ integers such that each element of the set is relatively prime to n, and no two different elements of the set are congruent modulo n.

For example, if $n=10$, $\varphi(10)=4$. Each element in the set $\{1, 3, 7, 9\}$ is relatively prime to 10, and no two different elements of the set are congruent modulo 10. Therefore, the set $\{1, 3, 7, 9\}$ is a reduced residue system modulo 10. For the same reason, the set $\{-3, -1, 1, 3\}$ is also a reduced residue system modulo 10.

Theorem 3.3.4 If a set $\{r_1, r_2, ..., r_{\varphi(n)}\}$ is a reduced residue system modulo n, and if n and a are coprime positive integers, then the set $\{ar_1, ar_2, ..., ar_{\varphi(n)}\}$ is also a reduced residue system modulo n.

For example, the set $\{1, 3, 7, 9\}$ is a reduced residue system modulo 10, and 3 and 10 are coprime positive integers. Then the set $\{3, 9, 21, 27\}$ is also a reduced residue system modulo 10.

Theorem 3.3.5 (Euler's Theorem, or Fermat–Euler Theorem). If n and a are coprime positive integers, then $a^{\varphi(n)}\equiv1(\mathrm{mod}\ n)$.

Proof. Suppose a set $\{r_1, r_2, ..., r_{\varphi(n)}\}$ is a reduced residue system whose element doesn't exceed n and is relatively prime to n. By **Theorem 3.3.4**, if n and a are coprime positive integers, then the set $\{ar_1, ar_2, ..., ar_{\varphi(n)}\}$ is also a reduced residue system modulo n. Therefore, the least positive residue system for $\{ar_1, ar_2, ..., ar_{\varphi(n)}\}$ is the set $\{r_1, r_2, ..., r_{\varphi(n)}\}$ in some order. If all elements in $\{ar_1, ar_2, ..., ar_{\varphi(n)}\}$ and $\{r_1, r_2, ..., r_{\varphi(n)}\}$ are multiplied together, $ar_1ar_2...ar_{\varphi(n)}\equiv r_1r_2...r_{\varphi(n)}$ (mod n). Therefore, $a^{\varphi(n)}r_1r_2...r_{\varphi(n)}\equiv r_1r_2...r_{\varphi(n)}$ (mod n). Because $r_1r_2...r_{\varphi(n)}$ and n are relatively prime numbers, then $a^{\varphi(n)}\equiv1$ (mod n).

For example, $\{1, 3, 7, 9\}$ is a reduced residue system whose element doesn't exceed 10 and is relatively prime to 10. 10 and 3 are coprime positive integers.

And {3, 9, 21, 27} is also a reduced residue system modulo 10. Therefore, the least positive residue system for {3, 9, 21, 27} is the set {1, 3, 7, 9} in some order. $3 \times 9 \times 21 \times 27 \equiv 1 \times 3 \times 7 \times 9$ (mod 10). $1 \times 3 \times 7 \times 9 (\text{mod } 10) = 9$. $n=10$, $a=3$, and $\varphi(10)=4$. $3^4 = 3^{\varphi(10)} \equiv 1 (\text{mod } 10)$.

Corollary 3.3.1. If n and a are coprime positive integers, then $a^{\varphi(n)+1} \equiv a (\text{mod } n)$.

Theorem 3.3.6 (Fermat's Little Theorem). If p is a prime number, a is a positive integer, and $GCD(a, p)=1$, then $a^{p-1} \equiv 1 (\text{mod } p)$. And if p is a prime and a is an positive integer, $a^p \equiv a (\text{mod } p)$.

For example, if $a=3$ and $p=5$, $3^4 \equiv 1 (\text{mod } 5)$. And if $a=6$ and $p=3$, $6^3 \equiv 6 (\text{mod } 3)$.

Definition 3.3.4 (Order of a Modulo n). Suppose a and n are relatively prime integers, where $a \neq 0$ and $n>0$. The least positive integer x such that $a^x \equiv 1 (\text{mod } n)$ is the order of a modulo n, and is denoted as $ord_n a$.

For example, suppose $a=3$ and $n=5$. $3^4 = 81 \equiv 1 (\text{mod } 5)$. Therefore $ord_5 3 = 4$.

Definition 3.3.5 (Primitive Root). Suppose a and n are relatively prime integers, where $n>0$. If $ord_n a = \varphi(n)$, then a is a primitive root modulo n, and n has a primitive root.

For example, $ord_5 3 = \varphi(5) = 4$. 3 is a primitive root of modulo 5, and 5 has a primitive root.

Theorem 3.3.7 If a positive integer n has a primitive root, then it has $\varphi(\varphi(n))$ different incongruent primitive roots.

3.3.1.1 Relatives

Given n, a positive integer, how many positive integers less than n are relatively prime to n? Two integers a and b are relatively prime if there are no integers $x>1$, $y>0$, $z>0$ such that $a=xy$ and $b=xz$.

Input

There are several test cases. For each test case, standard input contains a line with $n \leq 1,000,000,000$. A line containing 0 follows the last case.

Output

For each test case there should be a single line of output answering the question posed above.

Sample Input	Sample Output
7	6
12	4
0	

Source: Waterloo local 2002.07.01

IDs for Online Judges: POJ 2407, ZOJ 1906, UVA 10299

Analysis

Given a positive integer n, the number of positive integers less than n are relatively prime to n is the Euler phi-function $\varphi(n)$. n can be written as a product of primes: $n = p_1^{k_1} \times p_2^{k_2} \times ... \times p_r^{k_r}$. Therefore, $\varphi(n) = \varphi(p_1^{k_1}) \times \varphi(p_2^{k_2}) \times ... \times \varphi(p_r^{k_r})$, where $\varphi(p_i^{k_i}) = (p_i - 1) \times p_i^{k_i - 1}$, $1 \leq i \leq r$.

Program

```cpp
#include<iostream>
#include<cstdio>
#include<cstring>
#include<cmath>
#include<algorithm>
#include<cstdio>
#include<cstdlib>
using namespace std;
typedef long long ll;
bool u[50000];      //Prime sieve
ll su[50000],num;      //Prime list whose length is num
ll gcd(ll a,ll b){      //GCD(a, b)
    if(b==0)return a;
    return gcd(b,a%b);
}
void prepare(){      //Construct prime list su[] in [2, 50000]
    ll i,j,k;
    for(i=2;i<50000;i++)u[i]=1;
    for(i=2;i<50000;i++)
        if(u[i])
        for(j=2;j*i<50000;j++)
            u[i*j]=0;
    for(i=2;i<50000;i++)
        if(u[i])
            su[++num]=i;
}
ll phi(ll x)      // Euler phi-function φ(x)
{
    ll ans=1;
    int i,j,k;
    for(i=1;i<=num;i++)
    if(x%su[i]==0){      //the number of prime factor su[i] is j
        j=0;
```

```
            while(x%su[i]==0){++j;x/=su[i];}
            for(k=1;k<j;k++)ans=ans*su[i]%1000000007ll;
            ans=ans*(su[i]-1)%1000000007ll;
            if(x==1)break;
        }
    if(x>1)ans=ans*(x-1)%1000000007ll;
    return ans;    // return φ(x)
}
int main(){
    prepare();    // Construct prime list su[] in [2, 50000]
    int n,i,j,k;
    ll ans=1;
    while(scanf("%d",&n)==1&&n>0){
//Input test cases until 0
        ans=phi(n);    //calculate and output φ(n)
        printf("%d\n",(int)ans);
    }
}
```

3.3.1.2 Primitive Roots

We say that integer x, $0<x<p$, is a primitive root modulo odd prime p if and only if the set $\{(x^i \bmod p)|1\leq i\leq p-1\}$ is equal to $\{1, ..., p-1\}$. For example, the consecutive powers of 3 modulo 7 are 3, 2, 6, 4, 5, 1, and thus 3 is a primitive root modulo 7.

Write a program which given any odd prime $3\leq p<65536$ outputs the number of primitive roots modulo p.

Input

Each line of the input contains an odd prime numbers p. Input is terminated by the end-of-file separator.

Output

For each p, print a single number that gives the number of primitive roots in a single line.

Sample Input	Sample Output
23	10
31	8
79	24

Source: Jiayi@pku

ID for Online Judge: POJ 1284

Analysis

Based on the problem description, an integer x, $0<x<p$, is a primitive root modulo odd prime p if and only if the set $\{(x^i \bmod p)|1\leq i\leq p-1\}$ is equal to $\{1, ..., p-1\}$. If p has a primitive root, then it has $\varphi(\varphi(p))$ different primitive roots. Because p is a prime, $\varphi(\varphi(p))=\varphi(p-1)$.

Program

```cpp
#include<iostream>
#include<cstdio>
#include<cstring>
#include<cmath>
#include<algorithm>
#include<cstdio>
#include<cstdlib>
using namespace std;
typedef long long ll;
bool u[50000];      //prime sieve
ll su[50000],num;       //prime list whose length is num
void prepare(){     //Calculate prime list su[]
    ll i,j,k;
    for(i=2;i<50000;i++)u[i]=1;
    for(i=2;i<50000;i++)
        if(u[i])
            for(j=2;j*i<50000;j++) u[i*j]=0;
    for(i=2;i<50000;i++)
        if(u[i]) su[++num]=i;
}
ll phi(ll x)     // Euler phi-function φ(x)
{
    ll ans=1;
    int i,j,k;
    for(i=1;i<=num;i++)     //Enumerate each prime
    if(x%su[i]==0){     //if x has prime factor su[i], then
φ(s[i])=su[i]^{j-1}*(su[i]-1), and adjust φ(x)
        j=0;
        while(x%su[i]==0){++j;x/=su[i];}
        for(k=1;k<j;k++)ans=ans*su[i]%100000000711;
        ans=ans*(su[i]-1)%100000000711;
        if(x==1)break;
    }
```

```
    if(x>1)ans=ans*(x-1)%1000000007ll;      // φ(x)
    return ans;
}
int main(){
    prepare();      //construct prime list su[]
    int n,i,j,k;
    ll ans=1;
    while(scanf("%d",&n)==1){       //input test case n until EOF
        ans=phi(n-1);      //the number of primitive roots for n
        printf("%d\n",(int)ans);
    }
}
```

3.4 Problems

3.4.1 Prime Frequency

Given a string containing only alpha-numerals (0-9, A-Z and a-z), you have to count the frequency (the number of times the character is present) of all the characters and report only those characters whose frequency is a prime number. A prime number is a number which is divisible by exactly two different integers.

Some examples of prime numbers are 2, 3, 5, 7, 11, etc.

Input

The first line of the input is an integer T ($0 < T < 201$) that indicates how many sets of inputs are there. Each of the next T lines contains a single set of input.

The input of each test set is a string consisting of alpha-numerals only. The length of this string is positive and less than 2001.

Output

For each set of input, produce one line of output. This line contains the serial of output followed by the characters whose frequency in the input string is a prime number. These characters are to be sorted in lexicographically ascending order. Here "lexicographically ascending" means ascending in terms of the ASCII values. Look at the output for sample input for details. If none of the character frequency is a prime number, you should print "empty" (without the quotes) instead.

Sample Input	Sample Output
3	Case 1: C
ABCC	Case 2: AD
AABBBBBDDDDD	Case 3: empty
ABCDFFFF	

Source: Bangladesh National Computer Programming Contest

ID for Online Judge: UVA 10789

Hint

First, the offline method is used to calculate the prime sieve $u[]$ in [2, 2200]. Second, for each test case (a string), every character's frequency $p[]$ is calculated. Third, characters whose frequency is a prime number are sorted in lexicographically ascending order. If none of the character frequency is a prime number, "empty" is output.

3.4.2 Twin Primes

Twin primes are pairs of primes of the form (***p, p+2***). The term "twin prime" was coined by Paul Stäckel (1892–1919). The first few twin primes are (**3, 5**), (**5, 7**), (**11, 13**), (**17, 19**), (**29, 31**), (**41, 43**). In this problem you are asked to find out the *S*-th twin prime pair where *S* is an integer that will be given in the input.

Input

The input will contain less than **10001** lines of input. Each line contains an integers *S* (**1≤S≤100000**), which is the serial number of a twin prime pair. Input file is terminated by end of file.

Output

For each line of input, you will have to produce one line of output which contains the **S**-th twin prime pair. The pair is printed in the form (*p_1*,**<space>***p_2*). Here **<space>** means the space character (**ASCII 32**). You can safely assume that the primes in the **100000-th** twin prime pair are less than **20000000**.

Sample Input	Sample Output
1	(3, 5)
2	(5, 7)
3	(11, 13)
4	(17, 19)

Source: Regionals Warmup Contest 2002, Venue: Southeast University, Dhaka, Bangladesh

ID for Online Judge: UVA 10394

Hint

Suppose the sequence for twin primes is *ans*[], where *ans*[*i*] is the least prime for the *i*-th twin primes, 1≤*i*≤*num*.

The method for calculating *ans*[] is as follows:

First, the sieve method is used to calculate the prime sieve *u*[] for the interval [2, 20000000];
Second, each integer *i* is enumerated. If *i* and *i*+2 is a twin prime (*u*[*i*]&&*u*[*i*+2]), then *i* is added into the sequence for twin primes (*ans*[++*num*]=*i*);

Finally, for each test case *s*, the twin prime (*ans*[*s*], *ans*[*s*]+2) is output.

3.4.3 Less Prime

Let *n* be an integer, $100 \leq n \leq 10000$. Find the prime number *x*, $x \leq n$, so that $n - p \times x$ is maximum, where *p* is an integer such that $p \times x \leq n < (p+1) \times x$.

Input

The first line of the input contains an integer, *M*, indicating the number of test cases. For each test case, there is a line with a number *N*, $100 \leq N \leq 10000$.

Output

For each test case, the output should consist of one line showing the prime number that verifies the condition above.

Sample Input	Sample Output
5	2203
4399	311
614	4111
8201	53
101	3527
7048	

Source: III Local Contest in Murcia 2005

ID for Online Judge: UVA 10852

Hint

Because $n - p \times x$ is maximum (*x* is a prime number, *p* is an integer, $p \times x \leq n < (p+1) \times x$), *x* is such a prime number that *x*%*n* is maximal for all prime numbers less than *n*. The algorithm is as follows:

First, the prime list *su*[] for the interval [2, 11111] is calculated, where its length is *num*. Then, for each test case *n*, all prime numbers less than *n* are enumerated,

$tmp = \max\limits_{1\leq i\leq num} \{n\%su[i]\,|\,su[i] < n\}$. The prime number that verifies the condition above

is $su[k]$ that $tmp=n\%su[k]$.

3.4.4 Prime Words

A prime number is a number that has only two divisors: itself and the number one. Examples of prime numbers are: 1, 2, 3, 5, 17, 101, and 10007.

In this problem, you should read a set of words. Each word is composed only by letters in the range a-z and A-Z. Each letter has a specific value: the letter a is worth 1, letter b is worth 2, and so on until letter z, which is worth 26. In the same way, letter A is worth 27, letter B is worth 28, and letter Z is worth 52.

You should write a program to determine if a word is a prime word or not. A word is a prime word if the sum of its letters is a prime number.

Input

The input consists of a set of words. Each word is in a line by itself and has L letters, where $1\leq L\leq20$. The input is terminated by end of file (EOF).

Output

For each word you should print: **It is a prime word.**, if the sum of the letters of the word is a prime number; otherwise you should print: **It is not a prime word.**

Sample Input	Sample Output
UFRN	It is a prime word.
contest	It is not a prime word.
AcM	It is not a prime word.

Source: UFRN-2005 Contest 1

ID for Online Judge: UVA 10924

Hint

First, the offline method is used to calculate a prime list $u[]$ in the interval [2, 1010].

Second, a test case (a word whose length is n) is input, and the sum of letters in

the word is $X = \sum\limits_{i=1}^{n}(s[i]-'a'+1\,|\,s[i]\in\{'a'..'z'\}),s[i]-'A'+27\,|\,s[i]\in\{'A'..'Z'\}$.

If X is a prime number in [2, 1010], the word is a prime word; else it isn't a prime word.

3.4.5 Sum of Different Primes

A positive integer may be expressed as a sum of different prime numbers (primes), in one way or another. Given two positive integers n and k, you should count the number of ways to express n as a sum of k different primes. Here, two ways are considered to be the same if they sum up the same set of the primes. For example, 8 can be expressed as 3+5 and 5+3, but they are not distinguished.

When n and k are 24 and 3 respectively, the answer is two because there are two sets {2, 3, 19} and {2, 5, 17} whose sums are equal to 24. There are no other sets of three primes that sum up to 24. For $n=24$ and $k=2$, the answer is three, because there are three sets {5, 19}, {7, 17} and {11, 13}. For $n=2$ and $k=1$, the answer is one, because there is only one set {2} whose sum is 2. For $n=1$ and $k=1$, the answer is zero. As 1 is not a prime, you shouldn't count {1}. For $n=4$ and $k=2$, the answer is zero, because there are no sets of two different primes whose sums are 4.

Your job is to write a program that reports the number of such ways for the given n and k.

Input

The input is a sequence of datasets followed by a line containing two zeros separated by a space. A dataset is a line containing two positive integers n and k separated by a space. You may assume that $n \leq 1120$ and $k \leq 14$.

Output

The output should be composed of lines, each corresponding to an input dataset. An output line should contain one non-negative integer indicating the number of ways for n and k specified in the corresponding dataset. You may assume that it is less than 2^{31}.

Sample Input	Sample Output
24 3	2
24 2	3
2 1	1
1 1	0
4 2	0
18 3	2
17 1	1
17 3	0
17 4	1
100 5	55
1000 10	200102899
1120 14	2079324314
0 0	

Source: ACM Japan 2006

IDs for Online Judges: POJ 3132, ZOJ 2822, UVA 3619

Hint

Suppose *su*[] is the prime list in the interval [2, 1200]; $f[i][j]$ is the number of ways to express j as a sum of i different primes, $1 \leq i \leq 14$, and $su[i] \leq j \leq 1199$. Obviously, $f[0][0]=1$.

First, *su*[] is calculated. Its length is *num*.

Then, for a test case (two positive integers n and k), Dynamic Programming is used to compute the number of ways to express n as a sum of k different primes.

```
Enumerate each prime su[i] in su[] (1≤i≤num):
  Enumerate the number of different primes j in descending
order (j=14…1):
    Enumerate the sum of the first j primes p (p=1199…su[i]):
      Accumulate the number of ways that su[i] is as the j-th
prime f[j][p]+=f[j-1][p-su[i]];
```

Finally, $f[k][n]$ is the solution to the problem.

3.4.6 Gerg's Cake

Gerg is having a party, and he has invited his friends. **p** of them have arrived already, but **a** are running late. To occupy his guests, he tried playing some team games with them, but he found that it was impossible to divide the **p** guests into any number of equal-sized groups of more than one person.

Luckily, he has a backup plan—a cake that he would like to share between his friends. The cake is in the shape of a square, and Gerg insists on cutting it up into equal-sized square pieces. He wants to reserve one slice for each of the **a** missing friends, and the rest of the slices have to be divided evenly between the **p** remaining guests. He does not want any cake himself. Can he do it?

Input

The input will consist of several test cases. Each test case will be given as a non-negative integer **a** and a positive integer **p** as specified above, on a line. Both **a** and **p** will fit into a 32-bit signed integer. The last line will contain "−1 −1" and should not be processed.

Output

For each test case, output "Yes" if the cake can be fairly divided and "No" otherwise.

Sample Input	Sample Output
1 3	Yes
1024 17	Yes
2 101	No
0 1	Yes
−1 −1	

Source: 2005 ACM ICPC World Finals Warmup 2

ID for Online Judge: UVA 10831

Hint by the Problemsetter (http://www .algorithmist.com/index.php/Main_Page)

The summary of the problem is as follows. Given *a* and *p*, can a square cake be divided into $a+n \times p$ equal-sized pieces?

You have to test whether there is a solution to $x^2 = a + n \times p$, where *n* is an integer. Taking everything modulo *p*, we get $x^2 \equiv a \pmod{p}$. Now we use a trick to get to Fermat's Little Theorem: we take everything to the power $(p-1)/2$, so we get $x^{p-1} \equiv a^{(p-1)/2} \equiv 1 \pmod{p}$. So we only have to check whether $a^{(p-1)/2} \equiv 1 \pmod{p}$. If it is, there is a solution, and otherwise there isn't. This can easily be calculated in $O(\log p)$.

There are a few special cases, for example, $a \equiv 0 \pmod{p}$, $p=1$ and $p=2$.

3.4.7 Widget Factory

The widget factory produces several different kinds of widgets. Each widget is carefully built by a skilled widgeteer. The time required to build a widget depends on its type: the simple widgets need only three days, but the most complex ones may need as many as nine days.

The factory is currently in a state of complete chaos: recently, the factory has been bought by a new owner, and the new director has fired almost everyone. The new staff know almost nothing about building widgets, and it seems that no one remembers how many days are required to build each different type of widget. This is embarrassing when a client orders widgets and the factory cannot tell the client how many days are needed to produce the required goods. Fortunately, there are records that say, for each widgeteer, the date when he started working at the factory, the date when he was fired, and what types of widgets he built. The problem is that the record does not say the exact date of starting and leaving the job, only the day of the week. Nevertheless, even this information might be helpful in certain cases: for example, if a widgeteer started working on a Tuesday, built a Type 41 widget, and was fired on a Friday, then we know that it takes four days to build a Type 41 widget. Your task is to figure out from these records (if possible) the number of days that are required to build the different types of widgets.

Input

The input contains several blocks of test cases. Each case begins with a line containing two integers: the number $1 \leq n \leq 300$ of the different types, and the number $1 \leq m \leq 300$ of the records. This line is followed by a description of the m records. Each record is described by two lines. The first line contains the total number $1 \leq k \leq 10000$ of widgets built by this widgeteer, followed by the day of the week when he or she started working and the day of the week he or she was fired. The days of the week are given by the strings 'MON', 'TUE', 'WED', 'THU', 'FRI', 'SAT', and 'SUN'. The second line contains k integers separated by spaces. These numbers are between 1 and n, and they describe the different types of widgets that the widgeteer built. For example, the following two lines mean that the widgeteer started working on a Wednesday, built a Type 13 widget, a Type 18 widget, a Type 1 widget, again a Type 13 widget, and was fired on a Sunday.

4 WED SUN
13 18 1 13

Note that the widgeteers work seven days a week, and they were working on every day between their first and last day at the factory (if you like weekends and holidays, then do not become a widgeteer!).

The input is terminated by a test case with $n = m = 0$.

Hint: Huge input file, 'scanf' recommended to avoid TLE.

Output

For each test case, you have to output a single line containing n integers separated by spaces: the number of days required to build the different types of widgets. There should be no space before the first number or after the last number, and there should be exactly one space between two numbers. If there is more than one possible solution for the problem, then write "Multiple solutions." (without the quotes). If you are sure that there is no solution consistent with the input, then write "Inconsistent data." (without the quotes).

Sample Input	Sample Output
2 3	8 3
2 MON THU	
1 2	Inconsistent
	data.
3 MON FRI	
1 1 2	
3 MON SUN	
1 2 2	
10 2	
1 MON TUE	
3	
1 MON WED	
3	
0 0	

Source: ACM Central Europe 2005

IDs for Online Judges: POJ 2947, UVA 3529

 Hint

There are N types of widgets, and each type takes a fixed number of days (between three and nine) to be produced. In the factory there were several workers. For each of them, we know the information "He started on weekday X, produced c_1 widgets of type t_1, ..., c_k widgets of type t_k and finished on weekday Y." The task is to determine the production time for each of the widgets.

There may be inputs where there is no answer or more than one answer, and in these cases you just have to output a corresponding message.

Note that if we want to know the number of days D a widget takes to be completed, it is enough to determine (D modulo 7).

Each worker's information can be translated into a linear congruence modulo 7. The resulting set of equations can be solved using Gaussian elimination.

Note that all operations when solving the set of equations are done modulo 7. Seven (the number of days in a week) is a prime number. Thus Z_7 (the set {0, 1, 2, 3, 4, 5, 6} with addition and multiplication modulo 7) is a field. In other words, each number other than 0 has a multiplicative inverse, and thus we can divide in Z_7. E.g., in Z_7 $2{\times}4=1$, so instead of dividing a number by 4, we can multiply it by 2.

3.4.8 Count the Factors

Write a program that computes the number of different prime factors in a positive integer.

Input

The input tests will consist of a series of positive integers. Each number is on a line on its own. The maximum value is 1000000. The end of the input is reached when the number 0 is met. The number 0 shall not be considered as part of the test set.

Output

The program shall output each result on a line by its own, following the format given in the sample output.

Sample Input	Sample Output
289384	289384 : 3
930887	930887 : 2
692778	692778 : 5
636916	636916 : 4
747794	747794 : 3
238336	238336 : 3
885387	885387 : 2
760493	760493 : 2
516650	516650 : 3
641422	641422 : 3
0	

Source: 2004 Federal University of Rio Grande do Norte Classifying Contest-Round 2

ID for Online Judge: UVA 10699

 Hint

First, the prime list *su*[] in the interval [2, 1200] is calculated.

Then, for each test case, a positive integer x, the method by which the number of different prime factors k for x is calculated as follows:

Initially $k=0$. The prime list *su*[] is searched one by one. If *su*[i] is a prime factor for x (x% *su*[i]==0), then k++; and x/=*su*[i] is repeated until (x% *su*[i]≠0). If x>1 after all elements in *su*[] have been searched, k++.

3.4.9 Prime Land

Everybody in the Prime Land is using a prime base number system. In this system, each positive integer x is represented as follows: Let $\{p_i\}_{i=0}^{\infty}$ denote the increasing sequence of all prime numbers. We know that $x>1$ can be represented in only one way in the form of product of powers of prime factors. This implies that there is an integer k_x and uniquely determined integers $e_{k_x}, e_{k_x-1}, \ldots, e_1, e_0, (e_{k_x} > 0)$, that $x = p_{k_x}^{e_{k_x}} \times p_{k_x-1}^{e_{k_x-1}} \times \ldots \times p_1^{e_1} \times p_0^{e_0}$. The sequence $(e_{k_x}, e_{k_x-1}, \ldots, e_1, e_0)$ is considered to be the representation of x in the prime base number system.

It is really true that all numerical calculations in the prime base number system can seem to us a little bit unusual, or even hard. In fact, the children in Prime Land learn to add and to subtract numbers for several years. On the other hand, multiplication and division are very simple.

Recently, somebody has returned from a holiday in the Computer Land where small smart things called computers have been used. It turns out that they could be used to make addition and subtraction in the prime base number system much easier. It has been decided to make an experiment and let a computer do the operation "minus one".

Help people in the Prime Land and write a corresponding program for them.

For practical reasons, we will write here the prime base representation as a sequence of such p_i and e_i from the prime base representation above, for which $e_i > 0$. We will keep decreasing order with regard to p_i.

Input

The input file consists of lines (at least one), each of which, except the last, contains a prime base representation of just one positive integer greater than 2 and less or equal to 32767. All numbers in the line are separated by one space. The last line contains number 0.

Output

The output file contains one line for each but the last line of the input file. If x is a positive integer contained in a line of the input file, the line in the output file will contain $x-1$ in prime base representation. All numbers in the line are separated by one space. There is no line in the output file corresponding to the last "null" line of the input file.

Sample Input	Sample Output
17 1	2 4
5 1 2 1	3 2
509 1 59 1	13 1 11 1 7 1 5 1 3 1 2 1
0	

Source: ACM Central Europe 1997

IDs for Online Judges: POJ 1365, ZOJ 1261, UVA 516

Hint

First, a prime list in the interval [2, 32767] is calculated.

Then, for a test case (a prime base representation of a number x), x is calculated by multiplying $x = p_{k_x}^{e_{k_x}} \times p_{k_x-1}^{e_{k_x-1}} \times \ldots \ldots \times p_1^{e_1} \times p_0^{e_0}$.

Finally, the prime base representation of $x-1$ is output.

3.4.10 Prime Factors

An integer $g>1$ is said to be prime if and only if its only positive divisors are itself and one (otherwise, it is said to be composite). For example, the number 21 is composite; the number 23 is prime. Note that the decomposition of a positive number g into its prime factors, i.e., $g=f_1 \times f_2 \times \ldots \ldots \times f_n$ is unique if we assert $f_i>1$ that for all i and $f_i \leq f_j$ for $i<j$.

One interesting class of prime numbers are the so-called Mersenne primes which are of the form 2^p-1. Euler proved that $2^{31}-1$ is prime in 1772—all without the aid of a computer.

Input

The input will consist of a sequence of numbers. Each line of input will contain one number g in the range $-2^{31}<g<2^{31}$, but this number is different from -1 and 1. The end of input will be indicated by an input line having a value of zero.

Output

For each line of input, your program should print a line of output consisting of the input number and its prime factors. For an input number $g>0$, $g=f_1 \times f_2 \times \ldots \ldots \times f_n$, where each f_i is a prime number greater than unity (with $f_i \leq f_j$ for $i<j$), the format of the output line should be $g=f_1 \times f_2 \times \ldots \ldots \times f_n$. Where $g<0$, if $|g|=f_1 \times f_2 \times \ldots \ldots \times f_n$, the format of the output line should be $g=-1 \times f_1 \times f_2 \times \ldots \ldots \times f_n$.

Sample Input	Sample Output
−190	−190 = −1 x 2 x 5 x 19
−191	−191 = −1 x 191
−192	−192 = −1 x 2 x 2 x 2 x 2 x 2 x 2 x 3
−193	−193 = −1 x 193
−194	−194 = −1 x 2 x 97
.195	195 = 3 x 5 x 13
196	196 = 2 x 2 x 7 x 7
197	197 = 197
198	198 = 2 x 3 x 3 x 11
199	199 = 199
200	200 = 2 x 2 x 2 x 5 x 5
0	

Source: ACM East Central Region 1997

ID for Online Judge: UVA 583

Hint

First, a prime list in the interval $[2, \sqrt{2^{31}}]$ is calculated.

Then, for a input number x, if x is negative, the -1 coefficient should be added before factoring; this method is similar to the **3.4.8 Count the factors** method.

3.4.11 Perfect Pth Powers

We say that x is a perfect square if, for some integer b, $x=b^2$. Similarly, x is a perfect cube if, for some integer b, $x=b^3$. More generally, x is a perfect pth power if, for some integer b, $x=b^p$. Given an integer x, you are to determine the largest p such that x is a perfect pth power.

Input

Each test case is given by a line of input containing x. The value of x will have magnitude of at least 2 and be within the range of a (32-bit) *int* in C, C++, and Java. A line containing 0 follows the last test case.

Output

For each test case, output a line giving the largest integer p such that x is a perfect pth power.

Sample Input	Sample Output
17	1
1073741824	30
25	2
0	

Source: Waterloo local 2004.01.31

IDs for Online Judges: POJ 1730, ZOJ 2124

Hint

The positive integer x is represented as the product of powers of prime factors $x=p_1^{e1} p_2^{e2} * \cdots p_k^{ek}$. The largest integer p such that x is a perfect pth power is $p=GCD(e_1, e_2, \ldots, e_k)$.

3.4.12 *Factovisors*

The factorial function $n!$ is defined thus for n a non-negative integer:

```
0!=1
n!=n×(n-1)!  (n>0)
```

We say that a divides b if there exists an integer k such that $k×a=b$.

Input

The input to your program consists of several lines, each containing two non-negative integers, n and m, both less than 2^{31}.

Output

For each input line, output a line stating whether or not m divides $n!$, in the format shown below.

Sample Input	Sample Output
6 9	9 divides 6!
6 27	27 does not divide 6!
20 10000	10000 divides 20!
20 100000	100000 does not divide 20!
1000 1009	1009 does not divide 1000!

Source: 2001 Summer keep-fit 1

ID for Online Judge: UVA 10139

Hint

The non-negative integer m is represented as the product of powers of prime factors

$$m = \prod_{i=1}^{k} p_i^{e_i}.$$ m divides $n!$ if and only if $n!$ can be represented as the product of pow-

ers of prime factors $n! = \prod_{i=1}^{t} p_i^{'e_i'}$, where $\{p_1, p_2, \ldots, p_k\}$ is a subset for $\{p_1', p_2', \ldots, p_t'\}$

and the power for p_i in $\{p_1, p_2, \ldots, p_k\}$ is less than or equal to the power for p_i in $\{p_1', p_2', \ldots, p_t'\}$.

In order to avoid "Out Of Memory Error (OOME)", the power for p_i for $n!$ is

calculated directly from n: $e_i' = \sum_{j=1}^{k} \left\lfloor \frac{n}{p_i^j} \right\rfloor (p^{k+1} > n)$.

We should note: 0 can't divide $n!$; and m divides $n!$ is true if $m \leq n$.

3.4.13 Farey Sequence

The Farey Sequence *Fn* for any integer *n* with *n*≥2 is the set of irreducible rational numbers *a*/*b* with 0 <*a*<*b*≤*n* and *GCD*(*a*, *b*)=1 arranged in increasing order. The first few are as follows:

    F2 = {1/2}
    F3 = {1/3, 1/2, 2/3}
    F4 = {1/4, 1/3, 1/2, 2/3, 3/4}
    F5 = {1/5, 1/4, 1/3, 2/5, 1/2, 3/5, 2/3, 3/4, 4/5}

Your task is to calculate the number of terms in the Farey sequence *Fn*.

Input

There are several test cases. Each test case has only one line, which contains a positive integer *n* (2≤*n*≤10^6). There are no blank lines between cases. A line with a single 0 terminates the input.

Output

For each test case, you should output one line, which contains N(*n*)—the number of terms in the Farey sequence *Fn*.

Sample Input	Sample Output
2	1
3	3
4	5
5	9
0	

Source: POJ Contest, Author: Mathematica@ZSU

ID for Online Judge: POJ 2478

Hint

Based on the problem description, the Farey Sequence F_n for any integer *n* with *n*≥2 is the set of irreducible rational numbers *a*/*b* with 0<*a*<*b*≤*n* and *GCD*(*a*, *b*)=1 arranged in increasing order. Suppose F[*i*] is the number of terms in the Farey sequence F_i, and f_i' is the number of terms whose denominators are *i* in the Farey sequence F_i. Therefore,

$$F[i]=\begin{cases} f_i' & i=2 \\ F[i-1]+f_i' & 3\le i\le n \end{cases}.$$

For each term in the Farey sequence, F_i, its denominator i and numerator are relatively prime. Therefore, f_i' is Euler phi-function $\varphi(i)$. The offline method is used to calculate $F[]$. Then for each test case (a positive integer k), $F[k]$ is output.

3.4.14 Irreducible Basic Fractions

A *fraction m/n* is basic if $0<=m<n$ and it is *irreducible* if gcd(m, n)=1. Given a positive integer n, in this problem you are required to find out the number of *irreducible basic fractions* with denominator n.

For example, the set of all *basic fractions* with denominator 12, before reduction to lowest terms, is

$$\frac{0}{12}, \frac{1}{12}, \frac{2}{12}, \frac{3}{12}, \frac{4}{12}, \frac{5}{12}, \frac{6}{12}, \frac{7}{12}, \frac{8}{12}, \frac{9}{12}, \frac{10}{12}, \frac{11}{12}$$

Reduction yields

$$\frac{0}{12}, \frac{1}{12}, \frac{1}{6}, \frac{1}{4}, \frac{1}{3}, \frac{5}{12}, \frac{1}{2}, \frac{7}{12}, \frac{2}{3}, \frac{3}{4}, \frac{5}{6}, \frac{11}{12}$$

Hence there are only the following four *irreducible basic fractions* with denominator 12:

$$\frac{1}{12}, \frac{5}{12}, \frac{7}{12}, \frac{11}{12}$$

Input

Each line of the input contains a positive integer $n(<1000000000)$ and the input terminates with a value 0 for n (do not process this terminating value).

Output

For each n in the input, print a line containing the number of *irreducible basic fractions* with denominator n.

Sample Input	Sample Output
12	4
123456	41088
7654321	7251444
0	

Source: 2001 Regionals Warmup Contest

ID for Online Judge: UVA 10179

Hint

$\frac{m}{n}$ is *irreducible* if and only if gcd(m, n)=1. The number of m satisfying $n \leq m$ and $GCD(m, n)$=1 is $\varphi(n)$. Therefore, the number of *irreducible basic fractions* with denominator n is $\varphi(n)$.

3.4.15 LCM Cardinality

A pair of numbers has a unique **LCM** but a single number can be the **LCM** of more than one possible pairs. For example, **12** is the **LCM** of **(1,12)**, **(2,12)**, **(3,4)**, etc. For a given positive integer N, the number of different integer pairs with **LCM** that is equal to N can be called the **LCM** cardinality of that number N. In this problem, your job is to find out the **LCM** cardinality of a number.

Input

The input file contains at most **101** lines of inputs. Each line contains an integer N ($0 < N \leq 2 \times 10^9$). Input is terminated by a line containing a single zero. This line should not be processed.

Output

For each line of input except the last one, produce one line of output. This line contains two integers N and C. Here N is the input number and C is its cardinality. These two numbers are separated by a single space.

Sample Input	Sample Output
2	2 2
12	12 8
24	24 11
101101291	101101291 5
0	

Source: UVa Monthly Contest August 2005

ID for Online Judge: UVA 10892

Hint

For a given positive integer N, the number of different integer pairs with **LCM** is equal to N and can be called the **LCM** cardinality of that number N. Suppose A

and B are a pair of integers. A and B can be represented as the product of powers of prime factors, $A=\prod_i p_i^{a_i}$, and $B=\prod_i p_i^{b_i}$. The **LCM** for A and B is N,

$$N=LCM(A,B)=\prod_i p_i^{C_i},$$ where $\forall i$, $c_i=max\{a_i, b_i\}$. This is the insight that lets us solve the problem.

Suppose $f[i]$ is the **LCM** cardinality for the first i prime factors for N. For the first $i-1$ prime factors for N, there are two cases:

1. If $\forall j < i$, $c_j=a_j=b_j$. If $a_i=c_i$, then $b_i=0...c_i$, there are c_i+1 pairs of integers $(c_i,0)$, $(c_i,1)$, ..., (c_i,c_i);
2. Otherwise, there are $2\times c_i+1$ pairs of integers $(0,c_i)$, $(1,c_i)$, ..., (c_i-1,c_i), (c_i,c_i-1), ..., $(c_i,0)$, (c_i,c_i).

Therefore, $f[i]=(f[i-1]-1\times(2\times c_i+1)+c_i+1$.

3.4.16 GCD Determinant

We say that a set $S=\{x_1, x_2, ..., x_n\}$ is factor closed if, for any $x_i \in S$ and any divisor d of x_i. we have $d \in S$. Let's build a GCD matrix $(S)=(s_{ij})$, where $s_{ij}=GCD(x_i, x_j)$—the GCD of x_i and x_j. Given the factor closed set S, find the value of the determinant:

$$D_n=\begin{vmatrix} gcd(x_1,x_1) & gcd(x_1,x_2) & gcd(x_1,x_3) & \cdots & gcd(x_1,x_n) \\ gcd(x_2,x_1) & gcd(x_2,x_2) & gcd(x_2,x_3) & \cdots & gcd(x_2,x_n) \\ gcd(x_3,x_1) & gcd(x_3,x_2) & gcd(x_3,x_3) & \cdots & gcd(x_3,x_n) \\ \cdots & \cdots & \cdots & \cdots & \cdots \\ gcd(x_n,x_1) & gcd(x_n,x_2) & gcd(x_n,x_3) & \cdots & gcd(x_n,x_n) \end{vmatrix}$$

Input

The input file contains several test cases. Each test case starts with an integer n ($0<n<1000$), that stands for the cardinality of S. The next line contains the numbers of S: x_1, x_2, ..., x_n. It is known that each x_i is an integer, $0<x_i<2\times10^9$. The input data set is correct and ends with an end of file.

Output

For each test case, find and print the value Dn mod 1000000007.

Sample Input	Sample Output
2	1
1 2	12
3	4
1 3 9	
4	
1 2 3 6	

Source: ACM Southeastern European Regional Programming
  Contest 2008

IDs for Online Judges: POJ 3910, UVA 4190

Hint

Suppose a_i is the row (gcd(x_i, x_1) gcd(x_i, x_2) gcd(x_i, x_3) ... gcd(x_i, x_n)) for the matrix D_n, and a_{ij} represents gcd(x_i, x_j).

There is a linear transformation for the matrix D_n $a_b - \displaystyle\sum_{(d|b)\&\&(d\neq b)} a_d$. Each a_d satisfying $(d \mid b)\&\&(d\neq b)$ has been transformed before a_b is transformed.

After a_b has been transformed, $a_{ij} = \begin{cases} 0 & \gcd(x_i, x_j) < x_i \\ \varphi(x_i) & \gcd(x_i, x_j) = x_i \end{cases}$.

First, all x_i are sorted in ascending order. Second, the gcd matrix M is constructed as the problem description. Third, a linear transformation for the matrix is done as above. The matrix must be an upper triangular matrix, and each element for the diagonal line of the matrix is the Euler phi-function $\varphi(x_i)$ for the row's corresponding number x_i. Therefore, $\det(M) = \displaystyle\prod_{i=1}^{n} \varphi(x_i)$.

3.4.17 GCD and LCM Inverse

Given two positive integers a and b, we can easily calculate the GCD and the least common multiple (LCM) of a and b. But what about the inverse? That is: given GCD and LCM, finding a and b.

Input

The input contains multiple test cases, each of which contains two positive integers, the GCD and the LCM. You can assume that these two numbers are both less than 2^{63}.

Output

For each test case, output *a* and *b* in ascending order. If there are multiple solutions, output the pair with smallest *a+b*.

Sample Input	Sample Output
3 60	12 15

Source: POJ Achilles

ID for Online Judge: POJ 2429

Hint

For this problem, *LCM=LCM(a, b)*, *GCD=GCD(a, b)*, and *a×b=LCM×GCD* with smallest *a+b*.

First $N=\dfrac{LCM}{GCD}$ is calculated. If *N*==1, then the pair with smallest *a+b* is (*GCD*, *LCM*); else (*a*, *b*) is calculated.

Suppose $a=t\times GCD$, $b=\dfrac{LCM}{t}=\dfrac{N\times GCD}{t}$. Therefore, $a:b=t:\dfrac{N}{t}$. Obviously, *a+b* being the smallest is equivalent to $t+\dfrac{N}{t}$ being the smallest. The method for calculating *t* is as follows:

The positive integer *N* is represented as the product of powers of prime factors $N=\prod\limits_{i=1}^{k} p_i^{e_i}$. Array *a*[] is used to represent the product of powers of prime factors for *N*, where $a[i]=p_i^{e_i}(1\le i\le k)$.

The recursive function *dfs*(0,1,*N*) is used to calculate *t*.

```
void dfs(i, t', n){ //i is the pointer for a[], a:b=t', n is LCM/GCD

    if (i==m+1){    // a[] has been analyzed
        if ((minx==-1) || (t'+n/t' <minx)){
            minx= t'+n/t';
            t= t' ;
        }
        return;    //backtracking
    }
    dfs(i+1, t'*a[i], n);    //a:b=t'*a[i], the (i+1)-th prime
factor for N is analyzed
    dfs(i+1, t', n);    // a:b=t', the (i+1)-th prime factor
for N is analyzed
}
```

If $t^2>N$, then *t*=*N*/*t*. The pair with smallest *a+b* is (*t×GCD*, *LCM*/*t*).

Chapter 4

Practice for Combinatorics

Combinatorics is the branch of mathematics studying the enumeration, combination, and permutation of sets of elements and the mathematical relations that characterize their properties. This chapter focuses on the following topics:

- Generating Permutations;
- Enumeration of Permutations and Combinations;
- The Pigeonhole Principle and the Inclusion–Exclusion Principle;
- The Pólya Counting Formula.

4.1 Generating Permutations

In this section, experiments for generating the next permutation and all permutations based on lexicographic order are shown.

4.1.1 Generating the Next Permutation Based on Lexicographic Order

Lexicographic order refers to generating the next permutation based on the alphabetical order of their component elements. Suppose the current permutation is $(p)=p_1 \ldots p_{i-1} p_i \ldots p_n$. The method for generating the next permutation (q) based on lexicographic order is as follows:

Step 1: Find the longest suffix that is non-increasing by scanning the sequence from right to left. The element immediately to the left of the suffix is called "the first element." If there is no such element, the sequence is non-increasing and is the last permutation. That is, find such an index i that $i=\max\{j|p_{j-1}<p_j, p_j\geq p_{j+1}\}$, and p_{i-1} is the first element.

Step 2: Find the rightmost successor to the first element in the suffix. Because the first element is less than the head of the suffix, some elements in the suffix are greater than the first element. In the suffix, the rightmost successor to the first element is the smallest element greater than the first element. We call the element "the second element." That is, find such an index j that $j=\max\{k|k\geq i, p_{i-1}<p_k\}$.

Step 3: Swap p_{i-1} and p_j, and get a new sequence $p_1\ldots p_{i-2}\boxed{p_j}\,p_i\,p_{i+1}\ldots p_{j-1}\boxed{p_{i-1}}$ $p_{j+1}\ldots p_n$.

Step 4: Reverse the subsequence after the original index of the first element. The next permutation is $(q)=p_1\cdots p_{i-2}\,p_j\boxed{p_n\cdots p_{j+1}p_{i-1}p_{j-1}\cdots\cdot p_{i+1}\,p_i}\cdot$

Suppose the current permutation is $(p)=2763541$. Based on lexicographic order, the next permutation is $(q)=2764135$.

1. 2763541 : Find the first element, and $p_{i-1}p_i$ is 35.
2. 2763541 : Find the second element: 4.
3. 2764531 : Swap the first element and the second element.
4. 2764135 : Reverse the subsequence after the original index of the first element. And get the next permutation (q).

4.1.1.1 ID Codes

It is 2084 and the year of Big Brother has finally arrived, albeit a century late. In order to exercise greater control over its citizens and thereby counter a chronic breakdown in law and order, the government decides on a radical measure—all citizens are to have a tiny microcomputer surgically implanted in their left wrists. This computer will contain all sorts of personal information, as well as a transmitter which will allow people's movements to be logged and monitored by a central computer.

An essential component of each computer will be a unique identification code, consisting of up to 50 characters drawn from the 26 lowercase letters. The set of characters for any given code is chosen somewhat haphazardly. The complicated way in which the code is imprinted into the chip makes it much easier for the manufacturer to produce codes that are rearrangements of other codes, rather than to produce new codes with a different selection of letters. Thus, once a set of letters has been chosen, all possible codes derivable from it are used before changing the set.

For example, suppose it is decided that a code will contain exactly three occurrences of "a", two of "b", and one of "c"; then three of the allowable 60 codes under these conditions are:

abaabc
abaacb
ababac

These three codes are listed from top to bottom in alphabetic order. Among all codes generated with this set of characters, these codes appear consecutively in this order.

Write a program to assist in the issuing of these identification codes. Your program will accept a sequence of no more than 50 lowercase letters (which may contain repeated characters) and print the successor code if one exists, or print the message "No Successor" if the given code is the last in the sequence for that set of characters.

Input

Input will consist of a series of lines, each containing a string representing a code. The entire file will be terminated by a line consisting of a single #.

Output

Output will consist of one line for each code read, containing the successor code or the words "No Successor".

Sample Input	Sample Output
abaacb	ababac
cbbaa	No Successor
#	

Source: New Zealand Contest 1991

IDs for Online Judges: POJ 1146, UVA 146

 Analysis

1. The successor code is the next permutation based on lexicographic order. Therefore, the algorithm is as follows. Suppose the given code is $s_0 s_1 s_2 \ldots \ldots s_{l-1}$.
2. Find the index i that $i = \max\{j | s_{j-1} \geq s_j\}$.
3. If $i == 0$, then the given code is the last in the sequence for that set of characters, output "No Successor", and exit; else
4. On the right of "the first element", find the smallest character greater than it. That is, find such an index j that $j = \max\{k | s_{i-1} < s_k\}$;
5. Swap s_{i-1} and s_j, and get $s_0 \ldots s_{i-2} s_j s_i s_{i+1} \ldots s_{j-1} s_{i-1} s_{j+1} \ldots s_{l-1}$;
6. Reverse the substring after s_j, and get the successor code $(q) = s_0 \ldots s_{i-2} s_j s_{l-1} \ldots s_{j+1} s_{i-1} s_{j-1} \ldots s_{i+1} s_i$.

Program

```
# include <cstdio>
# include <cstring>
# include <cstdlib>
# include <iostream>
# include <string>
# include <cmath>
# include <algorithm>
using namespace std;
typedef long long int64;
char s[60];int l;    // identification codes whose length is l
int get() {    //If there is a successor code for s, output
the successor and return 1; else return 0
    int i=l-1;
    while (i>0&&s[i-1]>=s[i]) i--;    //find the first element
     if (!i) return 0;    //no successor
    int mp=i;    // find the second element
    for (int j=i+1;j<l;j++) {
       if(s[j]<=s[i-1])continue;
       if(s[j]<s[mp])mp=j;
    }
    swap(s[mp],s[i-1]);    // Swap s_{i-1} and s_{mp}
    sort(s+i,s+l);    // Reverse the suffix after the i-th
character
    return 1;
}
int main(){
    while (~scanf("%s",s)&&s[0]!='#'){    //Input
identification codes until '#'
        l=strlen(s);    //the length of the identification
code
        if(get())  printf("%s\n",s);  // output the successor
        else  printf("No Successor\n");
    }
    return 0;
}
```

Not only can lexicographic order generate the next permutation for $p_1 \ldots p_{i-1}$ $p_i \ldots p_n$, but it can also generate an r-combination of a set S of n elements $\{a_1, a_2, \ldots, a_n\}$, where $a_1 < a_2 < \ldots < a_n$. Suppose the current r-combination of a set S is $\{a_{k1}, a_{k2}, \ldots, a_{kr}\}$, where $1 \leq k_1 < k_2 < \ldots < k_r \leq n$. Obviously, the first r-combination of a set S

of n elements is $\{a_1, a_2,..., a_r\}$, and the last r-combination of a set S of n elements is $\{a_{n-r+1}, a_{n-r+2},..., a_n\}$.

If the current r-combination of a set $S\{a_{k1}, a_{k2},..., a_{kr}\}$ isn't $\{a_{n-r+1}, a_{n-r+2},..., a_n\}$, then the next r-combination is calculated as follows:

Suppose i is the maximal index kj that $a_{kj} < a_{n-kr+kj}$. Based on lexicographic order, the next r-combination is $\{a_{k1},..., a_{kj-1}, a_{kj+1},..., a_{kr}, a_{kr+1}\}$. Therefore, for an r-combination $\{a_{k1}, a_{k2},..., a_{kr}\}$, the algorithm for calculating the next r-combination is as follows:

1. $i=\max\{kj | a_{kj} < a_{n-kr+kj}\}$;
2. $a_i \leftarrow a_{i+1}$, where $kj \leq i \leq kr$.

4.1.2 Generating All Permutations Based on Lexicographic Order

Based on Section 4.1.1, the method for generating all permutations for a finite set with n elements is as follows:

First, sort the n elements in ascending order; the permutation is the first permutation. Then the method generating the next permutation based on lexicographic order is used repeatedly until the last permutation is generated.

4.1.2.1 Generating Fast, Sorted Permutation

Generating permutation has always been an important problem in computer science. In this problem, you will have to generate the permutation of a given string in ascending order. Remember that your algorithm must be efficient.

Input

The first line of the input contains an integer n, which indicates how many strings to follow. The next n lines contain n strings. Strings will only contain alpha-numerals and never contain any space. The maximum length of the string is 10.

Output

For each input string, print all the permutations possible in ascending order. Note that the strings should be treated as case-sensitive strings and no permutation should be repeated. A blank line should follow each output set.

Sample Input	Sample Output
3	ab
ab	ba
abc	
bca	abc
	acb
	bac
	bca
	cab
	cba
	abc
	acb
	bac
	bca
	cab
	cba

Source: TCL Programming Contest, 2001

ID for Online Judge: UVA 10098

 Analysis

Suppose the length of string *s* is *l*. Therefore, *l*! permutations in ascending order are required to output. The algorithm is as follows:

The first permutation is achieved by sorting string *s* in ascending order. For the current permutation *s*,

1. $i=\max\{j|s_{j-1}\geq s_j\}$;
2. $j=\max\{k|s_{i-1}<s_k\}$;
3. Swap s_{i-1} and s_j, and get $s_0\ldots s_{i-2}\ \underline{s_j}\ s_i\ s_{i+1}\ldots s_{j-1}\ \underline{s_{i-1}}\ s_{j+1}\ldots s_{l-1}$;
4. Reverse the substring after s_j, and get the next permutation $(q)=s_0\ldots s_{i-2}$ $s_j\ \underline{s_{l-1}\ldots s_{j+1}}\underline{s_{i-1}}\underline{s_{j-1}\ \ldots\ldots s_{i+1}}\ \underline{s_i}$;

The above process is repeated until *i*==0. All of the permutations possible in ascending order are generated.

 Program

```
# include <cstdio>
# include <cstring>
```

```
# include <cstdlib>
# include <iostream>
# include <string>
# include <cmath>
# include <algorithm>
using namespace std;
typedef long long int64;
char s[60];int l;      // the length of string s is l
int get(){     // If there is a successor for s, output the
successor and return 1; else return 0
      int i=l-1;      // find the first character
      while(i>0&&s[i-1]>=s[i])i--;
      if(!i) return 0;      // no successor
      int mp=i;      // find the second character
      for(int j=i+1;j<l;j++){
         if(s[j]<=s[i-1])continue;
         if(s[j]<s[mp])mp=j;
      }
      swap(s[mp],s[i-1]);      // Swap s_{i-1} and s_{mp}
      sort(s+i,s+l);      // Reverse the substring after s_i
      return 1;
}
int main(){
      int casen;scanf("%d",&casen);      //number of strings
      while(casen--){
         scanf("%s",s);      //current string
         l=strlen(s);      //the length of the current string
         sort(s,s+l);
         printf("%s\n",s);      //the first permutation
         while(get()) printf("%s\n",s);      //all permutations
         printf("\n");
         }
      return 0;
}
```

4.2 Enumeration of Permutations and Combinations

In this section, first, experiments for calculating numbers of permutations and combinations are shown; and then, experiments for Catalan numbers, Bell numbers, and Stirling numbers are shown.

4.2.1 Calculating Numbers of Permutations and Combinations

$P(n, r)$ is denoted as the number of r-permutations of an n-element set.
$$P(n, r) = \frac{n!}{(n-r)!}.$$

$C(n, r)$ is denoted as the number of r-combination of an n-element set.
$C(n, r) = \dfrac{n!}{r!(n-r)!}$ (or denote by $\begin{pmatrix} n \\ r \end{pmatrix}$).

In programs, optimization methods can be used to calculate $C(n, r)$.

Method 1:

$$C(n,r) = \frac{n!}{r!(n-r)!} = \frac{(n-r+1)\times(n-r+2)\times...\times n}{1\times 2\times...\times r} = \boxed{\frac{n-r+1}{r}}\times\boxed{\frac{n-r+2}{r-1}}\times...\boxed{\frac{n}{1}}.$$

Method 2: The formula for calculating binomial coefficients is used:

$$C(i, j) = C(i-1, j) + C(i-1, j-1), \text{ that is, } c[i][0] = 1, \text{ and}$$
$$c[i][j] = c[i-1][j] + c[i-1][j-1].$$

4.2.1.1 Binomial Showdown

In how many ways can you choose k elements out of n elements, not taking order into account? Write a program to compute this number.

Input

The input file will contain one or more test cases. Each test case consists of one line containing two integers n ($n>=1$) and k ($0<=k<=n$). Input is terminated by two zeros for n and k.

Output

For each test case, print one line containing the required number. This number will always fit into an integer; i.e., it will be less than 2^{31}.

Warning: Don't underestimate the problem. The result will fit into an integer— but whether all intermediate results arising during the computation will also fit into an integer depends on your algorithm. The test cases will go to the limit.

Sample Input	Sample Output
4 2	6
10 5	252
49 6	13983816
0 0	

Source: Ulm Local 1997

IDs for Online Judges: POJ 2249, ZOJ 1938

Analysis

Method 1 is used to solve the problem directly:

$$C(n,k) = \frac{n!}{k!(n-k)!} = \frac{(n-k+1)\times(n-k+2)\times...\times n}{1\times 2\times...\times k}$$

$$= \boxed{\frac{n-k+1}{k}}\times\boxed{\frac{n-k+2}{k-1}}\times...\boxed{\frac{n}{1}}.$$

Program

```cpp
# include <cstdio>
# include <cstring>
# include <cstdlib>
# include <iostream>
# include <string>
# include <cmath>
# include <algorithm>
using namespace std;
typedef long long int64;
int64 work(int64 n,int64 k){      //Calculate C(n, k)
        if(k>n/2) k=n-k;      // To reduce the amount of
calculation
        int64 a=1,b=1;
        for(int i=1;i<=k;i++){
                a*=n+1-i;
                b*=i;
                if(a%b==0)a/=b,b=1;
        }
        return a/b;      //return C(n, k)
}
int main()  {
        int n,k;
        while(~scanf("%d %d",&n,&k)&&n){      // Input test cases
                printf("%lld\n",work(n,k));      // Calculate and
output C(n, k)
        }
        return 0;
}
```

4.2.1.2 Combinations

Computing the exact number of ways that N things can be taken M at a time can be a great challenge when N and/or M become very large. Challenges are the stuff of contests. Therefore, you are to make just such a computation, given the following:

Given: $5 \leq N \leq 100$; $5 \leq M \leq 100$; $M \leq N$

Compute the **EXACT** value of: $C=N!/(N-M)!M!$

You may assume that the final value of C will fit in a 32-bit Pascal LongInt or a C long. For the record, the exact value of 100! is:

93,326,215,443,944,152,681,699,238,856,266,700,490,715,968,264,381,621,
468,592,963,895,217,599,993,229,915,608,941,463,976,156,518,286,253,697,920,
827,223,758,251,185,210,916,864,000,000,000,000,000,000,000,000

Input

The input to this program will be one or more lines each containing zero or more leading spaces, a value for N, one or more spaces, and a value for M. The last line of the input file will contain a dummy N, M pair with both values equal to zero. Your program should terminate when this line is read.

Output

The output from this program should be in the form:

N things taken M at a time is C exactly.

Sample Input	Sample Output
100 6	100 things taken 6 at a time is 1192052400 exactly.
20 5	20 things taken 5 at a time is 15504 exactly.
18 6	18 things taken 6 at a time is 18564 exactly.
0 0	

IDs for Online Judges: POJ 1306, UVA 369

 Analysis

Suppose $c[i][j]$ is $C(i, j)$.

Based on the formula calculating binomial coefficients, $c[i][j]=c[i-1][j]+c[i-1][j-1]$.

Initially, $c[i][0]=1$, $0 \leq i \leq 101$. Then, based on the above formula, $c[i][j]$ can be calculated, $1 \leq i \leq 100$, $1 \leq j \leq 100$. Finally, for each test case N, M, output $c[N][M]$.

Program

```
# include <cstdio>
# include <cstring>
# include <cstdlib>
# include <iostream>
# include <string>
# include <cmath>
# include <algorithm>
using namespace std;
typedef unsigned long long int64;
unsigned int c[110][110];    //c[i][j] is C(i, j)
void pp(){    //calculate c[][] using the formula
      for (int i=0;i<102;i++)   c[i][0]=1;
      for (int i=1;i<101;i++)
         for(int j=1;j<101;j++)   c[i][j]=c[i-1][j-1]+c[i-1]
[j];
}
int main(){
      pp();    //offline method is used to calculate c[][]
      int n,m;
      while (~scanf("%d %d",&n,&m) && (n||m))    //Input test
cases
         printf("%d things taken %d at a time is %u
exactly.\n",n,m,c[n][m]);    //output C(n, m)
      return 0;
}
```

4.2.1.3 Packing Rectangles

Four rectangles are given. Find the smallest enclosing (new) rectangle into which these four may be fitted without overlapping. By "smallest rectangle", we mean the one with the smallest area.

All four rectangles should have their sides parallel to the corresponding sides of the enclosing rectangle. Figure 4.1 shows six ways to fit four rectangles together. These six are the only possible basic layouts, since any other layout can be obtained

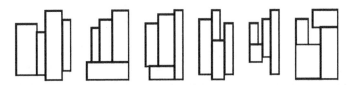

Figure 4.1

from a basic layout by rotation or reflection. Several different enclosing rectangles fulfilling the requirements may exist, all with the same area. You have to produce all such enclosing rectangles.

Input

Your program is to read from standard input. The input consists of four lines. Each line describes one given rectangle by two positive integers: the lengths of the sides of the rectangle. Each side of a rectangle is at least 1 and at most 50.

Output

Your program is to write to standard output. The output should contain one line more than the number of solutions. The first line contains a single integer: the minimum area of the enclosing rectangles. Each of the following lines contains one solution described by two numbers p and q with $p \leq q$. These lines must be sorted in ascending order of p, and the lines must all be different.

Sample Input	Sample Output
1 2	40
2 3	4 10
3 4	5 8
4 5	

Source: IOI 1995

ID for Online Judge: POJ 1169

 Analysis

1. **Calculating widths and heights for the enclosing rectangles.**
   The problem description shows six ways to fit four rectangles together. The key to the problem is to calculate the area of the enclosing rectangles for six ways. Suppose four rectangles which will be placed into the enclosing rectangle are represented as an array $t[0...3]$, and for rectangle $t[i]$, its height and width are $t[i].x$ and $t[i].y$ respectively, $0 \leq i \leq 3$.

   For each rectangle, there are two ways to place it into the enclosing rectangle: place it horizontally, or place it vertically. Obviously, if a rectangle's placement method is changed, we only need to exchange its height and width.

Based on the problem description, six ways to fit four rectangles together are analyzed as follows:

Case 1:

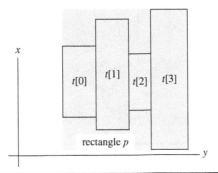

Figure 4.2

Four rectangles ($t[0]$, $t[1]$, $t[2]$, and $t[3]$) are placed in order as shown in Figure 4.2. For Case 1, the height and width for the enclosing rectangle p are as follows: $p.x=max\{t[0].x, t[1].x, t[2].x, t[3].x\}; p.y=t[0].y+t[1].y+t[2].y+t[3].y$.

Case 2:

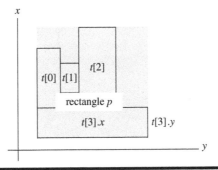

Figure 4.3

In the enclosing rectangle p, there are two parts: the upper part and the lower part, as shown in Figure 4.3. In the upper part, $t[0]$, $t[1]$, and $t[2]$ are placed; and in the lower part, $t[3]$ is placed. For case 2, the height and width for the enclosing rectangle p are as follows: $p.x=max\{t[0].x, t[1].x, t[2].x\}+t[3].y$, $p.y=max\{t[3].x, t[0].y+t[1].y+t[2].y\}$.

Case 3:

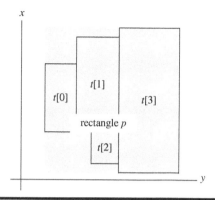

Figure 4.4

In the enclosing rectangle p, there are two parts: the left part and the right part, as shown in Figure 4.4. In the left part, $t[2]$ is placed below, $t[0]$ and $t[1]$ are placed up, and $t[2]$ and $t[1]$ are right-aligned. In the right part, $t[3]$ is placed. For case 3, the height and width for the enclosing rectangle p is as follows: $p.x=max\{max\{t[0].x, t[1].x\}+t[2].x, t[3].x)\}, p.y=max\{t[0].y+t[1].y, t[2].y\}+t[3].y$.

Case 4 and Case 5:

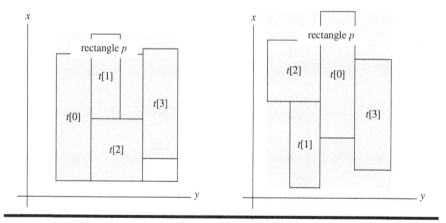

Figure 4.5

In the enclosing rectangle p, two rectangles, $t[1]$ and $t[2]$, are stacked together, and two other rectangles, $t[0]$ and $t[3]$, are placed alone, as shown in Figure 4.5. The height and width for the enclosing rectangle p is as follows: $p.x=max\{t[1].x+t[2].x, t[0].x, t[3].x\}, p.y=t[0].y+t[3].y+max(t[1].y, t[2].y\}$.

Case 6:

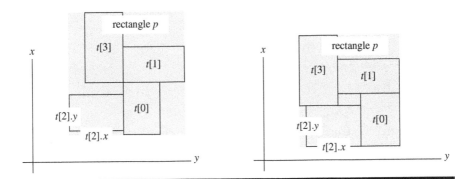

Figure 4.6

In the enclosing rectangle p, four rectangles are placed in two rows, and in each row there are two rectangles, where $t[1].x \leq t[3].x \leq t[0].x + t[1].x$ and $t[0].y \leq t[1].y$. In Figure 4.6, there are two different ways. All placements can be calculated through rotations and reflections for the two ways. The height and width for the enclosing rectangle p are as follows: $p.x = max\{t[0].x + t[1].x, t[2].y + t[3].x\}$, $p.y = max\{t[0].y + t[2].x, t[1].y + t[3].y\}$.

2. **Enumeration is used to calculate the minimal enclosing rectangle.**

   The algorithm is as follows:

   All possible permutations for four rectangles are enumerated (a, b, c, d) $(0 \leq a, b, c, d \leq 3, a \neq b \neq c \neq d)$, and $r[a...d]$ is stored in $t[0...3]$ in order. Suppose heights and widths for four rectangles are $r[i].x$ and $r[i].y$, respectively, $0 \leq i \leq 3$, and the placements for each rectangle are enumerated, where

$$v[i] = \begin{cases} 0 & \text{The rectangle is placed vertically.} \\ 1 & \text{The rectangle is placed horizontally.} \end{cases}, 0 \leq i \leq 3.$$

   If rectangle $t[i]$ is placed horizontally (i.e., $v[i]=1$, $0 \leq i \leq 3$), its height and width are exchanged (i.e., $t[i].x \leftrightarrow t[i].y$).

   There are four rectangles, and there are two placements for each rectangle. Therefore, there are $4! \times 24$ different $t[0...3]$. And for each $t[0...3]$, there are five enclosing rectangles (Case 1 to 5). Therefore, there are $4! \times 24 \times 5 = 1920$ areas of enclosing rectangles. We can calculate the minimum area of the enclosing rectangles as *min_area*.

   Suppose *soln*$[0...ps]$ stores the sequence of enclosing rectangles whose area is *min_area*. In the sequence, each element is described by two numbers *soln*$[i].x$ and *soln*$[i].y$ with *soln*$[i].x \leq$ *soln*$[i].y$. All elements are sorted in ascending order of *soln*$[i].x$, $0 \leq i \leq ps$.

Initially, *min_area*=∞, the rear pointer for *soln*[] *ps*=0. Then, each enclosing rectangle *p* in *r*[0...3] is enumerated:

If $p.x > p.y$, then $(p.x \leftrightarrow p.y)$;

If $p.x \times p.y < min\_area$, then $min\_area = p.x \times p.y$; and *p* is stored $(soln[0] \leftarrow p, ps=1)$;

If $p.x \times p.y = min\_area$, then *p* is added into *soln*[]$(soln[ps++] \leftarrow p)$;

If $p.x \times p.y > min\_area$, then *p* is omitted.

After the enumeration, *min_area* is the minimum area of the enclosing rectangles. All elements in array *soln* are sorted, *x* is the first key, and *y* is the second key. And all elements are different.

Program

```
#include <fstream>
#include <iostream>
#include <vector>
#include <algorithm>
using namespace std;
#define MAX 0x7fffffff
typedef struct      //rectangle
{
  int x;
  int y;
}rec;
int min_area = MAX;     //Initialization: the minimum area of
the enclosing rectangles
rec soln[1000];     //the sequence of enclosing rectangles
whose area is minimal, whose length is ps
int ps = 0;
rec r[4];     // Input 4 rectangles
rec t[4];     // 4 rectangles placed into the enclosing
rectangle
rec zero={0,0};     //Initialize height and width
int v[4];     // placements for each rectangle
inline void make(rec p)     // soln[] is adjusted based on the
current enclosing rectangle p
{
if(p.x>p.y)
{
  p.x = p.x ^ p.y; p.y = p.x ^ p.y; p.x = p.x ^ p.y;
}
```

```
if(min_area > p.x*p.y)
{
  min_area = p.x*p.y;
  ps = 0;
  soln[ps++] = p;
}
else if(min_area==p.x*p.y)
{
  soln[ps++] = p;
}
}
void search()   // Enumeration calculating the area of minimal
enclosing rectangles soln[ ]
{
int i;
for(int a=0;a<4;a++)        //all permutations for 4 rectangles
(a, b, c, d)
for(int b=0;b<4;b++)
for(int c=0;c<4;c++)
for(int d=0;d<4;d++)
{
  if(a != b)
  if(a != c)
  if(a != d)
  if(b != c)
  if(b != d)
  if(c != d)
  {
    for(v[0]=0;v[0]<2;v[0]++)     // Enumerating placements
(vertically or horizontally) for 4 rectangles
      for(v[1]=0;v[1]<2;v[1]++)
        for(v[2]=0;v[2]<2;v[2]++)
          for(v[3]=0;v[3]<2;v[3]++)
          {
            t[0]=r[a]; t[1]=r[b]; t[2]=r[c]; t[3]=r[d];
            for(i=0;i<4;i++)     //exchanging the height and width
for a rectangle
              if(v[i] == 1)
              {
                t[i].x = t[i].x ^ t[i].y; t[i].y = t[i].x ^ t[i].y;
t[i].x = t[i].x ^ t[i].y;
              }
            rec p=zero;     //Case 1
            p.x = max(t[0].x,max(t[1].x,max(t[2].x,t[3].x))); //the
height and width for p
            p.y = t[0].y + t[1].y + t[2].y + t[3].y;
            make(p);     // soln[] is adjusted based on the current
enclosing rectangle p
            if(p.x == 10 && p.y == 8) p=p;
            p = zero;     //Case 2
```

```
        p.x=max(t[0].x,max(t[1].x,t[2].x))+t[3].y;      // the
height and width for p
        p.y = max(t[0].y+t[1].y+t[2].y,t[3].x);
        make(p);      // soln[] is adjusted based on the current
enclosing rectangle p
        if(p.x == 10 && p.y == 8) p=p;
        p=zero;      //Case 3
        p.x=max(max(t[0].x,t[1].x)+t[2].x,t[3].x);      // the
height and width for p
        p.y = max(t[0].y+t[1].y,t[2].y)+t[3].y;
        make(p);      // soln[] is adjusted based on the current
enclosing rectangle p
        if(p.x == 10 && p.y == 8) p=p;
        p=zero;      //Case 4 and 5
        p.x=max(t[0].x,max(t[1].x+t[2].x,t[3].x));      // the
height and width for p
        p.y = t[0].y + max(t[1].y,t[2].y) + t[3].y ;
        make(p);      // soln[] is adjusted based on the current
enclosing rectangle p
        if(p.x == 10 && p.y == 8) p=p;
        if(t[0].y>t[1].y) continue;      //Case 6: If 4
rectangles can't satisfy t[1].x≤t[3].x≤t[0].x+t[1].x and t[0].
y≤t[1].y, continue to enumerate
        if(t[3].x > t[0].x+t[1].x) continue;
        if(t[3].x<t[1].x) continue;
        p = zero;      //Initialization
        p.x = max(t[0].x+t[1].x,t[2].y+t[3].x);      // the
height and width for p
        p.y = max(t[1].y+t[3].y,t[0].y+t[2].x);
        make(p);      // soln[] is adjusted based on the current
enclosing rectangle p
        if(p.x == 6 && p.y == 6) p=p;
        }
    }
  }
}
bool comp(rec a,rec b) //comparing enclosing rectangles a and
b (x is the 1st key, y is the 2nd key)
{
if(a.x<b.x) return 1;
     else if(a.x == b.x && a.y<b.y) return 1;
            else return 0;
}
bool comp2(rec a,rec b)      //determine whether enclosing
rectangles a and b are same
{
     return a.x==b.x && a.y==b.y;
}
int main()
{
```

```
for(int i=0;i<4;i++)      //Input heights and widths for
4 rectangles
{
  cin>>r[i].x>>r[i].y;
}
search();      //Calculating  soln[ ]
sort(&soln[0],&soln[ps],comp);
rec *t = unique(&soln[0],&soln[ps],comp2);
cout<<min_area<<endl;      //Output the minimum area of the
enclosing rectangles
for(rec *i=&soln[0];i!=t;i++)
  cout<<(*i).x<<" "<<(*i).y<<endl;
return 0;
}
```

4.2.2 Catalan Numbers, Bell Numbers and Stirling Numbers

4.2.2.1 Catalan Numbers

The Catalan sequence is the sequence $C_0, C_1, ..., C_n, ...$; where $C_0=1$, $C_1=1$, and $C_n = C_0 C_{n-1} + C_1 C_{n-2} + ... + C_{n-1} C_0$, $n \geq 2$.

Therefore, $C_n = \dfrac{C(2n, n)}{n+1}, n = 0, 1, 2, ...$; or $C_n = \dfrac{4n-2}{n+1} \times C_{n-1}$, $n > 1$.

The Catalan sequence is a frequent counting sequence. For example,

1. C_n is the number of stack-sortable permutations of $\{1, ..., n\}$.
2. C_n is the number of different ways that a convex polygon with $n+2$ sides can be cut into triangles by connecting vertices with non-crossing line segments.
3. C_n is the number of rooted binary trees with n nodes.

4.2.2.1.1 Game of Connections

This is a small but ancient game. You are supposed to write down the numbers 1, 2, 3, ..., $2n-1$, $2n$ consecutively in clockwise order on the ground to form a circle, and then, to draw some straight line segments to connect them into number pairs. Every number must be connected to exactly one another. And, no two segments are allowed to intersect.

It's still a simple game, isn't it? But after you've written down the $2n$ numbers, can you tell me in how many different ways you can connect the numbers into pairs?

Input

Each line of the input file will be a single positive number n, except the last line, which is a number -1. You may assume that $1 \leq n \leq 100$.

Output

For each n, print in a single line the number of ways to connect the $2n$ numbers into pairs.

Sample Input	Sample Output
2 3 −1	2 5

Source: ACM Shanghai 2004 Preliminary

IDs for Online Judges: POJ 2084, ZOJ 2424

 Analysis

Based on the problem description, there are n lines connecting $2n$ numbers into pairs. For each line, there are i pairs of numbers on the left, and there are $n-i-1$ pairs of numbers on the right. Suppose C_n is the number of ways to connect the $2n$ numbers into pairs. $C_0=1$, $C_1=1$, and $C_n=C_0C_{n-1}+C_1C_{n-2}+...+C_{n-1}C_0$, $n\geq2$. Therefore, C_n is a Catalan number.

The offline method is used to calculate the Catalan sequence C_0, C_1, ..., C_{120}. Because the range of Catalan numbers is out of the range of integers in programming languages, Catalan numbers are stored as high-precision numbers.

 Program

```
# include <cstdio>
# include <cstring>
# include <algorithm>
# include <iostream>
using namespace std;
struct BIGNUM{     //High-precision number
short s[200],l;    //the length of integer array s[] is l
}c[120];    // Catalan sequence, where c[i]=C_i

BIGNUM operator*(BIGNUM a,int b){     //a←a*b, where a is an
integer array, b is a integer
    for(int i=0;i<a.l;i++)   a.s[i]*=b;
    for(int i=0;i<a.l;i++){         //carry
```

```
                a.s[i+1]+=a.s[i]/10;
                a.s[i]%=10;
        }
while(a.s[a.l]!=0){
        a.s[a.l+1]+=a.s[a.l]/10;
        a.s[a.l]%=10;
        a.l++;
        }
return a;
}
BIGNUM operator/(BIGNUM a,int b){    //a←a/b, where a is an
integer array, b is an integer
        for(int i=a.l-1;i>0;i--){
        a.s[i-1]+=(a.s[i]%b)*10;
        a.s[i]/=b;
}
a.s[0]/=b;
while(a.s[a.l-1]==0) a.l--;       //number of digits
return a;
}
void print(BIGNUM a){     //output array a
        for(int i=a.l-1;i>=0;i--){
                printf("%d",a.s[i]);
        }
printf("\n");
}
int n;

int main(){
        c[0].l=1;c[0].s[0]=1;     //The first Catalan number C₀=1
        for(int i=0;i<=101;i++)     //Calculate Catalan sequence
```
$C_n = C_{n-1} * \dfrac{4*n-2}{n+1}$, offline method
```
                c[i+1]=(c[i]*(4*i+2))/(i+2);
        while(~scanf("%d",&n)){     //Input test cases
                if(n<0) break;
                print(c[n]);     //Output Cₙ
        }
return 0;
}
```

4.2.2.2 Bell Numbers and Stirling Numbers

Bell numbers, B_0, B_1, ..., B_n, ..., are numbers of partitions of a set, where B_n is the number of different ways to partition a set that has exactly n elements, or equivalently, the number of equivalence relations on it. Obviously, $B_0=1$,

$$B_1 = 1. \quad B_{n+1} = \sum_{k=0}^{n} C(n,k)B_k.$$

Stirling numbers of the first kind are the number of ways to arrange n objects into k cycles, where $S(n, 0)=0$, $S(1, 1)=1$, $S(n, k)=S(n-1, k-1)+(n-1)\times S(n-1, k)$.

Stirling numbers of the second kind are the number of ways to partition a set of n elements into k non-empty subsets. $S(n, n)=S(n, 1)=1$, $S(n, k)=S(n-1, k-1)+k\times S(n-1, k)$.

Obviously, $B_{n+1} = \sum_{k=1}^{n} S(n,k)$, where $S(n, k)$ is a Stirling number of the second kind.

Bell numbers and Stirling numbers of the second kind can be calculated through constructing Bell triangle a.

1. Put 1 on the first row. That is, $a[0, 0]=1$.
2. For the nth row, the leftmost number is the rightmost number on the $(n-1)$th row. That is, $a[n, 0]=a[n-1, n-1]$, $n\geq1$.
3. For the nth row, numbers not on the left column are sums of the number to the left and the number above the number to the left. That is, $a[n, m]=a[n, m-1]+a[n-1, m-1]$, m, $n\geq1$.

In a Bell triangle, the numbers on the left-hand side are the Bell numbers for that row (see Figure 4.7). That is, $B_i=a[i, 0]$, $i\geq0$. The sums of numbers on each row are Stirling numbers of the second kind.

4.2.2.2.1 Bloques

Little John has N blocks, all of them of different sizes. He is playing to build cities in the beach. A city is just a collection of buildings.

A single block over the sand can be considered as a building. Then John can construct higher buildings by putting a block above any other block. At most one block can be put immediately above any other block. However, he can stack several blocks together to construct a building. However, it's not allowed to put bigger blocks on top of smaller ones, since the stack of blocks may fall. A block can be specified by a natural number that represents its size.

1							
1	2						
2	3	5					
5	7	10	15				
15	20	27	37	52			
52	67	87	114	151	203		
203	255	322	409	523	674	877	
877	1080	1335	1657	2066	2589	3263	4140

Figure 4.7

The order among buildings doesn't matter. That is:

1 3
2 4

is the same configuration as:

3 1
4 2

Your problem is to compute the number of possible different cities using N blocks. We say that $\#(N)$ gives the number of different cities of size N. If $N=2$, for instance, there are only two possible cities:

City #1:

1 2

In this city, both blocks of size 1 and 2 are put over the sand.

City #2:

1
2

In this city a block of size 1 is over a block of size 2, and a block of size 2 is over the sand.
So, $\#(2)=2$.

Input

A sequence of non-negative integer numbers, each one in a different line. All of them but the last one are natural numbers. The last one is 0 and means the end. Each natural number is less than 900.

Output

For each natural number I in the input, you must write a line with the pair of numbers I, $\#(I)$.

Sample Input	Sample Output
2	2, 2
3	3, 5
0	

Source: Contest ACM-BUAP 2005

ID for Online Judge: UVA 10844

Analysis

The problem requires you to compute the number of possible different cities using N blocks, that is, the number of different ways to partition a set. Therefore, $\#(N)$ is a Bell number B_n.

The offline method is used to calculate Bell numbers B_0, B_1, ..., B_n ... in the range by constructing a Bell triangle. Because the range of Bell numbers is out of the range of integers in programming languages, Bell numbers are stored as high-precision numbers.

Program

```
# include <cstdio>
# include <cstring>
# include <cstdlib>
# include <iostream>
# include <string>
# include <cmath>
# include <algorithm>
using namespace std;
typedef unsigned long long int64;
int64 m=1e10;
struct Bigint{    //High-precision number
int l;int64 s[200];    //s[] stores a high-precision number,
each element stores a 10-digit decimal number, the length is l

void read(int64 x){    //integer x is represented by a high-
precision number s[]
    l=-1; memset(s,0,sizeof(s))
    do {
        s[++l]=x%m;
        x/=m;
    } while(x);
}
void print(){    // Output s[]
    printf("%llu",s[l]);    // s[l]: practical number of
digit, s[l-1]…s[0]: 10 digits
    for(int i=l-1;i>=0;i--)  printf("%010llu",s[i]);
}
} dp[2][1000],ans[1000];    //In a Bell triangle, the value
for (i, j) is dp[i&1][j]; the value for (i-1,j) is dp
[(i&1)^1][j], and the Bell number for i is ans[i+1]
```

```
Bigint operator+(Bigint a,Bigint &b){      //Addition for high-
precision numbers a←a+b
      a.l=max(a.l,b.l);int64 d=0;
      for(int i=0;i<=a.l;i++) {
            a.s[i]+=d+b.s[i];
            d=a.s[i]/m;a.s[i]%=m;
      }
      if(d)a.s[++a.l]=d;
      return a;
}
int n;
void getans(int id,int n){
      int i=id^1;
      for(int j=1;j<=n-1;i++)dp[id][j+1]=dp[i][j]+dp[id][j];
      }
void work(){     //Offline method: calculate Bell numbers
      dp[1][1].read(1);ans[2]=dp[0][1]=ans[1]=dp[1][1];      //
initialize Bell triangle: B₁=B₀=1
      for(int i=2;i<=900;i++){
            getans(i&1,i);
            dp[(i&1)^1][1]=ans[i+1]=dp[i&1][i];
      }
}
int main(){
      work();
      while(~scanf("%d",&n)&&n){
      printf("%d, ",n);  ans[n+1].print();      //output n and
its Bell number
      printf("\n");
      }
      return 0;
}
```

4.3 Applications of the Pigeonhole Principle and the Inclusion–Exclusion Principle

This section focuses on the Pigeonhole Principle and the Inclusion–Exclusion Principle.

4.3.1 Applications of the Pigeonhole Principle

Theorem 4.3.1. If $n+1$ objects are put into n containers, then at least one container must contain more than one object.

Theorem 4.3.2. If m objects are put into n containers, then at least one container

$$\text{must contain at least } k \text{ objects, where } k = \begin{cases} \dfrac{m}{n} & m \bmod n = 0 \\[3mm] \left\lfloor \dfrac{m}{n} \right\rfloor + 1 & m \bmod n \neq 0 \end{cases}.$$

The steps for applying the pigeonhole principle to solve problems are as follows:

1. Determine what objects and containers are based on problem descriptions;
2. Construct containers;
3. Apply the pigeonhole principle to solve problems.

4.3.1.1 Find a Multiple

The input contains N natural (i.e., positive integer) numbers ($N \leq 10000$). Each of these numbers is not greater than 15000. These numbers are not necessarily different (so it may happen that two or more of them will be equal). Your task is to choose a few of the given numbers ($1 \leq few \leq N$) so that the sum of chosen numbers is a multiple for N (i.e., $N \times k = $ [sum of chosen numbers] for some natural number k).

Input

The first line of the input contains the single number N. Each of the next N lines contains one number from the given set.

Output

In case your program decides that the target set of numbers cannot be found, it should print the single number 0 to the output. Otherwise, it should print the number of the chosen numbers in the first line followed by the chosen numbers themselves (on a separate line each) in arbitrary order.

If there are more than one set of numbers with the required properties, you should print to the output only one (preferably your favorite) of them.

Sample Input	Sample Output
5	2
1	2
2	3
3	
4	
1	

Source: Ural Collegiate Programming Contest 1999

IDs for Online Judes: POJ 2356, Ural 1032

Analysis

For this problem, we can prove this proposition.

For a sequence with N natural (i.e., positive integer) numbers $a_1,\ldots, a_N$, there exists a subsequence $a_l, \ldots, a_r$, $\sum_{i=l}^{r} a_i$ that can be divided exactly by N.

Proof. Suppose $B_i = \sum_{k=1}^{i} a_k$, $i = 1, 2,\ldots,N$.

If there exists a B_i exactly divisible by N, the proposition holds; else there are $N-1$ remainders for B_i divided by N; $i=1, 2, \ldots, N$.

Remainders are regarded as containers, and B_i are regarded as objects. There are N objects are put into $N-1$ containers. Based on the pigeonhole principle, there must exist B_j and B_i, $B_j\%N==B_i\%N$, $1\leq j < i \leq N$. Therefore, $(B_i-B_j)\%N==0$, that is, $\sum_{k=j+1}^{i} a_k$ can be divided exactly by N, so the proposition holds.

Program

```c
# include <stdio.h>
int a[10004],s[10004],mod[10004],n;
void print(int s,int t){    //Output a[s]..a[t]
printf("%d\n",t-s+1);      //the number of the chosen numbers
for(int i=s;i<=t;i++)      //the chosen numbers
printf("%d\n",a[i]);
}
int main(){
scanf("%d",&n);    //Input the number of positive integers N
for(int i=1;i<=n;i++){    // Remainders are regarded as
containers
   scanf("%d",a+i);    // the i-th positive integers
   s[i]=s[i-1]+a[i];    //the sum of the first i positive
integers
   if(s[i]%n==0){    //if the sum of the first i positive
integers can be divided exactly by N
      print(1,i);
      break;
```

```
        }else if(!mod[s[i]%n]){    //If the remainder never
appears, i is put into a container; else output the chosen
numbers
            mod[s[i]%n]=i;
        }else{
            print(mod[s[i]%n]+1,i);
            break;
            }
    }
return 0;
}
```

4.3.2 Applications of the Inclusion–Exclusion Principle

The inclusion–exclusion principle counts the number of elements in the union of finite sets. The inclusion–exclusion principle is as follows:

Suppose there are finite sets A_1, ..., A_n, and S is a finite universal set containing A_1, ..., A_n.

$$|A_1 \cup A_2 \cup ... \cup A_n| = \sum_{i=1}^{n} |A_i| - \sum_{1 \le i_1 < i_2 \le n} |A_{i_1} \cap A_{i_2}| + \sum_{1 \le i_1 < i_2 < i_3 \le n} |A_{i_1} \cap A_{i_2} \cap A_{i_3}| - ......$$

$$+ (-1)^{n-1} |A_1 \cap A_2 \cap ... \cap A_n|;$$

$$\overline{|A_1 \cup A_2 \cup ......A_n|} = |\overline{A_1} \cap \overline{A_2} ...... \cap \overline{A_n}| = |S| - |A_1 \cup A_2 \cup ...... \cup A_n|; \text{where } |A_i|$$

indicates the cardinality of a set A_i, $1 \le i \le n$.

When the inclusion–exclusion principle is used for A_1, ..., A_n, there are $C(n, 2)=n(n-1)/2$ two-set intersections, $C(n, 3)=n(n-1)(n-2)/3!$ three-set intersections, and so on.

4.3.2.1 Tmutarakan Exams

The University of New Tmutarakan trains first-class specialists in mental arithmetic. To enter the university, you should master arithmetic perfectly. One of the entrance exams at the Divisibility Department is the following. Examinees are asked to find k different numbers that have a common divisor greater than 1. All numbers in each set should not exceed a given number s. The numbers k and s are announced at the beginning of the exam. To exclude copying (the department is the most prestigious in town), each set of numbers is credited only once (to the person who submitted it first).

Last year, these numbers were $k=25$ and $s=49$ and, unfortunately, nobody passed the exam. Moreover, it was proven later by the best minds of the department that there do not exist sets of numbers with the required properties. To avoid embarrassment this year, the dean has asked for your help. You should find the number of sets of k different numbers, each of the numbers not exceeding s, which have a common divisor greater than 1. Of course, the number of such sets equals the maximum possible number of new students of the department.

Input

The input contains numbers k and s ($2 \leq k \leq s \leq 50$).

Output

You should output the maximum possible number of the department's new students if this number does not exceed 10000, which is the maximum capacity of the department; otherwise, you should output 10000.

Sample Input	Sample Output
3 10	11

Source: USU Open Collegiate Programming Contest March 2001 Senior Session

ID for Online Judge: Ural 1091

 Analysis

Every natural number $n \geq 2$ is a prime number or a product of prime numbers.

Every common divisor i ($2 \leq i \leq s$) is enumerated. In $1 \ldots s$ the number of numbers that have a common divisor i is $d = \left\lfloor \dfrac{s-i}{i} \right\rfloor + 1$. The number of k-combination of a d-element set is $C(d, k)$, and the number in each k-combination has the common divisor i.

If the common divisor i is a prime number, $C(d, k)$ is accumulated into the number of the department's new students;

If the common divisor i is a product of two prime numbers, in the number of the department's new students, $C(d, k)$ is counted twice. Based on the inclusion–exclusion principle, $C(d, k)$ must be subtracted from the number of the department's new students.

Because of the range of s, for this problem, we need not consider products of three prime numbers. Suppose ans is the maximum possible number of the department's new students.

For every number i in $[2s]$

    If (i is a prime number) $ans \mathrel{+}= C\left(\left\lfloor \dfrac{s-i}{i} \right\rfloor + 1, k\right)$

      Else if(i is a product of two prime numbers)

$ans\mathrel{-}= C\left(\left\lfloor \dfrac{s-i}{i} \right\rfloor + 1, k\right);$

  Output ans ($ans>10000$? 10000 : ans).

 Program

```cpp
# include <cstdio>
# include <algorithm>
# include <iostream>
using namespace std;
typedef long long int64;

bool pp[60];      //prime sieve
int64 c[60][60];      //c[n][m] is C(n, m)
int k,s;
void cal_prime() {      //calculate prime
pp[0]=pp[1]=1;
for(int i=2;i<=50;i++){
    if(pp[i])continue;
    for(int j=i*2;j<=50;j+=i)pp[j]=1;
    }
}

void cal_number(){      // The formula calculating binomial
coefficients is used to calculate c[][]
for(int i=0;i<=50;i++) c[i][0]=1;
for(int i=1;i<=50;i++)
    for(int j=1;i<=50;j++) c[i][j]=c[i-1][j]+c[i-1][j-1];
}

inline bool pxp(int a){      //determine whether a is a product
of two primes
    for(int i=2;i<=50;i++) if(a%i==0&&!pp[i]&&!pp[a/i]&&i!=a/i)
return true;
return false;
}
int work(){      // ans is the maximum possible number of the
department's new students
int64 ans=0;      //Initialization
```

```
for(int i=2;i<=s;i++){
```

if(!pp[i]){   //if *i* is a prime, $ans += C\left(\left\lfloor \dfrac{s-i}{i} \right\rfloor + 1, k\right)$

```
        int cnt=0;
        for(int j=i;j<=s;j+=i)cnt++;
        ans+=c[cnt][k];
```

}else if(pxp(i)){   //if *i* is a product of two prime

numbers, $ans -= C\left(\left\lfloor \dfrac{s-i}{i} \right\rfloor + 1,\ k\right)$

```
            int cnt=0;
            for(int j=i;j<=s;j+=i)cnt++;
            ans-=c[cnt][k];
        }
    }
    return ans>10000?10000:ans;    //return ans
}

int main(){
cal_prime();    //construct prime sieve p[]
cal_number();    // calculate c[][]
scanf("%d %d",&k,&s);    //calculate the solution
cout<<work()<<endl;
return 0;
}
```

Derangement: A derangement is a permutation of the elements of a set, such that no element appears in its original position. Suppose there are n elements a_1, a_2, ..., a_n, and the original position for a_i is the i-th position, $1 \le i \le n$. The number of derangements for n elements is $D_n = n!\left(1 - \dfrac{1}{1!} + \dfrac{1}{2!} - ... + (-1)^n \dfrac{1}{n!}\right)$.

Obviously, $D_1 = 0$, and $D_2 = 1$. For $n > 2$, the recursion formula for numbers of derangements is $D_n = (n-1)(D_{n-2} + D_{n-1})$, ($n = 3, 4, 5, ...$).

When n becomes large, the range of the number of derangements for n elements may be out of the range of integers in programming languages. Under such circumstances, the number of derangements for n elements are represented as a high-precision number.

4.3.2.2 Sweet Child Makes Trouble

Children are always sweet, but they can sometimes make you feel bitter. In this problem, you will see how Tintin, a five-year-old boy, creates trouble for his parents. Tintin is a joyful boy and is always busy doing something. But what he does is not always pleasant for his parents. He likes to play with household things like his father's wristwatch or his mother's comb. When he's finished playing, he places the item in some other place. Tintin is very intelligent and a boy with a very sharp memory. To make things worse for his parents, he never returns the things he has taken for playing to their original places.

Think about a morning when Tintin has managed to "steal" three household objects. Now, in how many ways he can place those things such that nothing is placed in their original place? Tintin does not like to give his parents that much trouble. So, he does not leave anything in a completely new place; he merely permutes the objects.

Input

There will be several test cases. Each will have a positive integer less than or equal to 800, indicating the number of things Tintin has taken for playing. Each integer will be in a line by itself. The input is terminated by a –1 (minus one) in a single line, which should not be processed.

Output

For each test case, print an integer indicating in how many ways Tintin can rearrange the things he has taken.

Sample Input	Sample Output
2	1
3	2
4	9
–1	

Source: The FOUNDATION Programming Contest 2004

ID for Online Judge: UVA 10497

 Analysis

Because Tintin never returns the things he has taken for playing to their original places, the problem requires you to calculate the number of derangements.

Because the number of things Tintin has taken for playing is a positive integer less than or equal to 800, a high-precision number is used to calculate the result. First, the offline method is used to calculate $D_1 \ldots D_{800}$. Then, for each test case n, output D_n directly.

Program

```
# include <cstdio>
# include <cstring>
# include <cstdlib>
# include <iostream>
# include <string>
# include <cmath>
# include <algorithm>
using namespace std;
typedef unsigned long long int64;
int64 m=1e10;      //High-precision number array s, each element
is a 10-digit decimal number
struct Bigint{     //Struct Bigint for high-precision
calculation
      int64 s[1000];int l; // High-precision number array s[],
its length is l
      Bigint(){l=0; memset(s,0,sizeof(s))}     //Initialization
      void operator *=(int x){  //s←s*x, where x is an integer
            int64 d=0;
            for(int i=0;i<=l;i++){
                  d+=s[i]*x;s[i]=d%m;
                  d/=m;
            }
            while(d){
            s[++l]=d%m;
            d/=m;
            }
      }
      void print(){
      printf("%llu",s[l]);     //output
      for(int i=l-1;i>=0;i--)
            printf("%010llu",s[i]);
  }
  void set(int64 a){     //integer a is transferred into high-
precision array s
      s[l]=a%m;a/=m;
      if(a)l++,s[l]=a%m;
      }
}dp[1000];     // dp[n] is D_n
Bigint operator+(Bigint b,Bigint&a){     //b←b+a, where b and a
are high-precision arrays
      int64 d=0;
      b.l=max(b.l,a.l);
      for(int i=0;i<=b.l;i++)
            {
```

```
                    b.s[i]+=d+a.s[i];
                    d=b.s[i]/m;b.s[i]%=m;
            }
        if(d)b.l++,b.s[b.l]=d;
        return b;
}
int n;
int main(){
        dp[1].set(0);dp[2].set(1);        // dp[1]=0, dp[1]=1
        for(int i=3;i<=800;i++)dp[i]=dp[i-2]+dp[i-1],
dp[i]*=(i-1);        //offline method to calculate dp[]
        while(~scanf("%d",&n)&&~n){        //input n
                dp[n].print();printf("\n");        //output Dn
        }
        return 0;
}
```

4.4 Applications of the Pólya Counting Formula

1. **Group and Permutation Group.**

   Definition 4.4.1 (Group). A group is a set G together with an operation, called the group law of G, that combines any two elements a and b to form another element, denoted as $a*b$ or ab. $(G,*)$, and this element must satisfy the following four requirements:

   1. Closure. For any a, $b \in G$, $a*b \in G$.
   2. Associativity. For any a, b, $c \in G$, $(a*b)*c = a*(b*c)$.
   3. Identity element. There exists an identity element e in G, such that for each $a \in G$, $e*a = a*e = a$.
   4. Inverse element. For each $a \in G$, there exists an element b in G, such that $a*b = b*a = e$, where e is the identity element.

      For example, $G = \{-1, 1\}$, and $(G,*)$ is a group.

      If G is a finite set, $(G,*)$ is a finite group; else $(G,*)$ is an infinite group.

   Definition 4.4.2 (Permutation Group). A permutation group is a group $(G,*)$ whose elements are permutations of $\{a_1, a_2, \ldots, a_n\}$ and $*$ is the composition of permutations.

      Pólya's theorem is based on permutation groups.

      There are $n!$ permutations for $\{a_1, a_2, \ldots, a_n\}$. If f is a permutation of $\{a_1, a_2, \ldots, a_n\}$, the permutation can be denoted by a 2-by-n array:

$$\begin{pmatrix} a_1 & a_2 & \ldots & a_n \\ f(a_1) & f(a_2) & \ldots & f(a_n) \end{pmatrix}.$$

For example, there are permutations f_1 and f_2 of $\{1, 2, 3, 4\}$,

$$f_1 = \begin{pmatrix} 1 & 2 & 3 & 4 \\ 3 & 1 & 2 & 4 \end{pmatrix}, \quad f_2 = \begin{pmatrix} 1 & 2 & 3 & 4 \\ 4 & 3 & 2 & 1 \end{pmatrix},$$

$$f_1 f_2 = \begin{pmatrix} 1 & 2 & 3 & 4 \\ 3 & 1 & 2 & 4 \end{pmatrix} \begin{pmatrix} 1 & 2 & 3 & 4 \\ 4 & 3 & 2 & 1 \end{pmatrix} = \begin{pmatrix} 1 & 2 & 3 & 4 \\ 2 & 4 & 3 & 1 \end{pmatrix}.$$

And $f_1 f_2$ is called the composition of permutations f_1 and f_2. Similarly,

$$f_2 f_1 = \begin{pmatrix} 1 & 2 & 3 & 4 \\ 4 & 3 & 2 & 1 \end{pmatrix} \begin{pmatrix} 1 & 2 & 3 & 4 \\ 3 & 1 & 2 & 4 \end{pmatrix} = \begin{pmatrix} 1 & 2 & 3 & 4 \\ 4 & 2 & 1 & 3 \end{pmatrix}.$$

Therefore, $f_1 f_2 \neq f_2 f_1$.

A permutation can be written in a product of cycles. For example,

$$\begin{pmatrix} 1 & 2 & 3 & 4 & 5 \\ 4 & 3 & 1 & 5 & 2 \end{pmatrix} = (1\ 4\ 5\ 2\ 3), \quad \begin{pmatrix} 1 & 4 & 5 & 2 & 3 \\ 5 & 1 & 4 & 2 & 3 \end{pmatrix} = (1\ 5\ 4)(2)(3) = (1\ 5\ 4),$$

and $\begin{pmatrix} 1 & 2 & 3 & 4 & 5 \\ 3 & 1 & 2 & 5 & 4 \end{pmatrix} = (1\ 3\ 2)(4\ 5).$

If there is a permutation $f=(1\ 2...n)$, then $f^n=(1)(2)...(n)=e$.

An even permutation is a permutation obtainable from an even number of two-element swaps. And an odd permutation is a permutation obtainable from an odd number of two-element swaps.

2. **Conjugacy Class.**

Suppose S_n is all permutations for $\{1, 2, ..., n\}$. For example, all permutations for $\{1, 2, 3, 4\}$ are $S_4=\{(1)(2)(3)(4), (12), (13), (14), (23), (24), (34), (123), (124), (132), (134), (142), (143), (234), (243), (1234), (1243), (1324), (1342), (1423), (1432), (12)(34), (13)(24), (14)(23)\}$.

A permutation P in S_n can be written as $P = \underbrace{(a_1 a_2 ... a_{k_1})(b_1 b_2 ... b_{k_2})...(h_1 h_2 ... h_{k_l})}_{l \text{ cycles}}$,

where $k_1+k_2+...+k_l=n$. Suppose C_k is the number of cycles whose order is k, $k=1...n$, and cycles whose order is k is denoted by $(k)^{C_k}$.

Therefore, S_n can be categorized into $(1)^{C_1}(2)^{C_2}...(n)^{C_n}$. If $C_i=0$, then $(i)^{C_i}$ can be omitted, $i=1...n$. Obviously, $\sum_{k=1}^{n} kc_k = n$. For example, in S_4, permutations with the same format are shown as follows:

There are three permutations for $(1)^0(2)^2(3)^0(4)^0$, or $(2)^2$: $(12)(34)$, $(13)(24)$, and $(14)(23)$;

There are eight permutations for $(1)^1(3)^1$: (123), (124), (132), (134), (142), (143), (234), and (243);

There are six permutations for $(1)^2(2)^1$: (12), (13), (14), (23), (24), and (34);

There is one permutation for $(1)^4$: (1)(2)(3)(4);

There are six permutations for $(4)^1$: (1234), (1243), (1324), (1342), (1423), and (1432).

Definition 4.4.3 (Conjugacy Class). In S_n, permutations with the same format are called conjugacy classes.

The number of conjugacy classes in S_n is equal to the number of integer partitions of n.

The number of permutations for a conjugacy class $(1)^{C_1}(2)^{C_2}...(n)^{C_n}$ is

$$\frac{n!}{c_1!...c_n!1^{c_1}2^{c_2}...n^{c_n}}.$$

For example, in S_4, numbers of permutations for all conjugacy classes are as follows:

In conjugacy class $(2)^2$ there are $\frac{4!}{2!\times2^2}=3$ permutations. In conjugacy class $(1)^1(3)^1$ there are $\frac{4!}{1!\times3}=8$ permutations. In conjugacy class $(1)^2(2)^1$ there are $\frac{4!}{2!\times2}=6$ permutations. In conjugacy class $(1)^4$ there are $\frac{4!}{4!}=1$ permutation. In conjugacy class $(4)^1$ there are $\frac{4!}{4}=6$ permutations.

Suppose G is a permutation group for $\{1, 2, ..., n\}$, and K is a number in $\{1, 2, ..., n\}$. Of course, G is a subgroup for S_n. The stabilizer of the number K, written Z_K, is the set of all permutations of G that leave K fixed.

For example, $G=\{e, (1\ 2), (3\ 4), (1\ 2)(3\ 4)\}$. $Z_1=\{e, (3\ 4)\}$; $Z_2=\{e, (3\ 4)\}$; $Z_3=\{e, (1\ 2)\}$; $Z_4=\{e, (1\ 2)\}$. Obviously, Z_K is a subgroup for G, K is a number in $\{1, 2, 3, 4\}$. For G, under the permutation, 1 can be permuted to 2, 2 can be permuted to 1; and 3 can be permuted to 4, 4 can be permuted to 3. But 1 or 2 can't be permuted to 3 or 4, and 3 or 4 can't be permuted to 1 or 2. Therefore, 1 and 2 are in one equivalence class, and 3 and 4 are in the other equivalence class.

Suppose G is a permutation group for $\{1, 2, ..., n\}$, and K is a number in $\{1, 2, ..., n\}$. Under the permutation, $\{1, 2, ..., n\}$ can be partitioned into several equivalence classes. The equivalence class that K belongs to is denoted as E_K.

3. **Burnside's Lemma and Pólya Counting Formula.**

Theorem 4.4.1 Suppose G is a permutation group for $\{1, 2, ..., n\}$, and K is a number in $\{1, 2, ..., n\}$. $|E_K|\times|Z_K|=|G|$.

For example, $G=\{e, (1\ 2), (3\ 4), (1\ 2)(3\ 4)\}$; $E_1=E_2=\{1, 2\}$, $E_3=E_4=\{3, 4\}$; $|E_1|=|E_2|=|E_3|=|E_4|=2$; $Z_1=Z_2=\{e, (3\ 4)\}$, $Z_3=Z_4=\{e, (1\ 2)\}$; $|Z_1|=|Z_2|=|Z_3|=|Z_4|=2$. Obviously, $|E_1|\times|Z_1|=|E_2|\times|Z_2|=|E_3|\times|Z_3|=|E_4|\times|Z_4|=4=|G|$.

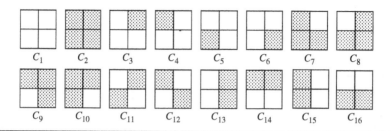

Figure 4.8

In S_4, even permutations $A_4=\{(1)(2)(3)(4), (1\ 2\ 3), (1\ 2\ 4), (1\ 3\ 2), (1\ 3\ 4), (1\ 4\ 2), (1\ 4\ 3), (2\ 3\ 4), (2\ 4\ 3), (1\ 2)(3\ 4), (1\ 3)(2\ 4), (1\ 4)(2\ 3)\}$. $E_1=\{1, 2, 3, 4\}$. $Z_1=\{e, (2\ 3\ 4), (2\ 4\ 3)\}$. Obviously, $|E_1|\times|Z_1|=4\times3=12=|A_4|$.

Suppose $G=\{\alpha_1, \alpha_2, \ldots, \alpha_m\}$ is a permutation group on $\{1, 2, \ldots, n\}$, where $\alpha_1=e$. α_k can be written as a product of cycles: $c_1(\alpha_k)$ is the number of cycles whose order is 1, $k=1, 2, \ldots, m$. For example, $G=\{e, (1\ 2), (3\ 4), (1\ 2)\ (3\ 4)\}$; $\alpha_1=e=(1)(2)(3)(4)$, $c_1(\alpha_1)=4$; $\alpha_2=(1\ 2)=(1\ 2)(3)(4)$, $c_1(\alpha_2)=2$; $\alpha_3=(3\ 4)=(1)(2)\ (3\ 4)$, $c_1(\alpha_3)=2$; $\alpha_4=(1\ 2)\ (3\ 4)$, $c_1(\alpha_4)=0$.

Burnside's Lemma. Suppose $G=\{\alpha_1, \alpha_2, \ldots, \alpha_m\}$ is a permutation group on $\{1, 2, \ldots, n\}$, and l is the number of equivalence classes under G.

$$l = \frac{1}{|G|}[c_1(\alpha_1)+c_1(\alpha_2)+\ldots+c_1(\alpha_m)].$$

Burnside's Lemma is used to count the number of nonequivalent colorings of a set X under the action of a group of permutations of X. For example, a square is divided into four little squares. Two colors are used to color four squares. There are 16 possible colorings, as shown in Figure 4.8.

If the above squares are rotated 90°, 180°, and 270° counterclockwise, there are three other permutations for 16 colorings.

1. Rotation by 0°: $P_1=(C_1)(C_2)(C_3)(C_4)(C_5)\ldots\ldots(C_{16})$, that is, $C_1(P_1)=16$;
2. Rotation by 90°: $P_2=(C_1)(C_2)(C_3C_4C_5C_6)(C_7C_8C_9C_{10})(C_{11}C_{12})(C_{13}C_{14}C_{15}C_{16})$, that is, $C_1(P_2)=2$;
3. Rotation by 180°: $P_3=(C_1)(C_2)(C_3C_5)(C_4C_6)(C_7C_9)(C_8C_{10})(C_{11})(C_{12})(C_{13}C_{15})$ $(C_{14}C_{16})$, that is, $C_1(P_3)=4$;
4. Rotation by 270°: $P_4=(C_1)(C_2)(C_3C_4C_5C_6)(C_7C_8C_9C_{10})(C_{11}C_{12})$ $(C_{13}C_{14}C_{15}C_{16})$, that is, $C_1(P_4)=2$.

Therefore, $G=\{P_1, P_2, P_3, P_4\}$, and $|G|=4$, the number of nonequivalent colorings $l = \frac{1}{4}(16+2+4+2) = 6$. The six corresponding nonequivalent colorings are as shown in Figure 4.9.

The four little squares can also be numbered 1, 2, 3, and 4 respectively (Figure 4.10).

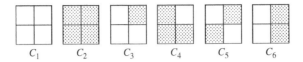

C_1 C_2 C_3 C_4 C_5 C_6

Figure 4.9

Figure 4.10

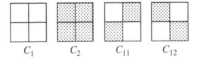

C_1 C_2

Figure 4.11

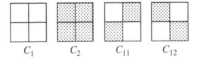

C_1 C_2 C_{11} C_{12}

Figure 4.12

If the above square is rotated 0°, 90°, 180°, and 270° counterclockwise, a permutation group $G=\{P_1, P_2, P_3, P_4\}$ is used to represent the rotations. The number of permutations $|G|=4$. Suppose $c(P_i)$ is the number of cycles for P_i, $i=1, 2, 3, 4$. Therefore, $P_1=(1)(2)(3)(4)$, $c(P_1)=4$; $P_2=(1\ 2\ 3\ 4)$, $c(P_2)=1$; $P_3=(1\ 3)$ $(2\ 4)$, $c(P_3)=2$; $P_4=(4\ 3\ 2\ 1)$, $c(P_4)=1$.

Suppose m is the number of colors. If each cycle is colored with same color in P_i, the number of colorings $m^{c(P_i)}$ is the number of colorings for G under permutation P_i. $2^{c(P_1)} = 2^4 = c_1(P_1) = 16$, $2^{c(P_2)} = 2^1 = c_1(P_2) = 2$, $2^{c(P_3)} = 2^2 = c_1(P_3) = 4$, and $2^{c(P_4)} = 2^1 = c_1(P_4) = 2$.

$P_4=(4\ 3\ 2\ 1)$, cycles whose order is 1 for P_4 are $(c_1)(c_2)$. That is, c_1 and c_2 are the four little squares are colored with the same color. (See Figure 4.11.)

$P_3=(1\ 3)(2\ 4)$, cycles whose order is 1 for P_3 are $(c_1)(c_2)(c_{11})(c_{12})$. That is, square 1 and square 3 are colored with the same color, and square 2 and square 4 are colored with the same color (see Figure 4.12).

Obviously, the number of nonequivalent colorings $l=\dfrac{1}{4}(2^4+2^1+2^2+2^1)=6$.

Based on that, the Pólya Counting Formula is as follows:

Pólya Counting Formula. Let G be a permutation group $\{P_1, P_2, \ldots\ldots, P_k\}$ of n elements. And m colors are used to color the n elements. Then the number of nonequivalent colorings $l = \dfrac{1}{|G|}(m^{c(P_1)} + m^{c(P_2)} + \cdots m^{c(P_k)})$, where $c(P_i)$ is the number of cycles for P_i, $i=1\ldots k$.

If there is a permutation group G of a set. Based on the number of permutations $|G|$, and the number of cycles $c(P_i)$ for each permutation P_i, the Pólya Counting Formula is used to calculate the number of produced equivalence classes.

4.4.1 Necklace of Beads

Beads of red, blue, or green colors are connected together into a circular necklace of n beads ($n<24$) (see Figure 4.13). If the repetitions that are produced by rotation around the center of the circular necklace or reflection to the axis of symmetry are all neglected, how many different forms of the necklace are there?

Input

The input has several lines, and each line contains the input data n. -1 denotes the end of the input file.

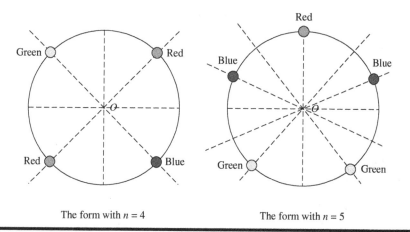

The form with $n = 4$	The form with $n = 5$

Figure 4.13

Output

The output should contain the output data: the number of different forms, in each line corresponding to the input data.

Sample Input	Sample Output
4	21
5	39
−1	

Source: ACM Xi'an 2002

IDs for Online Judges: POJ 1286, UVA 2708

 Analysis

Suppose a is the current permutation, where a_j is the number of beads in the jth position, $1 \leq j \leq n$.

Rotate around the center of the circular necklace and reflect to the axis of symmetry i times successively, $0 \leq i \leq n-1$.

1. The i-th times, a rotation: Bead j is permutated by bead $(j+i)\%n+1$, that is, $a[j]=a[(j+i)\%n+1]$, $1 \leq j \leq n$. Suppose c_i is the number of cycles for the i-th permutation. The number of colorings is 3^{c_i}.
2. The i-th times, a reflection: Bead j and bead $(n+1-j)$ exchange each other, that is, $a[j] \leftrightarrow a[n+1-j]$, $1 \leq j \leq n$. Suppose c_i' is the number of cycles for the i-th permutation. The number of colorings is $3^{c_i'}$.

Obviously, there are $2 \times n$ permutations. The Pólya Counting Formula is used to calculate the number of different forms: $l = \dfrac{1}{2n} \sum_{i=1}^{n} (3^{c_i} + 3^{c_i'})$.

 Program

```
# include <cstdio>
# include <cstring>
# include <cstdlib>
# include <iostream>
```

```cpp
# include <string>
# include <cmath>
# include <algorithm>
using namespace std;
typedef long long int64;

int n,vis[30],lab[30];     // lab[]: current permutation, bead
j is permutated by bead lab[j]; vis[j]: permutation flag for j
int64 qpow(int64 a,int64 b){     //calculate and return a^b
      int64 ans=1;
      while(b){
            if(b&1) ans*=a;
            a*=a;b>>=1;
      }
      return ans;
}

int getloop(){     // calculate and return the number of cycles
for the current permutation
      memset(vis,0,sizeof(vis));
      int cnt=0;
      for(int i=1;i<=n;i++) {
            if(vis[i])continue;     //calculate the cycle in
which i is
            cnt++;
            int j=i;
            do{
                  vis[j]=1;
                  j=lab[j];
            }while(!vis[j]);
      }
      return cnt;     //return the number of cycles for the
current permutation
}

void work(){     // calculate the number of different forms for
n beads
      if(!n){
            printf("0\b");
            return;
       }
      int64 ans=0;
      for(int i=0;i<n;i++){     // rotations and reflections
            for(int j=1;j<=n;j++) lab[j]=(j+i)%n+1;     // The
i-th times, a rotation
            ans+=qpow(3,getloop());     // the number of
colorings with 3 colors
            for(int j=1;j<=n/2;j++)swap(lab[j],lab[n+1-j]);
// The i-th times, a reflection
            ans+=qpow(3,getloop());     // the number of
colorings with 3 colors
```

```
        }
        ans/=(n*2);      // the number of different forms
        printf("%lld\n",ans);
}

int main(){
        while(~scanf("%d",&n)&&~n)work();      //Input n,
calculate and output
        return 0;
}
```

4.4.2 Toral Tickets

On the planet Eisiem, passenger tickets for the new means of transportation are planned to have the form of tores. Each tore is made of a single rectangular black rubber sheet containing $N \times M$ squares. Several squares are marked with white, thus encoding the ticket's source and destination.

When the passenger buys the ticket, the ticket booking machine takes the rubber sheet, marks some squares to identify the route of the passenger, and then provides it to the passenger. Next, the passenger must glue the ticket.

The ticket must be glued in as follows: First, two of its sides of greater length are glued together, forming a cylinder. Next, cylinder base circles, each of which has the length equal to the length of the short side of the original rubber sheet, are glued together. They must be glued in such a way that the cells, the sides of which are glued, first belonged to the same row of the sheet. Note that the inner and the outer part of the sheet can be distinguished.

The resulting tore is the valid ticket.

Note that if the original sheet is square, there are two topologically different ways to make a tore out of a rubber sheet.

Ticket material is so perfect, and gluing quality is so fine, that no one is able to find the seam, and this leads to some problems. First, the same tore can be obtained using different sheets. More than that, the same sheet can lead to tores that look a bit different.

Now, the transport companies of Eisiem wonder how many different routes they can organize, so that the following conditions are satisfied:

tickets for different routes are represented by different tores;
if some rubber sheet was marked to make the tore for some route, it cannot be used to make the tore for another route.

Help them to calculate the number of routes they can organize.

Input

The first line of the input file contains n and m $(1 \leq n, m \leq 20)$.

Output

Output the number of routes that Eisiem transport companies can organize.

Sample Input	Sample Output
2 2	6
2 3	13

Source: Petrozavodsk Summer Trainings 2003, 2003-08-23
(Andrew Stankevich's Contest #2)

IDs for Online Judges: ZOJ 2344, SGU 208

 Analysis

In the rectangular black rubber sheet, squares are numbered 1... $n\times m$ from top to down, and from left to right. For the rectangle, there are $n\times m$ classes of permutations, where the case that every square is moved left i squares circularly, and is moved down j squares circularly, is regarded as one class of permutations, $0 \le i \le n-1$, $0 \le j \le m-1$.

A class of permutations can also be classified into following permutations.

1. Rotation by 0°: square k is permutated by square $y \times n + x$;
2. Rotation by 180°: square k is permutated by square $(m-1-y) \times n + (n-1-x)$;

If the rectangle is a square ($m==n$), there are two other permutations:

1. Rotation by 90°: square k is permutated by square $(m-1-x) \times n + y$;
2. Rotation by 270°: square k is permutated by square $x \times n + (n-1-y)$;

where $x=(k\%n+i)\%n$, $y=(k/n+j)\%m$.

The number of cycles c_{ij}^1 and c_{ij}^2 under permutations 1 and 2 are calculated. If $m==n$, the number of cycles c_{ij}^3 and c_{ij}^4 under permutations 3 and 4 are also calculated.

Therefore, if $n \ne m$, there are $s=2 \times n \times m$ permutations; and if $m==n$, there are $s=4 \times n \times m$ permutations. Each square can be colored with white or black. The Pólya Counting Formula is used to calculate the number of routes that Eisiem transport companies can organize: $l = \dfrac{1}{s} \displaystyle\sum_{0 \le i \le n-1, 0 \le j \le m-1} (2^{c_{ij}^1} + 2^{c_{ij}^2} + (2^{c_{ij}^3} + 2^{c_{ij}^4} | \text{If } n == m))$.

Because the upper limit for *N* and *M* is 20, the number of routes may be out of the range for integers. Calculation of high-precision numbers should be used. In order to improve the time complexity, the offline method is also used.

Program

```
# include <cstdio>
# include <cstring>
# include <iostream>
# include <algorithm>
using namespace std;

struct BIGNUM{    // BIGNUM is used for calculation of high-
precision numbers
int s[200];   //high-precision number: s[] whose length is l
int l;
}ans,two[405];    // the number of routes Eisiem can organize
is ans; two[i] is 2^i

inline BIGNUM operator*(BIGNUM a,int b){     //a←a×b, where a
is a high-precision number, and b is an integer
  for(int i=0;i<a.l;i++)a.s[i]*=b;
  for(int i=0;i<a.l;i++){
  a.s[i+1]+=a.s[i]/10;
  a.s[i]%=10;
  }
  while(a.s[a.l]){      //carry
    a.s[a.l+1]+=a.s[a.l]/10;
    a.s[a.l]%=10;
    a.l++;
  }
}
return a;     //return a*b
}

inline BIGNUM operator+(BIGNUM a,BIGNUM b){      //a←a+b, where
a and b are high-precision numbers
a.l=max(a.l,b.l);
for(int i=0;i< a.l;i++)a.s[i]+=b.s[i];
for(int i=0;i< a.l;i++){
a.s[i+1]+=a.s[i]/10;
a.s[i]%=10;
}
while(a.s[a.l]){      //carry
  a.s[a.l+1]+=a.s[a.l]/10;
  a.s[a.l]%=10;
  a.l++;
}
}
```

```
return a;      //return a+b
}
inline BIGNUM operator/(BIGNUM a,int b){    //a←a/b, where a
is a high-precision number, and b is an integer
  for(int i=a.l-1;i>0;i--){
     a.s[i-1]+=(a.s[i]%b)*10;
     a.s[i]/=b;
  }
  a.s[0]/=b;
  while(!a.s[a.l-1])a.l--;
  return a;      // return a/b
}

void print(BIGNUM a){    //output high-precision number a
  for(int i=a.l-1;i>=0;i--){
     printf("%d",a.s[i]);
  }
  printf("\n");
}

void cal_two(){     // 2^i
  two[0].l=1;two[0].s[0]=1;
  for(int i=1;i<=400;i++)
  two[i]=two[i-1]*2;
}

int n,m,p[4][500],nm,vis[500];    //rectangular black rubber
sheet is n*m
int circle(int la){   //number of circles under permutation la
  int a=0;     //initialize number of circles
  memset(vis,0,sizeof(vis));
  for(int i=0;i<nm;i++) {    //Enumeration
    if(!vis[i])a++;    //if i isn't permutated, number of
circles+1
    vis[i]=1;     // set mark for permutation
    for(int j=p[la][i];!vis[j];j=p[la][j])    //elements in
the circle are set to the permutation mark
    vis[j]=1;
  }
  return a;     //return number of circles under permutation la
}

void work(){     // calculate the number of routes Eisiem can
organize
int div=0;
memset(ans.s,0,sizeof(ans.s));      //initialize the number of
routes 0
ans.l=0;
for(int i=0;i<n;i++)     //Enumeration
  for(int j=0;j<m;j++){
    for(int k=0;k<nm;k++){
```

```
        int x=(k%n+i)%n,y=(k/n+j)%m;
        p[0][k]=y*n+x;        // Rotation by 0°
        p[1][k]=(m-1-y)*n+(n-1-x);     // Rotation by 180°
          if(n==m){      //Square, Rotation by 90° and 270°
            p[2][k]=(m-1-x)*n+y;p[3][k]=x*n+(n-1-y);    }
          }
        div+=2;      //accumulation
        ans=ans+two[circle(0)];      //accumulation for the
number of circles for Rotation by 0°
        ans=ans+two[circle(1)];      // accumulation for the
number of circles for Rotation by 180°
        if(n==m){      //Square
          div+=2;
          ans=ans+two[circle(2)];      // accumulation for the
number of circles for Rotation by 90°
          ans=ans+two[circle(3)];      // accumulation for the
number of circles for Rotation by 270°
        }
      }
  ans=ans/div;
  print(ans);      //Output the result
}
int main(){
cal_two();
while(~scanf("%d %d",&n,&m)){      //Input test cases
  if(n<m)swap(n,m);
  nm=n*m;      //number of squares
  work();      // calculate and output the number of routes
Eisiem can organize
}
return 0;
}
```

4.4.3 Color

Beads of *n* colors are connected together into a circular necklace of *n* beads (*n* ≤1000000000). Your job is to calculate how many different types of the necklaces can be produced. You should know that the necklace might not use up all the *N* colors, and the repetitions that are produced by rotation around the center of the circular necklace are all neglected.

You only need to output the answer module as a given number *p*.

Input

The first line of the input is an integer *x* (*x*≤3500) representing the number of test cases. The following *x* lines each contains two numbers *n* and *P* (1≤*n*≤1000000000, 1≤*p*≤30000), representing a test case.

Output

For each test case, output one line containing the answer.

Sample Input	Sample Output
5 1 30000 2 30000 3 30000 4 30000 5 30000	1 3 11 70 629

Source: POJ Monthly, Lou Tiancheng

ID for Online Judge: POJ 2154

 Analysis

Method 1: Using the Pólya Counting Formula

Method 1 is analyzing each rotation, calculating the number of cycles, and using the **Pólya Counting Formula** to calculate the number of nonequivalent classes. For each rotation s, $a_i = a_{(i+k \times s)\%n}$, where a_i is the i-th bead, and these beads are in a cycle. Multiples of $s \bmod n$ are $0, d, 2 \times d, \ldots, n-d$, where $d = GCD(n, s)$. The number of cycles is $\dfrac{n}{\dfrac{n}{d}} = d$. Therefore, the number of different kinds of the necklace

$$ans = \frac{1}{n} \sum_{i=0}^{n-1} n^{GCD(n,i)}.$$

The time complexity is $O(n \times \log_2 n)$. Because the range of n is too large, optimization should be done.

Method 2: Euler Phi-Function $\varphi(n)$

The length of each cycle is enumerated. For all i such that $GCD(i, n) = k$, $\dfrac{i}{k}$ and $\dfrac{n}{k}$ are relative prime, and $\varphi\left(\dfrac{n}{k}\right)$ numbers and $\dfrac{n}{k}$ are relative prime (φ is Euler Phi-Function). The number of different kinds of the necklace is calculated as

$$ans = \frac{1}{n} \sum_{p \mid n} \varphi\left(\frac{n}{p}\right) \times n^p = \sum_{p \mid n} \varphi\left(\frac{n}{p}\right) \times n^{p-1}.$$

The time complexity for enumerating p is $O(\sqrt{n})$. And the time complexity for calculating the Euler Phi-Function is $O(\sqrt{\sqrt{n}})$. The time complexity for method 2 is $O(n^{\frac{3}{4}})$.

The given program uses **Method 2**.

Program

```
# include <cstdio>
# include <cstring>
# include <cstdlib>
# include <iostream>
# include <string>
# include <cmath>
# include <algorithm>
using namespace std;
typedef long long int64;
bool np[50000];       //Sieve
int prime[50000],pn,lim=50000;     //Prime list prime[], its
length pn, its upper limit lim
int n,p;
void pp(){     //calculate prime list prime[] in the interval
[2, lim-1]
      np[0]=np[1]=1;
      for(int i=2;i<lim;i++){
            if(np[i])       continue;
            prime[pn++]=i;
            for(int j=i*2;j<lim;j+=i)np[j]=1;
      }
}
int phi(int n){     // Euler Phi-Function φ(n)%p
      int ans=n;
      for(int i=0;i<pn&&prime[i]*prime[i]<=n;i++){     // Each
factor for n
            if(n%prime[i]!=0)continue;
            ans-=ans/prime[i];
            do{
               n/=prime[i];
               }while(n%prime[i]==0);
      }
      if(n!=1)ans-=ans/n;
      return ans%p;
}
int exp_m(int64 a,int b){     //calculate (a^b) %p
      int ans=1,x=a%p;
```

```
        while(b){
                if(b&1)ans=(ans*x)%p;
                x=(x*x)%p;
                b>>=1;
        }
        return ans;      // return (a^b)%p
}
int main(){
        int casen;
        pp();      //Calculating prime list
        scanf("%d",&casen);     //number of test cases
        while(casen--){
                int ans=0,i;
                scanf("%d %d",&n,&p);     //Input a test case
                for(i=1;i*i<n;i++){     //enumerate each factor i
```

$$\text{for } n, \quad ans = \left(\sum_{i^2 < n, n\%i=0} \varphi\left(\frac{n}{i}\right) * n^{i-1} + \varphi(i) * n^{\frac{n}{i}-1} \right) \% p$$

```
                if(n%i!=0)continue;
                ans+=(((phi(n/i)%p)*exp_m(n,i-1))+((phi(i)%p)*
exp_m(n,n/i-1)));
                ans%=p;
        }
        if(i*i==n){      //if n==i^2, then ans=(ans+φ(i)*n^{i-1})%p
                ans+=((phi(i)%p)*exp_m(n,i-1));
                ans%=p;      // the answer module a given number p
        }
        printf("%d\n",ans);      // output the answer
        }
        return 0;
}
```

4.5 Problems

4.5.1 Common Permutation

Given two strings of lowercase letters, *a* and *b*, print the longest string *x* of lower-case letters such that there is a permutation of *x* that is a subsequence of *a* and there is a permutation of *x* that is a subsequence of *b*.

Input

The input file contains several lines of input. Consecutive two lines make a set of input. That means, in the input file, lines **1** and **2** are a set of input, lines **3** and **4** are a set of input, and so on. The first line of a pair contains *a* and the second contains *b*. Each string is on a separate line and consists of at most **1000** lowercase letters.

Output

For each set of input, output a line containing *x*. If several *x* satisfy the criteria above, choose the first one in alphabetical order.

Sample Input	Sample Output
pretty women walking down the street	e nw et

Source: World Finals Warm-up Contest, University of Alberta Local Contest

ID for Online Judge: UVA 10252

 Hint

Given two strings of lowercase letters, *a* and *b*, the problem requires you to output the longest string *x* in alphabetical order such that there is a permutation of *x* that is a subsequence of *a* and there is a permutation of *x* that is a subsequence of *b*. The algorithm is as follows:

Suppose $S_1=a_1a_2...a_{l_a}$, and $S_2=b_1b_2...b_{l_b}$.

First, frequencies for each letter in S_1 and S_2 are calculated. Let $c_1[i]$ be the frequency for the *i*-th letter in S_1, $c_2[i]$ be the frequency for the *i*-th letter in S_2, where $1 \leq i \leq 26$, the first letter is "a", the second letter is "b", ..., and the 26th letter is "z".

Second, the common permutation for S_1 and S_2 is calculated. For each *i* ($1 \leq i \leq 26$), if the *i*-th letter appears in S_1 and S_2 (($c_1[i] \neq 0$)&&($c_2[i] \neq 0$)), the letter appears *k* (=min{$c_1[i], c_2[i]$}) times in the common permutation.

4.5.2 Anagram

You are to write a program that has to generate all possible words from a given set of letters.

Example: Given the word "abc", your program should—by exploring all different combination of the three letters—output the words "abc", "acb", "bac", "bca", "cab", and "cba".

In the word taken from the input file, some letters may appear more than once. For a given word, your program should not produce the same word more than once, and the words should be output in alphabetically ascending order.

Input

The input file consists of several words. The first line contains a number giving the number of words to follow. Each following line contains one word. A word consists of uppercase or lowercase letters from A to Z. Uppercase and lowercase letters are to be considered different.

Output

For each word in the input file, the output file should contain all different words that can be generated with the letters of the given word. The words generated from the same input word should be output in alphabetically ascending order. An uppercase letter goes before the corresponding lowercase letter.

Sample Input	Sample Output
3	Aab
aAb	Aba
abc	aAb
acba	abA
	bAa
	baA
	abc
	acb
	bac
	bca
	cab
	cba
	aabc
	aacb
	abac
	abca
	acab
	acba
	baac
	baca
	bcaa
	caab
	caba
	cbaa

Source: ACM Southwestern European Regional Contest 1995

IDs for Online Judges: POJ 1256, UVA 195

Hint

There are different strategies to solve this problem. The most efficient strategy is sorting the letters in the input word first, and then directly producing all possible anagrams without duplicates. A less efficient way is to first sort the letters in the input word, and then produce all possible permutations (correctly sorted) and eliminate all duplicates on the fly without storing more than one word. A completely inefficient way is to first produce all permutations and store them in memory, and then sort them and eliminate duplicates as the last step.

4.5.3 How Many Points of Intersection?

We have two rows. There are a dots on the top row and b dots on the bottom row. We draw line segments connecting every dot on the top row with every dot on the bottom row. The dots are arranged in such a way that the number of internal intersections among the line segments is maximized. To achieve this goal, we must not allow more than two line segments to intersect in a point. The intersection points on the top row and the bottom are not included in our count; we can allow more than two line segments to intersect on those two rows. Given the value of a and b, your task is to compute $P(a, b)$, the number of intersections in between the two rows. For example, in Figure 4.14, $a=2$ and $b=3$. This figure illustrates that $P(2, 3)=3$.

Input

Each line in the input will contain two positive integers $a(0<a≤20000)$ and $b(0<b≤20000)$. Input is terminated by a line where both a and b are zero. This case should not be processed. You will need to process at most 1200 sets of inputs.

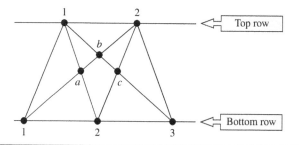

Figure 4.14

Output

For each line of input, print in a line the serial of the output, followed by the value of $P(a, b)$. Look at the output for sample input for details. You can assume that the output for the test cases will fit in 64-bit signed integers.

Sample Input	Sample Output
2 2	Case 1: 1
2 3	Case 2: 3
3 3	Case 3: 9
0 0	

Source: Bangladesh National Computer Programming Contest, 2004

ID for Online Judge: UVA 10790

 Hint

Line segments connecting two dots on the top row and dots on the bottom row will produce one intersection point. Based on the multiplication principle, $P(a, b)=C(a, 2)\times C(b, 2)$.

4.5.4 Permutations

We remind you that the *permutation* of some final set is a one-to-one mapping of the set onto itself. Less formally, that is a way to reorder elements of the set. For example, one can define a permutation of the set {1,2,3,4,5} as follows:

$$P(n) = \begin{pmatrix} 1 & 2 & 3 & 4 & 5 \\ 4 & 1 & 5 & 2 & 3 \end{pmatrix}$$

This record defines a permutation P as follows: $P(1)=4$, $P(2)=1$, $P(3)=5$, etc.

What is the value of the expression $P(P(1))$? It's clear that $P(P(1))=P(4)=2$. And $P(P(3))=P(5)=3$. One can easily see that if $P(n)$ is a permutation, then $P(P(n))$ is a permutation as well. In our example (believe us):

$$P(P(n)) = \begin{pmatrix} 1 & 2 & 3 & 4 & 5 \\ 2 & 4 & 3 & 1 & 5 \end{pmatrix}$$

It is natural to denote this permutation by $P^2(n)=P(P(n))$. In a general form the definition is as follows: $P(n)=P^1(n)$, $P^k(n)=P(P^{k-1}(n))$.

Among the permutations there is a very important one—that moves nothing:

$$E_N(n) = \begin{pmatrix} 1 & 2 & 3 & \dots & n \\ 1 & 2 & 3 & \dots & n \end{pmatrix}$$

It is clear that for every k the following relation is satisfied: $(E_N)^k = E_N$. The following less trivial statement is correct (we won't prove it here, but you may prove it to yourself incidentally):

Let $P(n)$ be some permutation of an N elements set. Then there exists a natural number k, so that $P^k = E_N$.

The least natural k such that $P^k = E_N$ is called *an order* of the permutation P. The problem that your program should solve is now formulated in a very simple manner: *"Given a permutation, find its order."*

Input

In the first line of the standard input, only a natural number N ($1 \leq N \leq 1000$) is contained, that is, a number of elements in the set that is rearranged by this permutation. In the second line, there are N natural numbers of the range from 1 up to N, separated by a space, that define a permutation—the numbers $P(1)$, $P(2)$,..., $P(N)$.

Output

You should write only a natural number to the standard output, that is an order of the permutation. You may consider that an answer shouldn't exceed 10^9.

Sample Input #1	Sample Output #1
5 4 1 5 2 3	6
Sample Input #2	Sample Output #2
8 1 2 3 4 5 6 7 8	1

Source: Ural State University Internal Contest October 2000 Junior Session

ID for Online Judge: POJ 2369

Hint

For the permutation $P(1)$, $P(2)$,..., $P(N)$, the numbers of elements in each cycle are calculated. Obviously, the least natural k such that $P^k = E_N$ is the Least Common Multiple (LCM) for these numbers.

4.5.5 *Coupons*

Coupons in cereal boxes are numbered 1 to n, and a set of one of each is required for a prize (a cereal box, of course). With one coupon per box, how many boxes on average are required to make a complete set of n coupons?

Input

Input consists of a sequence of lines each containing a single positive integer n, $1 \le n \le 33$, giving the size of the set of coupons. Input is terminated by end of file.

Output

For each input line, output the average number of boxes required to collect the complete set of n coupons. If the answer is an integer number, output the number. If the answer is not an integer, then output the integer part of the answer, followed by a space, and then by the proper fraction in the format shown below. The fractional part should be irreducible. There should be no trailing spaces in any line of output.

Sample Input	Sample Output
2	3
5	5
17	11 --
	12
	340463
	58 ------
	720720

Source: Math Lovers' Contest, *Source:* University of Alberta Local Contest

ID for Online Judge: UVA 10288

Hint

There are n coupons. Suppose that k coupons are collected, and E_K boxes are bought. The probability of getting a coupon in the next time is $\dfrac{n-k}{n}$. And the probability of getting two coupons two times is $\dfrac{n-k}{n} \times \dfrac{k}{n}, \cdots$, and so on. Therefore, there is a formula:

$$E_{k+1} = E_k + \frac{n-k}{n} \sum_{i=0}^{\infty} (i+1) \left(\frac{k}{n} \right)^i .$$

The formula $\sum_{k=0}^{\infty} kx^k = \dfrac{x}{(1-x)^2}$ (taking the derivative of two sides of the equa-

tion $\sum_{k=0}^{\infty} x^k = \dfrac{1}{(1-x)}$) is used to calculate the sum of $\dfrac{n-k}{n} \sum_{i=0}^{\infty} (i+1)\left(\dfrac{k}{n}\right)^i$ in the

above formula. $E_{k+1} = E_k + \dfrac{n}{n-k}$. Therefore, $E_n = n \times \sum_{i=1}^{n} \dfrac{1}{i}$.

4.5.6 Pixel Shuffle

Shuffling the pixels in a bitmap image sometimes yields random-looking images. However, by repeating the shuffling enough times, one finally recovers the original images. This should be no surprise, since "shuffling" means applying a one-to-one mapping (or permutation) over the cells of the image, which come in finite number, as shown in Figure 4.15.

Your program should read a number n, and a series of elementary transformations that define a "shuffling" φ of $n×n$ images. Then, your program should compute the minimal number m ($m>0$), such that m applications of φ always yield the original $n×n$ image.

For instance, if φ is counter-clockwise 90° rotation, then $m=4$, as shown in Figure 4.16.

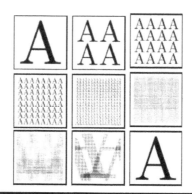

Figure 4.15

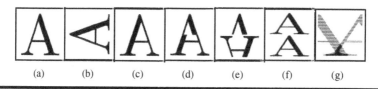

| (a) | (b) | (c) | (d) | (e) | (f) | (g) |

Figure 4.16

Input

Input consists of two lines, and the first line is number n ($2 \leq n \leq 2^{10}$, n even). The number n is the size of images. One image is represented internally by an $n \times n$ pixel matrix (a_i^j), where i is the row number and j is the column number. The pixel at the upper-left corner is at row 0 and column 0.

The second line is a non-empty list of at most 32 words, separated by spaces. Valid words are the keywords id, rot, sym, bhsym, bvsym, div, and mix, or a keyword followed by "-". Each keyword key designates an elementary transform (as defined by Figure 4.17), and key-designates the inverse of the transform key. For instance, rot- is the inverse of counterclockwise 90° rotation, that is, clockwise 90° rotation.

Transformations of image (a_i^j) into image (b_i^j)

id, identity. Nothing changes: $b_i^j = a_i^j$.

rot, counter-clockwise 90° rotation

sym, horizontal symmetry: $b_i^j = a_i^{n-1-j}$

bhsym, horizontal symmetry applied to the lower half of image: when $i \geq n/2$, then $b_i^j = a_i^{n-1-j}$ Otherwise $b_i^j = a_i^j$.

bvsym, vertical symmetry applied to the lower half of image ($i \geq n/2$)

div, division. Rows $0, 2, \ldots, n-2$ become rows $0, 1, \ldots n/2 - 1$, while rows $1, 3, \ldots n - 1$ become rows $n/2, n/2 + 1, \ldots n - 1$.

mix, row mix. Rows $2k$ and $2k + 1$ are interleaved. The pixels of row $2k$ in the new image are
$a_{2k}^0, a_{2k+1}^0, a_{2k}^1, a_{2k+1}^1 \cdots a_{2k}^{n/2-1}, a_{2k+1}^{n/2-1}$,
while the pixels of row $2k + 1$ in the new image are
$a_{2k}^{n/2}, a_{2k+1}^{n/2}, a_{2k}^{n/2+1}, a_{2k+1}^{n/2+1}, \ldots a_{2k}^{n-1}, a_{2k+1}^{n-1}$

Figure 4.17

Finally, the list k_1, k_2,..., k_p designates the compound transform $\varphi=k_1k_2...k_p$. For instance, "bvsym rot-" is the transform that first performs clockwise 90° rotation and then vertical symmetry on the lower half of the image.

Output

Your program should output a single line whose content is the minimal number m ($m>0$) such that φ^m is the identity. You may assume that, for all test input, you have $m <2^{31}$.

Sample Input 1	Sample Output 1
256 rot- div rot div	8
Sample Input 2	Sample Output 2
256 bvsym div mix	63457

Source: ACM Southwestern Europe 2005

IDs for Online Judges: POJ 2789, UVA 3510

 Hint

The problem statements define several operations on square images. Each of the operations is some simple permutation of the image's pixels. The input contains a sequence of operations. Your program should output the smallest positive K such that applying the whole sequence of operations K times always yields the original image.

The sequence of operations defines a (more complicated) permutation of the image's pixels. If you split this permutation into cycles, the answer is the LCM of the cycle lengths.

4.5.7 The Colored Cubes

All six sides of a cube are to be coated with paint. Each side is coated uniformly with one color. When a selection of n different colors of paint is available, how many different cubes can you make?

Note that any two cubes are only to be called "different" if it is not possible to rotate the one into such a position that it appears with the same coloring as the other.

Input

Each line of the input file contains a single integer *n* **(0<*n*<1000)** denoting the number of different colors. Input is terminated by a line where the value of *n*=**0**. This line should not be processed.

Output

For each line of input, produce one line of output. This line should contain the number of different cubes that can be made by using the matching number of colors.

Sample Input	Sample Output
1	1
2	10
0	

Source: 2004 ICPC Regional Contest Warmup 1

ID for Online Judge: UVA 10733

 Hint

All six sides of a cube are to be colored with paints. Each side is painted uniformly with one color. When a selection of *n* different colors of paint is available, how many different cubes can you make?

Two cubes are considered different if it is not possible to rotate one cube into a such position that it appears with the same coloring as the other (see Figure 4.18).

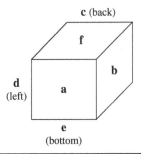

Figure 4.18

It's a pretty straightforward problem, if you know a bit of the Pólya-Burnside theory of counting.

First, you need to construct the permutation group of the cube's rotations. In simple terms, it's the set of ways (permutations) in which you can relabel the cube's faces, and get an equivalent cube (under rotations.)

The cube (with the initial labeling as shown in Figure 4.18) has 24 such ways, listed in the following table. The first column shows the final labeling of the cube, and the second one gives the corresponding permutation of faces.

Cube's arrangement	Permutation	Number of fixed points
abcdef	(a)(b)(c)(d)(e)(f)	n^6
adcbfe	(a)(bd)(c)(ef)	n^4
aecfdb	(a)(bedf)(c)	n^3
afcebd	(a)(bfde)(c)	n^3
badcfe	(ab)(cd)(ef)	n^3
bcdaef	(abcd)(e)(f)	n^3
bedfac	(abe)(cdf)	n^2
bfdeca	(abf)(cde)	n^2
cbadfe	(ac)(b)(d)(ef)	n^4
cdabef	(ac)(bd)(e)(f)	n^4
ceafbd	(ac)(be)(df)	n^3
cfaedb	(ac)(bf)(de)	n^3
dabcef	(adcb)(e)(f)	n^3
dcbafe	(ad)(bc)(ef)	n^3
debfca	(adf)(bec)	n^2
dfbeac	(ade)(bfc)	n^2
eafcbd	(aeb)(cfd)	n^2
ebfdca	(aecf)(b)(d)	n^3
ecfadb	(aed)(bcf)	n^2
edfbac	(ae)(bd)(cf)	n^3
faecdb	(afb)(ced)	n^2

fbedac	(afce)(b)(d)	n^3
fceabd	(afd)(bce)	n^2
fdebca	(af)(bd)(ce)	n^3

You can obtain all these permutations by first listing the most important ones—rotating around the X, Y, and Z axes, and then listing all their possible combinations.

A fixed point of a permutation is some coloring, such that the permutation results in a cube, which has the same coloring. If each face of the cube may be assigned one of n colors, and the permutation has c disjoint cycles, then it has n^c fixed points (the faces of each cycle have to be colored in the same color, there are c cycles, and n ways to choose colors for each).

By Burnside's Lemma, the total number of distinct colorings is equal to the arithmetic mean of the number of fixed points of permutations. That is, the answer to the problem is given by $\dfrac{1}{24}(n^6 + 3 \times n^4 + 12 \times n^3 + 8 \times n^2)$.

If you have never heard of the Pólya-Burnside theory, there are still some other methods to solve this problem.

For example, you could've guessed that the function, given that the number of colorings is a polynomial in n; obtain its values for small n by brute force, and use interpolation to find the polynomial's coefficients.

Here's another possible solution. Start by backtracking this subproblem: there are six available paints, the i-th of which must be used to color exactly $0 \leq n_i \leq 6$ sides of the cube (of course $n_1+n_2+\ldots+n_6=6$); how many colorings are possible? Then use well-known combinatorics (and probably, dynamic programming) to reduce the original problem to subproblems of this type.

Chapter 5

Practice for Greedy Algorithms

Greedy algorithms are used to solve optimization problems through a sequence of steps. At each step, greedy algorithms make the locally optimal choice in order to find a globally optimal solution. For some problems, greedy algorithms can yield a globally optimal solution, but for some problems, such as the traveling salesman problem (TSP), they can't.

This chapter organizes practices for greedy algorithms as follows:

- Practices for Greedy Algorithms;
- Greedy Choices Based on Sorted Data;
- Greedy Algorithms Used with Other Methods to Solve P-Problems.

5.1 Practices for Greedy Algorithms

Greedy algorithms are used to solve optimization problems through a sequence of steps, and to make the choice that looks best at each step. There are some famous greedy algorithms, such as Prim's algorithm and Kruskal's algorithm, used to find a minimum spanning tree for a weighted undirected graph; Dijkstra's algorithm, used to get single-source shortest paths between nodes in a graph; and Huffman coding.

There are two properties for optimization problems that can be solved by greedy algorithms:

1. Optimal substructures: Optimal solutions to problems consisting of a sequence of their optimal solutions to subproblems (necessity).
2. The property for greedy choices: Global optimal solutions to problems can be obtained by making a sequence of local optimal (greedy) choices (feasibility).

The following two experiments are practices for greedy algorithms.

5.1.1 Pass-Muraille

In modern-day magic shows, passing through walls is very popular, in which a magician performer passes through several walls in a predesigned stage show. The wall-passer (Pass-Muraille) has a limited wall-passing energy to pass through at most k walls in each wall-passing show. The walls are placed on a grid-like area. An example is shown in Figure 5.1, where the land is viewed from above. All the walls have unit widths, but different lengths. You may assume that no grid cell belongs to two or more walls. A spectator chooses a column of the grid. Our wall-passer starts from the upper side of the grid and walks along the entire column, passing through every wall on his way to get to the lower side of the grid. If he faces more than k walls when he tries to walk along a column, he would fail and would not present a good show. For example, in the wall configuration shown in Figure 5.1, a wall-passer with $k=3$ can pass from the upper side to the lower side by choosing any column except column 6.

Given a wall-passer with a given energy and a show stage, we want to remove the minimum number of walls from the stage so that our performer can pass through all the walls at any column chosen by spectators.

Input

The first line of the input file contains a single integer t ($1 \le t \le 10$), the number of test cases, followed by the input data for each test case. The first line of each test case contains two integers n ($1 \le n \le 100$), the number of walls, and k ($0 \le k \le 100$), the maximum number of walls that the wall-passer can pass through, respectively. After the first line, there are n lines each containing two (x, y) pairs representing coordinates of the two endpoints of a wall. Coordinates are non-negative integers less than or equal to 100. The upper-left of the grid is assumed to have coordinates $(0, 0)$. The second sample test case below corresponds to the land given in Figure 5.1.

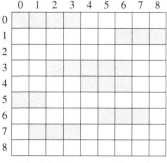

Shaded cells represent the walls

Figure 5.1

Output

There should be one line per test case containing an integer number which is the minimum number of walls to be removed, such that the wall-passer can pass through walls starting from any column on the upper side.

Sample Input	Sample Output
2	1
3 1	1
2 0 4 0	
0 1 1 1	
1 2 2 2	
7 3	
0 0 3 0	
6 1 8 1	
2 3 6 3	
4 4 6 4	
0 5 1 5	
5 6 7 6	
1 7 3 7	

Source: ACM Tehran 2002 Preliminary

IDs for Online Judges: POJ 1230, ZOJ 1375

Hint

Walls are parallel to X.

Analysis

All columns are scanned from left to right. Removing the minimum number of walls from the stage must guarantee removing the minimum number of walls in scanned columns. Therefore, the optimal solution to the problem consists of its optimal solutions to subproblems. The key to the problem is its greedy choice.

Suppose there are D walls in the current column. If $D \leq K$, we needn't remove any wall; and if $D > K$, $D - K$ walls must be removed. The greedy choice is as follows. For walls in the current column, the longest $D - K$ walls in unscanned columns are removed. Obviously, the greedy choice removes a minimum number of walls.

Program

```
#include<iostream>
using namespace std;
int t,n,k,x,y,x1,y2,max_x,max_y,sum_s=0;     //t: number of
test cases; n: number of walls; k: at most k walls can be
passed through; x,y,x1,y2: Coordinate; max_x,max_y: maximal
row and column Coordinate; sum_s: the minimum number of
removed walls
int map[105][105];
int main()
{
    scanf("%d",&t);     // number of test cases
    while(t--)     // all test cases are processed
    {
        memset(map,0,sizeof(map));
        max_x=0;     //Initialization
        max_y=0;
        sum_s=0;
        scanf("%d %d",&n,&k);
        for (int i=1;i<=n;i++)
        {
            scanf("%d %d %d %d",&x,&y,&x1,&y2);
            if (x>max_x)max_x=x;
            if (x1>max_x)max_x=x1;
            if(y>max_y)max_y=y;
            if (x<x1)
              {
                  for (int j=x;j<=x1;j++) map[j][y]=i;
              }
            else
            {
                  for (int j=x1;j<=x;j++) map[j][y]=i;

            }
        }
        for (int i=0;i<=max_x;i++)     //scan from left to right
        {
                int tem=0;     //calculate the number of walls
in the i-th column
                for (int j=0;j<=max_y;j++)  if (map[i][j]>0)
tem++;
                int offset=tem-k;
                if (offset>0)     // some walls are removed
                {
                    sum_s+=offset;
```

```
                    while(offset--)
                    {
                        int max_s=0,max_bh;
                        for (int k=0;k<=max_y;k++)    //search
                        {
                            if (map[i][k]>0)
//calculate length of wall in unscanned columns
                            {
                                int tem_s=0;
                                for (int z=i+1;z<=max_x;
z++)
                                    if (map[z][k]==map[i][k])
tem_s++;
                                        else  break;
                                if (max_s<tem_s)    //record
                                {
                                    max_s=tem_s; max_bh=k;
                                }
                            }
                        }
                        for (int a=i;a<=i+max_s;a++) map[a]
[max_bh]=0;    // some walls are removed
                    }
                }
        }
        printf("%d\n",sum_s);    //output the result
    }
    return 0;
}
```

5.1.2 Tian Ji: The Horse Racing

Here is a famous story from Chinese history.

> About 2300 years ago, General Tian Ji was a high official
> in the country Qi. He likes to play horse racing with the
> king and others.
>
> Both Tian and the king have three horses in different
> classes, namely, regular, plus, and super. The rule is to have
> three rounds in a match; each of the horses must be used in
> one round. The winner of a single round takes two hundred
> silver dollars from the loser.
>
> Being the most powerful man in the country, the king
> has such nice horses that in each class, his horse is better
> than Tian's. As a result, each time the king takes six hun-
> dred silver dollars from Tian.

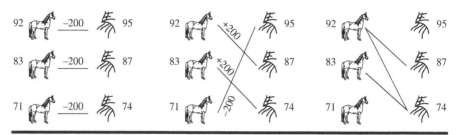

Figure 5.2

> Tian Ji was not happy about that, until he met Sun Bin, one of the most famous generals in Chinese history. Using a little trick that he learned from Sun, Tian Ji brought home two hundred silver dollars and such a grace in the next match.

It was a rather simple trick (Figure 5.2). Using his regular class horse race against the super class from the king, they will certainly lose that round. But then his plus beat the king's regular, and his super beat the king's plus. What a simple trick. And what do you think of Tian Ji, the high-ranked official in China?

Wherever Tian Ji lives nowadays, he will certainly laugh at himself. Even more, if he were sitting in the ACM contest right now, he may discover that the horse racing problem can be simply viewed as finding the maximum matching in a bipartite graph. Draw Tian's horses on one side, and the king's horses on the other. Whenever one of Tian's horses can beat one from the king, we draw an edge between them, meaning we wish to establish this pair. Then, the problem of winning as many rounds as possible is just to find the maximum matching in this graph. If there are ties, the problem becomes more complicated; he needs to assign weights 0, 1, or −1 to all the possible edges, and find a maximum weighted perfect matching.

However, the horse racing problem is a very special case of bipartite matching. The graph is decided by the speed of the horses—a vertex of higher speed always beats a vertex of lower speed. In this case, the weighted bipartite matching algorithm is too advanced a tool to deal with the problem.

In this problem, you are asked to write a program to solve this special case of matching problem.

Input

The input consists of up to 50 test cases. Each case starts with a positive integer n ($n \leq 1000$) on the first line, which is the number of horses on each side. The next n integers on the second line are the speeds of Tian's horses. Then the next n integers on the third line are the speeds of the king's horses. The input ends with a line that has a single "0" after the last test case.

Output

For each input case, output a line containing a single number, which is the maximum money Tian Ji will get, in silver dollars.

Sample Input	Sample Output
3	200
92 83 71	0
95 87 74	0
2	
20 20	
20 20	
2	
20 19	
22 18	
0	

Source: ACM Shanghai 2004

IDs for Online Judges: POJ 2287, ZOJ 2397 UVA 3266

 Analysis

The problem can be solved by several different methods. Maximum matching in a bipartite graph or dynamic programming can be used to solve the problem, but using a greedy algorithm to solve the problem is simple and efficient. The greedy algorithm is as follows:

First, the speeds of Tian's horses and the speeds of the king's horses are sorted in ascending order respectively. Suppose the sequence for speeds of Tian's current horses in ascending order is $A=a_1 \ldots a_n$; and the sequence for the speeds of the king's current horses are sorted in ascending order is $B=b_1 \ldots b_n$.

Second, greedy choices are as follows:

1. If Tian's current slowest horse is faster than the king's current slowest horse, that is, $a_1 > b_1$; then Tian's current slowest horse races against the king's current slowest horse, that is, a_1 is compared with b_1. Because b_1 is less than any elements in A and the king's current slowest horse can be defeated by any Tian's remainder horse, the king's current slowest horse is defeated by Tian's current slowest horse.

2. If Tian's current slowest horse is slower than the king's current slowest horse, that is, $a_1 < b_1$; then Tian's current slowest horse races against the king's current fastest horse, that is, a_1 is compared with b_n. Because a_1 is less than any elements in B and Tian's current slowest horse can be defeated by any king's remainder horse, Tian's current slowest horse is defeated by the king's current fastest horse.

3. If Tian's current fastest horse is faster than the king's current fastest horse, that is, $a_n > b_n$; then Tian's current fastest horse races against the king's current fastest horse, that is, a_n is compared with b_n. Because a_n is larger than any elements in B and Tian's current fastest horse can defeat any king's remainder horse, Tian's current fastest horse defeats the king's current fastest horse.

4. If Tian's current fastest horse is slower than the king's current fastest horse, that is, $a_n < b_n$; then Tian's current slowest horse races against the king's current fastest horse, that is, a_1 is compared with b_n. Because b_n is larger than any elements in A and the king's current fastest horse can defeat any Tian's remainder horse, the king's current fastest horse defeats Tian's current slowest horse.

5. If $(a_1 == b_1)$ and $(a_n > b_n)$, then it is suitable that Tian's current fastest horse races against the king's current fastest horse, that is, a_n is compared with b_n.

6. If $(a_n == b_n)$, then there exists an optimal solution that a_1 is compared with b_n.

The above process repeats until the horse racing ends. Tian's current fastest or slowest horse races against the king's current fastest or slowest horse each time based on the above greedy choices. Optimal solutions to subproblems constitute the global optimal solution to the problem.

Program

```
#include<cstdio>
#include<cstring>
#include<algorithm>
using namespace std;
int a[1010],b[1010];      //Speeds of Tian's horses and the
king's horses
int main()
{
    int n;
    while(scanf("%d",&n),n)      //number of Tian's horses (the
king's horses)
    {
        for(int i=1; i<=n; i++)  scanf("%d",&a[i]);      // Input
speeds of Tian's horses
        for(int i=1; i<=n; i++)  scanf("%d",&b[i]);      // Input
speeds of the king's horses
        sort(a+1,a+1+n);     //Sorting speeds in ascending
order
        sort(b+1,b+1+n);
        int tl=1,tr=n,ql=1,qr=n;     //Initialization
        int sum=0;
```

```
        while(tl<=tr)      // the horse racing doesn't end
        {
            if(a[tl]<b[ql])     // Tian's slowest horse is
slower than the king's slowest horse
            {
                qr--;tl++;sum=sum-200;
            }
            else if(a[tl]==b[ql])      // Speeds of the two
slowest horses are same
            {
                while(tl<=tr&&ql<=qr)
                {
                    if(a[tr]>b[qr])      //Tian's fastest horse
is faster than the king's fastest horse
                    {
                        sum+=200;tr--;qr--;
                    }
                    else     // Tian's slowest horse races
against the king's fastest horse
                    {
                        if(a[tl]<b[qr])   sum-=200;
                        tl++;qr--; break;
                    }
                }
            }
            else     // Tian's slowest horse is faster than the
king's slowest horse
            {
                tl++;ql++;sum=sum+200;
            }
        }
        printf("%d\n",sum);     //Output the result
    }
    return 0;
}
```

5.2 Greedy-Choices Based on Sorted Data

The key to a greedy algorithm is its greedy choices. Sometimes the greedy choices must be based on sorted data. First, data are sorted. Then greedy choices are made based on the sorted data.

5.2.1 The Shoemaker's Problem

A shoemaker has N jobs (orders from customers) which he must make. The shoemaker can work on only one job in each day. For each i-th job, the integer T_i ($1 \leq T_i \leq 1000$) indicates the time in days it takes the shoemaker to finish the job.

For each day of delay before starting to work for the i-th job, the shoemaker must pay a fine of S_i ($1 \leq S_i \leq 10000$) cents. Your task is to help the shoemaker, by writing a program to find the sequence of jobs with minimal total fine.

Input

The input begins with a single positive integer on a line by itself, indicating the number of the cases following, each of them as described below. This line is followed by a blank line, and there is also a blank line between two consecutive inputs.

The first line of input contains an integer N ($1 \leq N \leq 1000$). The next N lines each contain two numbers: the time and the fine of each task in order.

Output

For each test case, the output must follow the description below. The outputs of two consecutive cases will be separated by a blank line.

Your program should print the sequence of jobs with minimal fine. Each job should be represented by its number in input. All integers should be placed on only one output line and separated by one space. If multiple solutions are possible, print the first lexicographically.

Sample Input	Sample Output
1	2 1 3 4
4	
3 4	
1 1000	
2 2	
5 5	

Source: Second Programming Contest of Alex Gevak, 2000

ID for Online Judge: UVA 10026

 Analysis

"For each day of delay before starting to work for the i_{th} job, the shoemaker must pay a fine of S_i cents" means "For each day of delay after starting to work for the i_{th} job, the shoemaker must pay a fine of S_i/T_i cents". S_i/T_i is the measurement of influence for the i_{th} job, $1 \leq i \leq n$. Therefore, in order to pay a minimal fine, the job whose measurement of influence is higher must be finished earlier. The greedy algorithm is as follows:

The metric for jobs is their measurement of influence. The n jobs are sorted using their measurement of influence as the first key (in ascending order), and the numbers of jobs as the second key (in descending order). The sorted sequence is the sequence of jobs with minimal fine.

 Program

```
#include<iostream>
#include<cstdlib>
#include<cstdio>
#include<cmath>
#include<cstring>
#include<algorithm>
using namespace std;
const int maxN=1010;      // the upper limit of the number of
jobs
struct job
{
   double a;     // measurement of influence for the job
   int num;      // the number of a job
} p[maxN];       // the sequence of jobs with minimal fine
int n;
void init()
{
   double a1,a2;
   scanf("%d",&n);      // n jobs
   for (int i=1;i<=n;i++)       // the time and fine of each task
   {
      scanf("%lf%lf",&a1,&a2);
      p[i].a=a2/a1;p[i].num=i;      // Calculate S_i/T_i, and record
the number
   }
}
bool cmp(job x,job y)      // sort two jobs using their
measurement of influence as the first key (in ascending
order), and numbers of jobs as the second key (in descending
order)
{
   if ((x.a>y.a)||((x.a==y.a)&&(x.num<y.num))) return true;
   return false;
}
void work()
```

```
{
   sort(p+1,p+n+1,cmp);     // sort n jobs using their
measurement of influence as the first key (in ascending
order), and numbers of jobs as the second key (in descending
order)
   for (int i=1;i<n;i++) printf("%d ",p[i].num);     // Output
the result
   printf("%d\n",p[n].num);
}
int main()
{
   int t;
   scanf("%d",&t);     //the number of test cases
   for (int i=1;i<=t;i++)     // deal with each test case
   {
      if (i>1) printf("\n");
      init();
      work();
   }
   return 0;
}
```

5.2.2 Add All

The problem name reflects your task; just add a set of numbers. But you may feel that it is not interesting to write a C/C++ program just to add a set of numbers. Such a problem will simply question your erudition. So, let's add some flavor of ingenuity to it.

The addition operation requires cost now, and the cost is the summation of those two numbers to be added. So, to add 1 and 10, you need a cost of 11. If you want to add 1, 2 and 3, there are several ways:

$1+2=3$, cost $=3$	$1+3=4$, cost $=4$	$2+3=5$, cost $=5$
$3+3=6$, cost $=6$	$2+4=6$, cost $=6$	$1+5=6$, cost $=6$
Total $=9$	Total $=10$	Total $=11$

I hope you have already understood your mission, to add a set of integers so that the cost is minimal.

Input

Each test case will start with a positive number, N ($2 \leq N \leq 5000$) followed by N positive integers (all are less than 100000). Input is terminated by a case where the value of N is zero. This case should not be processed.

Output

For each case, print the minimum total cost of addition in a single line.

Sample Input	Sample Output
3 1 2 3 4 1 2 3 4 0	9 19

Source: UVa Regional Warmup Contest 2005

ID for Online Judge: UVA 10954

 Analysis

Initially there is a set of *n* positive numbers. Each time, two positive numbers are deleted from the set, and the sum of the two numbers is added into the set. The process repeats $n-1$ times. The final sum is the total cost of addition. The problem requires you to calculate the minimum total cost of the addition.

Obviously, in order to get the minimum total cost of addition, the greedy choice is to select two minimal positive numbers each time. Therefore a *min heap* is suitable to represent the set.

 Program

```cpp
#include<iostream>
#include<cstdio>
#include<cstdlib>
#include<cmath>
#include<cstring>
#include<algorithm>
using namespace std;
const int maxN=5010;    //the upper limit of the size of the set
int n,a[maxN];    // n: the size of the heap, a[]: min heap
void sift(int i)    // the subtree with root i is adjusted as
a min heap
{
```

```
   a[0]=a[i];
   int k=i<<1;
   while (k<=n)
   {
      if ((k<n)&&(a[k]>a[k+1])) k++;
      if (a[0]>a[k]) { a[i]=a[k];i=k;k=i<<1;} else k=n+1;
   }
   a[i]=a[0];
}
void work()      //Calculate and output the result
{
   for (int i=n >> 1;i;i--) sift(i);      // set up a min heap
   long long ans=0;
   while (n!=1)
   {
      swap(a[1],a[n--]);
      sift(1);      // adjust the heap
      a[1]+=a[n+1];
      ans+=a[1];
      sift(1);      // adjust the heap
   }
   cout << ans << endl;      //Output the result
}
int main()
{
   while (scanf("%d",&n),n)
   {
      for (int i=1;i<=n;i++) scanf("%d",&a[i]);      // Input n
positive numbers
      work();      // calculate and output the minimum total cost
of addition
   }
   return 0;
}
```

5.2.3 Wooden Sticks

There is a pile of n wooden sticks. The length and weight of each stick are known in advance. The sticks are to be processed by a woodworking machine in one-by-one fashion. It needs some time, called setup time, for the machine to prepare for processing a stick. The setup times are associated with cleaning operations and changing tools and shapes in the machine. The setup times of the woodworking machine are given as follows:

1. The setup time for the first wooden stick is one minute.
2. Right after processing a stick of length l and weight w, the machine will need no setup time for a stick of length l' and weight w' if $l \leq l'$ and $w \leq w'$. Otherwise, it will need one minute for setup.

You are to find the minimum setup time to process a given pile of n wooden sticks. For example, if you have five sticks whose pairs of length and weight are $(9 , 4)$, $(2 , 5)$, $(1 , 2)$, $(5 , 3)$, and $(4 , 1)$, then the minimum setup time should be two minutes since there is a sequence of pairs $(4 , 1)$, $(5 , 3)$, $(9 , 4)$, $(1 , 2)$, $(2 , 5)$.

Input

The input consists of T test cases. The number of test cases (T) is given in the first line of the input file. Each test case consists of two lines: The first line has an integer n, $1 \le n \le 5000$, that represents the number of wooden sticks in the test case, and the second line contains $2n$ positive integers l_1, w_1, l_2, w_2,......, l_n, w_n, each of magnitude at most 10000, where l_i and w_i are the length and weight of the ith wooden stick, respectively. The $2n$ integers are delimited by one or more spaces.

Output

The output should contain the minimum setup time in minutes, one per line.

Sample Input	Sample Output
3	2
5	1
4 9 5 2 2 1 3 5 1 4	3
3	
2 2 1 1 2 2	
3	
1 3 2 2 3 1	

Source: ACM Taejon 2001

IDs for Online Judges: POJ 1065, ZOJ 1025, UVA 2322

 Analysis

Right after processing a stick of length l and weight w, the machine will need no setup time for a stick of length l' and weight w' if $l \le l'$ and $w \le w'$. Otherwise, it will need one minute for setup. In order to reduce the setup time, the strategy for greedy choice is as follows:

For unprocessed sticks, the stick with minimal length is selected first. If there are more than one stick with minimal length, the stick with minimal weight is selected.

First, all sticks are sorted. A stick is represented as (l, w), where l is its length, and w is its weight. Sticks are sorted using l as the first key and w as the second key. That is, $(l1,w1)<(l2,w2)$, if $l1<l2 || (l1==l2 \&\& w1<w2)$.

After sorting sticks, the greedy choice is processed as follows:

Initially, setup time $c=0$, and stick 0 is as the first unprocessed stick in the sequence. Then the following steps repeat.

Step 1: In the sequence, all unprocessed sticks after stick 0 which can be processed without setup time are set as processed. That is to say, the machine will need no setup time for these sticks, if stick 0 is processed.

Step 2: Setup time c++.

Step 3: Search the first unprocessed stick in the sequence. If there is no unprocessed stick, then output the minimum setup time; else set the first unprocessed stick as stick 0, and return to Step 1.

 Program

```cpp
#include <iostream>
using namespace std;
const int N = 5000;
struct node{      // Struct of stick
    node& operator=(node &n){
            l=n.l, w=n.w, isUsed=n.isUsed;      //the length,
weight, flag that is processed or not for stick n
            return *this;
    }
    bool operator>(node &n){      //compare sticks
        return l>n.l || (l==n.l && w>n.w);
    }
    void swap(node &n){      //exchange sticks
        node tmp=*this;
        *this=n;
        n=tmp;
    }
    int l, w;
    bool isUsed;
}A[N];      //sequence of sticks A[ ]
int main()
{
    int t, n, i, j, k;
    cin >> t;      //number of test cases
    for(i=0;i<t;i++){      // test cases are processed one by one
        cin >> n;
```

```
        for(j=0;j<n;j++){     //Input length, weight for all
sticks
            cin >> A[j].l >> A[j].w;
            A[j].isUsed=false;
        }
        for(j=1;j<n;j++)      //Sorting A
            for(k=1;k<=n-j;k++)
                if(A[k-1] > A[k])
                    A[k-1].swap(A[k]);
        node cur = A[0];      // stick 0 is as the last
processed stick cur
        A[0].isUsed=true;
        int c=0;      //Initialize setup time
        while(true){
            for(j=1;j<n;j++)      //set sticks whose lengths and
weights are larger than the current stick as processed
                if(A[j].isUsed==false)
                    if(A[j].l >= cur.l && A[j].w >= cur.w){
                        A[j].isUsed=true;
                        cur = A[j];
                    }
            c++;      //setup time+1
            for(j=1;j<n;j++) if(A[j].isUsed==false){   //Search
the first unprocessed stick
                    cur = A[j];
                    A[j].isUsed=true;
                    break;
                }
            if(j==n) break;      //all sticks are processed
        }
        cout << c << endl;      // output the minimum setup time
    }
    return 0;
}
```

5.2.4 Radar Installation

Assume the coast is an infinite straight line. Land is on one side of the coast, and the sea is on the other. Each small island is a point located on the seaside. And any radar installation, located on the coast, can only cover d distance, so an island in the sea can be covered by a radius installation, if the distance between them is at most d.

We use the Cartesian coordinate system, defining the coast as the x-axis. The seaside is above the x-axis, and the land side is below. Given the position of each island in the sea, and given the distance of the coverage of the radar installation, your task is to write a program to find the minimal number of radar installations to cover all the islands. Note that the position of an island is represented by its x–y coordinates.

Input

The input consists of several test cases. The first line of each case contains two integers n ($1{\leq}n{\leq}1000$) and d, where n is the number of islands in the sea and d is the distance of coverage of the radar installation. This is followed by n lines, each containing two integers representing the coordinate of the position of each island. Then a blank line follows to separate the cases. The input is terminated by a line containing a pair of zeros.

Output

For each test case, output one line consisting of the test case number followed by the minimal number of radar installations needed. "−1" installation means no solution for that case.

Sample Input	Sample Output
3 2	Case 1: 2
1 2	Case 2: 1
−3 1	
2 1	
1 2	
0 2	
0 0	

Source: ACM Beijing 2002

IDs for Online Judge: POJ 1328, ZOJ 1360, UVA 2519

 Analysis

Each small island is represented as a segment on the coast. If a radar locates on the segment, the island can be covered by the radar. Suppose the Cartesian coordinate for the island is (x, y). If a radar locates on the coast from $(x{-}h, 0)$ to $(x{+}h, 0)$, where $h = \sqrt{d^2 - y^2}$, the island can be covered. Therefore, the island is represented as a segment from $(x{-}h, 0)$ to $(x{+}h, 0)$. It can be shown as in Figure 5.3.

Suppose there are n islands. First, n islands are represented as n segments. Second, right endpoints are as the first key (in ascending order), left endpoints are as the second key (in ascending order), and the n segments are sorted. Finally, all sorted segments are scanned one by one. If the current segment isn't covered by a radar, a radar locates at the right endpoint for the segment.

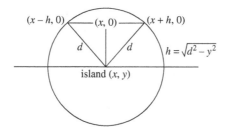

Figure 5.3

 Program

```cpp
#include <iostream>
#include <cstdio>
#include <algorithm>
#include <cmath>
using namespace std;
const int maxn = 1010;    //the upper limit of number of segments
struct tt {
        double l,r;     // left, right pointer
} p[maxn];      // the sequence of segments, where the i-th
island is represented as segment [p[i].l, p[i].r]
int n,d;      //n: number of islands, any radar covers d distance
bool flag;
void init( ) {      //Input positions of islands, and calculate
corresponding segments
        flag = true;
        int i;
        double x,y;
        for(i = 1 ; i <= n ; ++i){
                scanf("%lf%lf",&x,&y);
                if(d < y){      // if d<y, no solution
                    flag = false;
                }
                double h = sqrt(d*d - y*y);
                p[i].l = x - h;
                p[i].r = x + h;
        }
}
bool cmp (tt a, tt b){      //compare segment a and segment b
    if( b.r - a.r > 10e-7){
                return true;
        }
        if(abs(a.r - b.r) < 10e-7 && ( b.l - a.l > 10e-7)) {
```

```
                    return true;
        }
        return false;
}
void work( ) {      //Calculate and output the minimal number of
radar installations needed
if( d == -1){ printf("-1\n");  return ;  }
sort(p+1,p+1+n,cmp);     // Sorting segments
int ans = 0;    // Initialize the minimal number of radar
installations needed
double last = -10000.0;   //Initialize the position of radar
installation
int i;
for(i = 1 ; i <= n ; ++i){     // search segments one by one
    if(p[i].l <= last){     //there is a radar on the segment
            if(p[i].r <= last){
                    last = p[i].r;
            }
        continue;
        }
        ans++;     // a radar is installed on the right endpoint
        last = p[i].r;
}
printf("%d\n",ans);     //Output
}
int main(){
int counter = 1;
while(scanf("%d%d",&n,&d)!=EOF,n||d){     // Input test cases
    printf("Case %d: ",counter++);     // the number of test cases
    init();
    if(!flag){
        printf("-1\n");
    }else{
        work();
    }
}
return 0;
}
```

5.3 Greedy Algorithms Used with Other Methods to Solve P-Problems

In the real world, problems that we can solve can be classified into two classes:

P-Problems: P-Problems are polynomially solvable problems. That is, a P-Problem can be solved by an algorithm whose running time is bounded by a polynomial.
NP-Complete Problems: NP-Complete Problems cannot be solved in polynomial time.

In this section, practices for greedy algorithms used with other methods to solve P-Problems are shown.

5.3.1 Color a Tree

Bob is very interested in the data structure of a tree. A tree is a directed graph in which a special node is singled out, called the "root" of the tree, and there is a unique path from the root to each of the other nodes.

Bob intends to color all the nodes of a tree with a pen. A tree has N nodes, and these nodes are numbered 1, 2, ..., N. Suppose coloring a node takes one unit of time, and after finishing coloring one node, he is allowed to color another. Additionally, he is allowed to color a node only when its father node has been colored. Obviously, Bob is only allowed to color the root on the first try.

Each node has a "coloring cost factor", C_i. The coloring cost of each node depends both on C_i and the time when Bob finishes the coloring of this node. At the beginning, the time is set to 0. If the finishing time of coloring node i is F_i, then the coloring cost of node i is $C_i \times F_i$.

For example, a tree with five nodes is shown in Figure 5.4. The coloring cost factors of each node are 1, 2, 1, 2, and 4. Bob can color the tree in the order 1, 3, 5, 2, 4, with the minimum total coloring cost of 33.

Given a tree and the coloring cost factor of each node, please help Bob to find the minimum possible total coloring cost for coloring all the nodes.

Input

The input consists of several test cases. The first line of each case contains two integers N and R ($1 \le N \le 1000$, $1 \le R \le N$), where N is the number of nodes in the tree and R is the node number of the root node. The second line contains N integers, the i-th of which is C_i ($1 \le C_i \le 500$), the coloring cost factor of node i. Each of the next $N-1$ lines contains two space-separated node numbers V_1 and V_2, which are the endpoints of an edge in the tree, denoting that V_1 is the father node of V_2. No edge will be listed twice, and all edges will be listed.

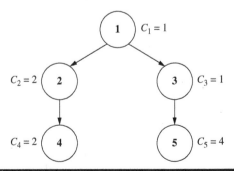

Figure 5.4

A test case of N=0 and R=0 indicates the end of input, and should not be processed.

Output

For each test case, output a line containing the minimum total coloring cost required for Bob to color all the nodes.

Sample Input	Sample Output
5 1	33
1 2 1 2 4	
1 2	
1 3	
2 4	
3 5	
0 0	

Source: ACM Beijing 2004

IDs for Online Judge: POJ 2054, ZOJ 2215, UVA 3138

 Analysis

For each node, the coloring cost is based on its coloring cost factor and the time at which Bob finishes coloring it. The coloring cost factor for each node is given. The key to the problem is determining the sequence coloring nodes.

Because Bob is allowed to color a node only when the node's father has been colored, the pointer pointing to its father for each node should be set up when edges are input. A DFS is used to calculate pointers pointing to its father for each node.

The sequence coloring nodes can be regarded as a merger process. For a father-child relationship (k, x), node x can be colored only after its father k is colored. If there are several children, the sequence coloring nodes should be determined.

Suppose $now[i]$ is the average for coloring cost factors for nodes which are merged into node i, and $cnt[i]$ is the number of nodes which are merged into node i. Initially $now[i]$= the coloring cost factor for node i, $cnt[i]$=1($1 \leq i \leq n$). After node x is colored, it is merged into node k, $now[k] = \dfrac{now[k] \times cnt[k] + now[x] \times cnt[x]}{cnt[k] + cnt[x]}$,

and $cnt[k]$=$cnt[k]$+$cnt[x]$. Such a merge process is performed n−1 times. Each time, the criteria for the merger is selecting a uncolored node whose now value is maximal. Obviously, it is a greedy strategy. The implementation process is as follows.

$n-1$ merger processes are run:

Selecting an unmerged node k (isn't the root) whose *now* value is maximal;

Setting the merger mark for node k;

Determining the sequence for coloring node k and its father f;

Searching node f which is the nearest for k and isn't merged based on the father pointer for k, and adjusting $now[f]$ and $cnt[f]$;

From the root, based on the coloring sequence, calculating the minimum total coloring cost required for Bob to color all the nodes.

$$ans = \sum_{i=1}^{n} i \times \text{the coloring cost factor for node } i \text{ in the coloring sequence.}$$

 Program

```cpp
#include<iostream>
#include<cstdlib>
#include<cstdio>
#include<cmath>
#include<cstring>
#include<algorithm>
using namespace std;
const int maxN=1100;    // the upper limit for the number of nodes
int root,n,fa[maxN],l[maxN],next[maxN],cnt[maxN],c[maxN],e[maxN]
[maxN];
// root: the root of the tree; n: the number of nodes; fa[ ]:
each node's father; next[ ]: the coloring sequence, where the
node is colored after node x is colored is node next[x]; cnt[ ]:
the number of merged nodes for each node; c[ ]: coloring cost
factor; e[ ][ ]: the adjacency matrix for the tree
double now[maxN];    //now[ ]: coloring costs for nodes after
merger
void init()    // Input coloring cost factors for n nodes and
edges, and construct e[ ][ ]
{
    int x,y;
    memset(e,0,sizeof(e));
    for (int i=1;i<=n;i++)  scanf("%d",&c[i]);
    for (int i=1;i<n;i++) {  scanf("%d%d",&x,&y);e[x][++e[x]
[0]]=y;e[y][++e[y][0]]=x;}
}
void dfs(int x)    //calculating the pointer pointing to its
father for each node
```

```
{
  int y;
  for (int i=1;i<=e[x][0];i++)      // for each child of x,
setting its father pointer x
  {
    y=e[x][i];
    if (fa[y]==0) { fa[y]=x;dfs(y);}
  }
}
void addedge(int x,int y)    //determine the coloring sequence
for x and y, that is, y is colored after x is colored
{
  while (next[x]) x=next[x];
  next[x]=y;
}
void work()     //calculate and output the minimum total
coloring cost
{
  memset(fa,0,sizeof(fa));     //initialization
  fa[root]=-1;
  dfs(root);   // Traverse the tree whose root is root, and
determine father-children relationships
  for (int i=1;i<=n;i++) now[i]=c[i];     // initialization
  bool flag[maxN];     // marks for merging nodes
  int k,f;
  double max;
  memset(flag,1,sizeof(flag));   memset(next,0,sizeof(next));
  for (int i=1;i<=n;i++) cnt[i]=1;
  for (int i=1;i<n;i++)     // n-1 merger processes
  {
    max=0;     // Selecting an unmerged node k (isn't the
root) whose now value is maximal
    for (int j=1;j<=n;j++) if ((j!=root)&&(flag[j])&&(max<now
[j])) { max=now[j];k=j;}
      f=fa[k];addedge(f,k);     // Determining the sequence
for coloring node k and its father f;
    while (!flag[f]) f=fa[f];     // Searching node f which is
the nearest for k and isn't merged based on the pointer
pointing to the father for k, that is, the father node for k
after merger
    flag[k]=false;     // Set the merger mark for node k
    now[f]=(now[f]*cnt[f]+now[k]*cnt[k])/(cnt[f]+cnt[k]);
// adjusting now[f]
    cnt[f]+=cnt[k];     // adjusting cnt[f]
  }
  int p=root,ans=0;     // calculate minimum total coloring cost
  for (int i=1;i<=n;i++)
  {
    ans+=i*c[p];p=next[p];
  }
```

```
    printf("%d\n",ans);      // output the minimum total coloring
cost
}
int main()
{
   while (scanf("%d%d",&n,&root),n+root)      //Input
   {
      init();
      work();      // calculate and output the minimum total
coloring cost
   }
   return 0;
}
```

5.3.2 Copying Books

Before the invention of book printing, it was very hard to make a copy of a book. All the contents had to be rewritten by hand by so-called scribers. The scriber was given a book, and after several months he finished creating a copy of it. One of the most famous scribers lived in the 15th century and his name was Xaverius Endricus Remius Ontius Xendrianus (Xerox). Anyway, the work was very annoying and boring. And the only way to speed it up was to hire more scribers.

Once upon a time, there was a theater ensemble that wanted to play famous antique tragedies. The scripts of these plays were divided into many books, and actors needed more copies of them, of course. So they hired many scribers to make copies of these books. Imagine you have m books (numbered 1, 2 ... m) that may have different numbers of pages $(p_1, p_2, ..., p_m)$, and you want to make one copy of each of them. Your task is to divide these books among k scribers, $k \leq m$. Each book can be assigned to a single scriber only, and every scriber must get a continuous sequence of books. That means, there exists an increasing succession of numbers $0 = b_0 < b_1 < b_2 < ... < b_{k-1} \leq b_k = m$ such that the i-th scriber gets a sequence of books with numbers between $b_{i-1}+1$ and b_i. The time needed to make a copy of all the books is determined by the scriber who was assigned the most work. Therefore, our goal is to minimize the maximum number of pages assigned to a single scriber. Your task is to find the optimal assignment.

Input

The input consists of N cases. The first line of the input contains only positive integer N. Then follow the cases. Each case consists of exactly two lines. At the first line, there are two integers m and k, $1 \leq k \leq m \leq 500$. At the second line, there are integers $p_1, p_2, ... p_m$ separated by spaces. All these values are positive and less than 10000000.

Output

For each case, print exactly one line. The line must contain the input succession p_1, p_2, ... p_m divided into exactly k parts, such that the maximum sum of a single part should be as small as possible. Use the slash character ('/') to separate the parts. There must be exactly one space character between any two successive numbers and between the number and the slash.

If there is more than one solution, print the one that minimizes the work assigned to the first scriber, and then to the second scriber, etc. But each scriber must be assigned at least one book.

Sample Input	Sample Output
2 9 3 100 200 300 400 500 600 700 800 900 5 4 100 100 100 100 100	100 200 300 400 500 / 600 700 / 800 900 100 / 100 / 100 / 100 100

Source: ACM Central European Regional Contest 1998

IDs for Online Judge: POJ 1505, ZOJ 2002, UVA 714

 Analysis

Binary search can be used to solve the problem. If the current maximum number of pages assigned to a single scriber x is feasible, we can reduce it; otherwise, we can increase it.

The key to the problem is to determine whether the current maximum number of pages assigned to a single scriber x is feasible or not. Because numbers of pages assigned to scribers are increasing from left to right, the greedy strategy is as follows. From back to front, every book is scanned, and the criteria that the current book can be assigned to the current scriber is that after the book is assigned to the scriber, the sum of numbers of pages assigned to the scriber isn't more than x, and every remainder scriber can be assigned at least one book. If the current book meets the criteria, the book is assigned to the current scriber; else the book is assigned to a new scriber, and the new scriber becomes the current scriber. A slash character ('/') is used to separate the two scribers' work.

Obviously, if k scribers can't finish copies for m books, then the current maximum number of pages assigned to a single scriber x isn't feasible; else the current maximum number of pages assigned to a single scriber x is feasible.

Binary search is used to find the minimal maximum number of pages assigned to a single scriber *min*. The above greedy algorithm is used to find the optimal assignment.

Program

```
#include<iostream>
#include<cstdlib>
#include<cstdio>
#include<cmath>
#include<cstring>
#include<algorithm>
using namespace std;
const int maxN=510;      //the upper limit of the number of
books
int n,m,a[maxN];      //n books, m scribers, a[]:the sequence of
books
long long sum;      // sum of pages
bool flag[maxN];      // flag to separate books

void init()      //Input the current test case
{
   sum=0;
   scanf("%d%d",&n,&m);      // Input numbers of books and scribers
   for (int i=1;i<=n;i++)      // Input the numbers of pages, and
sum
   {
      scanf("%d",&a[i]);sum+=a[i];
   }
}
bool judge(long long lmt)      // determine whether the current
maximum number of pages assigned to a single scriber lmt is
feasible or not
{   // determine whether the i-th book needs a new scriber or
not: isn't more than lmt, every remainder scriber can be
assigned at least one book
// from back to front
   memset(flag,0,sizeof(flag));
   int cnt=m;      // start from the mth scriber
   long long now=0;      //number of pages for the current
scriber
   for (int i=n;i;i--)      // scan books
   {
      if ((now+a[i]>lmt)||(i<cnt))      // large than lmt, or
every remainder scriber can't be assigned at least one book
      {
         now=a[i];cnt--;flag[i]=true;      // add a new scriber
         if (cnt==0) return false;      //need more scribers, for
lmt
      }
```

```
      else now+=a[i];      // accumulation
  }
  return true;      // lmt is feasible
}

void work()     //calculate and output the solution to the
current test case
{
  long long l=0,r=sum,mid;      // initial interval [l, sum],
middle pointer mid
  for (int i=1;i<=n;i++) if (l<a[i]) l=a[i];      // the maximal
number of pages in these books
  while (l!=r)      //Binary search in [l, r]
  {
    mid=(l+r)>>1;      // middle pointer mid
    if (judge(mid)) r=mid;else l=mid+1;      // if mid is
feasible, left subinterval; else right subinterval
  }
  judge(l);      //calculate

  for(int i=1;i<=n;i++)      // output
  {
    printf("%d",a[i]);
    if (i<n) printf(" ");
    if (flag[i]) printf("/ ");
  }
  printf("\n");
}
int main()
{
  int t;
  scanf("%d",&t);      // the number of test cases
  for (int i=1;i<=t;i++)      // deal with every test case
  {
    init();      // the i-th test case
    work();      //calculate and output the solution to the
i-th test case
  }
  return 0;
}
```

5.4 Problems

5.4.1 Stripies

Our chemical biologists have invented a new very useful form of life called stripies (in fact, they were first called in Russian "polosatiki", but the scientists had to invent an English name to apply for an international patent). The stripies are transparent amorphous amoebiform creatures that live in flat colonies in a jelly-like nutrient medium. Most of the time the stripies are moving. When two of them collide, a new stripie appears instead. Long observations made by our scientists enabled them to establish that the weight of the new stripie isn't equal to the sum of the weights of the two disappeared stripies that collided; nevertheless, they soon learned that when two stripies of weights m_1 and m_2 collide, the weight of the resulting stripie equals $2 \times \text{sqrt}(m_1 \times m_2)$. Our chemical biologists are very anxious to know to what limits the total weight of a given colony of stripies can decrease.

You are to write a program that will help them to answer this question. You may assume that three or more stripies never collide together.

Input

The first line of the input contains one integer N ($1 \leq N \leq 100$)—the number of stripies in a colony. Each of the next N lines contains one integer ranging from 1 to 10000—the weight of the corresponding stripie.

Output

The output must contain one line with the minimal possible total weight of the colony with the accuracy of three decimal digits after the point.

Sample Input	Sample Output
3	120.00
72	
30	
50	

Source: ACM Northeastern Europe 2001, Northern Subregion

IDs for Online Judge: POJ 1862, ZOJ 1543, Ural 1161

Hint

Suppose that the weights of n stripies are m_1, m_2, ..., m_n, respectively. After $n-1$ collisions, the total weight of the colony is as follows.

$$W = 2^{n-1}\left((m_1 m_2)^{\frac{1}{2^{n-1}}} m_3 \frac{1}{2^{n-2}}\dots m_n^{\frac{1}{2}}\right).$$

Obviously, if m_1, m_2, ..., m_n are sorted in ascending order, the total weight of the colony W is minimal.

5.4.2 The Product of Digits

Your task is to find the minimal positive integer number Q so that the product of the digits of Q is exactly equal to N.

Input

The input contains the single integer number N ($0 \leq N \leq 10^9$).

Output

Your program should print to the output only the number Q. If such a number does not exist, print −1.

Sample Input	Sample Output
10	25

Source: USU Local Contest 1999

IDs for Online Judge: Ural 1014

 Hint

The criteria for factorization of N is to produce factors as big as possible.

There are two special cases: If $N==0$, then $Q=0$; and if $N==1$, then $Q=1$.

Otherwise, the greedy strategy is used as follows. N is factorized from 9 to 2. First, factors 9 are produced, as many as possible; second, factors 8 are produced, as many as possible; ......; and so on. If the final result for the factorization is not 1, then there is no solution; else Q is the positive integer that lists the factors from small to large.

5.4.3 Democracy in Danger

In one of the countries of the Caribbean basin, all decisions were accepted by the simple majority of votes at the general meeting of citizens (fortunately, there were not many of them). One of the local parties, aspiring to come to power as lawfully as possible, got its way in putting into effect some reform of the election system.

The main argument was that the population of the island recently had increased, and it was no longer easy to hold general meetings.

The essence of the reform is as follows. From the moment of the reform coming into effect, all the citizens were divided into *K* (maybe not equal) groups. Votes on every question were to be held then in each group; moreover, the group was said to vote "for" if more than half of the group had voted "for"; otherwise, it was said to vote "against". After the voting in each group, a number of the group that had voted "for" and "against" was calculated. The answer to the question was positive if the number of groups that had voted "for" was greater than the half of the general number of groups.

At first the inhabitants of the island accepted this system with pleasure. But when the first delights dispersed, some negative properties became obvious. It appeared that supporters of the party that had introduced this system could influence the formation of groups of voters. Due to this, they had an opportunity to put into effect some decisions without a majority of voters voting "for" it.

Let's consider three groups of voters, containing five, five, and seven persons, respectively. Then it is enough for the party to have only three supporters in each of the first two groups. So it would be able to put into effect a decision with the help of only six votes "for" instead of the nine that would be necessary in the case of general votes.

You are to write a program which would determine according to the given partition of the electors the minimal number of supporters of the party, sufficient for putting into effect of any decision, with some distribution of those supporters among the groups.

Input

In the first line, only an odd integer *K*—a quantity of groups—is written ($1 \leq K \leq 101$). In the second line, there are written *K* odd integers, separated with a space. Those numbers define a number of voters in each group. The population of the island does not exceed 9999 persons.

Output

You should write a minimal quantity of supporters of the party that can put into effect any decision.

Sample Input	Sample Output
3 5 7 5	6

Source: Autumn School Contest 2000

IDs for Online Judge: Ural 1025

Hint

K groups are sorted in ascending order of the numbers of voters in groups. There are K groups. Therefore, the party needs $\left\lfloor \dfrac{K}{2} \right\rfloor + 1$ groups voting "for". If there are n voters in a group, the party needs $\left\lfloor \dfrac{n}{2} \right\rfloor + 1$ supporters in the group. Therefore, the minimal quantity of supporters of the party is that there are just over half supporters for the party in the first $\left\lfloor \dfrac{K}{2} \right\rfloor + 1$ groups.

5.4.4 Box of Bricks

Little Bob likes playing with his box of bricks. He puts the bricks one upon another and builds stacks of different heights. "Look, I've built a wall!", he tells his older sister Alice. "Nah, you should make all stacks the same height. Then you would have a real wall", she retorts. After a little consideration, Bob sees that she is right. So he sets out to rearrange the bricks, one by one, such that all stacks are the same height afterwards. But since Bob is lazy, he wants to do this with the minimum number of bricks moved, as shown in Figure 5.5. Can you help?

Input

The input consists of several data sets. Each set begins with a line containing the number n of stacks Bob has built. The next line contains n numbers, the heights h_i of the n stacks. You may assume $1 \leq n \leq 50$ and $1 \leq h_i \leq 100$.

The total number of bricks will be divisible by the number of stacks. Thus, it is always possible to rearrange the bricks such that all stacks have the same height.

The input is terminated by a set starting with $n=0$. This set should not be processed.

Output

For each set, first print the number of the set, as shown in the sample output. Then print the line "The minimum number of moves is k.", where k is the minimum number of bricks that have to be moved in order to make all the stacks the same height.

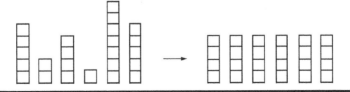

Figure 5.5

Output a blank line after each set.

Sample Input	Sample Output
6 5 2 4 1 7 5 0	Set #1 The minimum number of moves is 5.

Source: ACM Southwestern European Regional Contest 1997

IDs for Online Judge: POJ 1477, ZOJ 1251, UVA 591

Hint

$$\sum_{i=1}^{n} h_i$$

Suppose the average value of $avg = \dfrac{\displaystyle\sum_{i=1}^{n} h_i}{n}$. That is, avg is the heights of the n stacks

after the bricks are moved.

The criteria that bricks in the i-th stack should be moved is as follows. If $h_i > avg$, then $h_i - avg$ bricks should be moved in the stack. Therefore, the minimum number of bricks that have to be moved is $ans = \displaystyle\sum_{i=1}^{n} (h_i - avg | h_i > avg)$.

5.4.5 Minimal Coverage

Given several segments of line (in the X axis) with coordinates $[L_i, R_i]$, you are to choose the minimal number of them, such that they would completely cover the segment $[0, M]$.

Input

The first line is the number of test cases, followed by a blank line.

Each test case in the input should contain an integer M ($1 \leq M \leq 5000$), followed by pairs "L_i R_i"($|L_i|, |R_i| \leq 50000$, $i \leq 100000$), each on a separate line. Each test case of input is terminated by pair "0 0".

Each test case will be separated by a single line.

Output

For each test case, in the first line of output, your program should print the minimal number of line segments which can cover segment $[0, M]$. In the following lines, the coordinates of segments, sorted by their left end (L_i), should be printed in the same format as in the input. Pair "0 0" should not be printed. If $[0, M]$ cannot be covered by given line segments, your program should print "0" (without quotes).

Print a blank line between the outputs for two consecutive test cases.

Sample Input	Sample Output
2	0
1	1
−1 0	0 1
−5 −3	
2 5	
0 0	
1	
−1 0	
0 1	
0 0	

Source: USU Internal Contest March'2004

IDs for Online Judge: UVA 10020, Ural 1303

Hint

All segments are sorted in ascending order of left ends as the first key, and right ends as the second key ($(L_i \le L_{i+1} || ((L_i == L_{i+1}) \&\&(R_i < R_{i+1}))$, $1 \le i \le$ the number of segments -1).

The criteria for selecting segments is selecting a segment whose right end is the farthest among segments whose left ends are covered.

The greedy algorithm is as follows:

Suppose that *now* is the end position that the current segment covers; and *len* is the farthest position that a segment *k* whose left end is covered can reach. Initially *ans=now=len=*0.

Every segment in the sorted sequence is analyzed one by one:

```
if   ((Lᵢ ≤now)&&(len<Rᵢ))  {len= Rᵢ; k=i;}
if   ((Lᵢ₊₁ >now)&&(now<len))  {now=len; segment k is as a
new covered segment;}
       if (now≥m) output the result and exit;
If now<m after all segments are analyzed, then [0, M] cannot
be covered by given segments.
```

5.4.6 Annoying Painting Tool

Perhaps you wonder what an annoying painting tool is? First of all, the painting tool we speak of supports only black and white. Therefore, a picture consists of a rectangular area of pixels, which are either black or white. Second, there is only one operation that can change the color of pixels:

Select a rectangular area of *r* rows and *c* columns of pixels, which is completely inside the picture. As a result of the operation, each pixel inside the selected rectangle changes its color (from black to white, or from white to black).

Initially, all pixels are white. To create a picture, the operation described above can be applied several times. Can you paint a certain picture which you have in mind?

Input

The input contains several test cases. Each test case starts with one line containing four integers *n, m, r,* and *c*. ($1 \leq r \leq n \leq 100$, $1 \leq c \leq m \leq 100$), The following *n* lines each describe one row of pixels of the painting you want to create. The *i*-th line consists of *m* characters describing the desired pixel values of the *i*-th row in the finished painting ('0' indicates white, '1' indicates black).

The last test case is followed by a line containing four zeros.

Output

For each test case, print the minimum number of operations needed to create the painting, or −1 if it is impossible.

Sample Input	Sample Output
3 3 1 1	4
010	6
101	−1
010	
4 3 2 1	
011	
110	
011	
110	
3 4 2 2	
0110	
0111	
0000	
0 0 0 0	

Source: Ulm Local 2007

IDs for Online Judge: POJ 3363

Hint

The first thing to realize is that in an optimal solution, the painting operation is never applied more than once at the same position. Also, it doesn't matter in which order the operations are done; therefore, we can do the painting operations from top to bottom, and from left to right.

Using these ideas, we can easily check if a painting operation at some position is required or not. Since the pixel in the top left corner of a selected area for the painting operation will not be changed by later operations, we just check if it already has the required color. If its color still needs to be changed, we have to apply the painting operation.

After we have applied all the painting operations, we need to check the pixels in the rightmost $m-c$ columns and bottom $n-r$ rows to see if they have their required color. If one of these pixels doesn't have its required color, it is impossible to create the painting.

Since the size of the picture is at most 100×100, a naive implementation with $O(n^4)$ runs in time. There exists an optimal solution which runs in $O(n \times m)$. The idea is to store how many operations have been applied with the top left corner in one of the first i rows and j columns. With this stored data, it is possible to answer in constant time how many operations covering a pixel have been applied.

5.4.7 Troublemakers

Every school class has its troublemakers—those kids who can make the teacher's life miserable. On his own, a troublemaker is manageable, but when you put certain pairs of troublemakers together in the same room, teaching a class becomes very hard. There are *n* kids in Mrs. Shaida's math class, and there are *m* pairs of troublemakers among them. The situation has gotten so bad that Mrs. Shaida has decided to split the class into two classes. Help her do it in such a way that the number of troublemaker pairs is reduced by at least a half.

Input

The first line of input gives the number of cases, *N*. *N* test cases follow. Each one starts with a line containing *n* ($0 \leq n \leq 100$) and *m* ($0 < m < 5000$). The next *m* lines will contain a pair of integers *u* and *v* meaning that when kids *u* and *v* are in the same room, they make a troublemaker pair. Kids are numbered from 1 to *n*.

Output

For each test case, output one line containing "Case #*x*:" followed by *L*—the number of kids who will be moved to a different class (in a different room). The next line should list those kids. The total number of troublemaker pairs in the two rooms must be at most *m*/2. If that is impossible, print "Impossible." instead of *L* and an empty line afterwards.

Sample Input	Sample Output
2	Case #1: 3
4 3	1 3 4
1 2	Case #2: 2
2 3	1 2
3 4	
4 6	
1 2	
1 3	
1 4	
2 3	
2 4	
3 4	

Source: Abednego's Graph Lovers' Contest, 2006

IDs for Online Judge: UVA 10982

Hint

A graph is used to represent the problem, where kids in Mrs. Shaida's math class are represented as vertices, and there are edges between each pair of troublemakers. Mrs. Shaida splits the class into two classes, $s[0]$ and $s[1]$, where the number of kids in $s[1]$ is less than the number of kids in $s[0]$.

The method that Mrs. Shaida uses to split the class into two classes is as follows:

For kid i ($1 \leq i \leq n$), numbers of kids among kid 1 to kid $i-1$ who constitute a pair of troublemakers with kid i in $s[0]$ and $s[1]$ are calculated. If such a number in $s[1]$ is less than such a number in $s[0]$, the kid i is moved to $s[1]$; else the kid i stays in $s[0]$.

The greedy algorithm is as follows.

```
For (i=1; i≤n; i++)
    Calculate the numbers of vertices which connect with
vertice i in s[0] and s[1] from vertice 1 to vertice i-1;
    if (the number of such vertices in s[1]<the number of
such vertices in s[0])
        vertice i is moved to s[1];
Finally, vertices in s[1] corresponds to kids moved to a
different class (in a different room).
```

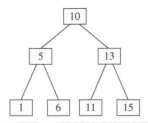

Figure 5.6

5.4.8 Constructing BST

BST (Binary Search Tree) is an efficient data structure for searching. In a BST, all the elements of the left subtree are smaller, and those of the right subtree are greater than the root. A typical example of BST is as shown in Figure 5.6.

Normally, we construct BST by successively inserting an element. In that case, the ordering of elements has great impact on the structure of the tree. Look at the following cases in Figure 5.7.

In this problem, you have to find the order of 1 to N integers such that the BST constructed by them has a height of at most H. The height of a BST is defined by the following relation:

1. A BST having no node has height 0.
2. Otherwise, it is equal to the maximum of the height of the left subtree and right subtree plus 1.

Again, several orders can satisfy the criterion. In that case, we prefer the sequence where smaller numbers come first. For example, for $N=4$, $H=3$, we want the sequence 1 3 2 4 rather than 2 1 4 3 or 3 2 1 4.

Input

Each test case starts with two positive integers N ($1 \leq N \leq 10000$) and H ($1 \leq H \leq 30$). Input is terminated by $N=0$, $H=0$. This case should not be processed. There can be at most 30 test cases.

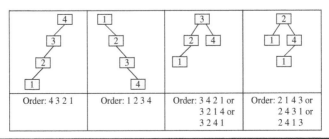

Figure 5.7

Output

The output of each test case should consist of a line starting with "Case #: " where # is the test case number. It should be followed by the sequence of N integers in the same line. There must not be any trailing space at the end of the line. If it is not possible to construct such a tree, then print "Impossible.". (without the quotes).

Sample Input	Sample Output
4 3	Case 1: 1 3 2 4
4 1	Case 2: Impossible.
6 3	Case 3: 3 1 2 5 4 6
0 0	

Source: ACM ICPC World Finals Warmup 1, 2005

IDs for Online Judge: UVA 10821

Hint

The problem requires you to output the Pre-order Traversal of a BST. Because smaller numbers come first in the sequence, the number for the root is as small as possible.

A BST with the height of at most H is constructed by the order of 1 to N integers. The number of nodes in its left subtree and right subtree is no more than $2^{H-1}-1$. The criteria for the number of the root is as follows:

If the right subtree is a full subtree, the number for the root is $N-(2^{H-1}-1)$; else the number for the root is 1.

Then the problem is transferred and a BST with the height of at most $H-1$ is constructed by the order of 1 to *root*−1 integers, the BST is as the left subtree; and a BST with the height of at most $H-1$ is constructed by the order of *root*+1 to N, the BST is as the right subtree.

Obviously the greedy algorithm is a recursive algorithm.

Program

```
#include<iostream>
#include<cstdio>
#include<cstdlib>
#include<cmath>
```

```
#include<cstring>
#include<algorithm>
using namespace std;
const int maxN=31;
int n,h,cnt,s[maxN];      //s[i]: the number of nodes for a full
binary treeth height i
//function work is an in order traversal for a BST with height
at most h, nodes from 1 to r
void work(int l,int r,int h)
{
  int m=max(l,r-s[h-1]);      //the number of root is as small
as possible
  printf("%d",m);
  if (++cnt<n) printf(" ");
  if (l<m) work(l,m-1,h-1);      // recursion for the left
subtree
  if (r>m) work(m+1,r,h-1);      // recursion for the right
subtree
}
int main()
{
  for (int i=0;i<=30;i++)  s[i]=(1<<i)-1;
  int t=0;
  while (scanf("%d%d",&n,&h),n+h)
  {
    cnt=0;
    printf("Case %d: ",++t);
    if (s[h]<n) printf("Impossible.");else work(1,n,h);
    printf("\n");
  }
  return 0;
}
```

5.4.9 Gone Fishing

John is going on a fishing trip. He has h hours available ($1 \leq h \leq 16$), and there are n lakes in the area ($2 \leq n \leq 25$), all reachable along a single, one-way road. John starts at lake 1, but he can finish at any lake he wants. He can only travel from one lake to the next one, but he does not have to stop at any lake unless he wishes to. For each $i=1, \ldots, n-1$, the number of five-minute intervals it takes to travel from lake i to lake $i+1$ is denoted t_i ($0 < t_i \leq 192$). For example, $t_3=4$ means that it takes 20 minutes to travel from lake 3 to lake 4. To help plan his fishing trip, John has gathered some information about the lakes. For each lake i, the number of fish expected to be caught in the initial five minutes, denoted as f_i ($f_i \geq 0$), is known. Each five minutes of fishing decreases the number of fish expected to be caught in the next five-minute interval by a constant rate of d_i ($d_i \geq 0$). If the number of fish expected to be caught in an interval is less than or equal to d_i, there will be no more fish left in the lake in the next interval. To simplify the planning, John assumes that no one else will be fishing at the lakes to affect the number of fish he expects to catch.

Write a program to help John plan his fishing trip to maximize the number of fish expected to be caught. The number of minutes spent at each lake must be a multiple of five.

Input

You will be given a number of cases in the input. Each case starts with a line containing n. This is followed by a line containing h. Next, there is a line of n integers specifying f_i ($1 \leq i \leq n$), then a line of n integers d_i ($1 \leq i \leq n$), and finally, a line of $n - 1$ integers t_i ($1 \leq i \leq n - 1$). Input is terminated by a case in which $n=0$.

Output

For each test case, print the number of minutes spent at each lake, separated by commas, for the plan achieving the maximum number of fish expected to be caught (you should print the entire plan on one line, even if it exceeds 80 characters). This is followed by a line containing the number of fish expected.

If multiple plans exist, choose the one that spends as long as possible at lake 1, even if no fish are expected to be caught in some intervals. If there is still a tie, choose the one that spends as long as possible at lake 2, and so on. Insert a blank line between cases.

Sample Input	Sample Output
2	45, 5
1	Number of fish expected: 31
10 1	
2 5	240, 0, 0, 0
2	Number of fish expected: 480
4	
4	115, 10, 50, 35
10 15 20 17	Number of fish expected: 724
0 3 4 3	
1 2 3	
4	
4	
10 15 50 30	
0 3 4 3	
1 2 3	
0	

Source: ACM East Central North America 1999

IDs for Online Judge: POJ 1042, UVA 757

Hint

Obviously, in the solution there is no turning back. That is, in John's fishing trip, if John fishes at a lake and leaves the lake, he can't go back to the lake.

Suppose John finishes the trip at lake *ed*. How can we calculate the maximum number of fish expected to be caught at lake *ed*?

The criteria for selecting a lake is as follows:

If the time is allowed, the lake in which there are a maximum number of fish is selected.

The greedy algorithm is as follows:

Initially, for lake *i*, the number of fish expected to be caught $f2[i]$ is the number of fish expected to be caught in the initial five minutes f_i, and the time that John fishes at the lake $tt[i]$ is 0, $1 \leq i \leq ed$. The time that John can fish is $h2 = h - \sum_{i=1}^{ed} t_i$, for there is no turning back in his fishing trip. The current number of fish to be caught $now=0$.

Then, for each terminal *ed*, repeat the following steps until $h2 \leq 0$:

```
Search a lake p in which there are a maximum number of fish,
that is, f2[p]= max {f2[i]};
              1≤i≤ed

         h2-=5;
         tt[p]+=5;
         now+=f2[p];
         the number of fish to be caught in lake p is adjusted
f2[p]=max(f2[p]-d_p, 0);
Finally, if (ans<now), then ans=now, and ans_tt[ ] is adjusted;
```

Obviously, after every terminal *ed* is enumerated ($1 \leq ed \leq n$), the number of minutes spent at each lake, and the plan for achieving the maximum number of fish expected to be caught can be computed. That is, *ans* is the maximum number of fish expected to be caught, and *ans_tt*[ ] is the number of minutes spent at each lake.

Program

```
#include<iostream>
#include<cstdio>
#include<cstdlib>
```

```
#include<cmath>
#include<cstring>
#include<algorithm>
using namespace std;
const int maxN=30;
int n,h,f[maxN],d[maxN],t[maxN];
int f2[maxN],tt[maxN],ans,ans_tt[maxN];     //f2 is the same as
f, tt is the time that John fishes at the lake; ans_tt is the
tt when ans is maximal
void init()
{
// Initialization
  ans=-1;
  memset(t,0,sizeof(t));
  memset(f,0,sizeof(f));
  memset(d,0,sizeof(d));
  memset(ans_tt,0,sizeof(ans_tt));
//Input
  scanf("%d",&h);h*=60;     //h is transferred into minutes
  for (int i=1;i<=n;i++) scanf("%d",&f[i]);
  for (int i=1;i<=n;i++) scanf("%d",&d[i]);
  for (int i=1;i<n;i++) {scanf("%d",&t[i]);t[i]+=t[i-1];}
}
//function work : calculate the maximum number of fish
expected to be caught at lake ed
void work(int ed)
{
  memcpy(f2,f,sizeof(f));
  memset(tt,0,sizeof(tt));
  int now=0,h2=h;     //now: the current number of fish to be
caught; h2: the time that John can fish
  h2-=t[ed-1]*5;     // the number of minutes spent from lake 1
to lake ed
  f2[0]=-1;
  while (h2>0)     // each while corresponds to five minutes of
fishing
  {
    int p=0;
    h2-=5;     // spend 5 minutes
    for (int i=1;i<=ed;i++)     // search a lake p in which
there are maximum number of fish
      if (f2[p]<f2[i]) p=i;
    tt[p]+=5;
    now+=f2[p];f2[p]=max(f2[p]-d[p],0);     // accumulation
  }
  if (ans<now) { ans=now;memcpy(ans_tt,tt,sizeof(tt));}
}
//output the result
void print()
{
```

```
  for (int i=1;i<=n;i++)
  {
    printf("%d",ans_tt[i]);
    if (i<n) printf(", ");
  }
  printf("\nNumber of fish expected: %d\n\n",ans);
}
int main()
{
  while (scanf("%d",&n),n)
  {
    init();
    for (int i=1;i<=n;i++)     // every terminal is enumerated
      work(i);
    print();
  }
  return 0;
}
```

Chapter 6

Practice for Dynamic Programming

Dynamic programming (DP) is used to solve optimization problems. DP breaks an optimization problem into a sequence of related subproblems, solves these subproblems just once, stores solutions to subproblems, and constructs an optimal solution to the problem, based on solutions to subproblems. The method for storing solutions to subproblems is called memorization. When the same subproblem occurs, its solution can be used directly.

There are two characteristics for a problem solved by DP:

1. Optimization. An optimal solution to a problem consists of optimal solutions to subproblems.
2. No aftereffect. A solution to a subproblem is only related to solutions to its direct predecessors.

In this chapter, DP experiments are organized as follows:

- Linear Dynamic Programming;
- Tree-Like Dynamic Programming;
- Dynamic Programming with State Compression.

6.1 Linear Dynamic Programming

6.1.1 Linear Dynamic Programming

Basic concepts for DP and the method for linear DP are as follows.

Stage k and State s_k: The solution to a problem is divided into k orderly and related stages. In a stage there are several states. State s_k is a state in stage k.

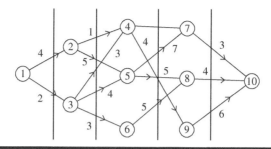

Figure 6.1

For example, Figure 6.1 shows a solution to a problem that is divided into five orderly and related stages. State 1 is called the initial state, in stage 1. State 10 is called the goal state, in stage 5. In stage 3 there are three states: state 4, state 5, and state 6.

Decision u_k and Available Decision Set $D_k(s_k)$: The choice from a state in stage $k-1$ (the current stage) to a state in stage k (the next stage) is called decision u_k. Normally, a state can be reachable through more than one decision from the last stage, and such decisions constitute an available decision set $D_k(s_k)$.

For example, there are two decisions reaching state 5: 2→5, 3→5, $D_3(5)=\{2, 3\}$. A decision sequence from the initial state to the goal state is called a strategy. For example, 1→3→5→8→10 is a strategy.

Successor Function and Optimization: A successor function is used to describe the transition from stage $k-1$ to stage k. The DP method is used to solve some optimization problems. Successor functions are used to find a solution with the optimal (minimum or maximum) value to a problem. A successor function can be formally defined as follows:

$$f_k(s_k) = \underset{u_k \in D_k(s_k)}{opt} \ g\big(f_{k-1}(T_k(s_k, u_k)), u_k\big);$$

where $T_k(s_k, u_k)$ is a state s_{k-1} in stage $k-1$ which relates to state s_k through decision u_k, and $f_{k-1}(T_k(s_k, u_k))$ is an optimal solution, $g(x, u_k)$ is a function for value x and decision u_k, that is, $g(f_{k-1}(T_k(s_k, u_k))$ is a function from state s_{k-1} to state s_k through decision u_k; *opt* means optimization; and $f_1(s_1)$ is an initial value. Because u_k is one decision in a decision set $D_k(s_k)$, all decisions are enumerated to get the optimal solution to s_k. From the initial state, successor functions are used to get the optimal solution f_n (goal state) to the problem finally.

If the stages are in linear order, linear DP is used to solve the problem.

```
for (every stage i is processed in linear order)

{
    for (every state j in stage i is enumerated (j∈S_i))
    { for (every state k in stage i-1 which is related to state
j is enumerated (k∈S_{i-1}))
```

```
    { calculate  f_i(j) =  opt  g(f_{i-1}(k) , u_k);  }
                          u_k ∈D_k(k)
  }
}
```

6.1.1.1 Brackets Sequence

Let us define a regular brackets sequence in the following way:

1. An empty sequence is a regular sequence.
2. If S is a regular sequence, then (S) and $[S]$ are both regular sequences.
3. If A and B are regular sequences, then AB is a regular sequence.

For example, all of the following sequences of characters are regular brackets sequences:

(), [], (()), ([]), () [], () [()]

And all of the following character sequences are not:

(, [,),)(, ([)], ([(

Some sequence of characters '(', ')', '[', and ']' is given. You are to find the shortest possible regular brackets sequence that contains the given character sequence as a subsequence. Here, a string $a_1 a_2 \ldots a_n$ is called a subsequence of the string $b_1 b_2 \ldots b_m$, if there exist such indices $1 \leq i_1 < i_2 < \ldots < i_n \leq m$, that $a_j = b_{i_j}$ for all $1 \leq j \leq n$.

Input

The input file contains at most 100 brackets (characters '(', ')', '[' and ']') that are situated on a single line without any other characters among them.

Output

Write to the output file a single line that contains some regular brackets sequence that has the minimal possible length and contains the given sequence as a subsequence.

Sample Input	Sample Output
([(	()[()]

Source: ACM Northeastern Europe 2001

IDs for Online Judges: POJ 1141, ZOJ 1463, Ural 1183, UVA 2451

Analysis

Suppose stage r is the length of subsequence, $1 \le r \le n$; and state i is the pointer pointing to the front of the current subsequence, $0 \le i \le n-r$. Based on i and r, the pointer j pointing to the rear of the current subsequence can be calculated, $j=i+r-1$. Suppose $dp[i, j]$ is the minimal number of characters that must be inserted into $s_i \ldots s_j$. Obviously, if the length of subsequence is 1, $dp[i, i]=1$, $0 \le i <$ strlen(s).

If $((s_i=='[')\&\&(s_j==']')||(s_i=='(')\&\&(s_j==')'))$, then the minimal number of characters that must be inserted into $s_i \ldots s_j$ is the minimal number of characters that must be inserted into $s_{i+1} \ldots s_{j-1}$, that is, $dp[i, j]=dp[i+1, j-1]$; otherwise $s_i \ldots s_j$ is divided into two parts, and we need to determine the pointer k ($i \le k < j$) so that $dp[i, j] = \min_{i \le k < j}(dp[i, k] + dp[k+1, j])$.

Based on the above, a memorized list $path[\ ][\ ]$ is used to store all solutions to subproblems:

$$
path[i][j] = \begin{cases} -1 & ((s_i == '[') \& \&(s_j == ']')) || ((s_i == '(') \& \&(s_j == ')')) \\ k & dp[i, j] = \min_{i \le k < j}(dp[i, k] + dp[k+1, j]) \end{cases}.
$$

After the memorized list $path[\ ][\ ]$ is calculated through DP, the regular brackets sequence that has the minimal possible length and contains the given sequence as a subsequence can be obtained through recursion.

Program

```
#include<cstdio>
#include<cstring>
const int N=100;
char str[N];      //Input String
int dp[N][N];
int path[N][N];
void oprint(int i,int j)      //output regular brackets sequence
containing subsequence str[i, j]
{
    if(i>j)
       return;
    if(i==j)      //there is only one character for subsequence
str[i, j]
       {
         if(str[i]=='['||str[i]==']')
```

```
            printf("[]");
          else
            printf("()");
      }
    else if(path[i][j]==-1)      // str[i] and str[j] are matched
brackets
        {
          printf("%c",str[i]);
          oprint(i+1,j-1);
          printf("%c",str[j]);
        }
      else      // otherwise
      {
        oprint(i,path[i][j]);
        oprint(path[i][j]+1,j);
      }
}
int main(void)
{
    while(gets(str))
      {
          int n=strlen(str);
          if(n==0)
            {
                printf("\n");
                continue;
            }
        memset(dp,0,sizeof(dp));
        for(int i=0;i<n;i++)
            dp[i][i]=1;
        for(int r=1;r<n;r++)      //Stage: r is the length of
subsequences
          {
              for(int i=0;i<n-r;i++)      //State: fronts of
subsequences are enumerated
                {
                    int j=i+r;      // rears of subsequences
                    dp[i][j]=0x7fffffff;      // Initialization
                    if((str[i]=='(' && str[j]==')') || (str[i]=='['
&& str[j]==']'))      // str[i] and str[j] are matched
                      {
                          dp[i][j]=dp[i+1][j-1];
                          path[i][j]=-1;
                      }
                    for(int k=i; k<j; k++)      // k is enumerated
                      {
                          if(dp[i][j]>dp[i][k]+dp[k+1][j])
                            {
                                dp[i][j]=dp[i][k]+dp[k+1][j];
                                path[i][j]=k;
```

```
                    }
                }
            }
        }
    oprint(0,n-1);        // Output the regular brackets sequence
    printf("\n");
    }
    return 0;
}
```

There are three classical problems solved by DP method: Subset Sum; Longest Common Subsequence (LCS); and Longest Increasing Subsequence(LIS).

6.1.2 Subset Sum

Suppose $S=\{x_1, x_2, ..., x_n\}$ is a set of non-negative integers, and c is a non-negative integer. The Subset Sum problem is to determine whether there is a subset of the given set with the sum equal to given c.

Coin counting is a classical problem for Subset Sum. Given a set of n non-negative integers $\{a_1, a_2, ..., a_n\}$ and a non-negative integer T, coin counting is to determine how many solutions to $k_1a_1+k_2a_2+...+k_n a_n=T$, where $k_1, k_2, ..., k_n$ are non-negative integers. DP can be used to solve the problem. Suppose $c(i, j)$ is the number of solutions to $k_1a_1+k_2a_2+...+k_ia_i=j$, $k_i>0$. Obviously the goal for coin counting is to calculate $c(n, T)$. In order to calculate $c(i, j)$, stage i is the first and i integers are used, $1\le i\le n$; states are $k_1a_1+k_2a_2+...+k_i a_i=j$, $a_i\le j\le T$. The successor function is as follows:

$$c(i, j) = \begin{cases} 1 & i = 0 \\ \displaystyle\sum_{k=1}^{i-1} c(k, j - a_i) & i \ge 1, j \ge a_i \end{cases}$$

The final solution is $c(n, T)$.

6.1.2.1 Dollars

New Zealand currency consists of $100, $50, $20, $10, and $5 notes and $2, $1, 50c, 20c, 10c and 5c coins. Write a program that will determine, for any given amount, in how many ways that amount may be made up. Changing the order of listing does not increase the count. Thus 20c may be made up in four ways: 1×20c, 2×10c, 10c+2×5c, and 4×5c.

Input

Input will consist of a series of real numbers no greater than $50.00 each on a separate line. Each amount will be valid, that is, it will be a multiple of 5c. The file will be terminated by a line containing zero (0.00).

Output

Output will consist of a line for each of the amounts in the input, each line consisting of the amount of money (with two decimal places and right-justified in a field of width 5), followed by the number of ways in which that amount may be made up, right-justified in a field of width 12.

Sample Input	Sample Output
0.20 2.00 0.00	0.20 4 2.00 293

Source: New Zealand Contest 1991

IDs for Online Judge: UVA 147

 Analysis

First, DP is used to calculate all solutions to the problem in the range. The 5c coin is the smallest coin. Other notes and coins for New Zealand currency are multiples for the 5c coin. Therefore, the 5c coin is used as the unit for notes and coins for New Zealand currency. Suppose $b[i]$ is the number of 5c coins for the i-th currency, $0 \leq i \leq 10$; $a[i, j]$ is the number of ways in which j 5c coins may be made up using the first i-th currencies, $0 \leq i \leq 10$, $0 \leq j \leq 6000$.

Obviously, the number of ways in which j 5c coins may be made up only using 5c coin is 1, that is, $a[0, j]=1$, $0 \leq j \leq 6000$. If the amount is equal to a coin or a note, there is a way that the amount may be made up using the coin or the note.

For 10 cents, there are two ways. 10 cents are made up using 5c coins or a 10c coin.

For 15 cents, the first way is that 15c cents are made up only using 5c coins. Then we calculate the number of ways in which 15 cents are made up using 5c coin and 10c coin (the way only using 5c coin needn't be considered). First a 10c coin is used (at least one 10c coin is used), and then a 5c coin is used. Therefore, there are two ways for 15 cents.

For 20 cents, the first case is that only 5c coins are used. For the second case, a 10c coin is used first (at least one 10c coin is used), and for the remaining 10 cents, there are two ways. The final case is that only the 20c coin is used. Therefore, there are four ways.

Based on the above, the number of ways in which j 5c coins may be made up using the first i-th currencies is based on the number of ways in which $j-b[i]$ 5c coins may be made up using the first $(i-1)$th currencies. That is,

$$a[i,j]=\sum_{k=0}^{i-1}a[k,j-b[i]]\Big|j\geq b[i].$$

Then, for each test case, the solution can be computed based on array a. For a real number n, the solution is $a[10, \lfloor n\times 20\rfloor]$.

The problem can also be solved by generation function.

 Program

```cpp
#include <iomanip>
#include <iostream>
using namespace std;
int main(void) {
    int b[] = {1, 2, 4, 10, 20, 40, 100, 200, 400, 1000, 2000};
//5c coin is used as the unit for notes and coins for
New Zealand currency
    long long a[6001] = {1};      // the number of ways in which n
5c coins may be made up using notes and coins for New Zealand
currency is a[n]
    //Off-line method, DP
    for (int i = 0; i < 11; i++){      // Enumerate all coins and
notes
        for (int j = b[i]; j < 6001; j++) {      // Enumerate
            a[j] += a[j - b[i]];
        }
    }
    cout << fixed << showpoint << setprecision(2);
    for (float fIn; cin >> fIn && fIn != 0; cout << endl) {
        cout << setw(6) << fIn << setw(17) << a[(int)(fIn * 20 +
0.5f)];
    }
    return 0;
}
```

6.1.3 Longest Common Subsequence (LCS)

For a sequence, elements in its subsequence appear in the same relative order, and are not necessarily contiguous. For example, for the string "abcdefg", "abc", "abg", "bdf", and "aeg" are all subsequences. And for strings "HIEROGLYPHOLOGY" and "MICHAELANGELO", string "HELLO" is a common subsequence.

Given two sequences of items, the Longest Common Subsequence (LCS) is to find the longest subsequence in both of them.

The LCS problem can be solved in terms of smaller subproblems. Given two sequences x and y, of length m and n respectively, the longest common subsequence z of x and y is found as follows:

Suppose sequence $x=<x_1, x_2, .., x_m>$, and the i-th prefix $x_i'==<x_1, x_2, .., x_i>$, $i=0,1,..,m$; sequence $y=<y_1, y_2, .., y_n>$, and the i-th prefix $y_i'==<y_1, y_2, .., y_i>$, $i=0,1,..,n$; and sequence $z=<z_1, z_2, .., z_k>$ is an LCS for x and y. For example, if $x=<A,B,C,B,D,A,B>$, then $x_4'=<A,B,C,B>$, and x_0' is an empty sequence.

Stage and state are pointer i for prefix of x and pointer j for prefix of y respectively. And x_{i-1} and y_{i-1} have been calculated through LCS. Decisions are made based on the following properties.

Property 1: If $x_m=y_n$, then $z_K=x_m=y_n$ and z_{k-1}' is an LCS for x_{m-1}' and y_{n-1}'.
Property 2: If $x_m \neq y_n$, then $z_K \neq x_m$, and z is an LCS for x_{m-1}' and y.
Property 3: If $x_m \neq y_n$, then $z_K \neq y_n$, and z is an LCS for x and y_{n-1}'.

Suppose $c[i, j]$ is the length of LCS for x_i' and y_j'.

$$c[i,j] = \begin{cases} 0 & i = 0 \text{ or } j = 0 \\ c[i-1, j-1]+1 & i, j > 0 \text{ and } x_i = y_j \\ \max\{c[i, j-1], c[i-1, j]\} & i, j > 0 \text{ and } x_i \neq y_j \end{cases}$$

The time complexity for calculating $c[i, j]$ is $O(n^2)$.

6.1.3.1 Longest Match

A newly opened detective agency is struggling with their limited intelligence to find out a secret information for passing technique among its detectives. Since they are new in this profession, they know well that their messages will easily be trapped and hence modified by other groups. They want to guess the intentions of other groups by checking the changed sections of messages. First, they have to get the length of the longest match. You are going to help them.

Input

The input file may contain multiple test cases. Each case will contain two successive lines of string. Blank lines and non-letter printable punctuation characters may appear. Each line of string will be no longer than 1000 characters. The length of each word will be less than 20 characters.

Output

For each case of input, you have to output a line starting with the case number right-justified in a field width of two, followed by the longest match, as shown in the sample output. In the case of at least one blank line for each input, output "Blank!". Consider the non-letter punctuation characters as white spaces.

Sample Input	Sample Output
This is a test. test Hello! The document provides late-breaking information late breaking.	1. Length of longest match: 1 2. Blank! 3. Length of longest match: 2

Source: TCL Programming Contest 2001

IDs for Online Judge: UVA 10100

Analysis

Consecutive letters in a string are regarded as a word. Words in two strings are gotten one by one, where words in the first string are stored in $T1.word[1]...T1.word[n]$, and words in the second string are stored in $T2.word[1]...T2.word[m]$.

Then every word is regarded as a "character". The LCS algorithm is used to calculate the Longest Common Subsequence (LCS). The length of the subsequence is the length of the longest match.

Program

```
#include<iostream>
#include<cstring>
#include<cstdio>
#include<string>
#include<algorithm>
#define N (1024)
using namespace std;
struct text{    // two successive lines of string
   int num;    // number of words
   string word[1024];    // words
}t1,t2;
string s1,s2;
int f[N][N];    //the number of matched words for the first
i-th words in s1 and the first j-th words in s2 is f[i, j]
void devide(string s,text &t)    // sequence of words t.word[]
whose length is t.num is taken out from s
{
```

```
    int l=s.size();      //the length of s
    t.num=1;
    for(int i=0;i<1000;i++) t.word[i].clear();
    for (int i=0;i<l;++i)
       if ('A'<=s[i] && s[i]<='Z' || 'a'<=s[i] && s[i]<='z'||'0'
<=s[i]&&s[i]<='9')
             t.word[t.num]+=s[i];
       else ++t.num;
    int now=0;
    for(int i=1;i<=t.num;i++)if(!t.word[i].empty())
t.word[++now]=t.word[i];
    t.num=now;
}
int main(void)
{
    int test=0;      //Initialization: the number of test cases
    while (!cin.eof())
    {
       ++test;
       getline(cin,s1);      // Input string s1
       devide(s1,t1);
       getline(cin,s2);      //Input string s2
       devide(s2,t2);
       printf("%2d. ",test);
       if(s1.empty() || s2.empty())
       {
          printf("Blank!\n");
          continue;
       }
       memset(f,0,sizeof(f));
       for (int i=1;i<=t1.num;++i)      // words in s1
          for (int j=1;j<=t2.num;++j)      //words in s2
          {      //Calculation
             f[i][j]=max(f[i-1][j],f[i][j-1]);
             if (t1.word[i]==t2.word[j])
                f[i][j]=max(f[i][j],f[i-1][j-1]+1);
          }
       printf("Length of longest match: %d\n",f[t1.num]
[t2.num]);      // Output result
    }
    return 0;
}
```

6.1.4 Longest Increasing Subsequence (LIS)

The Longest Increasing Subsequence (LIS) problem is to find the longest increasing subsequence of a given sequence. Given a real sequence $A=<a_1, a_2, ..., a_n>$, the Longest Increasing Subsequence for A is such a longest subsequence $L=<a_{k1}, a_{k2}, ..., a_{km}>$, where $k_1<k_2<...<k_m$ and $a_{k1}<a_{k2}<...<a_{km}$.

There are three DP methods to calculate LIS.

Method 1: A LIS problem is transformed into an LCS problem.

A LIS problem can be transformed into an LCS problem. Suppose $X=<b_1, b_2, \ldots, b_n>$ is a sorted sequence in ascending order for $A=<a_1, a_2, \ldots, a_n>$. Obviously the LCS for X and A is the LIS for A.

The time complexity for sorting A is $O(n\log_2(n))$. The time complexity for calculating the LCS for X and A is $O(n^2)$. Therefore, the time complexity for **Method 1** is $O(n\log_2(n)+n^2)$.

Method 2: DP method.

Suppose $f(i)$ is the length of the LIS for the subsequence in A whose rear is a_i. Obviously,

```
f(1)=1
f(i)=max {f(j)|a_j<a_i}+1
     1≤j≤i-1
f(n) is the length of the LIS for A. Obviously, the time
complexity using the DP method is O(n²).
```

Method 3: Dichotomy.

For **Method 2**, in order to calculate $f(i)$, the maximal $f(j)(j<i)$ must be found. Array B is used to store the rear for LIS of subsequences, that is, $B[f(j)]=a_j$. When $f(i)$ is calculated, dichotomy is used to find j in array B where $j<i$ and $B[f(j)]=a_j<a_i$. Then $B[f(j)+1]=a_i$.

Experiments for the three DP methods are as follows.

6.1.4.1 History Grading

Many problems in computer science involve maximizing some measure according to constraints. Consider a history exam in which students are asked to put several historical events into chronological order. Students who order all the events correctly will receive full credit, but how should partial credit be awarded to students who incorrectly rank one or more of the historical events?

Some possibilities for partial credit include:

1. One point for each event whose rank matches its correct rank;
2. One point for each event in the longest (not necessarily contiguous) sequence of events which are in the correct order relative to each other.

For example, if four events are correctly ordered 1 2 3 4, then the order 1 3 2 4 would receive a score of 2 using the first method (events 1 and 4 are correctly ranked) and a score of 3 using the second method (event sequences 1 2 4 and 1 3 4 are both in the correct order relative to each other).

In this problem, you are asked to write a program to score such questions using the second method.

Given the correct chronological order of n events 1, 2, ..., n as $c_1, c_2, ..., c_n$ where $1 \le c_i \le n$ denotes the ranking of event i in the correct chronological order and a sequence of student responses $r_1, r_2, ..., r_n$ where $1 \le r_i \le n$ denotes the chronological rank given by the student to event i; determine the length of the longest (not necessarily contiguous) sequence of events in the student responses that are in the correct chronological order relative to each other.

Input

The first line of the input will consist of one integer n indicating the number of events with $2 \le n \le 20$. The second line will contain n integers, indicating the correct chronological order of n events. The remaining lines will each consist of n integers with each line representing a student's chronological ordering of the n events. All lines will contain n numbers in the range [1..n], with each number appearing exactly once per line, and with each number separated from other numbers on the same line by one or more spaces.

Output

For each student ranking of events, your program should print the score for that ranking. There should be one line of output for each student ranking.

Sample Input 1	Sample Output 1
4	1
4 2 3 1	2
1 3 2 4	3
3 2 1 4	
2 3 4 1	

Sample Input 2	Sample Output 2
10	6
3 1 2 4 9 5 10 6 8 7	5
1 2 3 4 5 6 7 8 9 10	10
4 7 2 3 10 6 9 1 5 8	9
3 1 2 4 9 5 10 6 8 7	
2 10 1 3 8 4 9 5 7 6	

Source: Internet Programming Contest 1991

IDs for Online Judge: UVA 111

Analysis

Suppose *st*[ ] is the correct chronological order of *n* events, where *st*[*t*] is the *t*-th event in the chronological order; *ed*[ ] is the current student's chronological ordering of the *n* events, where *ed*[*t*] is the *t*-th event in the current student's chronological order.

Obviously, the Longest Common Subsequence (LCS) for *st*[ ] and *ed*[ ] is the Longest Increasing Subsequence (LIS) for *ed*[ ], where its length is the score for that ranking. **Method 1** is used to solve the problem.

 Program

```
#include<iostream>
#include<cstring>
#include<cstdio>
using namespace std;
int n;      //number of events
int f[30][30];
int st[30];      // st[t] is the t-th event in the chronological
order
int ed[30];      // ed[t] is the t-th event in the current
student's chronological order
int tmp[30];
int main(void)
{
  freopen("111.in","r",stdin);
  freopen("HG.out","w",stdout);
  scanf("%d",&n);      // Input number of events
  for(int i=1;i<=n;++i)      // Input the correct chronological
order of n events
  {
    cin >> tmp[i];
    st[tmp[i]]=i;
  }
  while(!cin.eof())      //Input students' chronological
ordering of the n events
  {
    for(int i=1;i<=n;++i)      // Input current student's
chronological ordering of the n events
    {
```

```
        cin >> tmp[i];
        ed[tmp[i]]=i;
    }
    if(cin.eof()) break;
    memset(f,0,sizeof(f));
    for(int i=1;i<=n;++i)      //Calculate the LCS for st[ ]
and ed[ ]
        for(int j=1;j<=n;++j)
        {
            f[i][j]=max(f[i-1][j],f[i][j-1]);
            if(st[i]==ed[j])
                f[i][j]=max(f[i][j],f[i-1][j-1]+1);
        }
    cout << f[n][n] << endl;    //Output the current
student's score
    }
    return 0;
}
```

6.1.4.2 Ski

Michael likes to ski. Skiing is really exciting for him. In order to get speed, the ski area must be down. When he skis down to the bottom, he has to walk up the hill again or wait for the lift to carry him. Michael wants to know the longest skidway in a ski area. The ski area is given by a two-dimensional array. Each digit of the array represents the height of the point. There is an example as follows:

1	2	3	4	5
16	17	18	19	6
15	24	25	20	7
14	23	22	21	8
13	12	11	10	9

From a point, he can ski to one of four adjacent points (up, down, left, and right), if and only if the height of an adjacent point is less than the height of the point. In the above example, a viable skidway is 24–17–16–1. Obviously 25–24–23–...–3–2–1 is the longest viable skidway.

Input

Row R and column C for the ski area are shown in the first line ($1 \leq R, C \leq 100$). Then there are R rows, and in each row there are C integers representing the height of points h, where $0 \leq h \leq 10000$.

Output

Output the length of the longest viable skidway that Michael can ski.

Sample Input	Sample Output
5 5 1 2 3 4 5 16 17 18 19 6 15 24 25 20 7 14 23 22 21 8 13 12 11 10 9	25

Source: SHTSC 2002 (Problemsetter: Yongji Zhou)

IDs for Online Judge: POJ 1088

 Analysis

The problem requires you to calculate the length of the longest viable skidway whose points are adjacent and in descending order. The skidway is the Longest Decreasing Subsequence, if heights are as keys. **Method 2** is used to solve the problem. Suppose $f[\ ][\ ]$ is visited marks, if point (x, y) is in the skidway, then $f[x][y]$=true; and $c[\ ][\ ]$ is the successor function, where $c[x][y]$ is the longest viable skidway which starts from (x, y):

$$c[x][y] = \begin{cases} 1 & \textit{Initialization} \\ \max\{c[xx][yy]+1\} & (xx, yy) \text{ is adjacent to } (x, y), \text{ unvisited, and lower.} \end{cases}$$

Because the start point isn't given in the problem, the length of the longest viable skidway $ans = \max_{(1 \le x \le n, 1 \le y \le m)}\{c[x][y]\}$.

 Program

```
#include<cstdio>
using namespace std;
int n,m,s1[5],s2[5],i,j,ans;      //size of the ski area is n*m;
the length of the longest viable skidway is ans
int a[105][105],c[105][105];      // adjacency matrix for ski
area a[ ][ ], state transition equation c[ ][ ]
bool f[105][105];      //visited mark
```

```
void work(int x,int y){      //recursively calculate the length
of the longest viable skidway c[x][y] starting from (x, y)
                int i,xx,yy;
                f[x][y]=true;       //Set visited mark for (x, y)
                c[x][y]=1;      // Initialization
                for (i=1;i<=4;i++){     // 4 adjacent points
                        xx=x+s1[i];yy=y+s2[i];     // (xx, yy):
adjacent point in direction i
                        if (a[xx][yy]<a[x]
[y]&&xx>0&&xx<=n&&yy>0&&yy<=m){      // (xx, yy) is in the area,
can be skied down from (x, y), isn't visited
                        if (!f[xx][yy]) work(xx,yy);
                        // adjustment
                        c[x][y]=c[x][y]>(c[xx][yy]+1)?c[x][y]:(c[xx]
[yy]+1);}
                }
        }
int main(){
    s1[1]=-1; s2[1]=0;
    s1[2]=1; s2[2]=0;
    s1[3]=0; s2[3]=-1;
    s1[4]=0; s2[4]=1;
    scanf("%d%d",&n,&m);      //numbers of rows and columns
    for (i=1;i<=n;i++)      // heights of points
        for (j=1;j<=m;j++)scanf("%d",&a[i][j]);
    ans=0;      // Initialization
    for (i=1;i<=n;i++)
     for (j=1;j<=m;j++)
        if (!f[i][j]) {work(i,j); ans=ans>c[i][j]?ans:c[i][j];}
    printf("%d\n",ans);      // output result
    return 0;
}
```

6.1.4.3 Wavio Sequence

A Wavio sequence is a sequence of integers. It has some interesting properties:

- Wavio is of odd length, i.e., $L=2\times n+1$.
- The first $(n+1)$ integers of Wavio sequence make a strictly increasing sequence.
- The last $(n+1)$ integers of Wavio sequence make a strictly decreasing sequence.
- No two adjacent integers are same in a Wavio sequence.

For example 1, 2, 3, 4, 5, 4, 3, 2, 0 is a Wavio sequence of length 9. But 1, 2, 3, 4, 5, 4, 3, 2, 2 is not a valid Wavio sequence. In this problem, you will be given a sequence of integers. You have to find out the length of the longest Wavio sequence which is a subsequence of the given sequence. Consider the given sequence as:

1 2 3 2 1 2 3 4 3 2 1 5 4 1 2 3 2 2 1.

Here the longest Wavio sequence is : 1 2 3 4 5 4 3 2 1. So, the output will be 9.

Input

The input file contains less than 75 test cases. The description of each test case is given below. Input is terminated by end of file.

Each set starts with a positive integer, $N(1 \leq N \leq 10000)$. In the next few lines there will be N integers.

Output

For each set of input, print the length of the longest Wavio sequence in a line.

Sample Input	Sample Output
10	9
1 2 3 4 5 4 3 2 1 10	9
19	1
1 2 3 2 1 2 3 4 3 2 1 5 4 1 2 3 2 2 1	
5	
1 2 3 4 5	

Source: The Diamond Wedding Contest: Elite Panel's 1st Contest 2003

IDs for Online Judge: UVA 10534

 Analysis

Suppose the sequence of integers is $A=a_1...a_n$; $LIS[k]$ is the length of the Longest Increasing Subsequence in $[a_1...a_k]$; and $LDS[k]$ is the length of the Longest Decreasing Subsequence in $[a_k...a_n]$.

First, **Method 3** is used to calculate the length of the Longest Increasing Subsequence in the prefix for A. If the prefix's rear is a_i, the length is $f[i]$, $1 \leq i \leq k$. Therefore, $LIS[k] = \max_{1 \leq i \leq k}\{f[i]\}$.

Second, **Method 3** is used to calculate the length of the Longest Decreasing Subsequence in the postfix for A. If the postfix's front is a_i, the length is $f[i]$, $k \leq i \leq n$. Therefore, $LDS[k] = \max_{k \leq i \leq n}\{f[i]\}$.

If k is the pointer pointing to the middle of a Wavio sequence, that is, the number of integers in the left half and the number of integers in the right half is $\min\{LIS[k], LDS[k]\}$. The length of the Wavio sequence is $ans[k]=2\times\min\{LIS[k], LDS[k]\}-1$.

The length of the longest Wavio sequence is $ans = \max_{1 \leq k \leq n}\{ans[k]\}$.

Program

```
#include<cstdio>
#include<cstring>
using namespace std;
const int MAXN = 10010,INF = 2147483647;
int N,A[MAXN],F[MAXN],G[MAXN],L[MAXN];     // N: the number of
integers, A[ ] is the given sequence, L[ ]: increasing
sequence, F[ ] is as LIS[ ], G[ ] is as LDS[ ]
int binary(int l,int r,int x)     // return the number of
elements in L[l, r] which are less than x
{
  int mid;
  l = 0; r = N;
  while (l<r)
  {
    mid = (l+r)>>1;
    if (L[mid+1]>=x) r = mid; else l = mid+1;
  }
  return l;
}
inline int min(int x,int y) { return (x<y) ? (x) : (y); }
// return min{x, y}
int main()
{
  int i,j,k,Ans;
  while (scanf("%d",&N) != EOF)
  {
    for (i=1;i<=N;i++) scanf("%d",A+i);     // Input N
integers
    for (i=1;i<=N;i++) L[i]=INF; L[0]=-INF-1;
// Initialization
    for (i=1;i<=N;i++)     //Right to left in array A
    {
      F[i]=binary(1,N,A[i])+1;
      if (A[i]<L[F[i]]) L[F[i]]=A[i];
    }
    for (i=1;i<=N;i++) L[i]=INF; L[0]=-INF-1;     //
Initialization
    for (i=N;i>=1;i--)     // Left to right in array A
    {
      G[i]=binary(1,N,A[i])+1;
      if (A[i] < L[G[i]]) L[G[i]]=A[i];
    }
    Ans=0;
```

```
    for (i=1;i<=N;i++)    // every element in A[ ] as the
middle, and adjust
        if ((k = min(F[i],G[i])) > Ans) Ans = k;
    printf("%d\n",Ans*2-1);    // Output the result
  }
  return 0;
}
```

6.2 Tree-Like Dynamic Programming

If the background or the relationships between stages for a DP problem are represented as a tree, tree-like DP can be used to solve such problems.

Tree-like DP is different from linear DP:

1. The calculation sequences are different. There are two calculation sequences for linear DP: forward and backward. The calculation sequence for tree-like DP is normally from leaves to the root, and the root is the solution.
2. The calculation methods are different. A traditional iteration method is used in linear DP. The recursive method is used in tree-like DP, for tree-like DP is normally implemented by memorized search.

In this section, two problems for tree-like DP are shown.

6.2.1 Binary Apple Tree

Let's imagine how an apple tree looks in the binary computer world. You're right, it looks just like a binary tree, i.e., any biparous branch splits up to exactly two new branches. We will enumerate by integers the root of a binary apple tree, points of branching, and the ends of twigs. In this way, we may distinguish different branches by their ending points. We will assume that the root of the tree always is numbered by 1, and all numbers used for enumerating are numbered in range from 1 to *N*, where *N* is the total number of all enumerated points. For instance, in Figure 6.2, *N* is equal to 5. Figure 6.2 is an example of an enumerated tree with four branches.

As you may know, it's not convenient to pick an apple from a tree when there are too many branches. That's why some of them should be removed from a tree. But you are interested in removing branches in order to achieve a minimal loss of

---

Figure 6.2

apples. So you are given numbers of apples on a branch and the number of branches that should be preserved. Your task is to determine how many apples can remain on a tree after the removal of excessive branches.

Input

The first line of input contains two numbers: N and Q ($2 \leq N \leq 100$; $1 \leq Q \leq N-1$). N denotes the number of enumerated points in a tree. Q denotes the number of branches that should be preserved. The next $N-1$ lines contain descriptions of branches. Each description consists of three integer numbers divided by spaces. The first two of them define a branch by its ending points. The third number defines the number of apples on this branch. You may assume that no branch contains more than 30000 apples.

Output

Output should contain only the number—the number of apples that can be preserved. And don't forget to preserve the tree's root.

Sample Input	Sample Output
5 2 1 3 1 1 4 10 2 3 20 3 5 20	21

Source: Ural State University Internal Contest '99 #2

IDs for Online Judge: Ural 1018

 Analysis

In this problem, the apple tree is a weighted binary tree. The problem requires you to get such a subtree with Q branches (i.e., $Q+1$ points) whose weight is maximal. For each internal point, there are three choices: pruning its left subtree, pruning its right subtree, or pruning some points in its left subtree and its right subtree; to get a subtree with maximal weight.

Suppose $g[x][k]$ is the maximal weight for the subtree with root x in which there are k points (including the weight of the branch from root x to its parent). For each leaf, DP is used in the order of post-order traversal. The successor function for DP is as follows:

If x is a leaf, then $g[x][k]=$ the weight of the edge from x to its parent; else all cases where $k-1$ nodes are distributed in its left subtree and its right subtree are enumerated, and the best case is found. That is,

$$
g[x][k] = \begin{cases}
0 & k = 0 \\
\text{the weight of the edge from } x \text{ to its parent} & x \text{ is a leaf} \\
\text{the weight of the edge from } x \text{ to its parent} + & x \text{ isn't a leaf} \\
\quad \max_{0 \le k \le k-1} \{g[\text{the left child for } x][i] + \\
\quad g[\text{the right child for } x][k-i-1]\}
\end{cases}
$$

DP is used bottom-up until the root. Finally, *ans*=$g[root][Q+1]$.

Program

```
#include <cstdio>
#include <cstdlib>
#include <cstring>
#define Max(a,b) ((a)>(b)?(a):(b))
#define N (256)
using namespace std;
int n,m,ne,x,y,z;    //n: number of points, m: amount of
preserve branches, that should be preserved, ne: the number of
a branch, (x,y): a branch, z: weight
int id[N],w[N],v[N],next[N],head[N],lch[N],rch[N],f[N];
int g[N][N];    // Successor Function
void add(int x,int y,int z)    //add branch (x, y) with weight
z into adjacency list
{
   id[++ne]=y; w[ne]=z; next[ne]=head[x]; head[x]=ne;
}
void dfs(int x)    //a binary tree is constructed from point x
{
   for (int p=head[x];p;p=next[p])    //search every branch p
connecting x
      if (id[p]!=f[x])
{
      if (!lch[x]) lch[x]=id[p]; else rch[x]=id[p];
      f[id[p]]=x;
```

```
        v[id[p]]=w[p];
        dfs(id[p]);
    }
}
int dp(int x,int k)      //from x, the subtree with k points and
maximal number of apples
{
    if (!k) return 0;       //subtree is empty
    if (g[x][k]>=0) return g[x][k];     // return the result
    if (!lch[x]) return (g[x][k]=v[x]);    // x is a leaf
    for (int i=0;i<k;++i)      // calculate the best case
        g[x][k]=Max(g[x][k],dp(lch[x],i)+dp(rch[x],k-i-1));
    g[x][k]+=v[x];
    return g[x][k];
}
int main()
{
    scanf("%d%d",&n,&m);
    for (int i=1;i<n;++i)
    {
        scanf("%d%d%d",&x,&y,&z);
        add(x,y,z);
        add(y,x,z);
    }
    dfs(1);
    memset(g,255,sizeof(g));
    printf("%d\n",dp(1,m+1));
    return 0;
}
```

6.2.2 Anniversary Party

The president of the Ural State University is planning an eightieth anniversary party. The university has a hierarchical structure of employees; that is, the supervisor relation forms a tree rooted at the president. Employees are numbered by integer numbers in a range from 1 to N. The personnel office has ranked each employee with a conviviality rating. In order to make the party fun for all attendees, the president does not want both an employee and his or her immediate supervisor to attend.

Your task is to make up a guest list with the maximal conviviality rating of the guests.

Input

The first line of the input contains a number N. $1 \le N \le 6000$. Each of the subsequent N lines contains the conviviality rating of the corresponding employee.

Conviviality rating is an integer number in a range from −128 to 127. After that, the supervisor relation tree goes. Each line of the tree specification has the form

    $<L><K>$

which means that the K-th employee is an immediate supervisor of the L-th employee. Input is ended with the line

    0 0

Output

The output should contain the maximal total rating of the guests.

Sample Input	Sample Output
7	5
1	
1	
1	
1	
1	
1	
1	
1 3	
2 3	
6 4	
7 4	
4 5	
3 5	
0 0	

Source: Ural State University Internal Contest October 2000 Students Session

IDs for Online Judge: POJ 2342, Ural 1039

 Analysis

The supervisor relation in Ural State University forms a tree rooted at the president. For each internal node u in the tree, there are two possible conviviality ratings of the subtree rooted at u:

1. Conviviality rating including node u;
2. Conviviality rating not including node u.

If the maximal conviviality rating of the subtree doesn't include node u, then the maximal conviviality rating of the subtree is the sum of the maximal conviviality ratings of all subtrees for node *u* (subtrees root at *u*'s children), and for such subtrees, their maximal conviviality ratings also have two cases: including their roots or not including their roots.

If the maximal conviviality rating of the subtree includes node *u* is the maximal conviviality rating of the subtree doesn't include node *u* plus conviviality rating for node *u*. Suppose $F[u][0]$ is the maximal conviviality rating of the subtree rooted at *u* which doesn't include node *u*; and $F[u][1]$ is the maximal conviviality rating of the subtree rooted at *u* which includes node *u*. Initially, $F[u][0]=0$, $F[u][1]=u$, $1\leq u\leq n$. Then from leaf nodes, based on post-order traversal, the successor function is calculated as follows:

$$F[u][0] = \sum_{v \in u\text{'s children}} \max\{F[v][0], F[v][1]\} ;$$

```
F[u] [1] = F[u] [1] ( the conviviality rating of u)+F[u] [0].
```

The successor function is calculated until *root*. Finally, *ans*=max{$F[root][0]$, $F[root][1]$}.

 Program

```
#include<cstdio>
#include<cstring>
using namespace std;
const int MAXN = 6010;      //the upper limit of the number
of nodes
int N,root,Ri[MAXN],F[MAXN][2],son[MAXN],bro[MAXN];
// successor function F[] []
bool is_son[MAXN];
void init()     //Input and construct the adjacency list for
the tree
{
     int i,j,k;
     scanf("%d",&N);      //number of employees
     for (i=1;i<=N;i++) scanf("%d",Ri+i);     // the
conviviality rating of employee
     memset(son,0,sizeof(son));
memset(is_son,0,sizeof(is_son));
     for (i=1;i<N;i++)
  {
     scanf("%d%d",&j,&k);     //k is the immediate supervisor
for j
     bro[j] =son[k]; son[k] = j;     //j is added into the
adjacency list for k
```

```
        is_son[j] = true;      // j has parent
    }
    for (i=1;i<=N;i++)
        if (!is_son[i]) root = i;
}
inline int max(int x,int y) { return (x>y)?(x):(y); }
void DP(int u)    // Dynamic Programming on a Tree, F[u][0]
and F[u][1] are calculated
{
    int v;
    F[u][0] = 0; F[u][1] = Ri[u];    // Initialization
    for (v=son[u]; v; v=bro[v])    //u's every subtree
    {
        DP(v);
        F[u][0]+=max(F[v][0],F[v][1]);
        F[u][1]+=F[v][0];
    }
}
void solve()    //Calculate the maximal total rating and
output
{
    DP(root);
    printf("%d\n",max(F[root][0],F[root][1]));
}
int main()
{
    init();
    solve();
    return 0;
}
```

6.3 Dynamic Programming with State Compression

In some problems, each constituent part for a state can be represented as 0 or 1, and states can be represented as strings for 0 and 1. For example, grids in a chessboard can be represented by a string. And states for a chessboard can be represented as strings. We call this state compression. DP with state compression can be implemented by bitwise operations.

6.3.1 Nuts for Nuts

Ryan and Larry decided that some nuts don't really taste so good, but they realized that there are some nuts located in certain places of the island.. and they love them! Since they're lazy, but greedy, they want to know the shortest tour that they can use to gather every single nut!

Can you help them?

Input

You'll be given x, and y, both less than 20, followed by x lines of y characters each as a map of the area, consisting solely of ".", "#", and "L". Larry and Ryan are currently located in "L", and the nuts are represented by "#". They can travel in all eight adjacent directions in one step. See below for an example. There will be at most 15 places where there are nuts, and "L" will only appear once.

Output

On each line, output the minimum number of steps starting from "L", gather all the nuts, and back to "L".

Sample Input	Sample Output
5 5	8
L....	8
#....	
#....	
.....	
#....	
5 5	
L....	
#....	
#....	
.....	
#....	

Source: UVA Local Qualification Contest, 2005

IDs for Online Judge: UVA 10944

Analysis

The places where Ryan and Larry and all nuts locate are as vertices. Their coordinates are recorded, where (x_0, y_0) is the place where Larry and Ryan are currently located, and (x_i, y_i) is the place where the i-th nut is located, $1 \leq i \leq n$. $Map[\ ][\ ]$ is used to represent distances between vertices, where $Map[i][j] = max\{|x_i - x_j|, |y_i - y_j|\}$ for vertex i and vertex j.

The state that nuts are gathered is represented as a n bit binary number (b_{n-1}, ..., b_0), where $b_i=0$ means the $(i+1)$-th nut isn't gathered, and $b_i=1$ means the $(i+1)$-th nut is gathered. Suppose j is the current state value that nuts are gathered, where nut i is the nut that is gathered finally, and the minimum number of steps is $f[i][j]$. Obviously the minimum number of steps that Ryan and Larry gather for every nut is $f[i][2^{i-1}]=map[0][i](1\leq i\leq n)$. Suppose the state the nuts are gathered currently is j, where the number of gathered nuts is i, and the minimum number of steps is $f[i][j]$. Obviously, the minimum number of steps that Ryan and Larry gather for each nut is $f[i][2^{i-1}]=map[0][i]$, $1\leq i\leq n$. The successor function is analyzed as follows:

Stage i; states are enumerated in ascending order, $0\leq i\leq 2^n-1$;
State j; The last gathered nut j in stage i is enumerated, $1\leq j\leq n$, i & $2^{j-1}\neq 0$;
Decision k: Nut k which isn't in stage i is enumerated ($1\leq k\leq n,i$ & $2^{k-1}==0$), and determine whether gathering nut k is better. If gathering nut k is better, $f[k][i+2^{k-1}]$ is adjusted, that is,

$$f[k][i+2^{k-1}]=min\{f[k][i+2^{k-1}],f[j][i]+map[j][k]\}$$

After n nuts are gathered, if nut i is the last gathered nut, the minimum number of steps to reach the position of nut i is $f[i][2^n-1]$, the number of steps back to "L" is $map[0][i]$. The number of steps is $f[i][2^n-1]+map[0][i]$.

Finally, all results are compared $i(1\leq i\leq n)$, the minimum number of steps starting from "L", gather all the nuts, and back to "L" is:

$$ans=\min_{1\leq i\leq n}\{f[i][2^n-1]+map[0][i]\}.$$

Program

```
#include <cstdio>
#include <cstring>
#define Max(a,b) ((a)>(b)?(a):(b))
#define Inf (1<<20)
#define N (30)
#define M (65536)

using namespace std;

int f[N][M];    // nuts are gathered currently is j, where the
number of gathered nuts is i, and the minimum number of steps
is f[i][j]
char s[N];    //current row
int map[N][N];    // distance between vertice i and vertex j
is map[i, j]
int x[N],y[N];    //The sequence of vertices' coordinates
int num,n,m,ans,maxz;    //num: number of nuts, n*m: the size
of the map, ans: the minimum amount of steps starting from
```

"L", gather all the nuts, and back to "L", *maxz*: the state that all nuts are gathered

```
int Abs(int x) { if (x>0) return x; return -x; }  //|x|
void Update(int &x,int y) { if (x>y) x=y; }   //x←max{x,y}

int main()
{
  while (scanf("%d%d",&n,&m)!=EOF)    //Input the size of the
map
  {
    num=0;
    for (int i=0;i<n;++i)    // Input every row, calculate
the number of nuts, set up the sequence of vertices'
coordinates, where (x[0], y[0]) is the position where Larry
and Ryan are currently located, (x[1...num], y[1...num]) are
positions for num nuts
    {
      scanf("%s",s);
      for (int j=0;j<m;++j)
      if (s[j]=='#') { x[++num]=i; y[num]=j; } else
      if (s[j]=='L') { x[0]=i; y[0]=j; }
    }
    if (!num) {printf("0\n"); continue; }
    for (int i=0;i<=num;++i)    //Calculate distances between
vertices
for (int j=0;j<=num;++j)
map[i][j]=Max(Abs(x[i]-x[j]),Abs(y[i]-y[j]));
    maxz=(1<<num)-1;    // Calculate the state that all nuts
are gathered
    for (int i=0;i<=maxz;++i)    // Initialize successor
function
        for (int j=0;j<=num;++j) f[j][i]=Inf;
      for (int i=1;i<=num;++i) f[i][1<<(i-1)]=map[0][i];
      for (int i=0;i<maxz;++i)    // states are enumerated
      {
          for (int j=1;j<=num;++j) if (i & (1<<(j-1)))    // The
last gathered nut j in stage i is enumerated
            for(int k=1;k<=num;++k)    // Nut k which isn't in
stage i is enumerated, and adjusted
              if (!(i & (1<<(k-1))))Update(f[k][i+(1<<(k-1))],
f[j][i]+map[j][k]);
      }
      ans=Inf;
      for (int i=1;i<=num;++i)    // Enumerate the last
gathered nut i, and adjust
        Update(ans,f[i][maxz]+map[i][0]);
      printf("%d\n",ans);    // Output the result
  }
  return 0;
}
```

Figure 6.3

6.3.2 Mondriaan's Dream

Squares and rectangles fascinated the famous Dutch painter Piet Mondriaan. One night, after producing the drawings in his "toilet series" (where he had to use his toilet paper to draw on, for all of his paper was filled with squares and rectangles), he dreamt of filling a large rectangle with small rectangles of width 2 and height 1 in varying ways (see Figure 6.3).

Expert as he was in this material, he saw at a glance that he will need a computer to calculate the number of ways to fill the large rectangle whose dimensions were integer values, as well. Help him, so that his dream won't turn into a nightmare!

Input

The input file contains several test cases. Each test case is made up of two integer numbers: the height h and the width w of the large rectangle. Input is terminated by $h=w=0$. Otherwise, $1<=h$, $w<=11$.

Output

For each test case, output the number of different ways the given rectangle can be filled with small rectangles of size 2 times 1. Assume that the given large rectangle is oriented, i.e., count symmetrical tilings multiple times (see Figure 6.4).

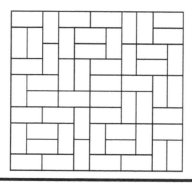

Figure 6.4

Sample Input	Sample Output
1 2	1
1 3	0
1 4	1
2 2	2
2 3	3
2 4	5
2 11	144
4 11	51205
0 0	

Source: Ulm Local 2000

IDs for Online Judges: POJ 2411, ZOJ 1100

 Analysis

Assume that you could calculate the number of different paintings for a rectangle with c columns and r rows where the first $r-1$ rows are completely filled and the last row has any of 2^c possible patterns. Then, by trying all variations of filling the last row where small rectangles may be spilled into a further row, you can calculate the number of different paintings for a rectangle with $r+1$ rows where the first r rows are completely filled and the last row again has any pattern.

This straightforwardly leads to a DP solution. All possible ways of filling a row, part of which may already be occupied and spilling into the next row and creating a new pattern, are generated by backtracking over a row. Viewing these as transitions from a pattern to another pattern, their number is given by the recursive equation $T_c = 2T_{c-1} + T_{c-2}$. Its solution is asymptotically exponential with a base of $sqrt(2)+1$, which is not a problem for $c <= 11$.

If both h and w are odd, the result is 0. Since the number of paintings is a symmetric function, the number of columns should be chosen as the smaller of the two input numbers whenever possible to improve runtime behaviour substantially.

Judges' test data includes all 121 legal combinations of h and w.

Since the size of the painting could be as large as 110, a simple backtracking solution won't do, not even with using five hours of contest time to precalculate the results. Once the DP algorithm is implemented, a quick review of the results should reveal that they don't fit into 32-bit *ints*. There are four ways to solve this problem: try *double* (which works actually), implement 64-bit arithmetics (only addition is needed), implement arbitrary precision arithmetics, or switch to Java and use *BigInteger*. A more efficient algebraic solution was not known to the judges.

Program

```c
#include <stdio.h>
static double cnt[12][1<<11];
static int trans[16384][2];
int rows, cols, ntrans;
/* there are ((sqrt(2)+1)^c - (sqrt(2)-1)^c) * (sqrt(2)+2) / 4
transitions
 * which is the solution to T_{c} = 2 * T_{c-1} + T_{c-2}
 */
void backtrack (int n, int from, int to)
{
  if (n > cols) return;
  if (n == cols)
  {
    trans[ntrans][0] = from;
    trans[ntrans][1] = to;
    ++ntrans;
    return;
  }
  backtrack (n+2, from<<2, to<<2);
  backtrack (n+1, from<<1, (to<<1)|1);
  backtrack (n+1, (from<<1)|1, to<<1);
}
int main ()
{
  int r, t;
  FILE* in = fopen ("dream.in", "r");
  while (fscanf (in, " %d %d ", &rows, &cols) == 2)
  {
    if (rows == 0 || cols == 0) break;
    if (rows < cols) { t = rows; rows = cols ; cols = t; }
    /* calculate map of possible transitions by linear
backtracking */
    ntrans = 0;
    backtrack (0, 0, 0);
    for (r=0 ; r<=rows ; r++)
      for (t=0 ; t<(1<<cols) ; t++)
        cnt[r][t] = 0;
    cnt[0][0] = 1;
    for (r=0 ; r<rows ; r++) /* the r topmost rows are already
filled */
      for (t=0 ; t<ntrans ; t++) /* perform all transitions */
        cnt[r+1][trans[t][1]] += cnt[r][trans[t][0]];
    printf ("%.0f\n", cnt[rows][0]);
  }
  return 0;
}
```

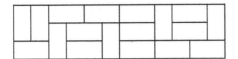

Figure 6.5

6.4 Problems

6.4.1 Tri Tiling

In how many ways can you tile a 3×n rectangle with 2×1 dominoes?

Figure 6.5 shows a sample tiling of a 3×12 rectangle.

Input

Input consists of several test cases followed by a line containing −1. Each test case is a line containing an integer $0 \le n \le 30$.

Output

For each test case, output one integer number giving the number of possible tilings.

Sample Input	Sample Output
2	3
8	153
12	2131
−1	

Source: Waterloo local 2005.09.24

IDs for Online Judges: POJ 2663, ZOJ 2547, UVA 10918

Hint

Suppose the state for column i is represented as a binary number j ($0 \le i \le n-1$, $0 \le j \le 7$), where 0 represents the square that is occupied by a domino, and 1 represents the square that isn't occupied by a domino. Obviously, the state for $(0, i)$ is $c=j\&1$; the state for $(1, i)$ is $b = \left\lfloor \dfrac{j}{2} \right\rfloor \&1$; and the state for $(2, i)$ is $a = \left\lfloor \dfrac{j}{4} \right\rfloor$.

Suppose $dp[i][j]$ is the number of possible tilings for the first i columns whose state is j. Obviously, $dp[0][0]=1$. From left to right, DP is used as follows.

```
If (1, i) and (2, i) are occupied by dominoes (!a&&!b==1),
then dp[i+1][!c]+=dp[i][j];
```

```
If (0, i) and (1, i) are occupied by dominoes (!b&&!c==1),
then dp[i+1][(!a)*4]+=dp[i][j];
dp[i+1][(!a)×4+(!b)×2+(!c)]+=dp[i][j];
```

Finally, $dp[n][0]$ is the solution to the problem.

6.4.2 Marks Distribution

In an examination. one student appeared in N subjects and has got total T marks. He has passed in all the N subjects where the minimum mark for passing in each subject is P. You have to calculate the number of ways the student can get the marks. For example, if $N=3$, $T=34$ and $P=10$, then the marks in the three subjects could be as follows:

	Subject 1	Subject 2	Subject 3
1	14	10	10
2	13	11	10
3	13	10	11
4	12	11	11
5	12	10	12
6	11	11	12
7	11	10	13
8	10	11	13
9	10	10	14
10	11	12	11
11	10	12	12
12	12	12	10
13	10	13	11
14	11	13	10
15	10	14	10

So there are 15 solutions. So F **(3, 34, 10)=15**.

Input

In the first line of the input, there will be a single positive integer K followed by K lines, each containing a single test case. Each test case contains three positive integers denoting *N*, *T*, and *P* respectively. The values of *N*, *T*. and *P* will be at most 70. You may assume that the final answer will fit in a standard 32-bit integer.

Output

For each input, print in a line the value of *F* (*N*, *T*, *P*).

Sample Input	Sample Output
2	15
3 34 10	15
3 34 10	

Source: 4th IIUC Inter-University Programming Contest, 2005

ID for Online Judge: UVA 10910

Hint

Suppose $dp[i][j]$ shows the number of ways the student passes i subjects and gets total j marks. Therefore, $dp[1][j]=1$, where $P{\leq}j{\leq}T$; and

$$dp[i][j] = \sum_{k=P}^{j-P} d[i-1][j-k] \big| j-k \geq P,$$ where $2{\leq}i{\leq}N$, $P{\leq}j{\leq}T$. Finally, $dp[N][T]$ is the solution to the problem.

6.4.3 Chocolate Box

Recently one of my friends, Tarik, became a member of the food committee of an ACM regional competition. He has been given *m* distinguishable boxes, and he has to put *n* types of chocolates in the boxes. The probability that one chocolate is placed in a certain box is $1/m$. What is the probability that one or more boxes are empty? At first he thought it was an easy task. But soon he found that it was much harder. So, he falls into great trouble and asks you to help him in this task.

Input

Each line of the input contains two integers *n* indicating the total number of distinguishable types of chocolate and *m* indicating the total number of distinguishable boxes (*m*≤*n*<100). A single line containing –1 denotes the end.

Output

For each of the cases, you should calculate the probability corrected to seven decimal places. The output format is shown below.

Sample Input	Sample Output
50 12	Case 1: 0.1476651
50 12	Case 2: 0.1476651
−1	

Source: The FOUNDATION Programming Contest 2004

ID for Online Judge: UVA 10648

 Hint

Suppose $dp[i][j]$ is the probability that j boxes have chocolates after the i-th chocolate is placed. Initially, $dp[1][1]=1$. And $dp[i][j]=dp[i-1][j] \times f(j)+dp[-1][j-1] \times f(m-j+1)$, where $f(x) = \dfrac{x}{m}$, represents the probability that one chocolate is placed in x boxes, $2 \le i \le n$, $1 \le j \le m$.

The solution is $1 - dp[n][m]$.

6.4.4 A Spy in the Metro

Secret agent Maria was sent to Algorithms City to carry out an especially dangerous mission. After several thrilling events, we find her in the first station of Algorithms City Metro, examining the time table. The Algorithms City Metro consists of a single line with trains running both ways, so its timetable is not complicated.

Maria has an appointment with a local spy at the last station of Algorithms City Metro. Maria knows that a powerful organization is after her. She also knows that while waiting at a station, she is at great risk of being caught. To hide in a running train is much safer, so she decides to stay in running trains as much as possible, even if this means traveling backward and forward. Maria needs to know a schedule with minimal waiting time at the stations that gets her to the last station in time for her appointment. You must write a program that finds the total waiting time in a best schedule for Maria.

The Algorithms City Metro system has N stations, consecutively numbered from 1 to N. Trains move in both directions: from the first station to the last station and from the last station back to the first station (see Figure 6.6). The time required

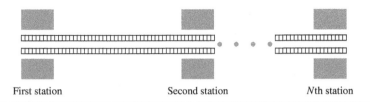

Figure 6.6

for a train to travel between two consecutive stations is fixed since all trains move at the same speed. Trains make a very short stop at each station, which you can ignore for simplicity. Since she is a very fast agent, Maria can always change trains at a station even if the trains involved stop in that station at the same time.

Input

The input file contains several test cases. Each test case consists of seven lines with information as follows.

- Line 1. The integer N ($2 \leq N \leq 50$), which is the number of stations.
- Line 2. The integer T ($0 \leq T \leq 200$), which is the time of the appointment.
- Line 3. $N-1$ integers: $t_1, t_2, \ldots t_{N-1}$ ($1 \leq t_i \leq 20$), representing the travel times for the trains between two consecutive stations: t_1 represents the travel time between the first two stations, t_2 the time between the second and the third station, and so on.
- Line 4. The integer M_1 ($1 \leq M_1 \leq 50$), representing the number of trains departing from the first station.
- Line 5. M_1 integers: $d_1, d_2, \ldots, d_{M_1}$ ($0 \leq d_i \leq 250$ and $d_i < d_{i+1}$), representing the times at which trains depart from the first station.
- Line 6. The integer M_2 ($1 \leq M_2 \leq 50$), representing the number of trains departing from the Nth station.
- Line 7. M_2 integers: $e_1, e_2, \ldots e_{M_2}$ ($0 \leq e_i \leq 250$ and $e_i < e_{i+1}$) representing the times at which trains depart from the Nth station.

The last case is followed by a line containing a single zero.

Output

For each test case, print a line containing the case number (starting with 1) and an integer representing the total waiting time in the stations for a best schedule, or the word "impossible" in case Maria is unable to make the appointment. Use the format of the sample output.

Sample Input	Sample Output
4	Case Number 1: 5
55	Case Number 2: 0
5 10 15	Case Number 3: impossible
4	
0 5 10 20	
4	
0 5 10 15	
4	
18	
1 2 3	
5	
0 3 6 10 12	
6	
0 3 5 7 12 15	
2	
30	
20	
1	
20	
7	
1 3 5 7 11 13 17	
0	

Source: ACM World Finals 2003

IDs for Online Judges: UVA 2728

Hint

First, the time that each train departing from the first station arrives at each station $x1[\ ][\ ]$ and the time that each train departing from the N-th station arrives at each station $x2[\ ][\ ]$ are calculated, where the time that the i-th train departing from the first stationt arrives at the j-th station is $x1[i][j]$:

$$
x1[i][j] = \begin{cases} \text{the time at which the i-th train departs} \\ \text{from the first station} \qquad\qquad\qquad j = 1 \\ \\ x1[i][j\text{-}1] + \text{the travel time between the (j-1)-th} \quad j > 1 \\ \text{station and the j-th station} \end{cases} ;
$$

and the time that the i-th train departing from the N-th stationt arrives at the j-th station is $x2[i][j]$:

$$
x2[i][j]=\begin{cases}
\text{the time at which the i-th train departs} & j=n\\
\text{from the N-th station} & \\
x2[i][j+1]+\text{the travel time between the (j+1)-th} & j<n\\
\text{station and the j-th station} &
\end{cases}
$$

States are minimal waiting times at each point in time for each station. Because for the previous point in time, the waiting time must also be minimal, DP is used to solve the problem.

Suppose $f[j][k]$ is the minimal waiting time that Maria arrives at the k-th station at point in time j. Obviously, $f[0][1]=0$;

Stage i: A stage is a point in time before the time of the appointment, $0\leq i\leq T-1$;

State k: each station is enumerated, $0\leq k\leq N$;

Decision: There are two kinds of decisions, forward and backward:

Forward: Each forward train j which arrives at the k-th station after point in time i is enumerated ($1\leq j\leq$ the number of trains departing from the first station, $i\leq x1[j][k]$). The minimal waiting time that train j arrives at the $(k+1)$-th station is $f[x1[j][k+1]][k+1]=\min\{f[x1[j][k+1]][k+1], f[i][k]+x1[j][k]-i\}$.

Backward: Each backward train j which arrives at the k-th station after point in time i is enumerated ($1\leq j\leq$ the number of trains departing from the N-th station, $i\leq x2[j][k]$). The minimal waiting time that train j arrives at the $(k-1)$-th station is $f[x2[j][k-1]][k-1]=\min\{f[x2[j][k-1]][k-1], f[i][k]+x1[j][k]-i\}$

Because the time that trains arrive at the $(k+1)$-th station or the $(k-1)$-th station may be after i, we need to adjust: $f[i+1][k]=\min\{f[i+1][k], f[i][k]+1\}$.

Obviously, if $f[T][N]$ is the initial value before DP, Maria is unable to make the appointment; otherwise, $f[T][N]$ is the total waiting time in the stations for a best schedule.

6.4.5 A Walk Through the Forest

Jimmy experiences a lot of stress at work these days, especially since his accident made working difficult. To relax after a hard day, he likes to walk home. To make things even nicer, his office is on one side of a forest, and his house is on the other. A nice walk through the forest, seeing the birds and chipmunks, is quite enjoyable.

The forest is beautiful, and Jimmy wants to take a different route every day. He also wants to get home before dark, so he always takes a path to make progress towards his house. He considers taking a path from A to B to be progress if there exists a route from B to his home that is shorter than any possible route from A. Calculate how many different routes through the forest Jimmy might take.

Input

Input contains several test cases followed by a line containing 0. Jimmy has numbered each intersection or joining of paths starting with 1. His office is numbered 1, and his house is numbered 2. The first line of each test case gives the number of intersections N, $1 < N \leq 1000$, and the number of paths M. The following M lines each contain a pair of intersections a b and an integer distance $1 \leq d \leq 1000000$ indicating a path of length d between intersection a and a different intersection b. Jimmy may walk a path any direction he chooses. There is at most one path between any pair of intersections.

Output

For each test case, output a single integer indicating the number of different routes through the forest. You may assume that this number does not exceed 2147483647.

Sample Input	Sample Output
5 6	2
1 3 2	4
1 4 2	
3 4 3	
1 5 12	
4 2 34	
5 2 24	
7 8	
1 3 1	
1 4 1	
3 7 1	
7 4 1	
7 5 1	
6 7 1	
5 2 1	
6 2 1	
0	

Source: Waterloo local 2005.09.24

IDs for Online Judges: POJ 2662, UVA 10917

Hint

First, a weighted graph is constructed. Each intersection is represented as a vertex. Each path is represented as an edge. And the length for a path is represented as the weight of the corresponding edge. Jimmy's office is as vertex 1, and Jimmy's house is as vertex 2.

Second, Dijkstra's algorithm is used to calculate the shortest path $dist[]$ from each vertex to vertex 2, where $dist[i]$ is the length of the shortest path form vertex i to vertex 2. Suppose $f[x]$ is the number of paths from vertex x to vertex 2:

$$f[x] = \begin{cases} 1 & x = 2 \\ \sum_{i=1}^{n}(f[i] \mid (i,x) \in E \, \& \, \& \, dist[i] < dist[x]) & x \neq 2 \end{cases}$$

Finally, $f[1]$ is the number of different routes through the forest.

6.4.6 *Common Subsequence*

A subsequence of a given sequence is the given sequence with some elements (possibly none) left out. Given a sequence $X=<x_1, x_2,..., x_m>$, another sequence $Z=<z_1, z_2,..., z_k>$ is a subsequence of X if there exists a strictly increasing sequence $<i_1, i_2, ..., i_k>$ of indices of X such that for all $j=1, 2, ..., k, x_{i_j}=z_j$. For example, $Z=<a, b, f, c>$ is a subsequence of $X=<a, b, c, f, b, c>$ with index sequence $<1, 2, 4, 6>$. Given two sequences X and Y, the problem is to find the length of the maximum-length common subsequence of X and Y.

Input

The program input is from the standard input. Each data set in the input contains two strings representing the given sequences. The sequences are separated by any number of white spaces. The input data are correct.

Output

For each set of data, the program prints on the standard output the length of the maximum-length common subsequence from the beginning of a separate line.

Sample Input		Sample Output
abcfbc	abfcab	4
programming	contest	2
abcd	mnp	0

Source: ACM Southeastern Europe 2003

IDs for Online Judges: POJ 1458, ZOJ 1733, UVA 2759

 Hint

The problem is an LCS problem.

6.4.7 Lazy Cows

Farmer John regrets having applied high-grade fertilizer to his pastures since the grass now grows so quickly that his cows no longer need to move around when they graze. As a result, the cows have grown quite large and lazy... and winter is approaching.

Farmer John wants to build a set of barns to provide shelter for his immobile cows, and he believes that he needs to build his barns around the cows based on their current locations since they won't walk to a barn, no matter how close or comfortable.

The cows' grazing pasture is represented by a $2 \times B$ ($1 \leq B \leq 15,000,000$) array of cells, some of which contain a cow and some of which are empty. N ($1 \leq N \leq 1000$) cows occupy the cells in this pasture:

	cow				cow	cow	cow	cow
	cow	cow	cow					

Ever the frugal agrarian, Farmer John would like to build a set of just K ($1 \leq K \leq N$) rectangular barns (oriented with walls parallel to the pasture's edges) whose total area covers the minimum possible number of cells. Each barn covers a rectangular group of cells in their entirety, and no two barns may overlap. Of course, the barns must cover all of the cells containing cows.

By way of example, in the picture above, if $K=2$. then the optimal solution contains a 2×3 barn and a 1×4 barn and covers a total of 10 units of area.

Input

Line 1: Three space-separated integers, N, K, and B.

Lines 2. *N*+1: Two space-separated integers in the range (1,1) to (2,*B*) giving the coordinates of the cell containing each cow. No cell contains more than one cow.

Output

Line 1: The minimum area required by the *K* barns in order to cover all of the cows.

Sample Input	Sample Output
8 2 9 1 2 1 6 1 7 1 8 1 9 2 2 2 3 2 4	10

Source: USACO 2005 USOpen Gold

ID for Online Judge: POJ 2430

Hint

This is a problem for DP with state compression. Suppose $dp[i][j][k]$ represents the best solution that the first i columns is covered by j barns, and the current state is k; where $k==1$ means only the first row is covered by a barn, $k==2$ means only the second row is covered by a barn, $k==3$ means the first row and the second row is covered by a barn, and $k==4$ means the first row and the second row are covered by two different barns.

6.4.8 Longest Common Subsequence

Given two sequences of characters, (Figure 6.7), print the length of the longest common subsequence of both sequences. For example, the longest common subsequence of the following two sequences:

abcdgh
aedfhr

is adh of length 3.

Sequence 1:

Sequence 2:

Figure 6.7

Input

The input consists of pairs of lines. The first line of a pair contains the first string and the second line contains the second string. Each string is on a separate line and consists of at most 1,000 characters.

Output

For each subsequent pair of input lines, output a line containing one integer number which satisfies the criteria stated above.

Sample Input	Sample Output
a1b2c3d4e	4
zz1yy2xx3ww4vv	3
abcdgh	26
aedfhr	14
abcdefghijklmnopqrstuvwxyz	
a0b0c0d0e0f0g0h0i0j0k0l0m0n0o0p0q0r0s0t0u0v0w0x0y0z0	
abcdefghijklmnzyxwvutsrqpo	
opqrstuvwxyzabcdefghijklmn	

Source: November 2002 Monthly Contest

ID for Online Judge: UVA 10405

 Hint

This problem is a classical LCS problem.

6.4.9 Make Palindrome

By definition, a palindrome is a string which is not changed when reversed. "MADAM" is a nice example of a palindrome. It is an easy job to test whether a given string is a palindrome or not. But it may not be so easy to generate a palindrome.

Here we will make a palindrome generator that will take an input string and return a palindrome. You can easily verify that for a string of length '*n*', no more than (*n*–1) characters are required to make it a palindrome. Consider "abcd" and its palindrome "abcdcba" or "abc" and its palindrome "abcba". But life is not so easy for programmers!! We always want optimal cost. And you have to find the minimum number of characters required to make a given string into a palindrome if you are allowed to insert characters at any position of the string.

Input

Each input line consists only of lowercase letters. The size of the input string will be at most 1000. Input is terminated by EOF.

Output

For each input, print the minimum number of characters and such a palindrome separated by one space in a line. There may be many such palindromes. Any one will be accepted.

Sample Input	Sample Output
abcd	3 abcdcba
aaaa	0 aaaa
abc	2 abcba
aab	1 baab
abababaabababa	0 abababaabababa
pqrsabcdpqrs	9 pqrsabcdpqrqpdcbasrqp

Source: The Real Programmers' Contest -2 -A BUET Sprinter Contest 2003

ID for Online Judge: UVA 10453

 Hint

First, the longest common subsequence of the string and its reverse are calculated. This will give you the optimal overlap in the palindrome. Then the rest of the characters are added into the string to make the shortest palindrome.

6.4.10 Vacation

You are planning to take some rest and to go on vacation, but you really don't know which cities you should visit. So, you ask your parents for help. Your mother says "My son, you MUST visit Paris, Madrid, Lisbon, and London. But it's only fun in this order." Then your father says: "Son, if you're planning to travel, go first to Paris, then to Lisbon, then to London and then, at last, go to Madrid. I know what I'm talking about."

Now you're a bit confused, as you didn't expect this situation. You're afraid that you'll hurt your mother if you follow your father's suggestion. But you're also afraid to hurt your father if you follow your mother's suggestion. But it can get worse, because you can hurt both of them if you simply ignore their suggestions!

Thus, you decide that you'll try to follow their suggestions in the best way that you can. So, you realize that the "Paris-Lisbon-London" order is the one which better satisfies both your mother and your father. Afterwards, you can say that you could not visit Madrid, even though you would've liked it very much.

If your father suggested the "London-Paris-Lisbon-Madrid" order, then you would have two orders, "Paris-Lisbon" and "Paris-Madrid", which would better satisfy both of your parents' suggestions. In this case, you could only visit two cities.

You want to avoid problems like this one in the future. And what if their travel suggestions were bigger? Probably you would not find the better way very easily. So, you decided to write a program to help you in this task. You'll represent each city by one character, using uppercase letters, lowercase letters, digits, and the space. Thus, you can have at most 63 different cities to visit. But it's possible that you'll visit some city more than once.

If you represent Paris with "a", Madrid with "b", Lisbon with "c", and London with "d", then your mother's suggestion would be "abcd" and your father's suggestion would be "acdb" (or "dacb", in the second example).

The program will read two travel sequences, and it must answer how many cities you can travel to such that you'll satisfy both of your parents and its maximum.

Input

The input will consist of an arbitrary number of city sequence pairs. The end of input occurs when the first sequence starts with an "#" character (without the quotes). Your program should not process this case. Each travel sequence will be on a line alone and will be formed by legal characters (as defined above). All travel sequences will appear in a single line and will have at most 100 cities.

Output

For each sequence pair, you must print the following message in a line alone:

Case #d: you can visit at most K cities.

Where d stands for the test case number (starting from 1) and K is the maximum number of cities you can visit such that you'll satisfy both your father's suggestion and your mother's suggestion.

Sample Input	Sample Output
abcd acdb abcd dacb #	Case #1: you can visit at most 3 cities. Case #2: you can visit at most 2 cities.

Source: 2001 Universidade do Brasil (UFRJ). Internal Contest Warmup

ID for Online Judge: UVA 10192

Hint

Your mother's suggestion is the first string, and your father's suggestion is the second string. The Longest Common Subsequence (LCS) for the two strings are cities that you'll visit to satisfy both your father's suggestion and your mother's suggestion.

6.4.11 Is Bigger Smarter?

Some people think that the bigger an elephant is, the smarter it is. To disprove this, you want to take the data on a collection of elephants and put as large a subset of this data as possible into a sequence so that the weights are increasing, but the IQs are decreasing.

The input will consist of data for a group of elephants, one elephant per line, terminated by the end-of-file. The data for a particular elephant will consist of a pair of integers: the first representing its size in kilograms and the second representing its IQ in hundredths of IQ points. Both integers are between 1 and 10000. The data will contain information for at most 1000 elephants. Two elephants may have the same weight, the same IQ, or even the same weight and IQ.

Say that the numbers on the i-th data line are $W[i]$ and $S[i]$. Your program should output a sequence of lines of data; the first line should contain a number n; the remaining n lines should each contain a single positive integer (each one representing an elephant). If these n integers are $a[1]$, $a[2]$,..., $a[n]$, then it must be the case that

$$W[a[1]]<W[a[2]]<...<W[a[n]]$$

and

$$S[a[1]]>S[a[2]]>...>S[a[n]].$$

In order for the answer to be correct, n should be as large as possible. All inequalities are strict: weights must be strictly increasing, and IQs must be strictly decreasing. There may be many correct outputs for a given input, but your program only needs to find one.

Sample Input	Sample Output
6008 1300	4
6000 2100	4
500 2000	5
1000 4000	9
1100 3000	7
6000 2000	
8000 1400	
6000 1200	
2000 1900	

Source: The "silver wedding" Contest 2001

ID for Online Judge: UVA 10131

Hint

It is a standard DP (Longest Increasing Subsequence) problem. First, n elephants are sorted. The weight is as the first key. And the IQ is as the second key. Then the Longest Increasing Subsequence (LIS) for the sorted sequence is calculated.

6.4.12 Stacking Boxes

Some concepts in mathematics and computer science are simple in one or two dimensions but become more complex when extended to arbitrary dimensions. Consider solving differential equations in several dimensions and analyzing the topology of an n-dimensional hypercube. The former is much more complicated than its one-dimensional relative, while the latter bears a remarkable resemblance to its "lower-class" cousin.

Consider an n-dimensional "box" given by its dimensions. In two dimensions the box (2,3) might represent a box with length two units and width three units. In three dimensions the box (4,8,9) can represent a box 4×8×9 (length, width, and height). In six dimensions it is, perhaps, unclear what the box (4,5,6,7,8,9) represents; but we can analyze the properties of the box, such as the sum of its dimensions.

In this problem you will analyze a property of a group of n-dimensional boxes. You are to determine the longest *nesting string* of boxes, that is a sequence of boxes $b_1, b_2, \ldots, b_k$ such that each box b_i nests in box b_{i+1} ($1 \leq i < k$).

A box $D=(d_1, d_2, ..., d_n)$ nests in a box $E=(e_1, e_2, ..., e_n)$ if there is some rearrangement of the d_i such that when rearranged, each dimension is less than the corresponding dimension in box E. This loosely corresponds to turning box D to see if it will fit in box E. However, since any rearrangement suffices, box D can be contorted, not just turned (see examples below).

For example, the box $D=(2,6)$ nests in the box $E=(7,3)$ since D can be rearranged as $(6,2)$ so that each dimension is less than the corresponding dimension in E. The box $D=(9,5,7,3)$ does NOT nest in the box $E=(2,10,6,8)$, since no rearrangement of D results in a box that satisfies the nesting property, but $F=(9,5,7,1)$ does nest in box E since F can be rearranged as $(1,9,5,7)$, which nests in E.

Formally, we define nesting as follows: box $D=(d_1, d_2, ..., d_n)$ nests in box $E=(e_1, e_2, ..., e_n)$ if there is a permutation π of $1..n$ such that $(d_{\pi(1)}, d_{\pi(2)}, ..., d_{\pi(n)})$ "fits" in $(e_1, e_2, ..., e_n)$, i.e., if $d_{\pi(i)} \leq e_i$ for all $d_{\pi(i)} \leq e_i$.

Input

The input consists of a series of box sequences. Each box sequence begins with a line consisting of the number of boxes k in the sequence followed by the dimensionality of the boxes, n (on the same line).

This line is followed by k lines, one line per box with the n measurements of each box on one line separated by one or more spaces. The line in the sequence ($1 \leq i \leq k$) gives the measurements for the box.

There may be several box sequences in the input file. Your program should process all of them and determine, for each sequence, which of the k boxes determine the longest nesting string and the length of that nesting string (the number of boxes in the string).

In this problem, the maximum dimensionality is 10 and the minimum dimensionality is 1. The maximum number of boxes in a sequence is 30.

Output

For each box sequence in the input file, output the length of the longest nesting string on one line, followed on the next line by a list of the boxes that comprise this string in order. The "smallest" or "innermost" box of the nesting string should be listed first, and the next box (if there is one) should be listed second, etc.

The boxes should be numbered according to the order in which they appeared in the input file (the first box is box 1, etc.).

If there is more than one longest nesting string, then any one of them can be output.

Sample Input	Sample Output
5 2	5
3 7	3 1 2 4 5
8 10	4
5 2	7 2 5 6
9 11	
21 18	
8 6	
5 2 20 1 30 10	
23 15 7 9 11 3	
40 50 34 24 14 4	
9 10 11 12 13 14	
31 4 18 8 27 17	
44 32 13 19 41 19	
1 2 3 4 5 6	
80 37 47 18 21 9	

Source: Internet Programming Contest 1990

ID for Online Judge: UVA 103

Hint

This problem is a Longest Increasing Subsequence problem. The problem requires you to check whether box *a* fits in box *b*.

First, for each box, its dimension (s_1, s_2, s_3, ..., s_n) is sorted such that $s_i \leq s_j$ for all $i < j$.
Second, boxes are sorted. For two boxes *a* and *b*, $a < b$ if $a_i \leq b_i$ for all *i*.
Finally, the Longest Increasing Subsequence algorithm is used.
The time complexity for the problem is $O(n^2)$.

6.4.13 Function Run Fun

We all love recursion! Don't we? Consider a three-parameter recursive function $w(a, b, c)$:

$$
w(a,b,c) = \begin{cases}
1 & a \leq 0 \text{ or } b \leq 0 \text{ or } c \leq 0 \\
w(20,20,20) & a > 20 \text{ or } b > 20 \text{ or } c > 20 \\
w(a,b,c-1) + w(a,b-1,c-1) - w(a,b-1,c) & a < b \text{ and } b < c \\
w(a-1,b,c) + w(a-1,b-1,c) + w(a-1,b,c-1) - w(a-1,b-1,c-1) & \text{otherwise}
\end{cases}
$$

This is an easy function to implement. The problem is, if implemented directly, for moderate values of *a*, *b*, and *c* (e.g., *a*=15, *b*=15, *c*=15), the program takes hours to run because of the massive recursion.

Input

The input for your program will be a series of integer triples, one per line, until the end-of-file flag of –1 –1 –1. Using the above technique, you are to calculate *w*(*a*, *b*, *c*) efficiently and print the result.

Output

Print the value for *w*(*a*, *b*, *c*) for each triple.

Sample Input	Sample Output
1 1 1	w(1, 1, 1) = 2
2 2 2	w(2, 2, 2) = 4
10 4 6	w(10, 4, 6) = 523
50 50 50	w(50, 50, 50) = 1048576
–1 7 18	w(–1, 7, 18) = 1
–1 –1 –1	

Source: ACM Pacific Northwest 1999

IDs for Online Judge: POJ 1579, ZOJ 1168

 Hint

A memorized search is used to solve the problem. Suppose *a*[][][] is the memorized list, where *a*[*x*][*y*][*z*] stores the result for *w*(*x*, *y*, *z*).

```
For w(x, y, z),
If (x≤0||y≤0||z≤0), return 1;
If (x>20||y>20||z>20), return w(20, 20, 20);
If (x<y&&y<z), then a[x][y][z] memorizes w(x, y, z-1)+w(x,
y-1, z-1)-w(x, y-1, z); else a[x][y][z] memorizes w(x-1, y,
z)+w(x-1, y-1, z)+w(x-1, y, z-1)-w(x-1, y-1, z-1).
```

6.4.14 To the Max

Given a two-dimensional array of positive and negative integers, a subrectangle is any contiguous subarray of size 1×1 or greater located within the whole array. The sum of a rectangle is the sum of all the elements in that rectangle. In this problem, the subrectangle with the largest sum is referred to as the maximal subrectangle.

As an example, the maximal subrectangle of the array:

$$
\begin{array}{rrrr}
0 & -2 & -7 & 0 \\
9 & 2 & -6 & 2 \\
-4 & 1 & -4 & 1 \\
-1 & 8 & 0 & -2
\end{array}
$$

is in the lower left corner:

$$
\begin{array}{rr}
9 & 2 \\
-4 & 1 \\
-1 & 8
\end{array}
$$

and has a sum of 15.

Input

The input consists of an $N \times N$ array of integers. The input begins with a single positive integer N on a line by itself, indicating the size of the square two-dimensional array. This is followed by N^2 integers separated by white space (spaces and new lines). These are the N^2 integers of the array, presented in row-major order; that is, all numbers in the first row, left to right, then all numbers in the second row, left to right, etc. N may be as large as 100. The numbers in the array will be in the range [−127,127].

Output

Output the sum of the maximal subrectangle.

Sample Input	Sample Output
4 0 −2 −7 0 9 2 −6 2 −4 1 −4 1 −1 8 0 −2	15

Source: ACM Greater New York 2001

IDs for Online Judges: POJ 1050, ZOJ 1074, UVA 2288

Hint

Suppose *max* is the sum of the maximal subrectangle, initially *max*=−10000; and array *m* is the input array.

First, array *m* is input. For row *i*, ma_i is the maximum for sums of continuous integers, $1 \leq i \leq N$. And after array *m* is input, the maximum *max* for all maximums for sums of continuous integers for each row is computed, $max = \max_{1 \leq i \leq N} \{ma_i\}$.

Second, from the first row, a *for* repetition statement deals with every row top-down. For the current row, integers for its below rows are added into its corresponding columns. The maximum for sums of continuous integers in the current row is calculated. That is, row by row, for each row, its below row is added into the row. And *max* is adjusted after a row is added into the current row, if it isn't the maximal value. After the *for* repetition, *max* is the sum of the maximal subrectangle.

6.4.15 Robbery

Inspector Robstop is very angry. Last night, a bank has been robbed and the robber has not been caught. And this has happened already for the third time this year, even though he did everything in his power to stop the robber: as quickly as possible, all roads leading out of the city were blocked, making it impossible for the robber to escape. Then, the inspector asked all the people in the city to watch out for the robber, but the only messages he received were of the form "We don't see him."

But this time, he has had enough! Inspector Robstop decides to analyze how the robber could have escaped. To do that, he asks you to write a program which takes all the information the inspector could get about the robber in order to find out where the robber has been at which time.

Coincidentally, the city in which the bank was robbed has a rectangular shape. The roads leaving the city are blocked for a certain period of time *t*, and during that time, several observations of the form "The robber isn't in the rectangle R_i at time T_i" are reported. Assuming that the robber can move at most one unit per time step, your program must try to find the exact position of the robber at each time step.

Input

The input contains the description of several robberies. The first line of each description consists of three numbers *W, H, t* ($1 \leq W, H, t \leq 100$) where *W* is the width, *H* is the height of the city, and *t* is the time during which the city is locked.

The next line contains a single integer *n* ($0 \leq n \leq 100$), the number of messages the inspector received. The next *n* lines (one for each of the messages) consist of five integers t_i, L_i, T_i, R_i, B_i each. The integer t_i is the time at which the observation has been made ($1 \leq t_i \leq t$), and L_i, T_i, R_i, B_i are the left, top, right, and bottom respectively of the (rectangular) area which has been observed. ($1 \leq L_i \leq R_i \leq W$, $1 \leq T_i \leq B_i \leq H$; the point (1, 1) is the upper left-hand corner, and (*W, H*) is the lower right-hand corner of the city.) The messages mean that the robber was not in the given rectangle at time t_i.

The input is terminated by a test case starting with $W=H=t=0$. This case should not be processed.

Output

For each robbery, first output the line "Robbery #*k*:", where *k* is the number of the robbery. Then, there are three possibilities:

1. If it is impossible that the robber is still in the city considering the messages, output the line "The robber has escaped."
2. If it is impossible that the robber is still in the city considering the messages, output the line "The robber has escaped." In all other cases, assume that the robber really is in the city. Output one line of the form "Time step *t*: The robber has been at *x*, *y*." for each time step, in which the exact location can be deduced. (*x* and *y* are the column and row of the robber in time step *t*.) Output these lines ordered by time *t*.
3. If nothing can be deduced, output the line "Nothing known." and hope that the inspector will not get even more angry.

Output a blank line after each processed case.

Sample Input	Sample Output
4 4 5	Robbery #1:
4	Time step 1: The robber has been at 4,4.
1 1 1 4 3	Time step 2: The robber has been at 4,3.
1 1 1 3 4	Time step 3: The robber has been at 4,2.
4 1 1 3 4	Time step 4: The robber has been at 4,1.
4 4 2 4 4	
10 10 3	Robbery #2:
1	The robber has escaped.
2 1 1 10 10	
0 0 0	

Source: ACM Mid-Central European Regional Contest 1999

IDs for Online Judges: POJ 1104, ZOJ 1144, UVA707

 Hint (given by the problemsetter)

We are told that there is a robber capable of moving one unit per time in a rectangular gridlike city. Furthermore, we are given subrectangles of the city that we know the robber was not in at different points in time. We have to determine where the robber could possibly be at each time slice in a given range.

Memoization is particularly well-suited for this problem. Maintain a three-dimensional table indexed by *width*, *height*, and *time* holding three possible values: yes the robber could be there, no the robber cannot be there, and uncomputed.

Initialize the table to uncomputed for all possible values. Then read the witness input and mark every rectangle given by them to vacant. Now we can use memoization to decide which paths lead to a valid city block after the time is over. Start from the time t, and work back to time 1. A given position (*width, height, time*) can be reached only if at least one of its at most five predecessors can be reached (*width*±1, *height*±1, *time*−1) and (*width, height, time*−1). Obviously, *time*=1 is the base case for the recursion. Don't be greedy; explore all five options, even if the first one works, since we not only want to determine if the position is feasible, but also if it is unique.

After trying all paths starting at the end time, we can then perform the output. If there are no places at the finishing time where the robber can be, output that the robber must have escaped. Otherwise, for each time that there is only one position, output that position. Finally, if nothing was printed, output "Nothing known."

6.4.16 Always on the Run

Screeching tires. Searching lights. Wailing sirens. Police cars everywhere. Trisha Quickfinger did it again! Stealing the "Mona Lisa" had been more difficult than planned, but being the world's best art thief means expecting the unexpected. So here she is, with the wrapped frame tucked firmly under her arm, running to catch the northbound metro to the Charles de Gaulle airport.

But even more important than actually stealing the painting is to shake off the police that will soon be following her. Trisha's plan is simple: for several days she will be flying from one city to another, making one flight per day. When she is reasonably sure that the police have lost her trail, she will fly to Atlanta and meet her "customer" (known only as Mr. P.) to deliver the painting.

Her plan is complicated by the fact that nowadays, even when you are stealing expensive art, you have to watch your spending budget. Trisha therefore wants to spend the least money possible on her escape flights. This is not easy, since airline prices and flight availability vary from day to day. The price and availability of an airline connection depends on the two cities involved and the day of travel. Every pair of cities has a "flight schedule" which repeats every few days. The length of the period may be different for each pair of cities and for each direction.

Although Trisha is a good at stealing paintings, she easily gets confused when booking airline flights. This is where you come in.

Input

The input contains the descriptions of several scenarios in which Trisha tries to escape. Every description starts with a line containing two integers n and k. n is the number of cities through which Trisha's escape may take her, and k is the number of flights she will take. The cities are numbered 1, 2, ..., n, where 1 is Paris, her starting point, and n is Atlanta, her final destination. The numbers will satisfy $2 \leq n \leq 10$ and $1 \leq k \leq 1000$.

Next you are given $n(n-1)$ flight schedules, one per line, describing the connection between every possible pair of cities. The first $n-1$ flight schedules correspond to the flights from city 1 to all other cities (2, 3, ..., n), the next $n-1$ lines to those from city 2 to all others (1, 3, 4, ..., n), and so on.

The description of the flight schedule itself starts with an integer d, the length of the period in days, with $1 \leq d \leq 30$. Following this are d non-negative integers, representing the cost of the flight between the two cities on days 1, 2, ..., d. A cost of 0 means that there is no flight between the two cities on that day.

So, for example, the flight schedule "3 75 0 80" means that on the first day the flight costs 75, on the second day there is no flight, on the third day it costs 80, and then the cycle repeats: on the fourth day the flight costs 75, there is no flight on the fifth day, etc.

The input is terminated by a scenario with the formula $n=k=0$.

Output

For each scenario in the input, first output the number of the scenario, as shown in the sample output. If it is possible for Trisha to travel k days, starting in city 1, each day flying to a different city than the day before, and finally (after k days) arriving in city n, then print "The best flight costs x.", where x is the least amount that the k flights can cost.

If it is not possible to travel in such a way, print "No flight possible.".

Print a blank line after each scenario.

Sample Input	Sample Output
3 6	Scenario #1
2 130 150	The best flight costs 460.
3 75 0 80	
7 120 110 0 100 110 120 0	Scenario #2
4 60 70 60 50	No flight possible.
3 0 135 140	
2 70 80	
2 3	
2 0 70	
1 80	
0 0	

Source: ACM Southwestern European Regional Contest 1997

IDs for Online Judges: POJ 1476, ZOJ 1250, UVA 590

Hint

A thief wants to find the cheapest way of travelling to a certain city in exactly k days. She must make exactly one flight each day.

Suppose *cost*[*a*][*b*] is the cost to travel to city *a* on *b* day. Then *cost*[*a*][*b*] could be calculated as the minimum of *cost*[*m*][*b*−1] + cost to travel from city *m* to city *a*.

6.4.17 Martian Mining

The NASA Space Center, Houston, is less than 200 miles from San Antonio, Texas (the site of the ACM Finals this year). This is the place where the astronauts are trained for Mission Seven Dwarfs, the next giant leap in space exploration. The Mars Odyssey program revealed that the surface of Mars is very rich in yeyenum and bloggium. These minerals are important ingredients for certain revolutionary new medicines, but they are extremely rare on Earth. The aim of Mission Seven Dwarfs is to mine these minerals on Mars and bring them back to Earth.

The Mars Odyssey orbiter identified a rectangular area on the surface of Mars that is rich in minerals. The area is divided into cells that form a matrix of *n* rows and *m* columns, where the rows go from east to west and the columns go from north to south. The orbiter determined the amount of yeyenum and bloggium in each cell. The astronauts will build a yeyenum refinement factory west of the rectangular area and a bloggium factory to the north. Your task is to design the conveyor belt system that will allow them to mine the largest amount of minerals.

There are two types of conveyor belts: the first moves minerals from east to west, and the second moves minerals from south to north. In each cell, you can build either type of conveyor belt, but you cannot build both of them in the same cell. If two conveyor belts of the same type are next to each other, then they can be connected. For example, the bloggium mined at a cell can be transported to the bloggium refinement factory via a series of south-north conveyor belts.

The minerals are very unstable, thus they have to be brought to the factories on a straight path without any turns. This means that if there is a south-north conveyor belt in a cell, but the cell north of it contains an east-west conveyor belt, then any mineral transported on the south-north conveyor belt will be lost (see Figure 6.8). The minerals mined in a particular cell have to be put on a conveyor

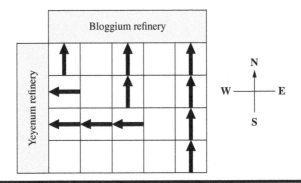

Figure 6.8

belt immediately, in the same cell (thus they cannot start the transportation in an adjacent cell). Furthermore, any bloggium transported to the yeyenum refinement factory will be lost, and vice versa.

Your program has to design a conveyor belt system that maximizes the total amount of minerals mined, i.e., the sum of the amount of yeyenum transported to the yeyenum refinery and the amount of bloggium transported to the bloggium refinery.

Input

The input contains several blocks of test cases. Each case begins with a line containing two integers: the number $1 \leq n \leq 500$ of rows, and the number $1 \leq m \leq 500$ of columns. The next n lines describe the amount of yeyenum that can be found in the cells. Each of these n lines contains m integers. The first line corresponds to the northernmost row; the first integer of each line corresponds to the westernmost cell of the row. The integers are between 0 and 1000. The next n lines describe in a similar fashion the amount of bloggium found in the cells.

The input is terminated by a block with $n=m=0$.

Output

For each test case, you have to output a single integer on a separate line: the maximum amount of minerals that can be mined.

Sample Input	Sample Output
4 4	98
0 0 10 9	
1 3 10 0	
4 2 1 3	
1 1 20 0	
10 0 0 0	
1 1 1 30	
0 0 5 5	
5 10 10 10	
0 0	

Source: ACM Central Europe 2005

IDs for Online Judges: POJ 2948, UVA 3530

Hint

Suppose the matrix that describes the amount of yeyenum is $A[\ ][\ ]$, and the matrix that describes the amount of bloggium is $B[\ ][\ ]$. $F[i][j]$ is the maximum amount of minerals that can be mined in the matrix whose upper-left corner is $(0,0)$ and lower-right corner is (i, j), $0 \le i \le n-1, 0 \le j \le m-1$.

$F[i][j]$ is calculated from top to down, and from left to right, $0 \le i \le n-1$, $0 \le j \le m-1$. That is, before $F[i][j]$ is calculated, $F[i-1][j]$ and $F[i][j-1]$ are calculated. At (i, j), the astronauts can build a south-north conveyor belt for bloggium, or they can build a east-west conveyor belt for yeyenum. Therefore,

$$F[i][j] = Max\{F[i][-1] + \sum_{k_0}^{t} B[k][j], F[i-1][j] + \sum_{k_0}^{j} A[i][k]\}.$$

Obviously, $F[n-1][m-1]$ is the maximum amount of minerals that can be mined.

6.4.18 String to Palindrome

In this problem you are asked to convert a string into a palindrome with a minimum number of operations. The operations are described below.

Here you'd have the ultimate freedom. You are allowed to:

■ Add any character at any position
■ Remove any character from any position
■ Replace any character at any position with another character

Every operation you do on the string would count for a unit cost. You'd have to keep that as low as possible.

For example, to convert "abccda" you would need at least two operations if we allowed you only to add characters. But when you have the option to replace any character, you can do it with only one operation. We hope you would be able to use this feature to your advantage.

Input

The input file contains several test cases. The first line of the input gives you the number of test cases, T ($1 \le T \le 10$). Then T test cases will follow, each in one line. The input for each test case consists of a string containing lowercase letters only. You can safely assume that the length of this string will not exceed 1000 characters.

Output

For each set of input, print the test case number first. Then print the minimum number of characters needed to turn the given string into a palindrome.

Sample Input	Sample Output
6	Case 1: 5
tanbirahmed	Case 2: 7
shahriarmanzoor	Case 3: 6
monirulhasan	Case 4: 8
syedmonowarhossain	Case 5: 8
sadrulhabibchowdhury	Case 6: 8
mohammadsajjadhossain	

Source: 2004-2005 ICPC Regional Contest Warmup 1

ID for Online Judge: UVA 10739

 Hint

Suppose $s_1...s_n$ is a string, and $f[i][j]$ is the minimum number of characters needed to turn $s_i...s_j$ into a palindrome, $1 \leq i \leq j \leq n$.

```
If s_i==s_j, then f[i][j]=f[i+1][j-1];
If s_i≠s_j, there are three possibilities:
        s_j is inserted into the i-th position, or s_i is
deleted, that is, f[i+1][j]+1;
        s_i is inserted into the j-th position, or s_j is
deleted, that is, f[i][j-1]+1;
        s_i is replaced by s_j, or s_j is replaced by s_i, that
is, f[i+1][j-1]+1;
and f[i][j]=min(f[i+1][j], f[i][j-1], f[i+1][j-1])+1.
```

The length of the current substring l is the current stage, $2 \leq l \leq n$. The front pointer i ($1 \leq i \leq n-l+1$) for the substring is the current state. The rear pointer is $j=i+l-1$. Based on the above successor function, $f[i][j]$ is calculated.

Finally, $f[1][n]$ is the minimum number of characters needed to turn the given string into a palindrome.

6.4.19 *String Morphing*

There is a special multiplication operator such that

Left \ Right	a	b	c
a	b	b	a
b	c	b	a
c	a	c	c

Thus *ab=b, ba=c, bc=a, cb=c,* ...

For example, you are given the string ***bbbba*** and the character ***a***,

```
(b(bb)) (ba) = (bb) (ba)   [as bb = b]
          = b(ba)     [as bb = b]
          = bc        [as ba = c]
          = a         [as bc = a]
```

By adding suitable brackets, *bbbba* can produce *a* according to the above multiplication table.

You are asked to write a program to show the morphing steps of a string into an expected character, or otherwise, output "None exist!" if the given string cannot be morphed as expected.

Input

The first line of the input file gives the number of test cases. Each case consists of two lines. The first line is the starting string, which has at most 100 characters. The second line is the target character. All characters in the input are within the range of *a–c*.

Output

For each test case, your output should consist of several lines, showing the morphing steps of a string into the character. In case there are more than one solution, always try to start the morphing from the left. Print a blank line between consecutive sets of output.

Sample Input	Sample Output
2	*bbbba*
bbbba	*bbba*
a	*bba*
bbbba	*bc*
a	*a*
	bbbba
	bbba
	bba
	bc
	a

Source: Second Programming Contest for Newbies 2006

ID for Online Judge: UVA 10981

Hint

First, the relationships between letters and numbers are as follows: $a=0$, $b=1$, $c=2$. The table for a special multiplication operator is shown in the following table.

Left \ Right	0	1	2
0	1	1	0
1	2	1	0
2	0	2	2

Suppose $F[i][j][t]$ shows whether the interval $[i, j]$ can can produce t or not. Obviously $F[i][i][str[i]-'a']$=true. Suppose Fm is used to store how the result is produced, where $Fm[i][j][t][0]$ stores the intermediate pointer producing t; the result for the left subinterval $[i, Fm[i][j][t][0]]$ is stored in $Fm[i][j][t][1]$, and the result for the right subinterval $[Fm[i][j][t][0]+1, j]$ is stored in $Fm[i][j][t][2]$, $1 \leq i \leq j \leq n$, $0 \leq t \leq 2$.

DP is used to calculate $F[][][]$ and $Fm[][][]$, where stage l is the length of the substring, $2 \leq l \leq n$; stage i is the front pointer for the current substring, $1 \leq i \leq n-l+1$; the rear pointer $j=i+l-1$; and decision k ($i \leq k \leq j-1$) is the intermediate pointer such that the left subinterval $[i, k]$ produces a and the right subinterval $[k+1, j]$ produces b, ($0 \leq a, b \leq 2$, $F[i][k][a]$&&$F[k+1][j][b]$=true). The result t ($t=mul[a][b]$) is stored. $Fm[i][j][t][0]=k$, $Fm[i][j][t][1]=a$, $Fm[i][j][t][2]=b$, and $F[i][j][t]$ = true.

Suppose t is the expected character. If $f[1][n][t]$==false, then output "None exist!"; else output the morphing steps of a string into the character based on Fm.

6.4.20 End Up with More Teams

The prestigious ICPC is here again. The coaches are busy selecting teams. This year, they have adopted a new policy. Contrary to the traditional selection process, where few individual contests are held and the top three are placed in one team and the next three in another and so on, this year the coaches decided to place members in such a way that the total number of *promising* teams is maximized. *Promising* teams are defined as a team having *ability points* of its members adding up to **20** or greater. Here the *ability point* of a member denotes his capability as a programmer, the higher the better.

Input

There will be as many as 100 cases in the input file. Each case of input has two lines. The first line contains a positive integer, where **n** indicates the number of contestants available for selection. The next line will contain **n** positive integers, each of which will be at most 30. End of input is indicated by a value of **0** for **n**.

Output

For every case of input, there will be one line of output. This line should contain the case number followed by the maximum number of *promising teams* that can be formed. Note that it is not mandatory to assign everyone in a team. In case you don't know, each team consists of exactly three members.

Constraint: $n \leq 15$

Sample Input	Sample Output
9 22 20 9 10 19 30 2 4 1 6 2 15 3 0	Case 1: 3 Case 2: 0

Source: IIUPC 2006

ID for Online Judge: UVA 11088

 Hint

Suppose S is a sequence of contestants' ability points, and $best(S)$ is the maximum number of *promising teams* that can be formed.

The problem is solved with its subproblem as such:

```
If (|S|<3)
  best(S)=0;
```

$$else\ best(S=\{a_1,a_2,...,a_n\}) = \max_{i,j,k}\left(best\left(S-a_i-a_j-a_k\right)+\begin{cases} 1 & a_i+a_j+a_k \geq 20 \\ 0 & otherwise \end{cases}\right).$$

6.4.21 Many a Little Makes a Mickle

A long string does not look so long if we can identify a few short substrings that were used (possibly more than once) in some permutation to construct the longer string. Your task is to find if a given (long) string can be made up by choosing some (shorter) strings from a given collection.

You should note that:

1. All the strings are composed of ASCII characters in the range 33 to 127.
2. Any of the short strings or their reversed forms can be used any number of times to construct the long string.
3. Each use of a short string or its reverse would be counted as one occurrence of that short string.

When you construct the longer string from these short strings, you should ensure that it is done by keeping the total occurrences of the short strings to a minimum.

For example, if we want to construct the string "aabbabbabbbb" from the set {"a", "bb", "abb"}, there can be many ways to achieve the goal. "a-abb-abb-abb-bb" and "a-abb-a-bba-bb-bb" are two such valid constructions. However, we would prefer "a-abb-abb-abb-bb" (five substrings) over "a-abb-a-bba-bb-bb" (six substrings) because it uses a lesser number of substrings. You would only need to find the minimum number of substrings that could be used to construct the given string.

Input

The first line of the the input contains S ($S<51$), the number of data sets. Then S number of data sets follows. The first line of each data set contains the long string, P ($0<length(P)<10001$). The next line contains the number of short strings, N ($0<N<51$) to choose from. Each of the next N lines contain the short string P_i ($0<length(P_i)<101$) [$i≥1,2,3?N$]. You can safely assume that there is no blank/empty line in the input file.

Output

For each data set print exactly one line of output.
Either Set S: C;
Or S: Not possible.
If it is possible to construct the string using the given strings, then print the first line; otherwise, print the second line. Here S is the serial of data set (sequentially from 1 to S) and C is the minimum number of times the substrings were used to construct P. For clarification see the sample output below.

Sample Input	Sample Output
2	Set 1: 5.
aabbabbabbbb	Set 2: Not possible.
3	
a	

Sample Input	Sample Output
bb	
abb	
ewu**bbacsecsc	
4	
ewu	
bba	
cse	
csc	

Source: Next Generation Contest 1

ID for Online Judge: UVA 10860

Hint

A graph is constructed as follows. Spaces between the characters are as vertices, and the characters are edges. Therefore, two vertices are connected if the string between them is located in given shorter strings (or the reverse). Then it's just a standard DP problem or a standard Shortest Path problem.

The graph is constructed in $O(n^2m)$, where n is the length of the longer string, and m is the length of the shorter string. The problem is strictly $O(n^2m)$ as that is the amount of data we're given.

6.4.22 Rivers

Nearly all of the Kingdom of Byteland is covered by forests and rivers. Small rivers meet to form bigger rivers, which also meet and, in the end, all the rivers flow together into one big river. The big river meets the sea near Bytetown.

There are n lumberjacks' villages in Byteland, each placed near a river. Currently, there is a big sawmill in Bytetown that processes all trees cut in the Kingdom. The trees float from the villages down the rivers to the sawmill in Bytetown. The king of Byteland decided to build k additional sawmills in villages to reduce the cost of transporting the trees down river. After building the sawmills, the trees need not float to Bytetown, but can be processed in the first sawmill they encounter down river. Obviously, the trees cut near a village with a sawmill need not be transported by river. It should be noted that the rivers in Byteland do not fork. Therefore, for each village, there is a unique way down river from the village to Bytetown.

The king's accountants calculated how many trees are cut by each village per year. You must decide where to build the sawmills to minimize the total cost of transporting the trees per year. River transportation costs one cent per kilometre, per tree.

Write a program that:

■ reads from the standard input the number of villages, the number of additional sawmills to be built, the number of trees cut near each village, and descriptions of the rivers,
■ calculates the minimal cost of river transportation after building additional sawmills,
■ writes the result to the standard output.

Figure 6.9 illustrates the example input data. Village numbers are given inside circles. Numbers below the circles represent the number of trees cut near villages. Numbers above the arrows represent the rivers' lengths.

The sawmills should be built in villages 2 and 3.

Input

The first line of input contains two integers: n—the number of villages other than Bytetown ($2 \leq n \leq 100$), and k—the number of additional sawmills to be built ($1 \leq k \leq 50$ and $k \leq n$). The villages are numbered 1, 2, . . . , n, while Bytetown has number 0.

Each of the following n lines contains three integers, separated by single spaces. Line i+1 contains:

W_i—the number of trees cut near village i per year ($0 \leq W_i \leq 10000$);
V_i—the first village (or Bytetown) down river from village i ($0 \leq V_i \leq n$);
D_i—the distance (in kilometres) by river from village i to V_i ($1 \leq D_i \leq 10000$);

It is guaranteed that the total cost of floating all the trees to the sawmill in Bytetown in one year does not exceed 2000000000 cents.

In 50 percent of test cases, n will not exceed 20.

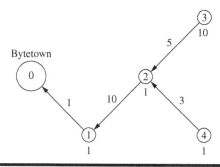

Figure 6.9

Output

The first and only line of the output should contain one integer: the minimal cost of river transportation (in cents).

Sample Input	Sample Output
4 2 1 0 1 1 1 10 10 2 5 1 2 3	4

Source: IOI 2005, Day 2

IDs for Online Judges: BZOJ 1812 http://www.lydsy.com/JudgeOnline/problem.php?id=1812

Hint

A directed graph is constructed as follows. n villages are represented as vertices (numbered from 1 to n), and the number of trees cut near village i per year is the weight for vertex i; there is an edge from a village to its first village (or Bytetown) down river, and the distance between the two villages is the weight for the edge; and Bytetown is vertex 0. The problem requires you to select k villages to build sawmills to reduce the cost of transporting the trees down river. That is, k villages constitute a set A. For each vertex i ($i \notin A$), there is a vertex j ($j \in A$) such that there is a path from vertex i to j, and the path is also the shortest path from vertex i to vertices in A, that is, in the path there are no other vertices in A. The length of the path × the weight of vertex i is the cost of transporting the trees for village i. The problem requires you to calculate the minimal cost of river transportation. Obviously, the problem is a problem for DP on a tree.

For vertex i, its parent pointer is $pa[i]$, its right child is $ch[i]$, and its left sibling is $b[i]$, $1 \le i \le n$.

The current vertex is cur, and its parent is r. In the subtree whose root is cur, l sawmills are built. The minimal cost of river transportation is $f[cur][r][l]$. A recursive function $dfs(cur, r, l, tot)$ calculates $f[cur][r][l]$, where tot is the length of the path from r to the nearest sawmill.

The end condition of recursion: If cur is a leaf ($cur==-1$), if there is no sawmill to be built ($l==0$), return 0; else return ∞.

If cur isn't a leaf, there are two choices;

1. At vertex cur a sawmill is built.
   The rest $l-1$ sawmills are built in the subtree for cur (the length of the path from vertex cur to the nearest sawmill is 0), and the subtree for cur's left

sibling (the length of the path from r to the nearest sawmill is still *tot*). The minimal cost of river transportation is

$$D1= \min_{0 \leq i \leq l-1} \left\{ dfs\left(ch[cur], cur, i, 0\right) + dfs\left(b[cur], r, l-1-i, tot\right) \right\};$$

2. At vertex *cur* there is no sawmill.
   The cost of river transportation from vertex *cur* to the nearest sawmill is $(tot+d[cur]) \times w[cur]$. l sawmills are built in the subtree for *cur* (the length of the path from vertex *cur* to the nearest sawmill is $tot+d[cur]$), and the subtree for *cur*'s left sibling (the length of the path from r to the nearest sawmill is still *tot*). The minimal cost of river transportation is:

$$D2=\min_{0 \leq i \leq s} \left\{ dfs\left(ch[cur], r, i, tot+d[cur]\right) + dfs\left(b[cur], r, l-i, tot\right) \right\}$$
$$+\left(tot+d[cur]\right) \times w[cur].$$

Obviously, $f[cur][r][l]=\min\{D1, D2\}$.
The solution to the problem is $dfs(ch[0], 0, k, 0)$.

6.4.23 Islands and Bridges

Given a map of islands and bridges that connect these islands, a Hamilton path, as we all know, is a path along the bridges such that it visits each island exactly once. On our map, there is also a positive integer value associated with each island. We call a Hamilton path the best triangular Hamilton path if it maximizes the value described below.

Suppose there are n islands. The value of a Hamilton path $C_1C_2...C_n$ is calculated as the sum of three parts. Let V_i be the value for the island C_i. As the first part, we sum over all the V_i values for each island in the path. For the second part, for each edge C_iC_{i+1} in the path, we add the product $V_i \times V_{i+1}$. And for the third part, whenever three consecutive islands $C_iC_{i+1}C_{i+2}$ in the path forms a triangle in the map, i.e., there is a bridge between C_i and C_{i+2}, we add the product $V_i \times V_{i+1} \times V_{i+2}$.

Most likely, but not necessarily, the best triangular Hamilton path you are going to find contains many triangles. It is quite possible that there might be more than one best triangular Hamilton path; your second task is to find the number of such paths.

Input

The input file starts with a number q ($q \leq 20$) on the first line, which is the number of test cases. Each test case starts with a line with two integers n and m, which are the number of islands and the number of bridges in the map, respectively. The next line contains n positive integers, the i-th number being the V_i value of island i. Each value is no more than 100. The following m lines are in the form $x\ y$, which indicates that there is a (two-way) bridge between island x and island y. Islands are numbered from 1 to n. You may assume there will be no more than 13 islands.

Output

For each test case, output a line with two numbers, separated by a space. The first number is the maximum value of a best triangular Hamilton path; the second number should be the number of different best triangular Hamilton paths. If the test case does not contain a Hamilton path, the output must be '0 0'.

Note: A path may be written down in the reverse order. We still think it is the same path.

Sample Input	Sample Output
2	22 3
3 3	69 1
2 2 2	
1 2	
2 3	
3 1	
4 6	
1 2 3 4	
1 2	
1 3	
1 4	
2 3	
2 4	
3 4	

Source: ACM Shanghai 2004

IDs for Online Judges: POJ 2288, ZOJ 2398, UVA 3267

Hint

A graph is constructed as follows. Islands are represented as vertices, bridges are represented as edges, and positive integers associated with islands are represented as weights associated with corresponding vertices. A state for a circuit is represented as a binary number with n digits $d_{n-1} \ldots d_0$. If vertex i is in the circuit, $d_{i+1}=0$; otherwise $d_{i+1}=1$. A circuit is marked by its last edge and its state. Suppose $f[\ ][\ ][\ ]$ and $way[\ ][\ ][\ ]$ are used to store the best triangular Hamilton path, where the last edge in the circuit is (i, j), and the the state value for the circuit is k. The value of the circuit whose state is k is $f[i][j][k]$, and the number of edges in the circuit is $way[i][j][k]$.

Queues $Q1[\ ]$, $Q2[\ ]$, and $Q3[\ ]$ are used to store the two vertices for the last edge and the state for the circuit respectively; and $IN[\ ][\ ][\ ]$ is used to store marks that the circuit exists.

Initially, $f[i][0][2^{i-1}]$ = the weight of vertex i; $way[i][0][2^{i-1}]=1$; $IN[i][0][2^{i-1}]$ = true; i, 0 and 2^{i-1} are stored in $Q1[\ ]$, $Q2[\ ]$, and $Q3[\ ]$ respectively, $1\le i\le n$.

BFS is used for states' transition and to calculate all circuits:

■ Delete fronts of queues (last edge (y, x) and state z), each unvisited vertex xt which is adjacent to vertex x ($(x,xt) \in E$, $z\ \&\ (2^{xt-1}) == 0$) is analyzed:
■ Edge (x, xt) is added into the circuit. The state for the circuit becomes $zt = z + 2^{xt-1}$. The value of the circuit is adjusted as $tmp=f[x][y][z]$+the weight of vertex xt+the weight of vertex x×the weight of vertex xt. If vertices y, x, and xt constitute a triangle ($y\&\&(y, xt)\in E$), $tmp=tmp$+the weight of vertex y× the weight of vertex x× the weight of vertex xt;
■ If the value of the current Hamilton path is maximal ($tmp>f[xt][x][zt]$), then $f[xt][x][zt]=tmp$, and the number of edges is noted ($way[xt][x][zt]=way[x][y][z]$). If ($IN[xt][x][zt]==$false), then edge (x, xt) and zt is added into queues $Q1[\ ]$, $Q2[\ ]$, and $Q3[\ ]$ respectively, and $IN[xt][x][zt]=$true;
■ If the value of the current Hamilton path is the same as $f[xt][x][zt]$ ($tmp==f[xt][x][zt]$), then $way[xt][x][zt]=way[xt][x][zt]+way[x][y][z]$;

Repeat the above process until queues are empty.

Obviously, all Hamilton paths are enumerated, and the maximum value of a best triangular Hamilton path is $max = \max\limits_{1\le i\le n, 0\le j\le n, i\ne j} \{f[i][j][2^n-1]\}$.

The number of different best triangular Hamilton paths is calculated as follows.

$$ans = \sum\limits_{1\le i\le n, 0\le j\le n, 1\ne j} (way[i][j][2^n-1]\,|\,f[i][j][2^n-1]\} = max).$$

If the number of vertices $n>1$, the number of different best triangular Hamilton paths is $ans/2$ (because of the symmetry in an undirected graph); if $n==1$, the number of different best triangular Hamilton paths is ans.

6.4.24 Hie with the Pie

The Pizazz Pizzeria prides itself in delivering pizzas to its customers as fast as possible. Unfortunately, due to cutbacks, they can afford to hire only one driver to do the deliveries. He will wait for one or more (up to ten) orders to be processed before he starts any deliveries. Needless to say, he would like to take the shortest route in delivering these goodies and returning to the pizzeria, even if it means passing the same location(s) or the pizzeria more than once on the way. He has commissioned you to write a program to help him.

Input

Input will consist of multiple test cases. The first line will contain a single integer n indicating the number of orders to deliver, where $1 \le n \le 10$. After this will be $n+1$ lines each containing $n+1$ integers indicating the times to travel between the pizzeria (numbered 0) and the n locations (numbers 1 to n). The j-th value on the i-th line indicates the time to go directly from location i to location j without visiting any other locations along the way. Note that there may be quicker ways to go from i to j via other locations, due to different speed limits, traffic lights, etc. Also, the time values may not be symmetric, i.e., the time to go directly from location i to j may not be the same as the time to go directly from location j to i. An input value of $n=0$ will terminate input.

Output

For each test case, you should output a single number indicating the minimum time to deliver all of the pizzas and return to the pizzeria.

Sample Input	Sample Output
3	8
0 1 10 10	
1 0 1 2	
10 1 0 10	
10 2 10 0	
0	

Source: ACM East Central North America 2006

IDs for Online Judges: POJ 3311, UVA 3725

Hint

Suppose the state for the path is represented as a binary number $D = d_n \ldots d_0$, where

$$d_i = \begin{cases} 1 & \text{Vertex } i \text{ is on the path} \\ 0 & \text{Vertex } i \text{ isn't on the path} \end{cases} \quad (0 \le i \le n);$$

$f[i][k]$ is the minimum time that the pizzeria (numbered 0) is the start, the state for the path is k, and i is the end ($0 \le i \le n$, $0 \le k \le 2^{n+1} - 1$).

First, the Floyd algorithm is used to calculate the shortest paths between any two vertices in the directed graph map[ ][ ]. Obviously, initially $f[i][2^{i-1}] = map[0][i]$. Then Dynamic Programming of State Compression is used to calculate $f[$][ ]:

All possible states of paths i are enumerated ($0 \le i \le 2^n$);

Vertices j and k are enumerated ($1{\le}j$, $k{\le}n$), where vertex j is in the state of the path $(i\&(2^{j-1})=1)$, and vertice k isn't in the state of the path $(i\&(2^{k-1})=0)$; $f[k][i+2^{k-1}]=\text{Min}\{f[k][i+2^{k-1}],\ f[j][i]+\text{map}[j][k]\}$ is calculated.

Obviously, the minimum time to deliver all of the pizzas and return to the pizzeria is $ans=\min\limits_{1\le i\le n}\{f[i][2^{n}-1]+map[i][0]\}$.

6.4.25 Tian Ji: The Horse Racing

Here is a famous story from Chinese history.

About 2300 years ago, General Tian Ji was a high official in the country Qi. He likes to play horse racing with the king and others.

Both Tian and the king have three horses in different classes, namely, regular, plus, and super. The rule is to have three rounds in a match; each of the horses must be used in one round. The winner of a single round takes two hundred silver dollars from the loser.

Being the most powerful man in the country, the king has such nice horses that in each class, his horse is better than Tian's. As a result, each time the king takes six hundred silver dollars from Tian.

Tian Ji was not happy about that, until he met Sun Bin, one of the most famous generals in Chinese history. Using a little trick he learned from Sun, Tian Ji brought home two hundred silver dollars in the next match.

It was a rather simple trick. Using his regular class horse race against the super class from the king, they will certainly lose that round. But then his plus beat the king's regular, and his super beat the king's plus (see Figure 6.10). What a simple trick. And what do you think of Tian Ji, the high-ranked official in China?

Where Tian Ji lives nowadays, he will certainly laugh at himself. Even more, if he were sitting in the ACM contest right now, he may discover that the horse racing problem can be simply viewed as finding the maximum matching in a bipartite graph. Draw Tian's horses on one side, and the king's horses on the other. Whenever one of Tian's horses can beat one from the king, we draw an edge between them, meaning we wish to establish this pair. Then, the problem of winning as many rounds as possible is just to find the maximum matching in this graph. If there are ties,

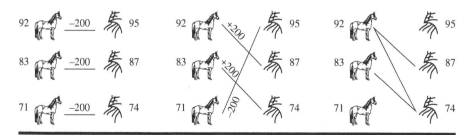

Figure 6.10

the problem becomes more complicated; he needs to assign weights 0, 1, or −1 to all the possible edges, and find a maximum weighted perfect matching.

However, the horse racing problem is a very special case of bipartite matching. The graph is decided by the speed of the horses—a vertex of higher speed always beats a vertex of lower speed. In this case, the weighted bipartite matching algorithm is too advanced a tool to deal with the problem.

In this problem, you are asked to write a program to solve this special case of matching problem.

Input

The input consists of up to 50 test cases. Each case starts with a positive integer n ($n \leq 1000$) on the first line, which is the number of horses on each side. The next n integers on the second line are the speeds of Tian's horses. Then the next n integers on the third line are the speeds of the king's horses. The input ends with a line that has a single "0" after the last test case.

Output

For each input case, output a line containing a single number, which is the maximum money Tian Ji will get, in silver dollars.

Sample Input	Sample Output
3	200
92 83 71	0
95 87 74	0
2	
20 20	
20 20	
2	
20 19	
22 18	
0	

Source: ACM Shanghai 2004

IDs for Online Judge: POJ 2287, ZOJ 2397, UVA 3266

Hint

The speeds of Tian's horses and the speeds of the king's horses are sorted in descending order respectively. If the king's horses participating in the horse racing are in this order, then Tian will let the current slowest horse or the current fastest horse participate in the horse racing each time. If the king's current fastest horse can defeat Tian's any current horse, it defeats Tian's current slowest horse. If the king's current fastest horse can be defeated, Tian's current fastest horse defeating it is suitable.

Suppose $f[i][j]$ is the maximum silver dollars Tian will get when Tian can use horse from number i to number j, and the king's current horse is horse $j-i+1$. Obviously, $f[1][n]$ is the solution to the problem.

```
f[i][j]=max(f[i+1][j]+cmp(a[i], b[j-i+1]), f[i]
[j-1]+cmp(a[j], b[j-i+1]))
```

where $a[\ ]$ is Tian's horse, $b[\ ]$ is the king's horse, and *cmp* is the result that the two horses race.

6.4.26 Batch Scheduling

There is a sequence of N jobs to be processed on one machine. The jobs are numbered from 1 to N, so that the sequence is 1, 2,..., N. The sequence of jobs must be partitioned into one or more batches, where each batch consists of consecutive jobs in the sequence. The processing starts at time 0. The batches are handled one by one starting from the first batch as follows. If a batch b contains jobs with smaller numbers than batch c, then batch b is handled before batch c. The jobs in a batch are processed successively on the machine. Immediately after all the jobs in a batch are processed, the machine outputs the results of all the jobs in that batch. The output time of a job j is the time when the batch containing j finishes.

A setup time S is needed to set up the machine for each batch. For each job i, we know its cost factor F_i and the time T_i required to process it. If a batch contains the jobs x, $x+1$,... , $x+k$, and starts at time t, then the output time of every job in that batch is $t+S+(T_x+T_{x+1}+...+T_{x+k})$. Note that the machine outputs the results of all jobs in a batch at the same time. If the output time of job i is O_i, its cost is $O_i \times F_i$. For example, assume that there are five jobs, and the setup time $S = 1$, $(T_1, T_2, T_3, T_4, T_5) = (1, 3, 4, 2, 1)$, and $(F_1, F_2, F_3, F_4, F_5) = (3, 2, 3, 3, 4)$. If the jobs are partitioned into three batches {1, 2}, {3}, {4, 5}, then the output times $(O_1, O_2, O_3, O_4, O_5) = (5, 5, 10, 14, 14)$ and the costs of the jobs are (15, 10, 30, 42, 56), respectively. The total cost for a partitioning is the sum of the costs of all jobs. The total cost for the example partitioning above is 153.

Input

Your program reads from standard input. The first line contains the number of jobs N, $1 \le N \le 10000$. The second line contains the batch setup time S which is an integer, $0 \le S \le 50$. The following N lines contain information about the jobs 1, 2,..., N in that order as follows. First on each of these lines is an integer T_i, $1 \le T_i \le 100$, the processing time of the job. Following that, there is an integer F_i, $1 \le F_i \le 100$, the cost factor of the job.

Output

Your program writes to standard output. The output contains one line, which contains one integer: the minimum possible total cost.

Sample Input	Sample Output
5	153
1	
1 3	
3 2	
4 3	
2 3	
1 4	

Source: IOI 2002

IDs for Online Judge: POJ 1180

Hint (given by the problemsetter)

This problem can be solved using DP. Let C_i be the minimum total cost of all partitionings of jobs $J_i, J_{i+1}, \ldots, J_n$ into batches. Let $C_i(k)$ be the minimum total cost when the first batch is selected as $\{J_i, J_{i+1}, \ldots, J_{k-1}\}$. That is,
$$C_i(k) = C_k + (S + T_i + T_{i+1} + \ldots + T_{k-1}) \times (F_i + F_{i+1} + \ldots + F_n).$$
Then we have that $C_i = \min \{C_i(k) | k = i+1, \ldots, n+1\}$ for $1 \leq i \leq n$, and $C_{n+1} = 0$.

The time complexity of the above algorithm is $O(n^2)$.

Investigating the property of $C_i(k)$, this problem can be solved in $O(n)$ time.

From $C_i(k) = C_k + (S + T_i + T_{i+1} + \ldots + T_{k-1}) \times (F_i + F_{i+1} + \ldots + F_n)$,
we have that for $i < k < l, C_i(k) \leq C_i(l)$

$$\Leftrightarrow C_l - C_k + (T_k + T_{k+1} + \ldots + T_{l-1}) \times (F_i + F_{i+1} + \ldots + F_n) \geq 0$$

$$\Leftrightarrow (C_k - C_l)/(T_k + T_{k+1} + \ldots + T_{l-1}) \leq (F_i + F_{i+1} + \ldots + F_n)$$

Let $g(k, l) = (C_k - C_l)/(T_k + T_{k+1} + \ldots + T_{l-1})$ and $f(i) = (F_i + F_{i+1} + \ldots + F_n)$

Property 1: Assume that $g(k, l) \leq f(i)$ for $1 \leq i < k < l$. Then $C_i(k) \leq C_i(l)$.

Property 2: Assume $g(j, k) \leq g(k, l)$ for some $1 \leq j < k < l \leq n$. Then for each i with $1 \leq i < j, C_i(j) \leq C_i(k)$ or $C_i(l) \leq C_i(k)$.

Property 2 implies that if $g(j, k) \leq g(k, l)$ for $j < k < l$, C_k is not needed for computing F_i. Using this property, a linear time algorithm can be designed, which is given in the following.

The algorithm calculates the values C_i for $i = n$ down to 1. It uses a queue-like list $Q = (i_r, i_{r-1}, \ldots, i_2, i_1)$ with tail i_r and head i_1 satisfying the following properties:

$$i_r < i_{r-1} < \ldots < i_2 < i_1 \text{ and } g(i_r, i_{r-1}) > g(i_{r-1}, i_{r-2}) > \ldots > g(i_2, i_1) \text{-------- (1)}$$

When C_i is calculated,

1. // Using $f(i)$, remove the unnecessary element at head of Q.

   ```
   If f(i)≥g(i₂, i₁), delete i₁ from Q since for all h≤i,
   f(h)≥f(i)≥g(i₂, i₁) and Cₕ(i₂)≤Cₕ(i₁) by Property 1.
   Continue this procedure until for some t≥1, g(iᵣ,
   iᵣ₋₁)>g(iᵣ₋₁,iᵣ₋₂)>.....>g(iₜ₊₁, iₜ)>f(i).
   Then by Property 1, Cᵢ(iᵥ₊₁)>Cᵢ(iᵥ) for v=t, ... , r-1 or
   r=t and Q contains only iₜ.
   Therefore, Cᵢ(iₜ) is equal to min{Cᵢ(k)|k=i+1, ... , n+1}.
   ```

2. // When inserting i at the tail of Q, maintain Q for the condition (1) to be satisfied.

   ```
   If g(i, iᵣ)≤g(iᵣ, iᵣ₋₁), delete iᵣ from Q by Property 2.
   Continue this procedure until g(i, iᵥ)>g(iᵥ, iᵥ₋₁).
   Append i as a new tail of Q.
   ```

Each i is inserted into Q and deleted from Q at most once. In each insertion and deletion, it takes a constant time. Therefore, the time complexity is $O(n)$.

Chapter 7

Practice for Advanced Data Structures

In this chapter, experiments for some frequently used data structures are discussed as follows:

- In linear lists, experiments for suffix arrays are shown;
- In trees, practices for segment trees are given;
- In graphs, some special graph algorithms are introduced.

7.1 Suffix Arrays

A string is a sequence of characters. A suffix for a string is a substring from a character in the string to the end of the string. A suffix array is a sorted array of all suffixes of a string, and is used in full text indices, data compression algorithms, and so on.

7.1.1 Doubling Algorithm Used to Calculate a Rank Array and a Suffix Array

Suppose S is a string, where its length is $length(S)$, the i-th character in S is $S[i]$, $S[i...j]$ is the substring from $S[i]$ to $S[j]$ in S, and $1 \le i \le j \le length(S)$. The suffix array of S is an array whose elements are suffixes from the i-th character, $1 \le i \le length(S)$, represented as $Suffix(S, i)$, that is, $Suffix(S, i)=S[i..length(S)]$. For convenience, for a string S, the suffix from the i-th character can be written as $Suffix(i)$. Figure 7.1 is an example.

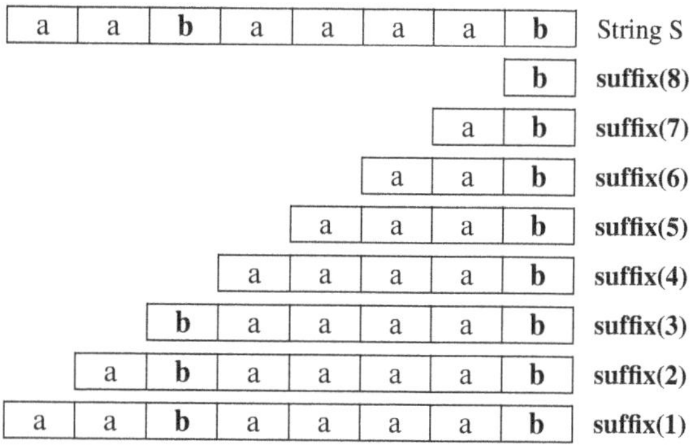

a	a	b	a	a	a	a	b	String S
							b	suffix(8)
						a	b	suffix(7)
					a	a	b	suffix(6)
				a	a	a	b	suffix(5)
			a	a	a	a	b	suffix(4)
		b	a	a	a	a	b	suffix(3)
	a	b	a	a	a	a	b	suffix(2)
a	a	b	a	a	a	a	b	suffix(1)

Figure 7.1

All suffixes for a string can be sorted in lexicographic order. For a string whose length is n, there are n different suffixes. A suffix array SA and a rank array $Rank$ are used to represent sorting the n suffixes.

Suffix Array SA: SA is an integer array storing a permutation of $1, 2, \ldots, n$, and $Suffix(SA[i]) < Suffix(SA[i+1])$, $1 \leq i < n$. For a string S, n suffixes are sorted in lexicographic order, and $SA[i]$ stores the starting position for the i-th suffix. That is, a suffix array SA represents which is the i-th suffix in lexicographic order.

Rank Array $Rank$: $Rank$ is an integer array with respect to SA. If $SA[i]=j$, then $Rank[j]=i$. That is, $Rank$ represents which position a suffix is in.

Therefore, calculating a suffix array SA is the inverse operation for calculating a rank array $Rank$, $Rank = SA^{-1}$. For example, for a string "aabaaaab", its suffix array SA and rank array $Rank$ are shown in Figure 7.2.

Doubling the algorithm is used to calculate a rank array $Rank$ for a string. In order to calculate $Rank$ conveniently, the least character which doesn't appear in the string is added to the end of the substring to make the length of the substring become an integer power of 2.

Doubling the algorithm is as follows. Substrings starting from every character with length 2^k are sorted in lexicographic order, $k \geq 0$. Power k is increased by 1 each time. That is, the length of sorted substrings is doubled each time. And each time, sorting substrings is based on the last $Rank$. Suppose the key xy is the value for the current $Rank$ for the substring whose starting position is i ($1 \leq i \leq n$), and whose length is 2^k, where

  x is the rank for its left substring, that is, its starting position is i, and its length is 2^{k-1}, that is, x is $Rank[i]$ for the last $Rank$;
  y is the rank for its right substring, that is, its starting position is $i+2^{k-1}$, and its length is 2^{k-1}, that is, y is $Rank[i+2^{k-1}]$ for the last $Rank$.

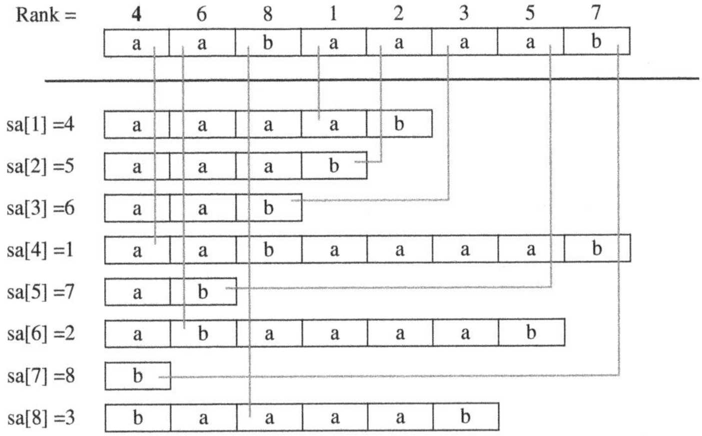

Array SA and Array Rank for "aabaaaab"

Figure 7.2

Values xy representing substrings whose length is 2^k constitute an array $xy[\ ]$. And *Rank* for substrings whose length is 2^k is calculated through sorting xy. When $2^k \geq n$, *Rank* is the rank array. For example, there is a string $S=$"aabaaaab".

$k=0$, substrings whose starting position is every character and length is $2^0=1$ are sorted. $Rank[1..8]=\{1, 1, 2, 1, 1, 1, 1, 2\}$.

$k=1$, substrings whose starting position is every character and length is $2^1=2$ are sorted. That is, based on the previous rank values x and y, key $xy[1..8]=\{11,12, 21,11,11,11,12,20\}$. And $Rank[1..8]=\{1, 2, 4, 1, 1, 1, 2, 3\}$.

$k=2$, substrings whose starting position is every character and length is $2^2=4$ are sorted. Key $xy[1..8]=\{14, 21, 41, 11, 12, 13, 20, 30\}$. And $Rank[1..8]=\{4, 6, 8, 1, 2, 3, 5, 7\}$.

$k=3$, substrings whose starting position is every character and length is $2^3=8$ are sorted. Key $xy[1..8]=\{42, 63, 85, 17, 20, 30, 50, 70\}$. And $Rank[1..8]=\{4, 6, 8, 1, 2, 3, 5, 7\}$.

The process for doubling the algorithm is shown in Figure 7.3.

The program segment *get_suffix_array()* calculating rank array *Rank*[ ] and suffix array *SA*[ ] is as follows:

```
    struct node{int now, next}d[maxn];      // linear list,
where d[ ].now is the sequence number for an element, and
d[ ].next is the successor pointer
    int val[maxn][2], c[maxn], Rank[maxn], SA[maxn], pos[maxn],
x[maxn];      // x[ ] is the string; val[ ][ ] are keys, where x
is val[ ][1], and y is val[ ][2]; c[ ] stores elements' front
```

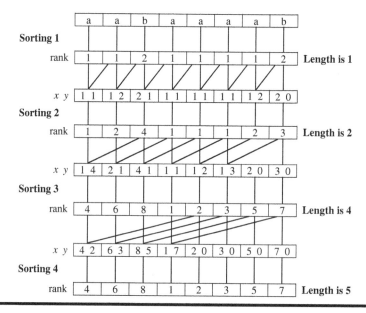

Figure 7.3

```
pointers in d[ ]; Rank[] stores suffixes' rank, SA[] stores
starting position for suffixes
    int n;    // the length of the string
void get_suffix_array( )    //Calculating Rank[ ] and SA[ ]
{
    int t = 1;    //Initialize the length of a substring
    while (t/2<=n){    //calculating Rank[] for substrings
whose length is t
        for (int i=1; i<=n; i++) {
        val[i][0]=Rank[i];  //left substring (start position i,
length t/2)
        val[i][1]=(((i+t/2<=n)?Rank[i+t/2]:0));//right
substring (start position i+t/2, length t/2)
        pos[i]=i;
        }
        radix_sort(1, n);    //val[][0] and val[][1] are
combined into xy, calculate Rank[] that substring's length is t
        t *= 2;
    }
    for (int i=1; i<=n; i++) SA[Rank[i]]=i;    //Calculate
SA[] based on Rank[]
}
// radix_sort(1, n) used to sort key xy
void radix_sort(int l, int r) // val[][0] and val[][1] are
combined into xy, calculate Rank[1...r] that substring's length
is t
```

```
{
    for (int k =1; k>=0;k --)
    {
        memset(c, 0, sizeof(c));
        for (int i=r; i>=1; i --)
           add_value(val[pos[i]][k], pos[i], i);
        int t = 0;
        for (int i =0; i<=20000; i ++)
           for (int j=c[i]; j; j=d[j].next) pos[++t]=d[j].now;
    }
    int t=0;
    for (int i=1; i<=n; i ++) {
        if (val[pos[i]][0]!=val[pos[i-1]][0]||val[pos[i]]
[1]!=val[pos[i-1]][1]) t++;
        Rank[pos[i]] = t;
    }
}
void add_value(int u, int v, int i)
{
    d[i].next=c[u]; c[u]=i;
    d[i].now=v;
}
```

The time complexity of doubling the algorithm is $O(n*\log_2 n)$.

7.1.2 The Longest Common Prefix

The algorithm calculating the longest common prefix is also important in processing strings.

Property 7.1.1 Suppose *height[i]* is the length of the longest common prefix for *suffix(SA[i−1])* and *suffix(SA[i])*, that is, the length of the longest common prefix for two adjacent suffixes in *SA*. For *j* and *k*, if *Rank[j]<Rank[k]*, there is the following property:

The length of the longest common prefix for *suffix(j)* and *suffix(k)* is the minimum for {*height[Rank[j]+1]*, *height[Rank[j]+2]*, *height[Rank[j]+3]*, . . . , *height[Rank[k]]*}.

For example, for a string "aabaaaab", the length of the longest common prefix for suffixes "abaaaab" and "aaab" is calculated as shown in Figure 7.4.

In Figure 7.4, the rank for suffix "abaaaab" is 6, that is, *SA[6]*=2, and *Rank[2]*=6; the rank for suffix "aaab" is 2, that is, *SA[2]*=5, and *Rank[5]*=2. The length of the longest common prefix for suffix "abaaaab" and suffix "aaab" is *min{height[3]*, *height[4]*, *height[5]*, *height[6]}*=*min*{2, 3, 1, 2}=1.

Calculating the longest common prefix for suffixes is to calculate the minimum in an interval. The longest common prefix for *suffix(j)* and *suffix(k)* is the minimum in an interval [*Rank[j]+1* ... *Rank[k]*].

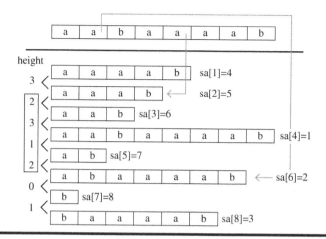

Figure 7.4

The key to the problem is how to calculate array *height*[] effectively. If *height*[2], *height*[3], ..., and *height*[n] are calculated one by one, the time complexity is $O(n^2)$. In order to calculate array *height*[] effectively, $h[i]$ is defined: $h[i]=height[Rank[i]]$.

Property 7.1.2 $h[i] \geq h[i-1]-1$.

Based on $h[1]$, $h[2]$,..., $h[n]$, array *height*[ ] can be calculated. Its time complexity is $O(n)$. The program segment is as follows:

```
void get_common_prefix()       //Calculating the array for the
longest common prefix height[ ]
{
    memset(h, 0, sizeof(h));
    for (int i=1; i<=n; i++) {   // Recursion: calculating h[ ]
      if (Rank[i]==1)
          h[i]=0;
      else{
          int now=0;
          if (i>1 && h[i-1]>1) now=h[i-1]-1;
          while(now+i<=n&&now+sa[Rank[i]-1]<=n&&x[now+i]==x[now
          +sa[Rank[i]-1]])
              now ++;
          h[i] = now;
      }
    }
    for (int i =1; i <= n; i ++) height[Rank[i]]=h[i];
//Based on h[ ], height[ ] is calculated
}
```

7.1.3 Application of Suffix Array

The reasons why suffix arrays can be widely used in string processing are as follows:

1. Based on the rank array *Rank*[ ] and the array for the longest common prefix *height*[ ], brute-force searches can be avoided and algorithms can be optimized;
2. The efficiency for calculating the rank array *Rank*[ ] and the array for the longest common prefix *height*[ ] can be improved. Calculating a rank array *Rank*[ ] and the array for the longest common prefix *height*[ ] can also be implemented as standard program segments.

In **7.1.3.1 Musical Theme**, *get_suffix_array*() is to calculate the rank array *Rank*[ ]. And in **7.1.3.2 Common Substrings**, *get_common_prefix*() is to calculate the array for the longest common prefix *height*[ ].

7.1.3.1 Musical Theme

A musical melody is represented as a sequence of *N* (1≤*N*≤20000) notes that are integers in the range 1..88, each representing a key on the piano. It is unfortunate but true that this representation of melodies ignores the notion of musical timing; but, this programming task is about notes and not about timings.

Many composers structure their music around a repeating "theme", which, being a subsequence of an entire melody, is a sequence of integers in our representation. A subsequence of a melody is a theme if it:

Is at least five notes long
Appears (potentially transposed—see below) again somewhere else in the piece of music
Is disjoint from (i.e., non-overlapping with) at least one of its other appearance(s)

Transposed means that a constant positive or negative value is added to every note value in the theme subsequence.
Given a melody, compute the length (number of notes) of the longest theme.
One-second time limit for this problem's solutions!

Input

The input contains several test cases. The first line of each test case contains the integer *N*. The following *N* integers represent the sequence of notes.
The last test case is followed by one zero.
Use scanf instead of cin to reduce the read time.

Output

For each test case, the output file should contain a single line with a single integer that represents the length of the longest theme. If there are no themes, output 0.

Sample Input	Sample Output
30 25 27 30 34 39 45 52 60 69 79 69 60 52 45 39 34 30 26 22 18 82 78 74 70 66 67 64 60 65 80 0	5

Source: LouTiancheng@POJ

IDs for Online Judges: POJ 1743

 Analysis

One application of suffix arrays is to compute the length of the longest disjoint repeating substrings in a string. First, we need to determine two substrings whose length is k are the same and disjoint. The length of the longest common prefix, array *height*[ ], is used to solve the problem. Sorted suffixes are divided into several groups, where in each group suffixes' *height* aren't less than a number. For example, there is a string "aabaaaab". If $k=2$, suffixes for "aabaaaab" are divided into four groups, as shown in Figure 7.5.

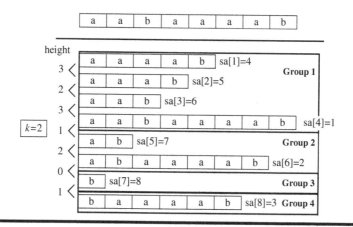

Figure 7.5

In group 1: *height*[2]=3, *height*[3]=2, and *height*[4]=3. In the group suffixes' *height* aren't less than 2. The difference between the maximum and the minimum for suffixes' *SA* is *SA*[3]−*SA*[4]=5.

In group 2: *height*[5]=1, and *height*[6]=2. In the group suffixes' *height* aren't less than 1. The difference between the maximum and the minimum for suffixes' *SA* is *SA*[5]−*SA*[6]=5.

In group 3: *height*[7]=0. In the group, the difference between the maximum and the minimum for suffixes' *SA* is 0.

In group 4: *height*[8]=1. In the group, the difference between the maximum and the minimum for suffixes' *SA* is 0.

Obviously, the two suffixes whose suffixes' *height* are less than k must be in a group. Then, for each group, we need to determine whether the difference between the maximum and the minimum for suffixes' *SA* is less than k or not. If the difference is less than k, then there exist two suffixes whose length of the longest disjoint repeating substrings is less than k; else there isn't such a pair of suffixes. For example, in group 1, there exist two suffixes whose length of the longest disjoint repeating substrings is less than 3 (*height*[2]=*height*[4]=3), and the difference between the maximum and the minimum for suffixes' *SA* is *SA*[3]−*SA*[4]=6−1=5>3. Therefore, the longest disjoint repeating substring is "aab", and it appears two times.

The algorithm is as follows:

First, a string a is input and pretreated. Because of transposition, the current number subtracts the previous number. A new string whose length is $n-1$ is generated.

Second, for the new string, the longest common prefix, array *height*[ ], is calculated.

Then, the longest repeating substring is calculated.

Finally, if the length of the longest repeating substrings is less than 5, there is no theme; else the length of the longest theme is the length of the longest repeating substring $s+1$, for the longest repeating substrings can't be adjacent, and can be overlapped. If the longest repeating substrings are adjacent in the new string, then they are overlapped in the original string.

The time complexity for the algorithm is $O(n*\log_2 n)$.

 Program

```
#include <iostream>
#include <cstdio>
#include <cmath>
#include <cstdlib>
#include <cstring>
#include <string>
```

```
#include <map>
#include <utility>
#include <vector>
#include <set>
#include <algorithm>
#define maxn 20010      //the upper limit of the length for the
sequence of notes
#define Fup(i,s,t) for (int i=s; i <=t; i ++) //Increasing
loop
#define Fdn(i,s,t) for (int i = s; i >= t; i --) //Descending
loop
#define Path(i,s) for (int i=s; i; i=d[i].next)   //Singly
Linked List d[]
using namespace std;
struct node {int now, next;}d[maxn];      // d[], where d[].now
is the sequence number for an element, and d[].next is the
successor pointer
int val[maxn][2], c[maxn], rank[maxn], sa[maxn], pos[maxn],
h[maxn], height[maxn], x[maxn];      //x[]: the sequence of
notes is transposed; val[][]: keys, where x is val[][0], y is
val[][1]; c[ ] stores elements' front pointers in d[ ]; Rank[]
stores suffixes' rank; SA[] stores the starting position for
suffixes; height[] is the array for the longest common prefix;
h[i]=height[Rank[i]]
int n;      // the length of the sequence of notes
void add_value(int u, int v, int i)    //add an element into d[]
{
    d[i].next = c[u]; c[u] = i;
    d[i].now = v;
}
void radix_sort(int l, int r)         // val[][0] and val[][1]
are combined into xy, calculate Rank[l...r] that substring's
length is t
{
    Fdn(k, 1, 0){
      memset(c, 0, sizeof(c));
      Fdn(i, r, 1) add_value(val[pos[i]][k], pos[i], i);
      int t = 0;
        Fup(i, 0, 20000)
            Path(j, c[i])
            pos[++ t] = d[j].now;
    }
    int t = 0;
    Fup(i, 1, n){
        if (val[pos[i]][0] != val[pos[i - 1]][0] ||
val[pos[i]][1] != val[pos[i - 1]][1])
            t ++;
        rank[pos[i]] = t;
    }
}
```

```
bool exist(int len)    //If there are disjoint repeating
substrings whose length is len, return 1; else return 0
{
    int Min = n + 1, Max = 0;    // the maximum and the
minimum for suffixes' SA is initialized
    Fup(i, 1, n)    //Rank is in ascending order
        if (height[i] < len){    //if height[i]<len, and the
difference between the maximum and the minimum for suffixes'
SA isn't less than len, return 1
            if (Max - Min >= len)
                return 1;
            Min = Max = sa[i];
        }else{    //adjust the maximum and the minimum for
suffixes' SA
            Min = min(Min, sa[i]);
            Max = max(Max, sa[i]);
        }
    if (Max - Min >= len)    //if the difference between the
maximum and the minimum for suffixes' SA isn't less than len,
return 1; else return 0
        return 1;
    return 0;
}
void get_suffix_array()    // get_suffix_array() is in 7.1.1
{
    int t = 1;
    while (t / 2 <= n){
        Fup(i, 1, n){
            val[i][0]=rank[i];
            val[i][1] = (((i + t / 2 <= n) ? rank[i + t / 2] :
0));
            pos[i] = i;
        }
        radix_sort(1, n);
        t *= 2;
    }
    Fup(i, 1, n) sa[rank[i]] = i;
}
void get_common_prefix()    // get_common_prefix() is in 7.1.2
{
    memset(h, 0, sizeof(h));
    Fup(i, 1, n){
        if (rank[i] == 1)
            h[i] = 0;
        else{
            int now = 0;
            if (i > 1 && h[i - 1] > 1)
                now = h[i - 1] - 1;
            while (now + i <= n && now + sa[rank[i] - 1] <= n
&& x[now + i] == x[now + sa[rank[i] - 1]])
```

```
                now ++;
           h[i] = now;
        }
    }
    Fup(i, 1, n) height[rank[i]] = h[i];
}
int binary_search(int l, int r)    //using binary search to
calculate the length of the longest disjoint repeating
substring
{
    while (l <= r){
        int mid = (l + r) / 2;
        if (exist(mid))    //If there exists a disjoint
repeating substring whose length is mid, search the left
interval; else search the right interval
            l = mid + 1;
        else
            r = mid - 1;
    }
    return r;    //return the length
}
void solve()    //compute and output the length of the longest
theme
{
    Fup(i, 1, n - 1)    //For two adjacent notes, subtract the
previous note from the current note, a new string is formed
        rank[i] = x[i]= x[i + 1] - x[i] + 88;
    n --;    //the length of the new string
    get_suffix_array();    //calculate Rank[]
    get_common_prefix();    // calculate height[]
    int ans = binary_search(0, n) + 1;    // using binary
search to calculate the length of the longest disjoint
repeating substring
    ans = ((ans < 5) ? 0 : ans);    // at least five notes
long
    printf("%d\n", ans);    //output the length of the longest
theme
}
int main()
{
    while (scanf("%d\n", &n), n > 0){
        Fup(i, 1, n)scanf("%d", &x[i]);    // the sequence of
notes
        solve();    // calculate output the length of the
longest theme
    }
    return 0;
}
```

7.1.3.2 Common Substrings

A substring of a string T is defined as:

$$T(i,k) = T_iT_{i+1}...T_{i+k-1}, 1 \leq i \leq i+k-1 \leq |T|$$

Given two strings A, B and one integer K, we define S, a set of triples (i, j, k):

$$S = \{(i, j, k)|k \geq K, A(i, k) = B(j, k)\}.$$

You are to give the value of $|S|$ for specific A, B and K.

Input

The input file contains several blocks of data. For each block, the first line contains one integer K, followed by two lines containing strings A and B, respectively. The input file is ended by $K=0$. $1 \leq |A|, |B| \leq 10^5$, $1 \leq K \leq min\{|A|, |B|\}$. Characters of A and B are all Latin letters.

Output

For each case, output an integer $|S|$.

Sample Input	Sample Output
2	22
aababaa	5
abaabaa	
1	
xx	
xx	
0	

Source: POJ Monthly, 2007.10.06, wintokk

ID for Online Judge: POJ 3415

 Analysis

The problem requires you to calculate the number of common substrings whose length isn't less than k for two strings A, B.

In the previous problem, array *height*[ ] is the length of the longest common prefix for two suffixes whose ranks are adjacent. In this problem, array *height*[ ] is the number of common substrings whose length isn't less than k for the longest common prefix for two suffixes whose ranks are adjacent. If *height*[i]$-k+1>0$, then there are *height*[*i*]−*k*+1 common substrings whose length is *k* for two suffixes whose ranks are *i* and *i*−1 respectively, and *height*[*i*]←*height*[*i*]−*k*+1; else there are no common substrings whose length is *k* for the two suffixes. Therefore, the idea for solving the problem is as follows.

Calculate the length of the longest common prefix for all suffixes for strings *A*, *B*, and accumulate the number of common substrings whose length isn't less than *k*.

The algorithm is as follows. Strings *A* and *B* adjoin. And a character (e.g., "$") which doesn't appear is inserted into the string to separate Strings *A* and *B*. Based on array *height*[ ], strings are divided into several groups. For each group, the number of common substrings whose length isn't less than *k* is calculated. For each suffix for *B*, calculate the number of common substrings whose length isn't less than *k* for the longest common prefix for all suffixes for *A*. And for each suffix for *A*, calculate the number of common substrings whose length isn't less than *k* for the longest common prefix for all suffixes for *B*.

 Program

```
#include <iostream>
#include <cstdio>
#include <cmath>
#include <cstdlib>
#include <cstring>
#include <string>
#include <map>
#include <utility>
#include <vector>
#include <set>
#include <algorithm>
#define maxn 200010
#define Fup(i, s, t) for (int i = s; i <= t; i ++)
#define Fdn(i, s, t) for (int i = s; i >= t; i --)
#define Path(i, s) for (int i = s; i; i = d[i].next)
using namespace std;
struct node {int now, next;}d[maxn];    // d[], where d[].now
is the sequence number for an element, and d[].next is the
successor pointer
int val[maxn][2], c[maxn], rank[maxn], sa[maxn], pos[maxn],
h[maxn], height[maxn], x[maxn], sta[maxn], num1[maxn],
num2[maxn];    //x[] is a combined array; val[][] are keys,
where x is val[][0], and y is val[][1]; c[] stores front
pointers for elements in d[]; Rank[], SA[] and height[] have
been defined; h[i]=height[Rank[i]]; h[i]=height[Rank[i]];
string S, s;    // two strings for a test case
```

```
int n, k;
void add_value(int u, int v, int i) // add an element into d[]
{
    d[i].next = c[u]; c[u] = i;
    d[i].now = v;
}
void radix_sort(int l, int r)     // val[][0] and val[][1] are
combined into xy, calculate Rank[l...r] that substring's length
is t
{
    Fdn(k, 1, 0){
        memset(c, 0, sizeof(c));
        Fdn(i, r, 1)
            add_value(val[pos[i]][k], pos[i], i);
        int t = 0;
        Fup(i, 0, 200000)
            Path(j, c[i])
            pos[++ t] = d[j].now;
    }
    int t = 0;
    Fup(i, 1, n){
        if (val[pos[i]][0] != val[pos[i - 1]][0] ||
val[pos[i]][1] != val[pos[i - 1]][1])
            t ++;
        rank[pos[i]] = t;
    }
}
void get_suffix_array()   //calculate Rank[] and SA[]
{
    int t = 1;
    while (t / 2 <= n){
        Fup(i, 1, n){
            val[i][0] = rank[i];
            val[i][1] = (((i + t / 2 <= n) ? rank[i + t / 2] :
0));
            pos[i] = i;
        }
        radix_sort(1, n);
        t *= 2;
    }
    Fup(i, 1, n)
        sa[rank[i]] = i;
}
void get_common_prefix()   //Calculating the array for the
longest common prefix height[ ]
{
    memset(h, 0, sizeof(h));
    Fup(i, 1, n){
        if (rank[i] == 1)
            h[i] = 0;
```

```
        else{
            int now = 0;
            if (i > 1 && h[i - 1] > 1)
                now = h[i - 1] - 1;
            while (now + i <= n && now + sa[rank[i] - 1] <= n
&& x[now + i] == x[now + sa[rank[i] - 1]])
                now ++;
            h[i] = now;
        }
    }
    Fup(i, 1, n)
        height[rank[i]] = h[i];
}
void get_ans()    //calculate the number of common substrings
whose length isn't less than k
{
    Fup(i, 2, n)
        height[i] -= k - 1;
    long long sum1 = 0, sum2 = 0, ans = 0;
    int top = 0;
    Fup(i, 2, n)
        if (height[i] <= 0){
            top = sum1 = sum2 = 0;
        }else{
            sta[++ top] = height[i];
            if (sa[i - 1] <= (int)S.size()){
                num1[top] = 1; num2[top] = 0;
                sum1 += (long long)sta[top];
            }else{
                num1[top] = 0; num2[top] = 1;
                sum2 += (long long)sta[top];
            }
            while (top > 0 && sta[top] <= sta[top - 1]){
                sum1 = sum1 - (long long)sta[top - 1] *
num1[top - 1] + (long long)sta[top] * num1[top - 1];
                sum2 = sum2 - (long long)sta[top - 1] *
num2[top - 1] + (long long)sta[top] * num2[top - 1];
                num1[top - 1] += num1[top];
                num2[top - 1] += num2[top];
                sta[top - 1] = sta[top];
                top --;
            }
            if (sa[i] <= (int)S.size())
                ans += sum2;
            else
                ans += sum1;
        }
    cout << ans << endl;
}
void init()    //Input the current test case (two strings) and
are combined into array x[]
```

```
{
    cin >> S >> s;
    n = (int)S.size() + s.size() + 1;
    string str = S + '$' + s;
    Fup(i, 1, n)
        x[i] = rank[i] = (int)str[i - 1];
}
void solve()    //calculate the number of common substrings
whose length isn't less than k
{
    get_suffix_array();
    get_common_prefix();
    get_ans();
}
int main()
{
    ios::sync_with_stdio(false);
    while (cin >> k, k > 0){
        init();
        solve();
    }
    return 0;
}
```

7.1.3.3 Checking the Text

Wind's birthday is approaching. In order to buy a really fantastic gift for her, Jiajia has to take a boring, yet money-making job—a text checker.

This job is very humdrum. Jiajia will be given a string of text consisting of English letters, and he must count the maximum number of letters that can be matched, starting from position two of the current text simultaneously. The matching proceeds from left to right, one character by one.

Even worse, sometimes the boss will insert some characters before, after, or within the text. Jiajia wants to write a program to do his job automatically, but this program should be fast enough, because there are only a few days before Wind's birthday.

Input

The first line of input file contains initial text.

The second line contains the number of commands n. And the following n lines describe each command. There are two formats of commands:

I *ch p*: Insert a character *ch* before the *p*-th. If *p* is larger than the current length of text, then insert at end of the text.

Q *i j*: Ask the length of matching started from the *i*-th and *j*-th character of the initial text, which doesn't include the inserted characters.

You can assume that the length of initial text will not exceed 50000, the number of I commands will not exceed 200, and the number of Q commands will not exceed 20000.

Output

Print one line for each Q command, containing the max length of matching.

Sample Input	Sample Output
abaab	0
5	1
Q 1 2	0
Q 1 3	3
I a 2	
Q 1 2	
Q 1 3	

Source: POJ Monthly, 2006.02.26, zgl & twb

ID for Online Judge: POJ 2758

 Analysis

Jiajia will be given a string of text consisting of English letters, and he must count the maximum number of letters that can be matched, starting from position two of the current text simultaneously. That is, given a string, the longest common prefix is required to calculate. Based on the definition of the longest common prefix, the longest common prefix for *suffix*(j) and *suffix*(k) (*Rank*[j]<*Rank*[k]) is *min*{*height*[*Rank*[j]+1], *height*[*Rank*[j]+2], ..., *height*[*Rank*[k]]}, 1≤j<k≤*length*(S).

Dynamic programming is used to calculate the minimal values of *height*[] for all subintervals. A two-dimensional array f is used to store results, where $f[i, j]$ stores the minimal *height* in the subinterval [j, j+2^i−1].

Therefore, for suffixes *suffix*[a] and *suffix*[b], the max length of matching is the minimal values of *height*[] for the rank interval [l, r], where l=*min*(*Rank*[a], *Rank*[b])+1, r=*max*(*Rank*[a], *Rank*[b]).

Suppose *cor*[k] is the current position for the character whose initial position is k; *dis*[k] is the distance between the character whose initial position is k, and the recently inserted character right; *opp*[i] is the initial position for the current i-th character.

1. If Ranks for *suffix*[*a*] and *suffix*[*b*] are same (*l*>*r*), the max length of matching is the length of the suffix *suffix*[*a*], that is, the length of string *s*−*cor*[*a*]+1;
2. If there is no inserted character in the max matching (the minimal values of *height*[] for the rank interval [*l*, *r*] is less than *dis*[*a*] and *dis*[*b*]), the max length of matching is the minimal values of *height*[] for the rank interval [*l*, *r*];
3. Otherwise, the max length of matching is *len*=min(*dis*[*a*], *dis*[*b*]) at least. Then the max length of matching *len* is calculated through a loop statement, and the condition for the loop statement is *cor*[*a*]+*len*≤the length of the string *s*&&*cor*[*b*]+*len*≤the length of the string *s*.

If (the (*cor*[*a*]+*len*−1)-th character in *s*≠the (*str*[*cor*[*b*]+*len*−1])-th character in *s*), then the max length of matching is *len*; else if the (*cor*[*a*]+*len*)-th character in *s* and the (*str*[*cor*[*b*]+*len*])-th character in *s* aren't inserted characters, then the max length of matching is *len*+ the max length of matching for *suffix*[*opp*[*cor*[*a*]+*len*] and *suffix* [*opp*[*cor*[*b*]+*len*]]; else *len*++, and continue to loop.

When the loop ends, *len* is the max length of matching.

 Program

```
#include <iostream>
#include <cstdio>
#include <cmath>
#include <cstdlib>
#include <cstring>
#include <string>
#include <map>
#include <utility>
#include <vector>
#include <set>
#include <algorithm>
#define maxn 50210    //the upper limit of the length of the
text
#define Fup(i, s, t) for (int i = s; i <= t; i ++)
//Increasing loop
#define Fdn(i, s, t) for (int i = s; i >= t; i --)
//Descending loop
#define Path(i, s) for (int i = s; i; i = d[i].next)
//Singly Linked List d[]
using namespace std;
struct node {int now, next;}d[maxn];    // d[], where d[].now
is the sequence number for an element, and d[].next is the
successor pointer
```

```
int f[maxn][20];      // f[i, j] stores the minimal height in
the subinterval [j, j+2^i-1]
int val[maxn][2], c[maxn], rank[maxn], sa[maxn], pos[maxn],
h[maxn], height[maxn], x[maxn], cor[maxn], dis[maxn],
opp[maxn];      //x[]: character array; val[][], x is val[][0],
y is val[][1]; Rank[], SA[] and height[] have been defined;
h[i]=height[Rank[i]]; cor[k] is the current position for the
character whose initial position is k; dis[k] is the distance
between the character whose initial position is k and the
recently inserted character right; opp[i] is the initial
position for the current i-th character
string str;
int n, k;      //length of string, number of commands
void add_value(int u, int v, int i)   //add an element into d[i]
{
    d[i].next = c[u]; c[u] = i;
    d[i].now = v;
}
void radix_sort(int l, int r)      //val[][0] and val[][1] are
combined to constitute xy, and calculate Rank[1..r] when the
length is t
{
    Fdn(k, 1, 0){    //sort y and x
      memset(c, 0, sizeof(c));
      Fdn(i, r, 1) add_value(val[pos[i]][k], pos[i], i);
      int t = 0;
      Fup(i, 0, 50000)
            Path(j, c[i])
            pos[++ t] = d[j].now;
    }
    int t = 0;
    Fup(i, 1, n){
        if (val[pos[i]][0] != val[pos[i - 1]][0] ||
val[pos[i]][1] != val[pos[i - 1]][1])
            t ++;
        rank[pos[i]] = t;
    }
}
void get_suffix_array()    //calculating Rank[] and SA[]
{
    int t = 1;      //initialize the length of the substring
    while (t / 2 <= n){      //while the string can be divided
into left and right substrings, Rank[] for substrings whose
length is t is calculated
        Fup(i, 1, n){
            val[i][0]=rank[i];      //rank for left substring
(starting position is i, length is t/2)
            val[i][1] = (((i + t / 2 <= n) ? rank[i + t / 2] :
0));    ///rank for right substring (starting position is
i+t/2, length is t/2)
```

```
                pos[i] = i;
        }
        radix_sort(1, n);    //val[][0] and val[][1] are
combined to xy, and calculate Rank[] with length t
        t *= 2;    //the length of substring *2
    }
    Fup(i, 1, n) sa[rank[i]] = i;    //SA[]
}
void get_common_prefix()    //Calculate the longest common
prefix height[]
{
    memset(h, 0, sizeof(h));
    Fup(i, 1, n){
        if (rank[i] == 1)
            h[i] = 0;
        else{
            int now = 0;
            if (i > 1 && h[i - 1] > 1)
                now = h[i - 1] - 1;
            while (now + i <= n && now + sa[rank[i] - 1] <= n
&& x[now + i] == x[now + sa[rank[i] - 1]])
                now ++;
            h[i] = now;
        }
    }
    Fup(i, 1, n) height[rank[i]] = h[i];    //calculate
height[] based on h[]
}
void get_RMQ()  //calculate f[][], f[i, j] stores the minimal
height in the subinterval [j, j+2^i-1]
{
    Fup(i, 1, n)f[i][0] = height[i];
    Fup(k, 1, (int)(log(n) / log(2)))    //length is
enumerated (integral power of 2)
        Fup(i, 1, n - (1 << k) + 1)
            f[i][k]=min(f[i][k-1],f[i+(1<<(k - 1))][k-1]);
}
int query(int a, int b)    //calculate the length of maximum
matching string for suffix[a] and suffix[b]
{
    int head = min(rank[a], rank[b])+1,
tail=max(rank[a],rank[b]);
    if (head > tail)
        return (int)str.size() - cor[a] + 1;
    int t = (int)(log(tail - head + 1) / log(2));
    int len = min(f[head][t], f[tail - (1 << t) + 1][t]);
    if (len < dis[a] && len < dis[b])return len;
    len = min(dis[a], dis[b]);
    while (cor[a] + len <= (int)str.size() && cor[b] + len <=
(int)str.size()){
```

```
        if(str[cor[a]+len-1]!=str[cor[b]+len-1])return len;
        if (opp[cor[a] + len] && opp[cor[b] + len])
            return len + query(opp[cor[a] + len], opp[cor[b] +
len]);
        len ++;
    }
    return len;
}
void insert(char ch, int pre)     //character ch is inserted
{
    int t = (int)str.size();     //length of str
    pre = min(t + 1, pre);     //inserted position
    str = str + ' ';     //space is added to the end of the
string
    Fdn(i, t, pre){
        str[i] = str[i - 1];
        opp[i + 1] = opp[i];     // opp[i] is the initial
position for the current i-th character
        if (opp[i])
            cor[opp[i]] = i + 1;     // cor[k] is the current
position for the character whose initial position is k
    }
    opp[pre] = 0;     //the current the pre-th character is the
inserted character
    str[pre - 1] = ch;     //Insertion
    Fdn(i, pre - 1, 1){
        if (!opp[i])break;
        dis[opp[i]] = min(dis[opp[i]], pre - i);
    }
}
void init()     //Input the initial text and commands
{
    cin >> str;     // the initial text
    n = (int)str.size();     //the length of the initial text
    Fup(i, 1, n){     //Initialization
        x[i] = rank[i] = (int)str[i - 1];
        cor[i] = i;opp[i] = i;
    }
    cin >> k;     //number of commands
}
void solve()     //commands are executed one by one
{
    get_suffix_array();     //calculate Rank[]
    get_common_prefix();     //calculate height[]
    get_RMQ();     // calculate the minimal height in the
subinterval
    memset(dis, 127, sizeof(dis));
    Fup(i, 1, k){     // commands are executed one by one
        char kind;
```

```
            cin >> kind;     // the format of a command
            if (kind == 'Q'){    //command Q
                int a, b;
                cin >> a >> b;
                int ans = query(a, b);    //calculate and output
the length of matching
                cout << ans << endl;
            }else{    //command I
                char ch;
                int pos;
                cin >> ch >> pos;
                insert(ch, pos);    // Insert a character ch
before the pos-th.
            }
        }
}
int main()
{
    ios::sync_with_stdio(false);
    init();    //Input initial text and commands
    solve();    //commands are executed
    return 0;
}
```

7.2 Segment Trees

We often meet some interval operations, such as calculating the length of the union of intervals or segments, and so on. A segment tree is a tree storing intervals or segments. Interval operations can be implemented based on segment trees. In this section, experiments for segment trees are given.

7.2.1 Segment Trees

A segment tree is a binary tree $T(a, b)$, where an interval $[a, b]$ represents the root for the binary tree. Suppose $L=b-a$. $T(a, b)$ is defined recursively as follows:

If $L>1$: Interval $\left[a, \left\lfloor \dfrac{a+b}{2} \right\rfloor\right]$ represents the left child for the root, and interval

$\left[\left\lfloor \dfrac{a+b}{2} \right\rfloor+1, b\right]$ is the right child for the root;

If $L=1$: The left child and the right child for $T(a, b)$ are leaves $[a]$ and $[b]$ respectively.

If $L=0$, that is, $a==b$: $T(a, b)$ is a leaf representing $[a]$, that is, an element a.

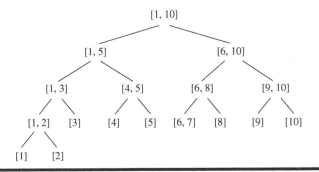

Figure 7.6

In Figure 7.6, there is a segment tree whose root is [1, 10].

Leaves are all data in the interval. An internal node can be regarded not only as an interval, but also as the midpoint for the interval.

An array $a[\ ]$ is used to store a segment tree. If node $a[i]$ represents an interval $[l, r]$, then its left child $a[2\times i+1]$ represents the left subinterval $\left[l, \left\lfloor \dfrac{l+r}{2} \right\rfloor\right]$, and its right child $a[2\times i+2]$ represents the right subinterval $\left[\left\lfloor \dfrac{l+r}{2} \right\rfloor+1, r\right]$. Therefore, each node stores not only an interval, but also some special data, for example, how many segments cover the interval, and so on.

Fundamental operations for a segment tree are as follows:

1. A segment tree is built;
2. A segment or an element is inserted in an interval;
3. A segment or an element is deleted from an interval;
4. A segment tree is updated.

1. **A segment tree is built for interval [*l*, *r*].**

   Based on dichotomy, interval [*l*, *r*] is divided into *tot* ($\geq 2\times\log_2(r-l)$) empty subintervals. These subintervals aren't covered by any segment. *tot* is a global variable, and shows how many nodes are used. Initially *tot*=0. A segment tree $T(l, r)$ is built as follows.

   ```
   void build_tree(int l, int r, int i)    //From node i, a
   segment tree is built for interval [l, r]
   {
      Data field for node i is initialized;
      if (l==r){   //There is only one element in the interval
   ```

```
      Set the sequence number for the leaf containing the
element;
    }
    int mid=(l+r) / 2;    // pointer pointing to the middle
of the interval
    build_tree(l, mid, i+i);    // A segment tree is built
for the left subinterval
    build_tree(mid+1, r, i+i+1);    // A segment tree is
built for the right subinterval
  }
```

2. **A segment or an element is inserted in an interval.**

Suppose R is the root for a segment tree $T(l, r)$, and R represents an interval $[l, r]$. A segment $[c, d]$ will be inserted into the segment tree.

If interval $[c, d]$ covers $[l, r]$ completely, that is, $((c \leq l) \& \&(r \leq d))$, then the number of covered segments in node R increases 1;

If interval $[c, d]$ doesn't cover the midpoint $\left(d \leq \left\lfloor \dfrac{l+r}{2} \right\rfloor \| \left\lfloor \dfrac{l+r}{2} \right\rfloor + 1 \leq c \right)$,

then the segment is inserted into the left subtree or right subtree for node R;

If interval $[c, d]$ covers the midpoint $\left(c \leq \left\lfloor \dfrac{l+r}{2} \right\rfloor \& \& d \geq \left\lfloor \dfrac{l+r}{2} \right\rfloor + 1 \right)$,

then the segment is inserted into the left subtree and right subtree for node R. Its time complexity is $O(\log_2 n)$.

If an element x is inserted in segment tree $T(l, r)$, binary search is used to find the position of the leaf containing x, and element x is interested into the leaf. Its time complexity is $O(\log_2 n)$.

3. **A segment or an element is deleted from an interval.**

Suppose R is the root for a segment tree $T(l, r)$ representing interval $[l, r]$. A segment $[c, d]$ will be deleted from the segment tree. The method is similar to inserting a segment into the segment tree. In order to guarantee that updating the segment tree is correct, a segment $[c, d]$ can be deleted only if there is at least one segment on the interval $[c, d]$.

The method for deleting an element is similar to the method for inserting an element. Obviously an element can be deleted only if the element has been inserted.

4. **A segment tree is updated.**

There are two methods for updating a segment tree:

1. Updating a single point in a segment tree, that is, a segment tree is updated after an element is inserted or deleted.
2. Updating a subinterval in a segment tree, that is, a segment tree is updated after a segment is inserted or deleted.

7.2.2 Updating a Single Point in a Segment Tree

In a segment tree, a leaf node is used to a represent an integer in the interval. Updating a single point in a segment tree means that a segment tree is updated after an element x is inserted into the interval or deleted from the interval. First, binary search is used to find the leaf containing x. Then, statuses for all nodes in path from the leaf to the root are adjusted, for these nodes contain element x.

7.2.2.1 Buy Tickets

Railway tickets were difficult to buy around the Lunar New Year in China, so we must get up early and join a long queue...

The Lunar New Year was approaching, but unluckily the Little Cat still had schedules going here and there. Now, he had to travel by train to Mianyang, Sichuan Province, for the winter camp selection of the national team of Olympiad in Informatics.

It was 1 a.m. and dark outside. A chill wind from the northwest did not scare off the people in the queue. The cold night gave the Little Cat a shiver. Why not find a problem to think about? That was better than freezing to death!

People kept jumping the queue. Since it was too dark all around, such moves would not be discovered even by the people adjacent to the queue-jumpers. "If every person in the queue is assigned an integral value and all the information about those who have jumped the queue and where they stand after queue-jumping is given, can I find out the final order of people in the queue?" thought the Little Cat.

Input

There will be several test cases in the input. Each test case consists of $N+1$ lines where N ($1 \leq N \leq 200{,}000$) is given in the first line of the test case. The next N lines contain the pairs of values Pos_i and Val_i in the increasing order of i ($1 \leq i \leq N$). For each i, the ranges and meanings of Pos_i and Val_i are as follows:

$Pos_i \in [0, i-1]$: The i-th person came to the queue and stood right behind the Pos_i-th person in the queue. The booking office was considered the 0th person and the person at the front of the queue was considered the first person in the queue.
$Val_i \in [0, 32767]$: The i-th person was assigned the value Val_i.

There are no blank lines between test cases. Proceed to the end of input.

Output

For each test case, output a single line of space-separated integers which are the values of people in the order they stand in the queue.

Sample Input	Sample Output
4	77 33 69 51
0 77	31492 20523 3890 19243
1 51	
1 33	
2 69	
4	
0 20523	
1 19243	
1 3890	
0 31492	

Source: POJ Monthly, 2006.05.28, Zhu Zeyuan

IDs for Online Judges: POJ 2828

Hint

Figure 7.7 shows how the Little Cat found out the final order of people in the queue described in the first test case of the sample input.

Analysis

Initially there is an empty sequence. N persons are interested into the sequence. Each person has a value. Values of persons are output in the order they stand in the queue finally.

For each test case, N pairs of values are dealt with in reverse order, in order to guarantee that the inserted position can't be changed. For example, for the second

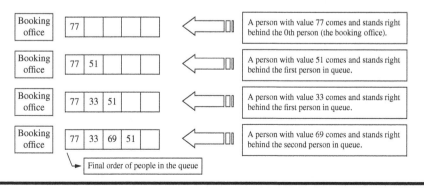

Figure 7.7

sample test case, the sequence of the sample input is 0 20523 1 19243 1 3890 0 31492. These four pairs of values are dealt with in reverse order. First, the fourth pair (*pos*[4], *val*[4])is dealt: *pos*[4]=0, *val*[4]=31492. *j*= *pos*[4]+1=1. The fourth person will be inserted into the "current" *j*-th empty position (the "current" first empty position). Second, the third pair (*pos*[3], *val*[3])is dealt: *pos*[3]=1, *val*[3]=3890. *j*=*pos*[3]+1=2. That is, the third person will be inserted into the current *j*-th empty position (the current second empty position). Third, the second pair (*pos*[2], *val*[2]) is dealt: *pos*[2]=1, *val*[2]=19243. *j*=*pos*[2]+1=2. That is, the third person will be inserted into the current *j*-th empty position (the current second empty position). Finally, the first pair (*pos*[1], *val*[1]) is dealt: *pos*[1]=0, *val*[1]=20523. *j*=*pos*[1]+1=1. That is, the first person will be inserted into the current first empty position. Therefore, for the second test case, the values of people in the order they stand in the queue are 31492 20523 3890 19243.

A segment tree is used to solve the problem. Nodes in the segment tree are used to store the number of empty positions in the corresponding interval. For each time, first the *pos*[*i*]-th empty position is searched, and then nodes' states for representing the *pos*[*i*]-th empty position are changed.

Initially, the state value for a node is the length of interval. Leaf nodes represent persons.

The process is as follows.

From the *n*-th person, each person's position is dealt with one by one. When the *i*-th person is inserted into the sequence, his position is the current *j*-th empty position (*j*=*pos*[*i*]+1, *i*=*n*...1, *pos*[*i*]<*i*), and then the number of the empty position is recursively calculated from the root of the segment tree.

If the number of empty positions in the left subtree ≥*j*, recursive search is on the left subtree; otherwise the *k*-th empty position in the right subtree is calculated, *k*=*j*−(number of empty positions in the left subtree). Repeat the above steps until leaf node *d* is found, where node *d* represents interval [*t*]. Then the *i*-th person's position is *t*.

Then the segment tree is adjusted. On the path from leaf node *d* to the root, the number of empty positions in each node −1.

Repeat the above steps until the first person's position is calculated.

 Program

```
#include <iostream>
#include <cstdio>
#include <cstring>
#include <string>
#include <map>
```

```
#include <utility>
#include <algorithm>
#define maxn 200100      //the upper limit of number of persons
#define Fup(i, s, t) for (int i = s; i <= t; i ++)
#define Fdn(i, s, t) for (int i = s; i >= t; i --)
using namespace std;
int pos[maxn], val[maxn], size[maxn * 3], ans[maxn],
point[maxn];       // pos[i] and val[i]: described in problem;
ans[k]: the i-th person in the queue; point[k]: the sequence
number for the leaf representing [k]; size[j]: number of empty
positions in node j
int n;      //number of persons
void build_tree(int l, int r, int i)      //From node i, a
segment tree is built for interval [l, r]
{
  size[i] = r - l + 1;      // number of empty positions in
node i
  if (l==r){        // There is only one element in the
interval. Set the sequence number for the leaf containing the
element
      point[l] = i;
      return;
  }
  int mid = (l + r) / 2;      // pointer pointing to the middle
of the interval
  build_tree(l, mid, i + i);       // A segment tree is built for
the left subinterval
  build_tree(mid + 1, r, i + i + 1);      // A segment tree is
built for the right subinterval
}
int require(int sum, int l, int r, int i)      //Calculate the
sequence number of the leaf for the sum-th empty position
{
  if (l == r)      // In the interval there is only one element,
return the element
      return l;
  int mid = (l + r) / 2;      // pointer pointing to the middle
  if (size[i + i] >= sum)      //number of empty positions in
the left subtree ≥ sum
      return require(sum, l, mid, i + i);
  return require(sum - size[i + i], mid + 1, r, i + i + 1);
}
void change(int i)      //Updating the segment tree, from leaf i
to the root, adjust number of empty positions
{
  while (i > 0){
    size[i] --;
    i = i / 2;
  }
}
```

```
void init()
{
  Fup(i, 1, n)      // Input test case
    scanf("%d%d\n", &pos[i], &val[i]);
}
void solve()    // calculate and output the values of people
in the order they stand in the queue
{
  memset(size, 0, sizeof(size));
  build_tree(1, n, 1);      //construct segment tree (1, n)
  Fdn(i, n, 1){    // n pairs of values are dealt with in
reverse order
    int t = require(pos[i] + 1, 1, n, 1);
    ans[t] = i;
    change(point[t]);      //updating segment tree
  }
  Fup(i, 1, n - 1)     // output the values of people in the
order they stand in the queue
    cout << val[ans[i]] << ' ';
  cout << val[ans[n]] << endl;
}
int main()
{
  while (scanf("%d\n", &n) == 1){
    init();      //Input
    solve();     //calculate and output the values of people
in the order they stand in the queue
  }
  return 0;
}
```

7.2.3 Updating a Subinterval in a Segment Tree

Updating a subinterval means that data in a subsequence are modified. The method is similar to updating a single point in a segment tree. When a subinterval is updated, a segment tree must be updated from bottom to top. In order to improve the efficiency, a label is used.

In each node, a label is used: If the interval that the node corresponds to is covered completely, then the node is labeled. If a labeled node is found during updating a subinterval, then its left child and right child are labeled, and the node's label is removed. The label's information is determined by updating a subinterval.

In this section, three kinds of experiments for updating a subinterval are shown.

1. Updating data uniformly and calculating data dynamically in a subinterval;
2. Calculating visible segments;
3. Updating and calculating disjoint segments.

7.2.3.1 Updating Data Uniformly and Calculating Data Dynamically in a Subinterval

An interval is represented as a segment tree. Updating data uniformly in a subinterval means that the same value is added to each number in the subinterval. Calculating data dynamically in a subinterval means that the sum of numbers in a subinterval is calculated, and so on. A node's information includes:

1. A label, the value of which is added to each number in the corresponding subinterval;
2. The calculation result of the corresponding subinterval.

7.2.3.1.1 A Simple Problem with Integers

You have N integers, A_1, A_2, ..., A_N, and you need to deal with two kinds of operations. One type of operation is to add some given number to each number in a given interval. The other is to ask for the sum of numbers in a given interval.

Input

The first line contains two numbers N and Q. $1 \leq N, Q \leq 100000$.

The second line contains N numbers, the initial values of A_1, A_2, ..., A_N. $-1000000000 \leq A_i \leq 1000000000$.

Each of the next Q lines represents an operation.

"C a b c" means adding c to each of A_a, A_{a+1}, ..., A_b. $-10000 \leq c \leq 10000$.
"Q a b" means querying the sum of A_a, A_{a+1}, ..., A_b.

Output

You need to answer all Q commands in order. One answer in a line.

Sample Input	Sample Output
10 5	4
1 2 3 4 5 6 7 8 9 10	55
Q 4 4	9
Q 1 10	15
Q 2 4	
C 3 6 3	
Q 2 4	

Hint: The sums may exceed the range of 32-bit integers.

Source: POJ Monthly, 2007.11.25, Yang Yi

IDs for Online Judges: POJ 3468

Analysis

A segment tree is used to solve the problem. Subintervals in the tree correspond to indexes of numbers, that is, $[l, r]$ corresponds to numbers $A_l, A_{l+1}, ..., A_r$. Obviously, leaves represent initial values for $A_l, A_{l+1}, ..., A_r$ from left to right. In each node, there are two attributes:

Attribute 1: The sum of current numbers in the subinterval s. Initially s is the sum of initial numbers in the subinterval.

Attribute 2: Label v, the increasement value for each number in the subinterval. If the operation is "C a b c", then, for all subintervals in $[a\ b]$, $c \times l$ is added to all sums s, where l is the length of the subinterval.

Each time, the label is used to update the segment tree. If node i isn't labeled, then return; else subintervals that the left child and right child correspond to are covered. Sums of current numbers in subintervals for the left child and right child are calculated. And the left child and right child are labeled v.

Suppose the root of the segment tree is i, and corresponds to an interval $[l, r]$.

Then the sum of current numbers in the subinterval $[tl, tr]$ is calculated:

```
If (tl>r| tr<l), return 0;
If [tl, tr] covers [l, r] completely (tl≤l&&r≤tr), return the
sum s for node i;
Node i is labeled to update the segment tree;
For the subinterval [tl, tr], the sum of current numbers in
the left subtree s₁ and the sum of current numbers in right
subtree s₂ are recursively calculated, and return s₁+s₂; v is
added to each number in the subinterval [tl, tr].
If (tl>r||tr<l), then return;
If [tl, tr] covers [l, r] completely (tl≤l&&r≤tr), then v is
added to each number in the subinterval [tl, tr], in node i, s
and v are updated: v*(r-l+1) is added to s, v is accumulated
in node i, and return;
For node i, the label method is used to update the segment
tree;
For node i, the sum of current numbers in the left subtree s₁
and the sum of current numbers in right subtree s₂ are
calculated;
In node i, s =s₁+s₂;
```

Program

```cpp
#include <iostream>
#include <cstdio>
#include <cmath>
#include <cstdlib>
#include <cstring>
#include <string>
#include <map>
#include <utility>
#include <set>
#include <algorithm>
#define maxn 100010     // the upper limit of the number of
numbers
using namespace std;
struct node {long long mark,sum;}tree[maxn*4];     // segment
tree, for node i, the sum of numbers is tree[i].sum, and the
label is tree[i].mark
int x[maxn];     // the sequence of initial numbers
int n, m;     //numbers of numbers and operations
void update(int l, int r, int i)     //label method is to
update a segment tree(i is the root, corresponding to an
interval [l, r])
{
    if (!tree[i].mark) return;     // if label i is labeled,
then return; else subintervals that left child and right child
correspond to are covered. Sums of current numbers in
subintervals for left child and right child are calculated.
And left child and right child are labeled v.
    int mid = (l + r) / 2;
    tree[i + i].sum += tree[i].mark * (long long)(mid - l + 1);
    tree[i + i + 1].sum += tree[i].mark * (long long)(r - mid);
    tree[i+i].mark+=tree[i].mark;
    tree[i+ i+1].mark += tree[i].mark;
    tree[i].mark = 0;     //removing the label for node i
}
long long query(int tl, int tr, int l, int r, int i)     // the
sum of current numbers in the subinterval [tl, tr] is
calculated. (i is the root of segment tree corresponding to
the interval [l, r], [tl, tr] is a subinterval for [l, r])
{
    if (tl > r || tr < l)
        return 0;
    if (tl <= l && r <= tr)     // If [tl, tr] covers [l, r]
completely, return the sum s for node i
        return tree[i].sum;
```

```
    update(l, r, i);    //label method is used to update the
segment tree (i is the root of segment tree corresponding to
the interval [l, r])
    int mid = (l + r) / 2;    //calculate sums of numbers that
[tl, tr] contains in left subtree and right subtree, and
return the sum of sums
    return query(tl, tr, l, mid, i + i) + query(tl, tr, mid +
1, r, i + i + 1);
}
void add_value(int tl, int tr, int l, int r, int i, int val)
//In segment tree (i is the root of segment tree corresponding
to the interval [l, r]), each number in the subinterval [tl,
tr] + val
{
    if (tl > r || tr < l)
        return;
    if (tl<=l && r<=tr){    // If [tl, tr] covers [l, r]
completely (tl≤l&&r≤tr), then val is added to each number in
the subinterval [l, r],
        tree[i].sum += val * (long long)(r - l + 1);
        tree[i].mark += val;    //label
        return;
    }
    update(l, r, i);    // Update the segment tree
    int mid = (l + r) / 2;
    add_value(tl, tr, l, mid, i + i, val);    //recursion for
left and right subtree
    add_value(tl, tr, mid + 1, r, i + i + 1, val);
    tree[i].sum = tree[i + i].sum + tree[i+ i+1].sum;
//accumulation
}
void build_tree(int l, int r, int i)    //construct a segment
tree (i is the root of segment tree corresponding to the
interval [l, r])
{
    if (l == r){    //leaf node
        tree[i].sum = x[l];
        return;
    }
    int mid = (l + r) / 2;    //midpoint
    build_tree(l, mid, i + i);    // left and right subtrees
    build_tree(mid + 1, r, i + i + 1);
    tree[i].sum = tree[i + i].sum + tree[i + i + 1].sum;
//accumulation
}
void solve()    //dealing with operations one by one
{
    memset(tree, 0, sizeof(tree));
    build_tree(1, n, 1);    //construct a segment tree
    scanf("\n");
```

```
    for (int i = 1; i <=m; i ++)     // dealing with operations
one by one
 {
        char ch;
        int l, r, v;
        scanf("%c", &ch);     //Input the i-th operation
        if (ch == 'Q'){     // 'Q' operation, input the
interval [l, r]
            scanf("%d%d\n", &l, &r);
            long long ans = query(l, r, 1, n, 1);
            printf("%lld\n", ans);
        }else{     //     'C' operation
            scanf("%d%d%d\n", &l, &r, &v);
            add_value(l, r, 1, n, 1, v);
        }
    }
}
int main()
{
    scanf("%d%d\n", &n, &m);     //numbers of numbers and
operations
    for (int i = 1; i <=n; i ++)     //n initial numbers
      scanf("%d", x + i);
    solve();     // Operations are dealt with one by one
    return 0;
}
```

7.2.3.2 Calculating Visible Segments

Segments are inserted into an interval one by one. And later segments can cover previous segments. Final visible segments are required to calculate. Labels for the segment tree are sequence numbers for covered subintervals.

   A segment tree is constructed based on the discretization on the segment coordinate.

7.2.3.2.1 Mayor's Posters

The citizens of Bytetown, Alberta, could not stand that the candidates in the mayoral election campaign have been placing their electoral posters in all places at their whim. The city council has finally decided to build an electoral wall for placing the posters and introduces the following rules:

   Every candidate can place exactly one poster on the wall.
   All posters are of the same height equal to the height of the wall; the width of a poster can be any integer number of bytes (byte is the unit of length in Bytetown).
   The wall is divided into segments and the width of each segment is one byte.
   Each poster must completely cover a contiguous number of wall segments.

They have built a wall 10000000 bytes long (such that there is enough place for all candidates). When the electoral campaign was restarted, the candidates were placing their posters on the wall, and their posters differed widely in width. Moreover, the candidates started placing their posters on wall segments already occupied by other posters. Everyone in Bytetown was curious whose posters will be visible (entirely or in part) on the last day before elections.

Your task is to find the number of visible posters when all the posters are placed, given the information about posters' size, their place, and order of placement on the electoral wall.

Input

The first line of input contains a number c giving the number of cases that follow. The first line of data for a single case contains number $1 \leq n \leq 10000$. The subsequent n lines describe the posters in the order in which they were placed. The i-th line among the n lines contains two integer numbers l_i and r_i, which are the number of the wall segment occupied by the left end and the right end of the i-th poster, respectively. We know that for each $1 \leq i \leq n$, $1 \leq l_i \leq r_i \leq 10000000$. After the i-th poster is placed, it entirely covers all wall segments numbered $l_i, l_{i+1}, \ldots, r_i$.

Output

For each input data set, print the number of visible posters after all the posters are placed.

Sample Input	Sample Output
1	4
5	
1 4	
2 6	
8 10	
3 4	
7 10	

Source: Alberta Collegiate Programming Contest 2003.10.18

IDs for Online Judges: POJ 2528

Figure 7.8 illustrates the case of the sample input.

Analysis

The wall is represented as an interval [0, 10000000]. One poster being placed on the wall can be regarded as a subinterval being colored. Placing the i-th

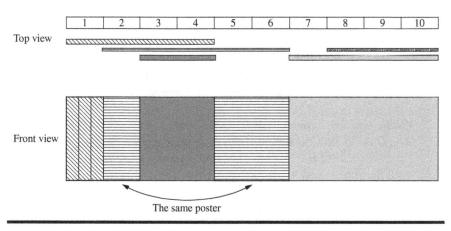

Front view

The same poster

Figure 7.8

poster can be regarded as the *i*-th subinterval being colored color *i*, $1 \le i \le n$. The final number of colors in the interval [0, 10000000] (previous colors can be covered by later colors) is the number of visible posters after all the posters are placed.

The problem is a basic problem for a segment tree. In the segment tree, each node stores its subinterval's color, where colorless is represented as 0, mixed color is represented as −1; otherwise, the color is represented as the number of color that the node is colored. Then a subinterval in the segment tree is updated each time. Because the wall is 10000000 bytes long, and $1 \le n \le 10000$. Discretization should be used. And it is not simple discretization.

The algorithm is as follows:

1. **Discretization.**

   The left boundaries, right boundaries, and middle positions for *n* posters are stored in array $x[1 \ldots 3 \times n]$. Then $x[\ ]$ is sorted to delete repeated coordinates. For the *i*-th poster, numbers of coordinates which aren't larger than its left boundary and right boundary are $l[i]$ and $r[i]$ respectively. Obviously, $l[i]$ and $r[i]$ constitute the *i*-th segment, and the color of the segment is *i*, $1 \le i \le n$.

   For example, there are three posters placed on the wall. Subintervals [1, 5], [1, 2] and [4, 5] are covered by the three posters. After the three posters are placed on the wall, there are three colors in the interval [0, 10000000]. For the first poster, numbers of coordinates which aren't larger than its left boundary and right boundary are 1 and 4. For the second poster, numbers of coordinates which aren't larger than its left boundary and right boundary are 1 and 2. And for the third poster, numbers of coordinates which aren't larger than its left boundary and right boundary are 3 and 4.

2. **Constructing a segment tree.**

    A segment is constructed, where the root is 1, and represents an interval [1...3*n*]. The label for a node is the number of color that the subinterval corresponds to. *n* segments are inserted into the segment tree one by one. And the segment tree is updated with the label method.

3. **Visible segments are recursively calculated.**

    If node *i* is labeled (the subinterval is covered by segments), and if the segment wasn't colored before, then set the label the color and return 1; else return 0 (in order to avoid repeated calculation);

    If node *i* is a leaf (the node isn't colored), return 0;

    Numbers of segments in the left subinterval and right subinterval are recursively calculated, and return the sum;

 Program

```
#include <iostream>
#include <cstdio>
#include <cstring>
#include <string>
#include <algorithm>
#define maxn 10010      // the upper limit of the number of
posters
using namespace std;
bool tab[maxn];      //tab[k]: the label that color k is used
int l[maxn], r[maxn], x[maxn*3], num[maxn*3], tree[maxn*12];
//For the i-th posters, numbers of coordinates which aren't
larger than its left boundary and right boundary are l[i] and
r[i] respectively; its left boundary, right boundary, and
middle position are x[3*i-2], x[3*i-1] and x[3*i]
respectively; after x[ ] is sorted, in x[1..j] the number of
non-repeating coordinates is num[j]; the label for node k in
the segment tree is tree[k], is the color for its subinterval
int c, n;      //c: number of test cases, n: number of posters

int binary_search(int sum)      //calculate different
coordinates in interval [0..sum]
{
   int l = 1, r = 3*n;
   while (r >= l){      //binary search is used to find the
sequence number r in x[ ] whose coordinate is sum
      int mid = (l + r) / 2;
      if (x[mid] <= sum)
        l = mid + 1;
```

```
      else
         r = mid - 1;
   }
   return num[r];     //the number of different coordinates in
x[1...r]
}
void update(int i)     //Update a segment with label method
{
   if (!tree[i])     // if label i isn't labeled, return
      return;
   tree[i+i]=tree[i+i+1]=tree[i];     //the label for node i is
given to its left and right child, and removed
   tree[i] = 0;
}
void change(int tl, int tr, int l, int r, int i, int co)
// In segment tree (root i, interval [l, r]), a subinterval
[tl, tr]) whose color is co is inserted
{
   if (tr < l || tl > r)
      return;
   if (tl<=l && r<=tr){     //[tl, tr] covers [l, r] completely
      tree[i] = co;
      return;
   }
   update(i);     // update the segment tree with label method
   int mid = (l + r) / 2;     // recursions for left and right
subtree
   change(tl, tr, l, mid, i+i, co);
   change(tl, tr, mid + 1, r, i + i + 1, co);
}
int require(int l, int r, int i)     //the number of visible
posters in the interval [l, r] (i is the root of its subtree)
{
   int mid = (l+r)/2;     //middle pointer
   if (tree[i]){     //i has been labeled, if the segment wasn't
colored before, then set the label to the color and return 1,
else return 0
      if (!tab[tree[i]]){
         tab[tree[i]] = 1;
         return 1;
      }
      return 0;
   }
   if (l == r)     //the current vertex isn't covered, return 0
      return 0;
   return require(l, mid,i+i)+require(mid+1,r,i+i+1);
//accumulate the number of visible posters in left and right
subintervals
}
void init()     // Discretization
```

```
{
  scanf("%d\n", &n);    //number of posters
  for (int i = 1; i <=n; i ++){    //posters' left and right
boundaries, for the i-th poster, its left and right boundaries
are x[3*i-2] and x[3*i-1], x[3*i] stores the middle position
    scanf("%d%d\n", l + i, r + i);
    x[i+ i+i-2] = l[i]; x[i+i+i-1]=r[i]; x[i+i+i]=(l[i] +
r[i])/2;
  }
  sort(x + 1, x + 3 * n + 1);    //sort x[]
  memset(num, 0, sizeof(num));
  for (int i=1;i<=3*n;i++){    //calculate num[], where num[i]
is the number of coordinates in x[1...i]
    num[i] = num[i - 1];
    if (x[i] != x[i - 1]) num[i] ++;
  }
  for (int i=1; i<=n; i++){    //calculate coordinates for
left and right boundaries of each poster
    l[i] = binary_search(l[i]);
    r[i] = binary_search(r[i]);
  }
}
void solve()    //calculate the number of visible posters
{
  memset(tree, 0, sizeof(tree));
  for (int i = 1; i<=n; i++)    //insert subintervals in the
segment tree
    change(l[i], r[i], 1, 3 * n, 1, i);
  memset(tab, 0, sizeof(tab));
  int ans = require(1,3*n,1);    // calculate and output the
number of visible posters
  printf("%d\n", ans);
}
int main()
{
  scanf("%d\n", &c);    //number of test cases
  for (int i = 1; i<=c; i++) {
    init();    // calculate the number of visible posters
    solve();
  }
  return 0;
}
```

7.2.3.3 Updating and Calculating Disjoint Segments

Given a segment whose length is l, if there are subintervals whose number of empty positions is no less than l in the segment tree, then the segment can be inserted. Normally there is a priority for such subintervals. For deletion operation, if there exists an "occupied interval" for the deleted segment in the segment tree, the segment can be deleted.

The label for a node includes:

1. The mark for the corresponding subinterval: There are three kinds of marks—occupied, empty, and partly occupied;
2. The longest empty subinterval in the corresponding subinterval for the node: The start position *pos*, and the length *lm*;
3. The length of the rightmost empty subinterval *ls* in the node's left child and the length of the leftmost empty subinterval *rs* in the node's right child, that is, the length of the subinterval crossing the left and right subintervals for the node is *ls+rs*.

7.2.3.3.1 Hotel

The cows are journeying north to Thunder Bay in Canada to gain cultural enrichment and enjoy a vacation on the sunny shores of Lake Superior. Bessie, ever the competent travel agent, has named the Bullmoose Hotel on famed Cumberland Street as their vacation residence. This immense hotel has N ($1 \leq N \leq 50,000$) rooms all located on the same side of an extremely long hallway (all the better to see the lake, of course).

The cows and other visitors arrive in groups of size D_i ($1 \leq D_i \leq N$) and approach the front desk to check in. Each group i requests a set of D_i contiguous rooms from Canmuu, the moose staffing the counter. He assigns them some set of consecutive room numbers $r..r+D_i-1$ if they are available; or, if no contiguous set of rooms is available, politely suggests alternate lodging. Canmuu always chooses the value of r to be the smallest possible.

Visitors also depart the hotel from groups of contiguous rooms. Checkout i has the parameters X_i and D_i which specify the vacating of rooms $X_i ..X_i +D_i-1$ ($1 \leq X_i \leq N-D_i+1$). Some (or all) of those rooms might be empty before the checkout.

Your job is to assist Canmuu by processing M ($1 \leq M < 50,000$) checkin/checkout requests. The hotel is initially unoccupied.

Input

* Line 1: Two space-separated integers: N and M;
* Lines 2: $M+1$: Line $i+1$ contains a request expressed as one of two possible formats:
   1. Two space-separated integers representing a check-in request: 1 and D_i;
   2. Three space-separated integers representing a checkout: 2, X_i, and D_i.

Output

* Lines 1.....: For each check-in request, output a single line with a single integer r, the first room in the contiguous sequence of rooms to be occupied. If the request cannot be satisfied, output 0.

Sample Input	Sample Output
10 6	1
1 3	4
1 3	7
1 3	0
1 3	5
2 5 5	
1 6	

Source: USACO 2008 February Gold

IDs for Online Judges: POJ 3667

 Analysis

For each node, there are three kinds of marks: occupied, empty, and partly occupied. There are two types of operations:

Operation 1: Search the position for the foremost empty subinterval whose length is n;

Operation 2: Set the mark for a subinterval empty.

For each operation, the segment tree needs to be updated. The label method is used to update the segment tree. The label for a node includes:

mark: the state for the node's corresponding subinterval (0: undetermined; 1: empty; 2: occupied);

ls: the length of the rightmost empty subinterval in the node's left child;

rs: the length of the leftmost empty subinterval in the node's right child;

ms: the length of the longest empty subinterval in the node's corresponding subinterval: and the start position for the subinterval is *pos*;

The three operations (update, query, and modification for a subinterval) for a segment tree whose root is i, and corresponding subinterval is $[l, r]$ are as follows:

1. **Update (the label method is used).**

```
if (mark for node i == 0) return;    // "undetermined"
if (mark for node i == 1){    //the subinterval [l, r]
for node i is empty, r-l+1 empty positions are divided
equally to the left and right subtrees, set marks for the
left and right subtrees "empty",
```

ls, *rs* and *ms* for the left child are $\left\lfloor \dfrac{l-r+2}{2} \right\rfloor$, its *pos* is

l; *ls*, *rs* and *ms* for right child are $\left\lfloor \dfrac{l-r+1}{2} \right\rfloor$, and its *pos*

is $\left\lfloor \dfrac{l-r}{2} \right\rfloor+1$; *mark* for the left and right child is 1;

```
          }else{      // the subinterval [l, r] for node i is
occupied, 0 empty positions, set marks for the left and
right subtrees "occupied"
          ls, rs and ms for the left child are 0, its pos is
l; ls, rs and ms for right child are 0, and its pos is
```

$\left\lfloor \dfrac{l-r}{2} \right\rfloor+1$; *mark* for the left and right child is 2;

```
          }
    Set the state for node i 0;      //Set the state for
node i "undetermined"
```

2. **Query.**

   For node *i* (corresponding to the subinterval [*l*, *r*]), search whether there exist empty subintervals whose length is *d*. If there exist such subintervals, return the left pointer for the foremost empty subinterval.

```
The label method is used to update the segment tree;
        if ( ms for node i<d), return failure;
        if (ms for node i==d), return pos for node i;
        if (ms for the left subtree≥d), recursive query for
the left subtree;
        if (rs for the left child + ls for the right
```

child ≥ *d*), return $\left(\left\lfloor \dfrac{l+r}{2} \right\rfloor - rs \text{ for the left child } +1 \right)$;

```
        recursive query for the right subtree;
```

3. **Modification.**

   A segment [*tl*, *tr*] is inserted into or deleted from a segment tree whose root is *i*, represent an interval [*l*, *r*].

```
    if ([tl, tr] isn't in [l, r]) return;
    if ([tl, tr] covers [l, r] completely) {
        if ( Insertion){      //After insertion, mark for
node i is "occupied"
            ls, rs, and ms for node i is set 0, pos is
set l, and mark is set 2;
        }else{      //After deletion mark for node i is
"empty"
```

```
                    ls, rs, and ms for node i is set r-l+1, pos
        is set 1, and mark is set 1;
                }
            return;
        }
        Label method is used to update the segment tree;
        Recursive modification for the left subtree;
        Recursive modification for the right subtree;
        ls for node i is set ls for its left child;      // ls,
        rs, ms and pos for node i is adjusted
        if (its left subtree is "empty") ls for node i += ls
        for its right child;
        rs for node i is set rs for its right child;
        if (its right child is "empty") rs for node i += rs
        for its left child;
        ms for node i =max(rs for its left child + ls for its
        right child, ms for its left child, ms for its right
        child);
        if ( ms for node i == ms for its left child)    //the
        longest empty subinterval is in the left subinterval
            pos for node i= pos for its left child;
        else
            if(ms for node i == rs for its left child + ls
        for its right child) // the longest empty subinterval
        crosses the left and right subintervals
```

$$pos \text{ for node } i = \left\lfloor \frac{l+r}{2} \right\rfloor - rs \text{ for its left child } +1;$$

```
            else pos for node i = pos for its right child;
        //the longest empty subinterval is in the right
        subinterval
```

 Program

```cpp
#include <iostream>
#include <cstdio>
#include <cstring>
#include <string>
#include <map>
#include <utility>
#include <set>
#include <algorithm>
#define maxn 80010
using namespace std;
struct node {int ls, rs, ms, pos, mark;}tree[4*maxn];
//segment tree, where the label for node i: tree[i].mark: the
```

state for the node's corresponding subinterval (0:
undetermined; 1: empty; 2: occupied), *tree*[*i*].*ls*: the length
of the rightmost empty subinterval in the node's left child;
tree[*i*].*rs*: the length of the leftmost empty subinterval in
the node's right child; *tree*[*i*].*ms*: the length of the longest
empty subinterval in the node's corresponding subinterval: and
the start position for the subinterval is *tree*[*i*].*pos*

```
int n, m;       //number of rooms and requests
void build_tree(int l, int r, int i)     //construct an "empty"
segment tree
{
    tree[i].ls=tree[i].rs=tree[i].ms=r-l+1;     // the
subinterval [l, r] for node i is empty
    tree[i].pos = l;
    if (l == r)     //left node
        return;
    int mid = (l + r) / 2;     //Intermediate pointer
    build_tree(l, mid, i + i);     //left subtree and right
subtree
    build_tree(mid + 1, r, i + i + 1);
}
bool all_space(int l,int r,int i)     // if subinterval [l, r]
for node i is empty, return 1; else return 0
{
    if (tree[i].ls==r-l+ 1)     //label "empty"
        return 1;
    return 0;
}
void update(int l, int r, int i)     // Update
{
    if (!tree[i].mark)     //the interval for node i is
"undetermined"
        return;
    if (tree[i].mark == 1){     //interval [l, r] for node i is
empty, then left and right subtrees have r-l+1 empty rooms,
left and right subtrees are empty state
        int len = r - l + 1;
        tree[i + i].ls = tree[i + i].rs = tree[i + i].ms =
(len + 1) / 2;
        tree[i + i].pos = l;
        tree[i + i + 1].ls = tree[i + i + 1].rs = tree[i + i +
1].ms = len /2;
        tree[i + i + 1].pos = (l + r) / 2 + 1;
        tree[i + i].mark = tree[i + i + 1].mark = 1;
    }else{     // interval [l, r] for node i is "occupied",
left and right subtrees are occupied
        tree[i + i].ls = tree[i + i].rs = tree[i + i].ms = 0;
        tree[i + i].pos = l;
        tree[i + i + 1].ls = tree[i + i + 1].rs = tree[i + i +
1].ms = 0;
```

```
            tree[i + i + 1].pos = (l + r) / 2 + 1;
            tree[i + i].mark = tree[i + i + 1].mark = 2;
    }
    tree[i].mark = 0;      //  node i "undetermined"
}
int query(int d, int l, int r, int i) //Query. If there exist
empty subintervals whose length is d in the segment tree (root
i, interval [l, r]), return the left pointer for the empty
subinterval, else return 0.
{
    update(l, r, i);
    if (tree[i].ms < d)     // there is no empty subinterval
whose length is d
        return 0;
    if (tree[i].ms==d)      // if (ms for node i==d) return pos
for node i
        return tree[i].pos;
    int mid = (l + r)/2;    //Intermediate pointer
    if (tree[i+i].ms>=d)     // if (ms for the left subtree≥d)
recursion for the left subtree
        return query(d, l, mid, i + i);
    if (tree[i + i].rs + tree[i + i + 1].ls >= d)    //the
length for empty interval covering the intermediate pointer
≥d, return its left pointer
        return mid - tree[i + i].rs + 1;
    return query(d, mid + 1, r, i + i + 1);     // recursion
for the right subtree;
}
void change(int tl, int tr, int l, int r, int i, bool flag)
//Modification. Insert or delete a segment [tl, tr] into or
from a segment tree (root i, interval [l, r])
{
    if (tl > r || tr < l)      //[tl, tr] isn't in [l, r]
        return;
    if (tl <= l && r <= tr){     // [tl, tr] covers [l, r]
        if (flag){      //Insertion
            tree[i].ls = tree[i].rs = tree[i].ms = 0;
            tree[i].pos = l;
            tree[i].mark = 2;      //the interval for node i is
occupied
        }else{      //delete
            tree[i].ls = tree[i].rs = tree[i].ms = r - l + 1;
            tree[i].pos = l;
            tree[i].mark = 1;      // the interval for node i is
empty
        }
        return;
    }
    update(l, r, i);
    int mid = (l + r) / 2;     // Intermediate pointer
```

```
        change(tl, tr, l, mid, i + i, flag);      //left subtree
        change(tl, tr, mid + 1, r, i + i + 1, flag);      //right
subtree
        tree[i].ls = tree[i + i].ls;
        if (all_space(l, mid, i+i))      // left subtree is empty
            tree[i].ls += tree[i + i + 1].ls;
        tree[i].rs=tree[i+i+1].rs;
        if (all_space(mid+1, r,i+i+1))      // right subtree is empty
            tree[i].rs += tree[i + i].rs;
        tree[i].ms=max(tree[i+i].rs+tree[i+i+1].ls,max(tree[i+i].
ms,tree[i+i+1].ms));
        if (tree[i].ms == tree[i + i].ms)
            tree[i].pos = tree[i + i].pos;
        else
            if (tree[i].ms == tree[i + i].rs + tree[i + i + 1].ls)
                tree[i].pos = mid - tree[i + i].rs + 1;
            else
                tree[i].pos = tree[i + i + 1].pos;
}
int main()
{
    scanf("%d%d\n", &n, &m);      //number of rooms and requests
        memset(tree, 0, sizeof(tree));
        build_tree(1, n, 1);      //construct an "empty" segment
tree
        for (int i =1; i <=m; i ++) {      //requests are dealt
with one by one
            int kind;
            scanf("%d", &kind);      //the type of requests
            if (kind == 1){      //check in
                int d;
                scanf("%d\n", &d);      //number of check-in rooms
                int ans=query(d,1,n,1);      //whether there exists
an empty interval whose length is d, and return the left
pointer for the interval (return 0 if there isn't)
                printf("%d\n", ans);
                if (ans)      // there exists an empty interval
whose length is d, segment [ans, ans+d-1] is inserted into the
segment tree
                    change(ans, ans+d-1,1,n,1,1);
            }else{      //check out
                int x, d;
                scanf("%d%d\n", &x, &d);      // d contiguous rooms
are checked out from position x
                change(x, x+d-1,1, n,1,0);      //segment [x, x+d-1]
is deleted
            }
        }
    return 0;
}
```

7.3 Graph Algorithms

In this section, practices for Euler graphs, Hamiltonian graphs, Maximum Independent Sets, Articulation Points, Bridges, and Biconnected Components are shown.

7.3.1 Euler Graphs

A circuit in a graph G containing all edges is called an Euler circuit of G. And the graph G is called an Euler graph. Similarly, a trail in a graph G containing all edges is called an Euler trail.

Theorem 7.3.1. A non-trivial connected graph G has an Euler circuit if and only if each vertex has even degree.

Proof. Suppose G has an Euler circuit $x_1 x_2 \ldots x_m$, $x_1 = x_m$. And x_i occurs k times in the sequence $x_1 x_2 \ldots x_m$, $1 \le i \le m-1$. Then $d(x_i) = 2k$. Therefore, each vertex has even degree.

Because G is connected and each vertex has even degree, there is a circuit C in G and the circuit C can be obtained by DFS. If C isn't the Euler circuit, in C there must be a vertex v_k whose degree is larger than the number of edges connected by v_k in C. From v_k a circuit C' whose edges aren't in C can be obtained through DFS. If $C \cup C' = G$, $C \cup C'$ is an Euler circuit. Else by the same reason, in $C \cup C'$ there must be a vertex v_k' whose degree is larger than the number of edges connected by v_k' in $C \cup C'$. And from v_k' a circuit C' whose edges aren't in $C \cup C'$ can be obtained through DFS. Then C' is added into $C \cup C'$, and so on until the Euler circuit is computed.

Obviously, the proof for necessity is also the algorithm getting the Euler circuit.

Theorem 7.3.2. A connected graph has an Euler trail from a vertex x to a vertex y ($x \ne y$) if and only if x and y are the only vertices of odd degree.

Its proof is similar to the proof for **Theorem 7.3.1**.

7.3.1.1 Johnny's Trip

Little Johnny has a new car. He decided to drive around the town to visit his friends. Johnny wanted to visit all his friends, but there were many of them. In each street he had one friend. He started thinking how to make his trip as short as possible. Very soon he realized that the best way to do it was to travel through each street of town only once. Naturally, he wanted to finish his trip at the same place where he had started, at his parents' house.

The streets in Johnny's town were named by integer numbers from 1 to n, $n < 1995$. The junctions were independently named by integer numbers from 1 to m, $m \le 44$. No junction connects more than 44 streets. All junctions in the town had different numbers. Each street was connecting exactly two junctions. No two streets in the town had the same number. He immediately started to plan his round trip. If there was more than one such round trip, he would have chosen the one which, when written down as a sequence of street numbers, is lexicographically the smallest. But Johnny was not able to find even one such round trip.

Help Johnny and write a program which finds the desired shortest round trip. If the round trip does not exist, the program should write a message. Assume that Johnny lives at the junction ending where the street appears first in the input with a smaller number. All streets in the town are two-way. There exists a way from each street to another street in the town. The streets in the town are very narrow, and there is no possibility to turn back the car once he enters a street.

Input

Input file consists of several blocks. Each block describes one town. Each line in the block contains three integers x, y, z, where $x>0$ and $y>0$ are the numbers of junctions that are connected by the street number z. The end of the block is marked by the line containing $x=y=0$. At the end of the input file there is an empty block, $x=y=0$.

Output

Output one line of each block containing the sequence of street numbers (single members of the sequence are separated by spaces) describing Johnny's round trip. If the round trip cannot be found, the corresponding output block contains the message "Round trip does not exist."

Sample Input	Sample Output
1 2 1	1 2 3 5 4 6
2 3 2	
3 1 6	Round trip does not exist.
1 2 5	
2 3 3	
3 1 4	
0 0	
1 2 1	
2 3 2	
1 3 3	
2 4 4	
0 0	
0 0	

Source: ACM Central European Regional Contest 1995

IDs for Online Judges: POJ 1041, UVA 302

Analysis

The problem requires you to calculate the Euler circuit for which the sequence of street numbers is lexicographically the smallest for a graph. The algorithm is as follows:

1. An undirected graph is constructed when a town is input. Degrees for nodes, the smallest number for nodes S, and the number of edges n are calculated.
2. If there exists a node whose degree is odd, there is no Euler circuit.
3. DFS is used to find an Euler circuit from node S. In order to find the Euler circuit in which the sequence of street numbers is lexicographically the smallest for a graph, for the set of unvisited edges incident to the current node, the edge with the smallest street number is selected. Because of recursion, the computed Euler circuit is in reversed order.
4. The Euler circuit is output in reversed order.

 Program

```cpp
#include <iostream>
#include <cstdio>
#include <cmath>
#include <cstdlib>
#include <cstring>
#include <string>
#include <map>
#include <utility>
#include <vector>
#include <set>
#include <algorithm>
#define maxn 2000    // The upper limit of the number of edges
#define maxm 50    // The upper limit of the number of vertices
using namespace std;
struct node{int s,t;}r[maxn];    //the sequence of edges,
where the i-th edge is (r[i].s, r[i].t)
bool vis[maxn];    //visited marks for edges vis[ ]
int deg[maxm], s[maxn];    // degrees of nodes deg[ ], the
sequence of edges for the Euler circuit s[ ]
int n, S, stop;    //the number of edges n, the smallest number
for nodes S, the number of edges in the Euler circuit stop
bool exist()    // If there exists a node whose degree is odd,
return 0; else return 1
{
```

```
    for (int i = 1; i <= maxm-1; i ++)
        if (deg[i] % 2 == 1) return 0;
    return 1;
}
void dfs(int now)      //Calculate the Euler circuit from now
{
        for (int i = 1; i <= n; i ++)      //Search an unvisited
edge connecting now
            if (!vis[i] && (r[i].s == now || r[i].t == now)){
                vis[i] = 1;      //the i-th edge
                dfs(r[i].s + r[i].t - now);
                s[++ stop] = i;      //add the i-th edge into the
Euler circuit
            }
}
int main()
{
    ios::sync_with_stdio(false);
    int x, y, num;      //(x, y) is an edge, the number of edge
is num
    while (cin>>x>>y, x>0){      //Repeat input the first edge
(x, y) in the current test case until end
        S = min(x, y); n = 0;      //Initialization
        memset(deg, 0, sizeof(deg));
        cin >> num;      //the sequence number for edge (x, y)
        r[num].s = x; r[num].t = y;      //two nodes for the
num-th edge
        deg[x] ++; deg[y] ++;      //degree for nodes x and y
        n = max(n, num);
        while (cin >> x >> y, x > 0){      // input edge (x, y)
            S = min(S, min(x, y));
            cin >> num;
            r[num].s=x; r[num].t=y;
            deg[x] ++; deg[y] ++;
            n = max(n, num);
        }
        if (exist()){      //If degrees for all nodes are even,
calculate the Euler circuit
            stop = 0;
            memset(vis,0,sizeof(vis));      //all edges are
unvisited
            dfs(S);      //from S, calculate the Euler circuit
            for (int i=stop;i>=2;i --) cout << s[i] << ' ';
//Output the Euler circuit
            cout << s[1] << endl;
        }else      //there exists odd nodes
            cout << "Round trip does not exist." << endl;
    }
    return 0;
}
```

Theorem 7.3.3. A directed graph is Eulerian if and only if every graph vertex has equal in-degree and out-degree.

7.3.1.2 Catenyms

A catenym is a pair of words separated by a period such that the last letter of the first word is the same as the last letter of the second. For example, the following are catenyms:

dog.gopher
gopher.rat
rat.tiger
aloha.aloha
arachnid.dog

A compound catenym is a sequence of three or more words separated by periods such that each adjacent pair of words forms a catenym. For example:

aloha.aloha.arachnid.dog.gopher.rat.tiger

Given a dictionary of lowercase words, you are to find a compound catenym that contains each of the words exactly once.

Input

The first line of standard input contains t, the number of test cases. Each test case begins with $3 \le n \le 1000$—the number of words in the dictionary. n distinct dictionary words follow; each word is a string of between 1 and 20 lowercase letters on a line by itself.

Output

For each test case, output a line giving the lexicographically least compound catenym that contains each dictionary word exactly once. Output "***" if there is no solution.

Sample Input	Sample Output
2 6 aloha arachnid dog gopher rat	aloha.arachnid.dog.gopher.rat.tiger ***

Sample Input	Sample Output
tiger 3 oak maple elm	

Source: Waterloo local 2003.01.25

Ids for Online Judges: POJ 2337, ZOJ 1919

 Analysis

The key to the problem is data modeling: What are represented as vertices and what are represented as arcs?

A dictionary is represented as a digraph G, where all letters are represented as vertices, that is, 'a' corresponds to 1, ..., and 'z' corresponds to 26; and each word is represented as an arc (u, v), where u is the number of the first letter for the word, and v is the number of the last letter for the word. Two corresponding words become a catenym if and only if the last letter of the first word is the same as the first letter of the second word. Therefore, the problem requires you to calculate an Euler path in the digraph G.

The algorithm is as follows:

1. A digraph G is constructed when a dictionary is input. The in-degree and the out-degree for each vertex and the root of the union-find set containing the vertex are calculated;
2. Arcs are sorted in lexicographical order;
3. Search vertices in ascending order: If there are two vertices belonging to different union-find sets, graph G isn't weakly connected, and there is no Eulerian directed path; else
4. Search vertices in ascending order. And determine whether there is an Euler path in the graph or not:
   If there is a vertex in which the difference between its in-degree and its out-degree is larger than 1, there is no Euler path;
   If every vertex's in-degree is the same as its out-degree, the vertex s with the smallest number is as the starting point for the Euler path;
   If there are only two vertices and their out-degrees and in-degrees differ by 1, the vertex s whose out-degree is larger than its in-degree is as the starting point for the Euler path;
   Else there is no Eulerian directed path.
5. DFS is used to calculate the Eulerian directed path from s.

Program

```
#include <iostream>
#include <cstdio>
#include <cmath>
#include <cstdlib>
#include <cstring>
#include <string>
#include <map>
#include <utility>
#include <vector>
#include <set>
#include <algorithm>
#define maxn 1010
using namespace std;
struct node{int u,v;string name;}road[maxn];      //edges, the
i-th edge is (road[i].u, road[i].v), and the word is road[i].
name
bool app[30], use[maxn];    //marks for vertices and edges are
app[] and use[]
int ind[30], oud[30], anc[30], s[maxn];    //in-degree and
out-degree for vertex i are ind[i] and oud[i], the root for
its union-find set is anc[i], directed Euler path is s[]
int n, S, stop, t;    //number of edges n, the starting point
and length for the directed Euler path are S and stop, number
of test cases is t
bool cmp(const node &a, const node &b)    //Lexicographic order
{
    return a.name < b.name;
}
int get_father(int x)    //return the root for the union-find
set containing x
{
    if (!anc[x])    //x doesn't belong to any union-find set,
return x
        return x;
    anc[x] = get_father(anc[x]);    //calculating the root for
the union-find set containing x
    return anc[x];
}
int change(char ch)    //letter ch is transferred as its
corresponding number
{
    return (int)ch - (int)'a' + 1;
}
```

```
bool exist_euler_circuit()      //determine whether there exists
an Euler path, if there is an Euler, the starting point S is
calculated
{
    int t = 0;
    for (int i=1; i<=26; i++)      //for each vertex in the
graph
        if (app[i]){
          if (t == 0) t = get_father(i);
            if (get_father(i)!= t)
                  return 0;
        }
    int sum = 0;      //Initialization
    S = 0;
    for (int i = 1; i <=26; i ++)      // for each vertex in the
graph
        if (app[i]){
              if (ind[i] != oud[i]){      //in-degree and out-
degree for vertex i are different
                  if (abs(ind[i] - oud[i])>1) return 0;
// the difference between its in-degree and its out-degree is
larger than 1, there is no Euler path, and return 0
                  sum ++;      //accumulate the number of vertices
which its out-degree and its in-degree differs by 1
                  if (oud[i]>ind[i]) S=i;      // if its out-
degree is larger than its in-degree, the node S is as the
starting point for the Euler path
              }
        }
    if (sum == 0)      //in-degree and out-degree for each
vertex are the same, there is a cycle, the starting vertex is
the vertex s whose sequence number is the least
        for (int i = 1; i <=26; i ++)
            if (app[i]){
                S = i;
                break;
            }
    return 1;
}
void dfs(int now)      // from vertex now, calculate the Euler
path s[]
{
    for (int i = 1; i <=n; i ++)      //search unvisited edges
from now
        if (!use[i] && road[i].u == now){
            use[i] = 1;
            dfs(road[i].v);
            s[++ stop] = i;
        }
}
```

```
void init()     // input a dictionary, construct a directed
graph
{
    cin >> n;     //number of words
    memset(ind, 0, sizeof(ind));     //in-degree and out-degree
    memset(oud, 0, sizeof(oud));
    memset(anc, 0, sizeof(anc));     //union-find set
    memset(app, 0, sizeof(app));
    for (int i = 1; i <=n; i ++){     //input words and
construct a directed graph
        cin >> road[i].name;     //the i-th word
        road[i].u = change(road[i].name[0]);     //the i-th
edge
        road[i].v = change(road[i].name[(int)road[i].name.
size() - 1]);
        app[road[i].u] = app[road[i].v] = 1;
        int u=get_father(road[i].u),v=get_father(road[i].v);
//roots for union-find sets
        if (u != v) anc[u] = v;     // union-find sets are
combined
        oud[road[i].u] ++; ind[road[i].v] ++;     // in-degree
and out-degree
    }
}
void solve()     // calculate and output Euler path
{
    sort(road + 1, road + n + 1, cmp);     //sort degrees in
Lexicographic order
    if (!exist_euler_circuit()){     //there is no Euler path
        cout << "***" << endl;
        return;
    }
    stop = 0;     //Initialize the length of Euler path
    memset(use, 0, sizeof(use));
    dfs(S);     //calculate the Euler path s[] from S
    for (int i = stop; i >= 2; i --)     //Output
        cout << road[s[i]].name << '.';
    cout << road[s[1]].name << endl;
}
int main()
{
    ios::sync_with_stdio(false);
    cin >> t;     //number of test cases
    for (int i = 1; i <=t; i ++) {     //test cases
        init();     // input a dictionary, construct a directed
graph
        solve();     //calculate and output Euler path
    }
    return 0;
}
```

7.3.2 Traveling Salesman Problem and Tournaments

In a graph, a Hamiltonian path is a path that contains each vertex exactly once, and a Hamiltonian circuit is a circuit that contains each vertex exactly once. A graph that contains a Hamiltonian path is called a traceable graph. A graph that contains a Hamiltonian cycle is called a Hamiltonian graph.

Suppose $G(V, E)$ is a connected graph with n vertices, $n \geq 3$, and no loops and multiple edges; $v \in V$, $|V|=n$, and $deg(v)$ is the degree of v.

Theorem 7.3.4 Graph G has a Hamiltonian circuit if, for any two vertices u and v of G that aren't adjacent, $deg(u)+deg(v) \geq n$. G has a Hamiltonian path if, for any two vertices u and v of G that aren't adjacent, $deg(u)+deg(v) \geq n-1$.

Corollary. Graph G has a Hamiltonian circuit if each vertex has a degree greater than or equal to $n/2$.

The Travelling Salesman Problem (TSP) is such a problem: "Given a weighted complete graph, what is the shortest possible route that visits each vertex exactly once and returns to the original vertex?" It is an NP-hard problem.

A tournament is a directed graph without loops, in which every pair of vertices is connected by a single uniquely arc.

Theorem 7.3.5 In a tournament, there is a directed Hamiltonian path.

In this section, there are three kinds of problems.

Case 1: In a graph there are a few vertices. Brute-force search can be used to solve the traveling salesman problem, although its time complexity is $O(n! \times n)$.

Case 2: State compression is used in solving the traveling salesman problem when there are a few vertices in a graph.

Case 3: A Hamiltonian path is calculated in a tournament. The time complexity is $O(n^2)$.

7.3.2.1 Getting in Line

Computer networking requires that the computers in the network be linked.

This problem considers a "linear" network in which the computers are chained together so that each is connected to exactly two others, except for the two computers on the ends of the chain, which are connected to only one other computer. A picture is shown in Figure 7.9. Here the computers are the black dots, and their locations in the network are identified by planar coordinates (relative to a coordinate system not shown in the picture).

Distances between linked computers in the network are shown in feet in Figure 7.9. For various reasons, it is desirable to minimize the length of cable used.

Your problem is to determine how the computers should be connected into such a chain to minimize the total amount of cable needed. In the installation being constructed, the cabling will run beneath the floor, so the amount of cable used to join two adjacent computers on the network will be equal to the distance between

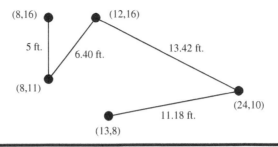

Figure 7.9

the computers plus 16 additional feet of cable to connect from the floor to the computers and provide some slack for ease of installation.

Figure 7.10 shows the optimal way of connecting the computers shown above, and the total length of cable required for this configuration is (4+16)+(5+16)+(5.83+16)+(11.18+16)=90.01 feet.

Input

The input file will consist of a series of data sets. Each data set will begin with a line consisting of a single number, indicating the number of computers in a network. Each network has at least two and at most eight computers. A value of 0 for the number of computers indicates the end of input.

After the initial line in a data set specifying the number of computers in a network, each additional line in the data set will give the coordinates of a computer in the network. These coordinates will be integers in the range 0 to 150. No two computers are at identical locations and each computer will be listed once.

Output

The output for each network should include a line which tells the number of the network (as determined by its position in the input data), and one line for each

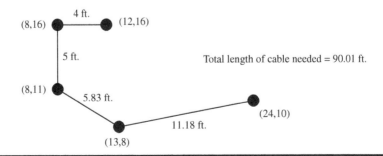

Figure 7.10

length of cable to be cut to connect each adjacent pair of computers in the network. The final line should be a sentence indicating the total amount of cable used.

In listing the lengths of cable to be cut, traverse the network from one end to the other. (It makes no difference at which end you start.) Use a format similar to the one shown in the sample output, with a line of asterisks separating output for different networks and with distances in feet printed to two decimal places.

Sample Input	Sample Output
6 5 19 55 28 38 101 28 62 111 84 43 116 5 11 27 84 99 142 81 88 30 95 38 3 132 73 49 86 72 111 0	*** Network #1 Cable requirement to connect (5,19) to (55,28) is 66.80 feet. Cable requirement to connect (55,28) to (28,62) is 59.42 feet. Cable requirement to connect (28,62) to (38,101) is 56.26 feet. Cable requirement to connect (38,101) to (43,116) is 31.81 feet. Cable requirement to connect (43,116) to (111,84) is 91.15 feet. Number of feet of cable required is 305.45. *** Network #2 Cable requirement to connect (11,27) to (88,30) is 93.06 feet. Cable requirement to connect (88,30) to (95,38) is 26.63 feet. Cable requirement to connect (95,38) to (84,99) is 77.98 feet. Cable requirement to connect (84,99) to (142,81) is 76.73 feet. Number of feet of cable required is 274.40. *** Network #3 Cable requirement to connect (132,73) to (72,111) is 87.02 feet. Cable requirement to connect (72,111) to (49,86) is 49.97 feet. Number of feet of cable required is 136.99.

Source: ACM/ICPC World Finals 1992

ID for Online Judge: UVA 216

 Analysis

A weighted graph is constructed as follows. Computers are represented as vertices. Euclidean distances between computers are as weights of edges connecting the two computers. Because each vertex's degree is $n-1$, there must be Hamilton paths in the graph. The problem requires you to calculate the Hamilton path with minimal length. Because the upper limit of the number of vertices is eight, DFS can be used to solve the problem.

Program

```
#include <iostream>
#include <cstdio>
#include <cmath>
#include <cstdlib>
#include <cstring>
#include <string>
#include <map>
#include <utility>
#include <vector>
#include <set>
#include <algorithm>
#define maxn 10
using namespace std;
bool vis[maxn];     //visited marks for vertices
int x[maxn], y[maxn], ans[maxn], t[maxn];     //computers'
coordinates x[] and y[], the shortest Hamiltonian path ans[],
the current path t[]
double dis[maxn][maxn];     //distance between vertices
double Min;     //the length of the shortest path
int n, casenum;     //number of vertices n, number of test
cases casenum
int sqr(int x)     //return x²
{
    return x * x;
}
void dfs(int sum, int now, double s)     //calculate the
Hamiltonian path from the current state (there are sum
vertices in the current path, the length of the current path
is s, the last vertex in the current path is now)
{
    if (sum == n){     // the Hamiltonian path
        if (s < Min){     // the current Hamiltonian path is
the shortest
            Min = s;
            for (int i = 1; i <=n; i ++) ans[i] = t[i];
        }
        return;     //backtracking
    }
    for (int i = 1; i <=n; i ++)     //search unvisited
vertices
        if (!vis[i]){
            vis[i] = 1;  //Set vertex i visited mark, (now, i)
is added into the path
```

```
                t[sum + 1] = i;
                dfs(sum + 1, i, s + dis[now][i]);
                vis[i] = 0;      // Set vertex i unvisited mark
        }
}
void init()     // Input computers' coordinates, construct
distance matrix
{
    for (int i = 1; i <=n; i ++)      // Input computers'
coordinates
        cin >> x[i] >> y[i];
    memset(dis, 0, sizeof(dis));
    for (int i = 1; i <=n; i ++)     //distances between vertices
        for (int j= 1; j<=n; j ++)
            dis[i][j] = sqrt(sqr(x[i] - x[j]) + sqr(y[i]
- y[j])) + 16;
}
void solve()     // calculate and output the shortest
Hamiltonian path
{
    cout << "********************************************
*********" << endl;
    cout << "Network #" << ++ casenum << endl;
    Min = 1e10;     //Initialization
    dfs(0, 0, 0.0);     // calculate the shortest Hamiltonian
path
    for (int i = 1; i <=n-1; i ++)      //Output the optimal way
of connecting the computers
        cout << "Cable requirement to connect (" << x[ans[i]]
<< "," << y[ans[i]] << ") to (" << x[ans[i + 1]] << "," <<
y[ans[i + 1]] << ") is " << dis[ans[i]][ans[i + 1]] << "
feet." << endl;
    cout << "Number of feet of cable required is " << Min <<
"." << endl;
}
int main()
{
    ios::sync_with_stdio(false);
    cout << fixed;
    cout.precision(2);
    while (cin >> n, n > 0){     //number of computers
        init();     //Input computers' coordinates, construct
distance matrix
        solve();     //calculate and output the shortest
Hamiltonian path
    }
    return 0;
}
```

7.3.2.2 Nuts for Nuts

So, Larry and Ryan decided that some nuts don't really taste so good, they realized that there are some nuts located in certain places of the island and they love them! Since they're lazy, but greedy, they want to know the shortest tour that they can use to gather every single nut!

Can you help them?

Input

You'll be given x and y, both less than 20, followed by x lines of y characters each as a map of the area, consisting sorely of ".", "#", and "L". Larry and Ryan are currently located in "L", and the nuts are represented by "#". They can travel in all eight adjacent directions in one step. See below for an example. There will be at the most 15 places where there are nuts, and "L" will only appear once.

Output

On each line, output the minimum amount of steps starting from "L", gather all the nuts, and back to "L".

Sample Input	Sample Output
5 5	8
L....	8
#....	
#....	
.....	
#....	
5 5	
L....	
#....	
#....	
.....	
#....	

Source: UVa Local Qualification Contest 2005

ID for Online Judge: UVA 10944

Larry and Ryan will go south for a nut, then south again for another nut, then south twice for another nut, and then back where they are.

 Analysis

Nuts are represented as vertices. Nuts are numbered 1…k from top to down and from left to right. A k-digit binary number is used to represent whether nuts are gathered or not. If the *i*-th nut is gathered, the (*i*-1)-th digit is 1, else the (*i*-1)-th digit is 0. Initially the *k*-digit binary number is 0. And finally the *k*-digit binary number is 2^k-1. Suppose the current position for Larry and Ryan is (x, y) and the current gathered nuts is z. The current state is represented as (x, y, z). Suppose a queue *q* is used to store states, and a hash table *hash* is to avoid repeated states.

Initially, the starting position for Larry and Ryan (l_x, l_y) and the current gathered nuts 0 are added into queue *q* as the initial state, and *hash*[the initial state]=1. Then BFS is used until queue *q* is empty.

The front is popped from *q* and extended in eight directions to produce new states. If a new state isn't in *hash*, then it is added into *q*, and *hash*[the new state]=1. If the new state is the goal state $(l_x, l_y, 2^k-1)$, Ryan and Larry have gathered all the nuts, and returned back to "L".

The amount of steps starting from "L", gather all the nuts, and back to "L" is calculated during the BFS.

Program

```
#include <iostream>
#include <cstdio>
#include <cmath>
#include <cstdlib>
#include <cstring>
#include <string>
#include <map>
#include <utility>
#include <vector>
#include <set>
#include <algorithm>
#define maxn 22     //upper limit for the size of the map
using namespace std;
const int dx[9] = {0, 0, -1, -1, -1, 0, 1, 1, 1};
// Horizontal displacement and vertical displacement
const int dy[9] = {0, 1, 1, 0, -1, -1, -1, 0, 1};
struct node {int x, y, get;}q[10000000];    //queue, where the
current position is (q[].x, q[].y), and q[].get represents
gathered nuts
bool hash[maxn][maxn][32768];    //Hash table, where hash[i]
[j][k] represents arriving at (i, j) and gathering k is the
current gathered nuts
int land[maxn][maxn];  // If (i, j) is the i-th nut from top to
down and from left to right, land[i][j]=2^i; else land[i][j]=0
int n, m, sum, Sx, Sy;    //the size for the map is (n, m);
the starting position for Larry and Ryan is (Sx, Sy)
void init()    //Input the map
```

```
{
    memset(land, 0, sizeof(land));
    sum = 1;
    for (int i = 1; i <=n; i ++){    // If (i, j) is the i-th
nut from top to down and from left to right, land[i][j]=2^i;
else land[i][j]=0
        char ch;
        cin.get(ch);
        for (int j = 1; j <=m; i ++) {
            cin.get(ch);
            switch (ch){
                case 'L': land[i][j]=0; Sx = i; Sy = j; break;
                case '#': land[i][j]=sum; sum *= 2; break;
                case '.': land[i][j] = 0; break;
            }
        }
    }
    for (int i = 0; i <=n+1; i ++)    //boundary value -1
        land[i][0] = land[i][m + 1] = -1;
    for (int i = 1; i <=m+1; i ++)
        land[0][i] = land[n + 1][i] = -1;
}
void solve()    //calculate and output the minimum amount of
steps
{
    memset(hash, 0, sizeof(hash));    //initialize Hash table
    hash[Sx][Sy][0] = 1;    //Hash value for the starting
position
    int head = 1, tail = 1, move = 0;    //Initialization
    q[1].x = Sx; q[1].y = Sy;
    q[1].get = 0;
    bool flag = 0;
    if (sum == 1) flag = 1;    //no nut
    while (head <= tail && !flag){    //queue is not empty,
no Hamiltonian Circuit
        int t = tail;    //the rear for the queue
        for (int i = head; i <= tail; i ++) {    // elements
in the queue
            int tx = q[i].x, ty = q[i].y;    //the current
element
            for (int j = 1; j <=8;j ++) {    // 8 directions
are searched
                int val=land[tx+dx[j]][ty+dy[j]];
                if (val >= 0 && !hash[tx+dx[j]][ty+dy[j]]
[q[i].get | val])    //add into the queue
                { t ++;
                    q[t].x = tx + dx[j]; q[t].y = ty + dy[j];
                    q[t].get = q[i].get | val;
                    hash[tx+dx[j]][ty+dy[j]][q[i].get|val]=1;
//Hash value
```

```
                if (q[t].x==Sx && q[t].y==Sy && q[t].
get==sum-1) /
                    flag = 1;
            }
        }
    }
    head =tail+1; tail=t;
    move ++;      //number of steps +1
    }
    cout << move << endl;    // output the minimum number of
steps
}
int main()
{
    ios::sync_with_stdio(false);
    while (cin >> n >> m){      //sizes of maps
        init();     // Input the map
        solve();    // calculate and output the minimum number
of steps
    }
    return 0;
}
```

Theorem 7.3.6 A tournament has a Hamiltonian path.

Proof. In a tournament there is a path. A vertex which is not in the path can be inserted into the path. Suppose there is a path $a_1 \to a_2 \to \dots a_i \dots a_{n-1} \to a_n$ in the tournament. A vertex a_{n+1} which isn't in the path can be inserted into the path:

Case 1: If (a_{n+1}, a_1) is an arc, then a_{n+1} is inserted into the path, and the path becomes $a_{n+1} \to a_1 \to a_2 \to \dots a_i \dots a_{n-1} \to a_n$;

Case 2: If there are arcs (a_i, a_{n+1}), $1 \le i \le n-1$, and a_{i+1} is the first vertex that there is an arc (a_{n+1}, a_{i+1}), then a_{n+1} is inserted into the path and the path becomes $a_1 \to a_2 \to \dots a_i \to a_{n+1} \to a_{i+1} \dots a_{n-1} \to a_n$;

Case 3: There is no such a vertex a_i in the path that (a_{n+1}, a_i) is an arc, $1 \le i \le n$. There must be an arc (a_n, a_{n+1}). Then a_{n+1} is inserted into the path and the path becomes $a_1 \to a_2 \to \dots a_i \dots a_{n-1} \to a_n \to a_{n+1}$.

Therefore, a tournament has a Hamiltonian path.

Obviously, the proof is also the algorithm for getting a Hamiltonian path in a tournament.

7.3.2.3 Task Sequences

Tom has received a lot of tasks from his boss, which are boring to deal with by hand. Fortunately, Tom got a special machine called an Advanced Computing Machine (ACM) to help him.

ACM works in a really special way. The machine can finish one task in a short time; after it has finished one task, it should smoothly move to the next one;

otherwise, the machine will stop automatically. You must start it up again to make it continue working. Of course, the machine cannot move arbitrarily from one task to another. So each time before it starts up, one task sequence should be well scheduled. Specially, a single task also can be regarded as a sequence. In the sequence, the machine should be able to smoothly move from one task to its successor (if a successor exists). After the machine has been started up, the machine always works according to the task sequence, and stops automatically when it finishes the last one. If all the tasks have not been finished, the machine has to start up again and works according to a new sequence. Of course, the finished tasks can't be scheduled again.

For some unknown reason, it was guaranteed that for any two tasks i and j, the machine can smoothly move from i to j or from j to i or both. Because the startup process is quite slow, Tom would like to schedule the task sequences properly, so that all the tasks can be completed with a minimal number of startup times. It is your task to help him achieve this goal.

Input

The input contains several test cases. For each test case, the first line contains only one integer n, ($0 < n \le 1,000$), representing the number of tasks Tom has received. Then n lines follow. Each line contains n integers, 0 or 1, separated by white spaces. If the j-th integer in the i-th line is 1, then the machine can smoothly move from task i to task j; otherwise. the machine can not smoothly move from task i to task j. The tasks are numbered from 1 to n.

Input is terminated by end of file.

Output

For each test case, the first line of output is only one integer k, the minimal number of startup times needed. And $2k$ lines follow, to describe the k task sequences. For each task sequence, the first line should contain one integer m, representing the number of tasks in the sequence. And the second line should contain m integers, representing the order of the m tasks in the sequence. Two consecutive integers in the same line should be separated by just one white space. Extra spaces are not allowed. There may be several solutions, and any appropriate one is accepted.

Sample Input	Sample Output
3	1
0 1 1	3
1 0 1	2 1 3
0 0 0	

Source: ACM Asia Guangzhou 2003

IDs for Online Judges: POJ 1776, ZOJ 2359, UVA 2954

Analysis

A directed graph $G(V, E)$ is used to represent the problem. Tasks are represented as vertices, and relationships for any two tasks are represented as arcs. For any two tasks i and j, the machine can smoothly move from i to j or from j to i or both. Therefore, the directed graph is a tournament. Because a tournament has a Hamiltonian path, the minimal number of startup times is 1. The algorithm calculating the Hamiltonian path in a tournament is shown in the proof for **Theorem 7.3.5**.

Vertex 1 is as the first vertex in the Hamiltonian path. Other vertices are inserted into the Hamiltonian path one by one. Suppose the current inserted vertex is vertex k. Vertices in the current Hamiltonian path are searched one by one, and the current vertex is vertex i.

```
If (k, i) ∉ E, t=i, that is, (t, k)∈ E;
If (k, i) ∈ E, then
   if vertex i is the first vertex in the current
Hamiltonian path, (k, i) is inserted into the current
Hamiltonian path, and vertex k is as the first vertex in the
current Hamiltonian path;
   else (t, k) and (k, i) are inserted into the current
Hamiltonian path;
```

If all vertices in the current Hamiltonian path have been searched, then (t, k) is inserted into the current Hamiltonian path.

 Program

```
#include <iostream>
#include <cstdio>
#include <cmath>
#include <cstdlib>
#include <cstring>
#include <string>
#include <map>
#include <utility>
#include <vector>
#include <set>
#include <algorithm>
#define maxn 1010
```

```
#define Path(i, s) for (int i = s; i; i = next[i])
using namespace std;
int pic[maxn][maxn];    // adjacency matrix
int next[maxn];    //pointers
int n;    //number of vertices
void init()    //construct an adjacency matrix
{
    memset(pic, 0, sizeof(pic));    //initialization
    string str;
    getline(cin, str);    //a blank line
    for (int i = 1; i <=n; i ++) {
        getline(cin, str);    //the i-th row
        for (int j= 1; j <=n;j ++)    // the i-th row for the
adjacency matrix
            pic[i][j] = str[(j - 1) * 2] - '0';
    }
}
void solve()    // calculate and output the Hamiltonian Path
{
    int head = 1, t;    //Initialization
    memset(next, 0, sizeof(next));
    for (int k = 2; k<=n; k++){    //vertex 2 ... vertex n are
inserted into the Hamiltonian Path
        bool flag = 0;    //vertice k isn't inserted
        for (int i = head; i; i = next[i])    //vertex i:
vertices in the current Hamiltonian Path
            if (pic[k][i]){    //vertex k and vertex i are
connected
                if (i==head) head=k;    //vertex i is the
first vertex in the Hamiltonian Path
                else next[t]=k;
                next[k] = i;    //(k, i) is inserted
                flag = 1;    // vertex k is inserted
                break;
            }else  t = i;
        if (!flag)    //(t, k) is inserted into the
Hamiltonian Path
            next[t] = k;
    }
    cout<<'1'<<endl<<n<<endl;    // output the minimal number
of startup times needed and the number of vertices in the
Hamiltonian Path n
    for (int i=head; i; i=next[i]){    //output the
Hamiltonian Path
        if (i != head) cout << ' ';
        cout << i;
    }
    cout << endl;
}
int main()
```

```
{
    ios::sync_with_stdio(false);
    while (cin >> n){      // the number of tasks (vertices)
        init();     //construct an adjacency matrix
        solve();    //calculate and output the Hamiltonian Path
    }
    return 0;
}
```

7.3.3 Maximum Independent Sets

In a graph $G(V, E)$, I is a subset of vertices, that is, $I \subseteq V$. If, for every two vertices in I, there is no edge connecting the two vertices, I is an independent set for G. A maximal independent set is such an independent set that if any other vertex is added to the set, the set isn't an independent set. A maximum independent set is an independent set of the largest possible size for a given graph G. This size is called the independence number of G, and denoted $\beta(G)$.

The Eight Queens Chess Problem is a problem for placing eight queens on the board so that no one queen can be taken by any other. The problem can be represented as a graph. In the board, each square is represented as a vertex. There are 64 vertices in the graph. If two placed queens can attack each other, there is an edge connecting the two corresponding vertices. Therefore, solving the Eight Queens Chess Problem is calculating the maximum independent set for the graph.

In a graph $G(V, E)$, if K is such a subset of vertices that each edge of the graph is incident to at least one vertex of the set, then K is a vertex cover. A minimum vertex cover is a vertex cover of the smallest possible size. This size is called the cover number of G, and denoted $\alpha(G)$. $\beta(G)+\alpha(G)=|V|$.

In a graph $G(V, E)$, a clique C is such a subset of the vertices that every two distinct vertices are adjacent. A maximum clique of a graph is such a clique that there is no clique with more vertices. The clique number $\omega(G)$ of a graph G is the number of vertices in a maximum clique in G.

The opposite of a clique is an independent set. Therefore, a maximum clique for a graph is a maximum independent set for its complement graph. And a maximum independent set for graph is a maximum clique for its complement graph.

Obviously, calculating an independent set for a graph can be implemented through calculating a maximum clique for its complement graph. When a graph is input, its complement graph can be constructed.

Suppose $f[i]$ is the number of vertices for the maximum clique for the subgraph induced by vertex i..vertex n; $get[i][]$ stores the number of adjacent vertices of the i-th vertex v for vertex $v+1$ .. vertex n in the current clique; *max* is the maximal number of vertices for current cliques; and $dfs(s, t)$ is used to calculate $f[ ]$, where s is the number of vertices in the clique, t is the number of adjacent vertices of the *s*-th vertex *v* for vertex *v*+1 .. vertex *n* in the current clique. Initially, vertex *i* is put into a clique, *s*=1, *t* is the number of vertices in *get*[1][]. The algorithm for *dfs*(*s*, *t*) is as follows:

```
    If s<max, then max=s, the current clique is the optimal
solution, and return;
    Enumerate get[s][i](1≤i≤t):
        For the current adjacent vertex v, if s+f[v]<max, then
it can't form the maximum clique, and return;
        Vertex v is as the (s+1)-th vertex;
        Calculate the number of vertices t' that are adjacent
to vertex v in get[s][i+1...t], and store these vertices in
get[s+1][];
        dfs(s+1, t');
```

Based on *dfs*(*s*, *t*), the main algorithm is as follows:

```
    max=0;
    Enumerate vertices in descending order (vertex i=n...1):
        Vertex i is the first vertex in the clique;
        Calculate the number of vertices adjacent to vertex i
from vertex i+1 to vertex n, and these adjacent vertices are
stored into get[1][];
        dfs(1, t);
        f[i]=max;
    Output max(the number of vertices for the optimal
solution);
```

7.3.3.1 Graph Coloring

You are to write a program that tries to find an optimal coloring for a given graph, as shown in Figure 7.11. Colors are applied to the nodes of the graph and the only available colors are black and white. The coloring of the graph is called optimal if a maximum of nodes is black. The coloring is restricted by the rule that no two connected nodes may be black.

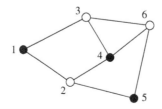

Figure 7.11   An optimal graph with three black nodes.

Input

The graph is given as a set of nodes denoted by numbers $1 \ldots n$, $n \leq 100$, and a set of undirected edges denoted by pairs of node numbers (n_1, n_2), $n_1 != n_2$. The input file contains m graphs. The number m is given on the first line. The first line of each graph contains n and k, the number of nodes and the number of edges, respectively. The following k lines contain the edges given by a pair of node numbers, which are separated by a space.

Output

The output should consists of $2m$ lines, two lines for each graph found in the input file. The first line should contain the maximum number of nodes that can be colored black in the graph. The second line should contain one possible optimal coloring. It is given by the list of black nodes, separated by a blank.

Sample Input	Sample Output
1	3
6 8	1 4 5
1 2	
1 3	
2 4	
2 5	
3 4	
3 6	
4 6	
5 6	

Source: ACM Southwestern European Regional Contest 1995

ID for Online Judge: POJ 1419, UVA 193

 Analysis

The coloring is restricted by the rule that no two connected nodes may be black. The coloring of the graph is called optimal if a maximum of nodes is black. Therefore, the problem requires you to calculate a maximum independent set for the graph.

When a graph is input, its complement graph is constructed. Then a maximum clique for the complement graph is calculated. The maximum clique for the complement graph is a maximum independent set for the graph. The cardinal number of the maximum independent set for the graph is the maximum number of nodes that can be colored black in the graph. And the maximum independent set for the graph is one possible optimal coloring.

For each node i ($i=n...1$), node i is as the first node for the current clique. Then, for node j ($j= i+1...n$), if node j and node i are adjacent, node j is put into a set, and cliques are calculated with the above method.

Obviously, finally the maximum clique is calculated when the loop ends.

 Program

```
#include <iostream>
#include <cstdio>
#include <cmath>
#include <cstdlib>
#include <cstring>
#include <string>
#include <map>
#include <utility>
#include <vector>
#include <set>
#include <algorithm>
#define maxn 105      //The upper limit for the number of nodes
using namespace std;
bool pic[maxn][maxn];     // adjacency matrix for complement
graph
int get[maxn][maxn];      // get[k][]: nodes adjacent to the
k-th node in the current clique
int node[maxn], ans[maxn], dp[maxn];    //node[]: current
clique; ans[]: maximum clique; dp[i]: the number of nodes for
the maximum clique storing node i... node n
int n, m, t, Max;      //number of nodes n, number of edge m,
the number of nodes for the current clique Max
void dfs(int now, int sum)    // the maximum clique is
calculated from the current state (the number of nodes for
the current clique now, number of edges connecting the last
node sum)
{
    if (sum == 0){     // clique, that is, a complete subgraph
        if (now>Max){    //adjust the number of nodes for the
maximum clique
            Max = now;
            for (int i=1; i<=Max; i ++) ans[i]=node[i];
        }
        return;
```

```
    }
    for (int i=1; i<=sum; i ++) {      //Enumeration
        int v=get[now][i], t=0;       //the other node v for the
i-th edge, the number of edges connecting v t
        if (now+dp[v]<=Max)return;
          for (int j=i+1;j<=sum; j++)      //v is added into the
clique
            if (pic[v][get[now][j]]) get[now+1][++t]=get[now]
[j];
        node[now+1]=v;
        dfs(now+1, t);
    }
}
void init()     //Input edges, construct the complement graph
{
    cin >> n >> m;     //numbers of nodes and edges
    memset(pic, true, sizeof(pic));     //initialize the
complement graph
      for (int i = 1; i <= m; i ++){     // Input edges,
construct the complement graph
        int a, b;
        cin >> a >> b;
        pic[a][b]=pic[b][a]=0;
    }
}
void solve()     // calculate the maximum clique for the
complement graph (the maximum independent set for the original
graph)
{
    Max = 0;     // the number of nodes for the current clique
Max
    for (int i = n; i >= 1; i --){     //node i is as the first
node for the current clique
        int sum = 0;
        for (int j=i+1; j<=n; j++)     // if node j (j=i+1…n)
and node i are adjacent, node j is put into get[1][]
            if (pic[i][j]) get[1][++sum]=j;
        node[1] = i;     // node i is as the first node for the
current clique
        dfs(1, sum);     //number of nodes for complete
subgraph for node i...n Max
        dp[i] =Max;
    }
    cout << Max << endl;     // number of nodes for the maximum
clique
    for (int i=1; i<=Max-1;i++)     // nodes for the maximum
clique
        cout << ans[i] << ' ';
    cout << ans[Max] << endl;
}
```

```
int main()
{
    ios::sync_with_stdio(false);
    cin >> t;      //number of test cases
    for (int i = 1; i <= t; i ++) {
        init();    //Input edges, construct the complement
graph
        solve();    // calculate the maximum clique for the
complement graph (the maximum independent set for the original
graph)
    }
    return 0;
}
```

7.3.4 Articulation Points, Bridges, and Biconnected Components

An articulation point in a connected graph is such a vertex that it would break the graph into two or more pieces if it is removed. A bridge in a connected graph is such an edge that it would break the graph into two or more pieces if it is removed. A cut, vertex cut, or separating set of a connected graph G is a set of vertices whose removal renders G disconnected. The connectivity or vertex connectivity $\kappa(G)$ (where G is not a complete graph) is the size of a minimal vertex cut. A graph is called k-connected or k-vertex-connected if its vertex connectivity is k or greater. The edge cut of G is a group of edges whose total removal renders the graph disconnected. The edge-connectivity $\lambda(G)$ is the size of a smallest edge cut, and the local edge-connectivity $\lambda(u, v)$ of two vertices u, v is the size of a smallest edge cut disconnecting u from v. Again, local edge-connectivity is symmetric. A graph is called k-edge-connected if its edge connectivity is k or greater. The vertex-connectivity and the edge-connectivity of a graph show the connectivity of a graph.

A connected component of a graph G is a connected subgraph of G that is not a proper subgraph of another connected subgraph of G. In an unconnected graph, how many connected components without a cut vertex can be computed? Such connected components are called biconnected components. A connected subgraph without a cut vertex is also called a block.

Function *low* is used to get cut vertices and bridges of a connected graph, and biconnected components of a graph. Suppose *pre[v]* is the sequence number of vertex v in DFS traversal. That is, *pre[v]* is the time that vertex v is visited. Function *low[u]* is the *pre[v]* of vertex v which is the earliest visited ancestor of u and u's descendants. That is,

$$low[u] = \min_{(u,s),(u,w) \in E} \{pre[u], low[s], pre[w]\}, \text{ where } s \text{ is a child of } u,$$

and (u, w) is a back edge.

A vertex itself is considered as one of its ancestors. Therefore $low[u]=pre[u]$ or $low[u]=pre[w]$ can hold. $low[u]$ is calculated as follows:

$$
low[u] = \begin{cases}
pre[u] & u \text{ is visited for the first time in DFS} \\
\min\{low[u], pre[w]\} & (u,w) \text{ is a back edge} \\
\min\{low[u], low[s]\} & \text{all edges related to u's children are inspected}
\end{cases}
$$

In the algorithm, $low[u]$ is changed until the DFS subtree whose root is u, and array low and array pre for u and its descendants are produced.

In DFS, edges can be classified into four types:

Branch edge T: Edge (u, v) is a branch edge, if it is the first time that v is visited in DFS.

Back edge B: Edge (u, v) is a back edge, if u is a descendant of v, and v has been visited, but all descendants of v haven't been visited.

Forward edge F: Edge (u, v) is a forward edge, if v is a descendant of u, all descendants of v have been visited, and $pre[u]<pre[v]$.

Cross edge C: All other edges (u, v). That is, u and v have no ancestor-descendant relationship in a DFS tree, or u and v are in different DFS trees. All descendants of v have been visited and $pre[u]>pre[v]$.

1. **Function low is used to get cut vertices in a connected graph.**

    We determine whether a vertex is a cut vertex or not based on the two following properties (see Figure 7.12).

    Property 1: If vertex U isn't a root, U is a cut vertex if and only if there exists a child s of U, $low[s]\geq pre[U]$. That is to say, there is no back edge from s and its descendants to U's ancestors.

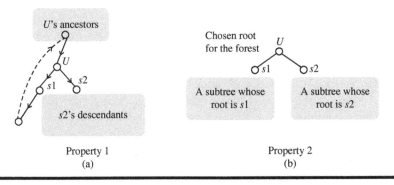

Property 1
(a)

Property 2
(b)

Figure 7.12

In Figure 7.12(a), although in the subtree whose root is *s*1 there is a back edge to *U*'s ancestor, there is no back edge to *U*'s ancestor from *s*2 or *s*2's descendants. If *U* is removed, the graph is not connected.

In an undirected graph, there are only branch edges and back edges. We can calculate *low* and *pre* through DFS, and find whether **Property 1** holds or not. The process is as follows:

```
    If (v, w) is a branch edge T (pre[w]==-1), and if
there is no back edge from w or w's descendants to v's
ancestors (low[w]≥pre[v]), then vertex v is a cut vertex,
and low[v]=min{low[v], low[w]}.
    If (v, w) is a back edge B (pre[w]!=-1), then
low[v]=min{low[v], pre[w]}.
void fund_cut_point(int v)     // DFS starts from v to
calculate a cut vertex in an undirected graph
{ int w;
  low[v]=pre[v]= ++d;     // Initialization
  for ( w∈the set of adjacent vertices for v) &&(w!=v)//
Search edge (v, w) for vertex v
  { if (pre[w]==-1)     //If (v, w) is branch edge T, w is
called recursively. If w and its descendants can't return
to v's ancestors, v is a cut vertex, calculate low[v]
        { fund_cut_point(w);     //w's all children's
related edges
          if (low[w]>=pre[v])     // v is a cut vertex
                v is a cut vertex;
          low[v]=min{ low[v], low[w]};
        };
    else low[v]=min{ low[v], pre[w]};     // If (v, w) is a
back edge, calculate low[v]
  }
}
```

Property 2: If *U* is selected as the root, then *U* is a cut vertex if and only if it has more than one child as in Figure 7.12(b).

In Figure 7.12(b), root *U* has two subtrees whose roots are *s*1 and *s*2 respectively, and there is no cross edge *C* between the two trees (in an undirected graph, there is no cross edge *C*). Therefore, the graph isn't connected after vertex *U* is deleted, and vertex *U* is a cut vertex.

Based on the above two properties, the algorithm for calculating cut vertices is as follows:

```
for(i = 0; i < n; i ++)          //Initialization
    pre[i] =-1;
low[s]=pre[s]=d=0;     // vertex s: start vertex
p=0;     // the number of children for vertex s
```

```
for (each w∈adj[s])  p++;
if (p>1)
  s is a cut vertex and exit;     //Property 2
fund_cut_point(s);     // Property 1
```

2. **Function *low* is used to get the bridge in a connected graph.**

In an undirected graph, edge (*u*, *v*) is a bridge if and only if (*u*, *v*) is not in any simple circuit.

The method for determining whether an edge is a bridge or not is as follows. Edge (*u*, *v*) is a branch edge discovered by DFS. If there is no back edge connecting *v* and its descendants to *u*'s ancestors; that is, *low*[*v*]>*pre*[*u*] or *low*[*v*]==*pre*[*v*]; then deleting (*u*, *v*) leads to *u* and *v* unconnected. Therefore, edge (*u*, *v*) is a bridge.

In Figure 7.13(a), DFS is used, a DFS tree is set up, as shown in Figure 7.13(b), and *pre* and *low* for all vertices are shown in Figure 7.13(c). Obviously for v_5, v_7, and v_{12}, *low*[*v*]==*pre*[*v*], and (v_0, v_5), (v_6, v_7), and (v_{11}, v_{12}) satisfy *low*[*v*]>*pre*[*u*] for edge (*u*, *v*). These edges are bridges in Figure 7.13(a).

In an undirected graph, there are only branch edges and back edges. DFS can be used to calculate *low* and *pre* for vertices (initial values for *pre*[ ] are −1), and calculate bridges in the undirected graph. The method is as follows:

```
    If (v, w) is a branch edge (pre[w]==-1), and if there
is no back edge from w or w's descendants to u's
ancestors, ((low[w]==pre[w])||(low[w]>pre[v])), then (v,
w) is a bridge, and low[v]=min{low[v], low[w]}.
```

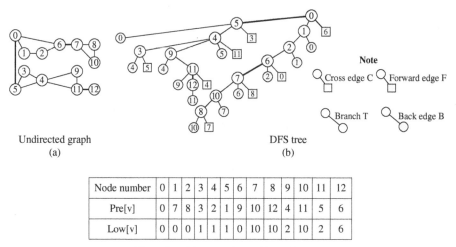

Undirected graph (a)	DFS tree (b)

Node number	0	1	2	3	4	5	6	7	8	9	10	11	12
Pre[v]	0	7	8	3	2	1	9	10	12	4	11	5	6
Low[v]	0	0	0	1	1	1	0	10	10	2	10	2	6

The nodes of the pre value and low value
(c)

Figure 7.13

```
     If (v, w) is a back edge (pre[w]!=-1), then
low[v]=min{low[v], pre[w]}.
void fund_bridge (v);     // DFS to find bridges from
vertex v
{ int w;
  low[v]= pre[v]=++d;
  for (each w∈ the set of adjacent vertices for v)
&(w!=v)  // Search edge(v, w)
  { if (pre[w]==-1)                       // if (v, w) is a
branch edge
    { fund_bridge (w);
        if ((low[w]== pre[w])||( low[w]> pre[v]))
            (v, w) is a branch edge;
        low[v]=min{ low[v], low[w]};
    };
    else low[v]=min{ low[v], pre[w]};     // if (v, w) is
a back edge
  }
}
```

3. **Function *low* is used to get biconnected components.**

   A biconnected component is a connected component without a cut vertex. Biconnected components of a graph are partitions of edges of the graph, that is, every edge must be in a block, and two different blocks don't contain common edges. In Figure 7.14, vertex *b* is a common vertex for block 3 and block 4, vertex *c* is a common vertex for block 3 and block 1, and vertex *e* is a common vertex for block 2 and block 4. The three vertices are cut vertices for the graph. The graph isn't connected when one of the three vertices is deleted.

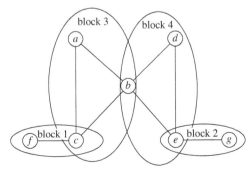

Cut vertices *b*, *c*, and *e* are common vertices for two blocks

Figure 7.14

The key to finding a block in an undirected graph is to find a cut vertex. DFS is used to get *low* and *pre* (initial values for *pre*[ ] are −1) and calculate blocks in the undirected graph. The process is as follows:

```
    For vertex v, u is the parent for v: if u is the
root, (u, v) is the first edge for the block; else
suppose f is u's parent. If u is deleted, v and f aren't
connected, then {f, u, v} isn't biconnected, (u, v) is
the first edge for the new block; else (u, v) and (f, u)
are in the same block. A stack is used to store vertices
in the current block. Suppose that
    st is a stack, sp is the pointer pointing to the top
of the stack;
    r is the number of blocks in the graph;
    ans is used to store blocks, where all vertices for
the t-th block are stored in ans[t][0]...ans[t][k], and
ans[t][k+1]=-1 (end mark for block t, 1≤t≤r);
void dfs(v)     //calculate block ans containing vertex v
{    st[sp++] = v;    //v is pushed into the stack
    pre[v]=low[v] =++d;    // set pre and low for v
    for (each w∈ the set of adjacent vertices for v)
&(w!=v)    //search adjacent edge (v, w) for v
         { if (pre[w]==-1) {    //(v, w) is a branch edge T
              dfs(w);
              if (low[w]< low[v])    // all children's
related edges for w have been checked, low[v]=min{
low[w], low[v]}
                   low[v]=low[w];
              if (low[w]>=pre[v]) {    //w and its
descendants can't return to an ancestor earlier than v,
then v is a cut vertex, the block is sent to ans[r]
                   k = 0;
                   st[sp] = -1;
                   ans[r][0] = v;    // vertex v enters
ans[r]
                   while (st[sp] != w)    // vertices
above w enter ans[r]
                        ans[r][++k] = st[--sp];
                   ans[r][++k] = -1;    // end mark for
ans[r]
                   if (k>2)    //if number of vertices
in the block > 2, accumulation
                        r++;
              }
         } else if (pre[w]< low[v])    //(v, w) is
back edge B, low[v]=min{ pre[w], low[v]}
                   low[v]= pre[w];
    }
}
```

7.3.4.1 Network

A Telephone Line Company (TLC) is establishing a new telephone cable network. They are connecting several places numbered by integers from 1 to N. No two places have the same number. The lines are bidirectional and always connect two places together, and in each place the lines end in a telephone exchange. There is one telephone exchange in each place. From each place, it is possible to reach every other place through lines; however, it need not be a direct connection, it can go through several exchanges. From time to time, the power supply fails at a place and then the exchange does not operate. The officials from TLC realized that in such a case, it can happen that besides the fact that the place with the failure is unreachable, this can also cause some other places to be unable to connect to each other. In such a case, we will say the place (where the failure occurred) is critical. Now the officials are trying to write a program for finding the number of all such critical places. Help them.

Input

The input file consists of several blocks of lines. Each block describes one network. In the first line of each block, there is the number of places $N<100$. Each of the next at most N lines contains the number of a place followed by the numbers of some places to which there is a direct line from this place. These at most N lines completely describe the network, i.e., each direct connection of two places in the network is contained at least in one row. All numbers in one line are separated by one space. Each block ends with a line containing just 0. The last block has only one line with $N=0$.

Output

The output contains for each block except the last in the input file one line containing the number of critical places.

Sample Input	Sample Output
5	1
5 1 2 3 4	2
0	
6	
2 1 3	
5 4 6 2	
0	
0	

Source: ACM Central Europe 1996

Ids for Online Judges: POJ 1144, ZOJ 1311, UVA 315

You need to determine the end of one line. In order to make it easy to determine, there are no extra blanks before the end of each line.

Analysis

A graph is constructed as follows. Places are represented as vertices. Lines between two places are represented as edges. Obviously, critical places are articulation points in a graph. The problem requires you to calculate the number of articulation points in a graph.

A Tarjan algorithm is used to recursively calculate *dfn*[ ] and *low*[ ] for vertices, and calculate the number of articulation points in a graph based on two properties.

Program

```cpp
#include <iostream>
#include <cstdio>
#include <cmath>
#include <cstdlib>
#include <cstring>
#include <string>
#include <map>
#include <utility>
#include <vector>
#include <set>
#include <algorithm>
#define maxn 110    //The upper limit of the number of vertices
using namespace std;
bool use[maxn];      //marks for articulation points
int pic[maxn][maxn];     // adjacency matrix
int dfn[maxn], low[maxn];     //dfn and low for vertices
int din, n, ans, s;     //visiting sequence din, number of
vertices n, number of articulation points ans, number of
children for the root s
void tarjan(int u)     //calculate the number of articulation
points from vertex u
{
    dfn[u] = low[u] = ++ din;
    for (int i = 1; i <=n; i ++)     //enumerate every adjacent
vertex for vertex u
        if (pic[u][i]){
            if (!dfn[i]){     //if (u, i) is a branch edge or a
cross-edge
```

```
                    tarjan(i);
                    low[u]=min(low[u], low[i]);      //adjust low
for u
                    if (low[i]>=dfn[u] && !use[u]){     //there are
no back edges for i or descendants for i to u's ancestors
                        if (u > 1){     //if u isn't the root, u is
an articulation point
                            ans ++;
                            use[u] = true;
                        }else     //u is the root, the number of
children +1
                            s ++;
                    }
            }else     //(u, i) is a back edge, adjust low for u
                low[u] = min(low[u], dfn[i]);
        }
}
void init()     //Input a graph, and construct an adjacency
matrix
{
    int u, v;     //two adjacent vertices
    memset(pic, 0, sizeof(pic));     //initialization
    while (cin >> u, u > 0){
        char ch;
        do{
            cin >> v;
            cin.get(ch);
            pic[u][v] = pic[v][u] = 1;     // two adjacent
vertices
        }while (ch != '\n');
    }
}
void solve()     //calculate and output articulation points
{
    memset(dfn, 0, sizeof(dfn));     //Initialization
    memset(low, 0, sizeof(low));
    memset(use, 0, sizeof(use));
    ans = din = s= 0;
    tarjan(1);     //calculate the number of articulation
points from the root
    if (s > 1) ans ++;     // if the root has more than one
child, the root is an articulation point
    cout << ans << endl;     // Output the number of
articulation points in a graph
}
int main()
{
    ios::sync_with_stdio(false);
    while (cin >> n, n > 0){     //Input the number of vertices
```

```
        init();     //Input a graph and construct its adjacency
matrix
        solve();    //calculate and output the number of
articulation points in a graph
    }
    return 0;
}
```

7.3.4.2 Road Construction

It's almost summer time, and that means that it's almost summer construction time! This year, the good people who are in charge of the roads on the tropical island paradise of Remote Island would like to repair and upgrade the various roads that lead between the various tourist attractions on the island.

The roads themselves are also rather interesting. Due to the strange customs of the island, the roads are arranged so that they never meet at intersections, but rather pass over or under each other using bridges and tunnels. In this way, each road runs between two specific tourist attractions, so that the tourists do not become irreparably lost.

Unfortunately, given the nature of the repairs and upgrades needed on each road, when the construction company works on a particular road, it is unusable in either direction. This could cause a problem if it becomes impossible to travel between two tourist attractions, even if the construction company works on only one road at any particular time.

So, the Road Department of Remote Island has decided to call upon your consulting services to help remedy this problem. It has been decided that new roads will have to be built between the various attractions in such a way that in the final configuration, if any one road is undergoing construction, it would still be possible to travel between any two tourist attractions using the remaining roads. Your task is to find the minimum number of new roads necessary.

Input

The first line of input will consist of positive integers n and r, separated by a space, where $3 \le n \le 1000$ is the number of tourist attractions on the island, and $2 \le r \le 1000$ is the number of roads. The tourist attractions are conveniently labelled from 1 to n. Each of the following r lines will consist of two integers, v and w, separated by a space, indicating that a road exists between the attractions labelled v and w. Note that you may travel in either direction down each road, and any pair of tourist attractions will have at most one road directly between them. Also, you are assured that in the current configuration, it is possible to travel between any two tourist attractions.

Output

One line, consisting of an integer, which gives the minimum number of roads that we need to add.

Sample Input 1	Sample Output 1
10 12	2
1 2	
1 3	
1 4	
2 5	
2 6	
5 6	
3 7	
3 8	
7 8	
4 9	
4 10	
9 10	
Sample Input 2	Sample Output 2
3 3	0
1 2	
2 3	
1 3	

Source: Canadian Computing Competition 2007

ID for Online Judge: POJ 3352

 Analysis

Remote Island is represented as a graph. Let tourist attractions be vertices, and roads be edges. Because any two tourist attractions are connected, the graph is a connected graph. Adding roads means adding edges in the graph. The goal for adding a road is "if any one road is undergoing construction, it would still be possible to travel between any two tourist attractions using the remaining roads." That is to say, a biconnected graph is constructed by adding roads with the minimum number. The algorithm is as follows.

First, all bridges are calculated. Second, all bridges are removed. And connected components are biconnected components. All biconnected components are represented as vertices, and all bridges are added back. The new graph is a tree, and its edge-connectivity is 1.

Then the number of vertices whose degree is 1 is calculated. Suppose the number of leaves is *leaf*. In order to make the tree become a biconnected graph, at least $\left\lfloor \dfrac{leaf+1}{2} \right\rfloor$ edges are added into the tree.

There are two lemmas.

Lemma 1: If there exists an edge (i, j), vertex i and vertex j are in a biconnected component if and only if $low[i]=low[j]$.

Lemma 2: There are n leaves in a tree. The tree will become a biconnected graph after adding at least $\left\lceil \dfrac{n}{2} \right\rceil$ edges.

The following algorithm is based on **Lemma 1** and **Lemma 2**. Suppose $e[][]$ is an adjacency list, $e[i][0]$ is the number of edges connecting vertex i, and the other vertex for the j-th edge is $e[i][j]$, $1 \le e[i][0] \le n-1$, and $1 \le j \le e[i][0]$.

1. Calculating $low[]$;
2. Calculating degrees for vertices in the contacted tree;
3. The number of vertices whose degree is 1 is calculated, denoted as *leaf*. In order to make the tree become a biconnected graph, at least $\left\lfloor \dfrac{leaf+1}{2} \right\rfloor$ edges are added into the tree.

Program

```cpp
# include <cstdio>
# include <cstring>
# include <cstdlib>
# include <vector>
# define vi vector<int>
# define pb push_back
using namespace std;
const int maxn=1010;      //the upper limit of the number of
vertices
vi e[maxn];      //an adjacency list for a graph
int dfsn[maxn],low[maxn],Time,deg[maxn];      // deg[]: degrees
for vertices in a tree, Time: visited time
int n,m;
void dfs(int a,int fa){      //calculate low from branch (fa, a)
    int q;dfsn[a]=low[a]=++Time;
    for(int p=0;p< e[a].size();p++)
        if(!dfsn[q=e[a][p]])
            dfs(q,a),low[a]=min(low[a],low[q]);
        else  if(q!=fa)low[a]=min(low[a],dfsn[q]);
```

```
}
void work(){
  for(int i=1;i<=n;i++) e[i].clear();    //Initialization
  for(int i=0;i<m;i++){    // adjacency list e is constructed
    int a,b;
    scanf("%d %d",&a,&b);
    e[a].pb(b);e[b].pb(a);
  }
  Time=0;    // Initialization
  memset(dfsn,0,sizeof(dfsn));
  memset(deg,0,sizeof(deg));
  dfs(1,-1);    //calculate low
  for(int i=1;i<=n;i++)        // Calculating degrees for
vertices in the contacted tree
    for(int p=0;p< e[i].size();p++) if(low[e[i][p]]!=low[i])
deg[low[i]]++;
  int cnt=0;    //number of leaves
  for(int i=1;i<=n;i++) if(deg[i]==1)cnt++;
  printf("%d\n",(cnt+1)/2);    // the minimum number of roads
}
int main(){
  while(~scanf("%d %d ",&n,&m))  work();
  return 0;
}
```

7.4 Problems

7.4.1 Long Long Message

Little Cat is majoring in physics in the capital of Byterland. A piece of sad news comes to him these days: his mother is ill. Being worried about spending so much on railway tickets (Byterland is such a big country, and he has to spend 16 hours on the train to get to his hometown), he decided only to send SMS messages to his mother.

Little Cat belongs to a family that is not rich, so he frequently visits the mobile service center to check how much money he has spent on SMS. Yesterday, the computer of the service center was broken, and printed two very long messages. The brilliant Little Cat soon found out the following:

1. All characters in messages are lowercase Latin letters, without punctuation and spaces.
2. All SMS has been appended to each other—(*i*+1)-th SMS comes directly after the *i*-th one—that is why those two messages are quite long.
3. His own SMS has been appended together, but possibly a great many redundant characters appear leftwards and rightwards due to the broken computer. For example, if his SMS is "motheriloveyou", either long message printed by that machine would possibly be one of "hahamotheriloveyou", "motheriloveyoureally", "motheriloveyouornot", "bbbmotheriloveyouaaa", etc.

4. For these broken issues, Little Cat has printed his original text twice (so there are two very long messages). Even though the original text remains the same in two printed messages, the redundancy characters on both sides would possibly be different.

You are given those two very long messages, and you have to output the length of the longest possible original text written by Little Cat.

Background: The SMS in the Byterland mobile service are charged in dollars-per-byte. That is why Little Cat is worrying about how long could the longest original text be.

Why ask you to write a program? There are four reasons:

1. Little Cat is so busy these days with physics lessons;
2. Little Cat wants to keep what he said to his mother a secret;
3. POJ is such a great Online Judge;
4. Little Cat wants to earn some money from POJ, and try to persuade his mother to see the doctor.

Input

Two strings with lowercase letters on two of the input lines individually. The number of characters in each one will never exceed 100000.

Output

A single line with a single integer number—what is the maximum length of the original text written by the little cat?

Sample Input	Sample Output
yeshowmuchiloveyoumydearmotherreallyicannotbelieve ityeaphowmuchiloveyoumydearmother	27

Source: POJ Monthly, 2006.03.26, Zeyuan Zhu, "Dedicated to my great beloved mother."

ID for Online Judge: POJ 2774

 Hint

Given two strings, the problem requires you to calculate the length of the longest common substring.

Any substring in a string is a prefix for a suffix in the string. Calculating the longest common substring for strings *A* and *B* is calculating the longest common

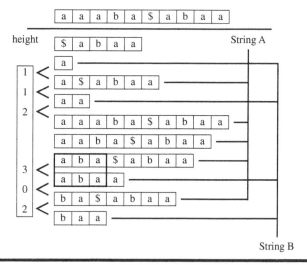

Figure 7.15

prefix for suffixes for strings A and B. It is inefficient to enumerate all suffixes for strings A and B. String B is adjoined to the end of string A, and a character which doesn't appear is inserted between A and B. For example, A="aaaba", B= "abaa", B is adjoined to the end of A, and a character '$' which doesn't appear is inserted between A and B. The longest common prefix for suffixes for the new string is calculated as Figure 7.15.

In Figure 7.15, "aa" is the longest common prefix for *suffix*(2) and *suffix*(9), and "aa" is a suffix for B and isn't a suffix for A; "aba" is the longest common prefix for *suffix*(3) and *suffix*(7), and "aba" is a suffix for A and isn't a suffix for B. The maximal value for *height*[] may not be the length of the longest common substring, for the two suffixes may be in the same string. Therefore, "aba" is the longest common substring for strings A and B.

The time complexity for the algorithm is $O(|A|+|B|)$.

7.4.2 *Milk Patterns*

Farmer John has noticed that the quality of milk given by his cows varies from day to day. On further investigation, he discovered that although he can't predict the quality of milk from one day to the next, there are some regular patterns in the daily milk quality.

To perform a rigorous study, he has invented a complex classification scheme by which each milk sample is recorded as an integer between 0 and 1,000,000 inclusive, and has recorded data from a single cow over N ($1 \le N \le 20,000$) days. He wishes to find the longest pattern of samples which repeats identically at least K ($2 \le K \le N$) times. This may include overlapping patterns—1 2 3 2 3 2 3 1 repeats 2 3 2 3 twice, for example.

Help Farmer John by finding the longest repeating subsequence in the sequence of samples. It is guaranteed that at least one subsequence is repeated at least K times.

Input

Line 1: Two space-separated integers: N and K;
Line 2: $N+1$: N integers, one per line, the quality of the milk on day i appears on the i-th line.

Output

Line 1: One integer, the length of the longest pattern which occurs at least K times.

Sample Input	Sample Output
8 2	4
1	
2	
3	
2	
3	
2	
3	
1	

Source: USACO 2006 December Gold

ID for Online Judge: POJ 3261

 Hint

Given a sequence of integers whose length is N, and an integer K, you are required to calculate the length of the longest repeating subsequences repeated at least K times in the sequence, and the subsequences can be overlapping.

The problem is a typical problem for suffix arrays. And dichotomy is also used in solving the problem.

7.4.3 Count Color

You have chosen Problem Solving and Program Design as an optional course, and you are required to solve all kinds of problems. Here, we get a new problem.

There is a very long board with length L centimeters, and L is a positive integer, so we can evenly divide the board into L segments, and they are labeled by 1, 2, ... L from left to right, each is 1 centimeter long. Now we have to color the board—one segment with only one color. We can do the following two operations on the board:

1. "C A B C" Color the board from segment A to segment B with color C.
2. "P A B" Output the number of different colors painted between segment A and segment B (including).

In our daily life, we have very few words to describe a color (red, green, blue, yellow...), so you may assume that the total number of different colors T is very small. To make it simple, we express the names of colors as color 1, color 2, ... color T. At the beginning, the board was painted in color 1. Now the rest of the problem is left to you.

Input

The first line of input contains L ($1 \leq L \leq 100000$), T ($1 \leq T \leq 30$) and O ($1 \leq O \leq 100000$). Here O denotes the number of operations. Following O lines, each contains "C A B C" or "P A B" (here A, B, and C are integers, and A may be larger than B) as an operation defined previously.

Output

Output the results of the output operation in order; each line contains a number.

Sample Input	Sample Output
2 2 4	2
C 1 1 2	1
P 1 2	
C 2 2 2	
P 1 2	

Source: POJ Monthly, 2006.03.26, dodo

ID for Online Judge: POJ 2777

Hint

Initially, the board is colored with color 1. Then update operations and query operations are dealt with one by one.

Update operations: Color a subinterval with a color;
Query operations: Output the number of different colors painted in a subinterval.

Obviously, the problem is a typical problem for visible segments. Its solution is the same as the solution to **7.2.3.2 Mayor's Posters**. Because the upper limit of the number of colors is 30, a bitwise operation can be used to improve the efficiency.

7.4.4 Who Gets the Most Candies?

N children are sitting in a circle to play a game.

The children are numbered from 1 to *N* in clockwise order. Each of them has a card with a non-zero integer on it in his/her hand. The game starts from the *K*-th child, who tells all the others the integer on his card and jumps out of the circle. The integer on his card tells the next child to jump out. Let *A* denote the integer. If *A* is positive, the next child will be the *A*-th child to the left. If *A* is negative, the next child will be the (−*A*)-th child to the right.

The game lasts until all children have jumped out of the circle. During the game, the *p*-th child jumping out will get *F*(*p*) candies where *F*(*p*) is the number of positive integers that perfectly divide *p*. Who gets the most candies?

Input

There are several test cases in the input. Each test case starts with two integers *N* ($0 < N \leq 500000$) and *K* ($1 \leq K \leq N$) on the first line. The next *N* lines contain the names of the children (consisting of at most 10 letters) and the integers (non-zero with magnitudes within 10^8) on their cards in increasing order of the children's numbers, a name, and an integer separated by a single space in a line with no leading or trailing spaces.

Output

Output one line for each test case containing the name of the luckiest child and the number of candies he/she gets. If ties occur, always choose the child who jumps out of the circle first.

Sample Input	Sample Output
4 2	Sam 3
Tom 2	
Jack 4	
Mary -1	
Sam 1	

Source: POJ Monthly, 2006.07.30, Sempr

ID for Online Judge: POJ 2886

 Hint

The key to the problem is: after the *i*-th child jumps out of the circle, who is the (*i*+1)-th child jumping out of the circle? A segment is used to represent children. A child jumping out of the circle can be implemented by updating a single point in a segment tree.

First, we calculate the numbers of factors for each integer in $[1, N]$. For example, Mike is the sixth child who jumps out of the circle. He will get four candies. Four is the number of positive integers that perfectly divide into 6 (factors for 6 are 1, 2, 3, and 6). It can be calculated in $O(N\log(N))$.

Suppose the i-th child jumps out of the circle, and his position is *now* (in the circle there are $n-i+1$ children, before he/she jumps out of the circle). After the i-th child jumps out of the circle, there are $n-i$ children in the circle. The i-th child jumping out of the circle is implemented by deleting the *now*-th element in the corresponding intervals. Suppose the root for the segment is i, and the interval that vertex i corresponds to is $[l, r]$; the algorithm is as follows:

```
if (l == r && now == 1)  return the vertex's sequence
number for element l;
    if (the now-th element is in the left subinterval)
calculate the sequence number for the now-th element is in the
left subinterval recursively;
        else{  now←num - number of elements in the left
subinterval;
            calculate the sequence number for the now-th
element is in the right subinterval recursively;
        }
```

After finding the sequence number for the vertex for the *now*-th element, the number of elements for vertices in the path from the vertex to the root -1.

7.4.5 Help with Intervals

LogLoader, Inc. is a company specialized in providing products for analyzing logs. While Ikki is working on graduation design, he is also engaged in an internship at LogLoader. Among his tasks, one is to write a module for manipulating time intervals, which have confused him a lot. Now he badly needs your help.

In discrete mathematics, you have studied several basic set operations, namely union, intersection, relative complementation, and symmetric difference, which naturally apply to the specialization of sets as intervals. For your quick reference, they are summarized in the table below:

Operation	Notation	Definition
Union	$A \cup B$	$\{x : x \in A \text{ or } x \in B\}$
Intersection	$A \cap B$	$\{x : x \in A \text{ and } x \in B\}$
Relative complementation	$A - B$	$\{x : x \in A \text{ but } x \notin B\}$
Symmetric difference	$A \oplus B$	$(A - B) \cup (B - A)$

Ikki has abstracted the interval operations emerging from his job as a tiny programming language. He wants you to implement an interpreter for him. The language maintains a set S, which starts out empty and is modified as specified by the following commands:

Command	Semantics
U T	$S \leftarrow S \cup T$
I T	$S \leftarrow S \cap T$
D T	$S \leftarrow S - T$
C T	$S \leftarrow T - S$
S T	$S \leftarrow S \oplus T$

Input

The input contains exactly one test case, which consists of between 0 and 65,535 (inclusive) commands of the language. Each command occupies a single line and appears like

$$X \; T$$

where X is one of "U", "I", "D", "C", and "S", and T is an interval in one of the forms $(a,b),(a,b],[a,b)$ and $[a,b](a,b \in \mathbf{Z}, 0 \le a \le b \le 65,535)$, which take their usual meanings. The commands are executed in the order they appear in the input.

End of file (EOF) indicates the end of input.

Output

Output the set S as it is after the last command is executed as the union of a minimal collection of disjoint intervals. The intervals should be printed on one line separated by single spaces and appear in increasing order of their endpoints. If S is empty, just print "empty set" and nothing else.

Sample Input	Sample Output
U [1,5] D [3,3] S [2,4] C (1,5) I (2,3]	(2,3)

Source: PKU Local 2007 (POJ Monthly, 2007.04.28), frkstyc

ID for Online Judge: POJ 3225

Hint

In the problem, four set operations, union, intersection, relative complementation, and symmetric difference, are given. Initially a set S is empty. After a sequence of set operations, S is the union of a minimal collection of disjoint intervals.

A segment tree is used to represent the universal set. "An interval is in the set" is represented as 1, "An interval isn't in the set" is represented as 0, and "Some parts for an interval are in the set" is represented as −1. Because there are open intervals, half-open intervals, and closed intervals, in the segment there are not only intervals, but also points, that is, the number of vertices should be doubled.

Two operations are used to simplify set operations.

Change(l, r, c): An interval [l, r] is added into the set, or taken out from the set (c==1, added; and c==0, taken out)

Reverse(l, r): An interval [l, r] is reversed. If the interval is in the set, then it is taken out; else it is added into the set.

Operation 'U' corresponds to *Change*(l, r, 1);

Operation 'I' corresponds to *Change*(1, l−1, 0) and *Change*(r+1, n, 0);

Operation 'D' corresponds to *Change*(l, r, 0);

Operation 'C' corresponds to *Change* (0, l−1, 0); *Change*(r+1, n, 0); *Reverse*(l, r);

Operation 'S' corresponds to *Reverse*(l, r);

7.4.6 Horizontally Visible Segments

There are a number of disjoint vertical line segments in the plane. We say that two segments are horizontally visible if they can be connected by a horizontal line segment that does not have any common points with other vertical segments. Three different vertical segments are said to form a triangle of segments if each two of them are horizontally visible. How many triangles can be found in a given set of vertical segments?

Your task is to write a program which for each data set:

reads the description of a set of vertical segments,
computes the number of triangles in this set,
writes the result.

Input

The first line of the input contains exactly one positive integer d equal to the number of data sets, 1≤d≤20. The data sets follow.

The first line of each data set contains exactly one integer n, 1≤n≤8000, equal to the number of vertical line segments. Each of the following n lines consists of exactly three non-negative integers separated by single spaces: y_i', y_i'', x_i—the y-coordinate of the beginning of a segment, y-coordinate of its end, and its x-coordinate, respectively. The coordinates satisfy $0 \le y_i' < y_i'' \le 8000$, $0 \le x_i \le 8000$. The segments are disjoint.

Output

The output should consist of exactly d lines, one line for each data set. Line i should contain exactly one integer equal to the number of triangles in the i-th data set.

Sample Input	Sample Output
1	1
5	
0 4 4	
0 3 1	
3 4 2	
0 2 2	
0 2 3	

Source: ACM Central Europe 2001

ID for Online Judges: POJ 1436, ZOJ 1391, UVA 2441

 Hint

The solution to the problem is similar to the solution to **7.4.3 Count Color**. The interval $[l, r]$ on the Y-axis is regarded as a segment tree, and the projection for a vertical line on the Y-axis is regarded as a segment. The number of triangles is calculated by enumerating segments from left to right.

7.4.7 Crane

ACM has bought a new crane (crane—jeřáb). The crane consists of n segments of various lengths, connected by flexible joints. The end of the i-th segment is joined to the beginning of the $i+1$-th one, for $1 \le i < n$. The beginning of the first segment is fixed at point with coordinates $(0, 0)$ and its end at point with coordinates $(0, w)$, where w is the length of the first segment. All of the segments lie always in one plane, and the joints allow arbitrary rotation in that plane. After a series of unpleasant accidents, it was decided that the software that controls the crane must contain a piece of code that constantly checks the position of the end of the crane and stops the crane if a collision should happen.

Your task is to write a part of this software that determines the position of the end of the n-th segment after each command. The state of the crane is determined by the angles between consecutive segments. Initially, all of the angles are straight, i.e., 180°. The operator issues commands that change the angle in exactly one joint.

Input

The input consists of several instances, separated by single empty lines.

The first line of each instance consists of two integers $1 \leq n \leq 10,000$ and $c \geq 0$ separated by a single space—the number of segments of the crane and the number of commands. The second line consists of n integers $l_1, \ldots, l_n$ ($1 \leq l_i \leq 100$) separated by single spaces. The length of the i-th segment of the crane is l_i. The following c lines specify the commands of the operator. Each line describing the command consists of two integers s and a ($1 \leq s < n$, $0 \leq a \leq 359$) separated by a single space—the order to change the angle between the s-th and the $s+1$-th segment to a degrees (the angle is measured counterclockwise from the s-th to the $s+1$-th segment).

Output

The output for each instance consists of c lines. The i-th of the lines consists of two rational numbers x and y separated by a single space—the coordinates of the end of the n-th segment after the i-th command, rounded to two digits after the decimal point.

The outputs for each two consecutive instances must be separated by a single empty line.

Sample Input	Sample Output
2 1	5.00 10.00
10 5	
1 90	−10.00 5.00
	−5.00 10.00
3 2	
5 5 5	
1 270	
2 90	

Source: CTU Open 2005

ID for Online Judge: POJ 2991

Hint

A segment tree is used to represent the problem. The root for the segment tree is the interval [1, n] representing n segments. Each node in the segment tree represents an interval [*l*, *r*]. In a node there are two pointers, where its left pointer points to the coordinates of the starting point for segment *l*, and its right pointer points to the coordinates of the end point for segment *r*. Obviously, after a command is executed, the right pointer of the root is the coordinates of the end of the *n*-th segment.

7.4.8 Is It a Tree?

A tree is a well-known data structure that is either empty (null, void, nothing) or is a set of one or more nodes connected by directed edges between nodes, satisfying the following properties:

> There is exactly one node, called the root, to which no directed edges point.
> Every node except the root has exactly one edge pointing to it.
> There is a unique sequence of directed edges from the root to each node.

For example, consider Figure 7.16, in which nodes are represented by circles and edges are represented by lines with arrowheads. The first two of these are trees, but the last is not.

In this problem, you will be given several descriptions of collections of nodes connected by directed edges. For each of these, you are to determine if the collection satisfies the definition of a tree or not.

Input

The input will consist of a sequence of descriptions (test cases) followed by a pair of negative integers. Each test case will consist of a sequence of edge descriptions followed by a pair of zeros. Each edge description will consist of a pair of integers; the first integer identifies the node from which the edge begins, and the second integer identifies the node to which the edge is directed. Node numbers will always be greater than zero.

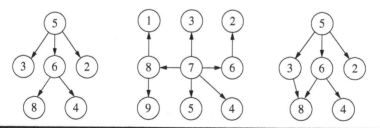

Figure 7.16

Output

For each test case display the line "Case *k* is a tree." or the line "Case *k* is not a tree.", where *k* corresponds to the test case number (they are sequentially numbered starting with 1).

Sample Input	Sample Output
6 8 5 3 5 2 6 4 5 6 0 0	Case 1 is a tree.
	Case 2 is a tree.
8 1 7 3 6 2 8 9 7 5 7 4 7 8 7 6 0 0	Case 3 is not a tree.
3 8 6 8 6 4 5 3 5 6 5 2 0 0	
−1 −1	

Source: ACM 1997 North Central Regionals

IDs for Online Judges: POJ 1308, ZOJ 1268, UVA 615

Hint

The problem is solved based on the definition of a tree.

When edges are input, in-degrees and out-degrees for nodes are calculated. If there exists a node whose in-degree is larger than 1, or the number of nodes whose in-degree is 0 is larger than 1, the case is not a tree. After all edges are input, if there is no above case, the case is a tree.

7.4.9 The Postal Worker Rings Once

Graph algorithms form a very important part of computer science and have a lineage that goes back at least to Euler and the famous *Seven Bridges of Königsberg* problem. Many optimization problems involve determining efficient methods for reasoning about graphs.

This problem involves determining a route for a postal worker so that all mail is delivered while the postal worker walks a minimal distance, so as to rest his weary legs.

Given a sequence of streets (connecting given intersections), you are to write a program that determines the minimal cost tour that traverses every street at least once. The tour must begin and end at the same intersection.

The "real-life" analogy concerns a postal worker who parks a truck at an intersection and then walks all streets on the postal delivery route (delivering mail) and returns to the truck to continue with the next route.

The cost of traversing a street is a function of the length of the street (there is a cost associated with delivering mail to houses and with walking even if no delivery occurs).

In this problem, the number of streets that meet at a given intersection is called the *degree* of the intersection. There will be at most two intersections with an odd degree. All other intersections will have an even degree, i.e., an even number of streets meeting at that intersection.

Input

The input consists of a sequence of one or more postal routes. A route is composed of a sequence of street names (strings), one per line, and is terminated by the string *"deadend"* which is NOT part of the route. The first and last letters of each street name specify the two intersections for that street, and the length of the street name indicates the cost of traversing the street. All street names will consist of lowercase alphabetic characters.

For example, the name *foo* indicates a street with intersections *f* and *o* of length 3, and the name *computer* indicates a street with intersections *c* and *r* of length 8. No street name will have the same first and last letter, and there will be at most one street directly connecting any two intersections. As specified, the number of intersections with odd degree in a postal route will be at most two. In each postal route, there will be a path between all intersections, i.e., the intersections are connected.

Output

For each postal route, the output should consist of the cost of the minimal tour that visits all streets at least once. The minimal tour costs should be output in the order corresponding to the input postal routes.

Sample Input	Sample Output
One	11
two	114
three	
deadend	
mit	
dartmouth	
linkoping	
tasmania	
york	
emory	
cornell	
duke	
kaunas	
hildesheim	
concord	
arkansas	
williams	
glasgow	
deadend	

Source: Duke Internet Programming Contest 1992

ID for Online Judge: UVA 117

Hint by the Problemsetter (http://www.algorithmist.com)

This problem reduces to a graph, by looking at each first or last character as a vertex, and the street name as an edge. We can reduce this problem to an Eulerian Path or an Eulerian Cycle problem, since each vertex will have an even number of degrees (except for at most two vertices).

Even though at first glance, it seems like it might need the Chinese Postman algorithm, but since each vertex will have an even number of degrees (except for at most two vertices), we can use the simpler Eulerian Path/Eulerian Cycle algorithm instead. If all vertices are of even degrees, then you're done, since the solution is simply the Eulerian Cycle—the sum of the weights of all the edges. Otherwise, we will have to calculate the Eulerian Path, and then you will have to find the shortest path between the two odd-degree vertices. This can be done with any of the Shortest Path algorithms.

7.4.10 Euler Circuit

An Euler circuit is a graph traversal starting and ending at the same vertex and using every edge exactly once. Finding an Euler circuit in an undirected or directed graph is a fairly easy task, but what about graphs where some of the edges are directed and some undirected? An undirected edge can only be traveled in one direction. However, sometimes any choice of direction for the undirected edges may not yield an Euler circuit.

Given such a graph, determine whether an Euler circuit exists. If so, output such a circuit in the format specified below. You can assume that the underlying undirected graph is connected.

Input

The first line in the input contains the number of test cases, at most **20**. Each test case starts with a line containing two numbers, V and E: the number of vertices ($1 \leq V \leq 100$) and edges ($1 \leq E \leq 500$) in the graph. The vertices are numbered from 1 to V. Then follow E lines specifying the edges. Each such line will be in the format *a b type* where *a* and *b* are two integers specifying the endpoints of the edge. *type* will either be the character **"U"**, if the edge is undirected, or **"D"**, if the edge is directed. In the latter case, the edge starts at *a* and ends at *b*.

Output

If an Euler circuit exists, output an order in which the vertices can be traversed on a single line. The vertex numbers should be delimited with a single space, and the start and end vertex should be included both at the beginning and the end of the sequence. Since most graphs have multiple solutions, any valid solution will be accepted. If no solution exists, output the line "No Euler circuit exists". Output a blank line between each test case.

Sample Input	Sample Output
2 6 8 1 3 U 1 4 U 2 4 U 2 5 D 3 4 D 4 5 U 5 6 D 5 6 U 4 4 1 2 D 1 4 D 2 3 U 3 4 U	1 3 4 2 5 6 5 4 1 No Euler circuit exists

Source: 2004 ICPC Regional Contest Warmup 1

ID for Online Judge: UVA 10735

Hint by the Problemsetter (http://www.algorithmist.com)

Given a graph *G*, which contains both directed edges and undirected edges, find a closed path in it, in which each edge is included exactly once.

Recall that, when an Euler tour exists in a directed graph: the underlying undirected graph is connected, and the in-degree of each vertex is equal to the out-degree.

In this problem, some of the graph's edges may be undirected. If we can orient them in such a way that the in-degree of each vertex will be equal to its out-degree, then the problem will be reduced to finding a tour in a directed graph. Such an orientation can be found by solving the following bipartite matching problem.

Construct a bipartite graph *H*. In one partition, put all edges (both directed and undirected) of *G*, and the other partition contains *G*'s vertices. For every edge, we have to know which of its two endpoints is the head. So, connect every object (edge of *G*) in the first partition with its *G*'s endpoints in the second partition.

We'll be finding a matching in this graph. If an undirected edge *e* = (*u*, *v*) of *G* will be matched with *v*, it means, that in the final directed graph, the edge *e* will go from vertex *u* to vertex *v*.

Each matched H's edge of (e, v) will contribute to the in-degree of vertex v in the directed graph, and unmatched edge (e, u) contributes to the out-degree of u.

Since we want to make the in-degree and out-degree of each vertex equal, each vertex must have an equal number of matched and unmatched edges in H. Additionally, each directed edge has to be matched with its respective head from G.

After finding a matching in H, satisfying the outlined constraints, we can assign direction to each undirected G's edge and find that the Euler tour is the resulting directed graph with any standard algorithm. If a matching doesn't exist, there will be no Euler tour in the original graph.

7.4.11 The Necklace

My little sister had a beautiful necklace made of colorful beads. Two successive beads in the necklace shared a common color at their meeting point. Figure 7.17 shows a segment of the necklace.

But, alas! One day, the necklace was torn and the beads were scattered all over the floor. My sister did her best to recollect all the beads from the floor, but she is not sure whether she was able to collect all of them. Now, she has come to me for help. She wants to know whether it is possible to make a necklace using all the beads she has in the same way her original necklace was made, and if so, in which order the beads must be put.

Please help me write a program to solve the problem.

Input

The input contains T test cases. The first line of the input contains the integer T.

The first line of each test case contains an integer N ($5 \le N \le 100$), giving the number of beads my sister was able to collect. Each of the next N lines contains two integers describing the colors of a bead. Colors are represented by integers ranging from 1 to 50.

Output

For each test case in the input, first output the test case number as shown in the sample output. Then, if you apprehend that some beads may be lost, just print the sentence "some beads may be lost" on a line by itself. Otherwise, print N lines with a single bead description on each line. Each bead description consists of two

Figure 7.17

integers giving the colors of its two ends. For $1 \leq i \leq N-1$, the second integer on line i must be the same as the first integer on line $i+1$. Additionally, the second integer on line N must be equal to the first integer on line 1. Since there are many solutions, any one of them is acceptable.

Print a blank line between two successive test cases.

Sample Input	Sample Output
2	Case #1
5	some beads may be lost
1 2	
2 3	Case #2
3 4	2 1
4 5	1 3
5 6	3 4
5	4 2
2 1	2 2
2 2	
3 4	
3 1	
2 4	

Source: ACM Shanghai 2000, University of Valladolid New Millenium Contest

IDs for Online Judges: UVA 10054, UVA 2036

Hint

A graph is constructed as follows: each color is represented as a node, and each bead is represented as an edge. The problem requires you to determine whether the graph is an Euler graph or not.

The problem is similar to **7.3.1.2 Catenyms**.

7.4.12 Dora Trip

Nobita is in great trouble. Today he failed to hand in his homework again, so he was heavily punished at school. Learning that, his mother is furious, and therefore assigns him many tasks to do—to buy vegetables at the market, to collect a parcel at the post office, and a lot more. Nobita certainly does not want to see his teacher on his way, nor would he like to meet Jyian, the tough bully. As usual, he asks Doraemon for help.

"Oh no!" cried Doraemon. "My door is broken, and my small propellers have all run out of batteries..." Well, that means Nobita has got to go without Doraemon's magic tools. "Ah, I still have this. It may well be useful." From his fourth-dimensional pocket, Doraemon takes out a map of their living area. He then marks on it the places where Nobita has to visit by asterisks ('*'), and where Jyian or his teacher may appear by crosses ('X'). Now Nobita's job is simple—he has to find the shortest route, through which he would not visit any of the "crosses", and he could finish the maximum number of the jobs (if not all) given by his mum. What he needs is just a computer program that works out the path.

Imagine that you are Nobita and write the program.

Input

The input file contains no more than **20** test cases. The details of each set are given as follows:

The first line of each case contains two integers *r* and *c* (1≤*r*,*c*≤20), which are the number of rows and columns of the map respectively. The next *r* lines, each with *c* characters, give the map itself. For each character, a space " " stands for an open space; a hash mark "#" stands for an obstructing wall; the capital letter "S" stands for the position of Nobita's house, which is where his journey is to start and end; the capital letter "X" stands for a dangerous place; and an asterisk "*" stands for a place he has to visit. The perimeter of the map is always closed, i.e., there is no way to get out from the coordinate of the "S". The number of places that Nobita has to visit is at most **10**.

The input file is terminated by a null case where *r* = *c* = **0**. This case should not be processed.

Output

For each test case, if Nobita cannot visit any target places at all, just print the line "**Stay home!**". Otherwise, your program should output the lexicographically smallest shortest path so that the number of target places that Nobita visits is maximized. Use the letters 'N', 'S', 'E', and 'W' to denote north, south, east and west respectively. Note that by "north" we mean facing upwards. You can be sure that the length of a correct output path will never exceed 200.

Sample Input	Sample Output
5 5	WWSSEEWWNNEE
#####	EEWW
# S#	Stay home!
# XX#	
# *#	

Sample Input	Sample Output
##### 5 5 ##### #* X# ###X# #S *# ##### 5 5 ##### #S X# # X# # #*# ##### 0 0	

Source: Programming Contest for Newbies 2005

ID for Online Judge: UVA 10818

Hint

The problem is a Traveling Salesperson problem, that is, the problem is NP-complete. Because the number of places that Nobita has to visit is at most **10**, a simple search suffices.

The problem is similar to **7.3.2.2 Nuts for Nuts**. Suppose Nobita's current position is (x, y) and the sequence of nodes that Nobita goes through is z. (x, y, z) constitutes a state. Initially, the position of Nobita's house (S_x, S_y) and $z=0$ is as the initial state and added into a queue q. Then BFS is used. And hash technology is also used to avoid repetitions.

7.4.13 Blackbeard the Pirate

Blackbeard the Pirate has stashed up to 10 treasures on a tropical island, and now he wishes to retrieve them. He is being chased by several authorities, however, and so would like to retrieve his treasure as quickly as possible. Blackbeard is no fool; when he hid the treasures, he carefully drew a map of the island which contains the position of each treasure and the positions of all obstacles and hostile natives that are present on the island.

Given a map of an island and the point where he comes ashore, help Blackbeard determine the least amount of time necessary for him to collect his treasure.

Input

The input consists of a number of test cases. The first line of each test case contains two integers h and w giving the height and width of the map, respectively, in miles. For simplicity, each map is divided into grid points that are a mile square. The next h lines contain w characters, each describing one square on the map. Each point on the map is one of the following:

@ The landing point where Blackbeard comes ashore.

~ Water. Blackbeard cannot travel over water while on the island.

A large group of palm trees; these are too dense for Blackbeard to travel through.

. Sand, which he can easily travel over.

* A camp of angry natives. Blackbeard must stay at least one square away or risk being captured by them, which will terminate his quest. Note this is one square in any of eight directions, including diagonals.

! A treasure. Blackbeard is a stubborn pirate and will not leave unless he collects all of them.

Blackbeard can only travel in the four cardinal directions; that is, he cannot travel diagonally. Blackbeard travels at a nice slow pace of one mile (or square) per hour, but he sure can dig fast, because digging up a treasure incurs no time penalty whatsoever.

The maximum dimension of the map is 50 by 50. The input ends with a case where both h and w are 0. This case should not be processed.

Output

For each test case, simply print the least number of hours Blackbeard needs to collect all his treasure and return to the landing point. If it is impossible to reach all the treasures, print out −1.

Sample Input	Sample Output
7 7	10
~~	32
~#!###~	
~...#.~	
~~....~	
~~.@~~	
.~~~~~~	
...~~~.	
10 10	

Sample Input	Sample Output
~~	
~~!!!###~~	
~##...###~	
~#....*##~	
~#!..**~~~	
~~....~~~~	
~~....~~~	
~~..~..@~~	
~#!.~~~~~~	
~~	
0 0	

Source: A Special Contest 2005

ID for Online Judge: UVA 10937

 Hint

The problem is also a Traveling Salesperson problem. The solution to the problem is the same as the solution to **7.3.2.2 Nuts for Nuts**. BFS, hash technology, and state compression are used to solve the problem.

Chapter 8

Practice for Computational Geometry

Computational geometry is the study of geometric algorithms for solving geometric problems. This chapter focuses on solving the following geometric problems:

1. Points, Line Segments, and Plans;
2. Calculating the Area for Union of Rectangles by Sweep Line Algorithms;
3. Intersection of Half-Planes;
4. Convex Hulls and Finding the Farthest Pair of Points.

8.1 Points, Line Segments, and Plans

In Euclidean space, a point is represented as a two-dimensional coordinate (x, y). Suppose there are two points $P_1(P_1=(x_1, y_1))$ and $P_2(P_2=(x_2, y_2))$, and there is a line segment from P_1 to P_2. Such a line segment is called a directed line segment, denoted by $\overrightarrow{P_1 P_2}$, where P_1 is the start point, P_2 is the end point, and the length of the line segment (i.e., its Euclidean distance) $|\overrightarrow{P_1 P_2}| = \sqrt{(x_1 - x_2)^2 + (y_1 - y_2)^2}$. If P_1 is the origin $(0, 0)$, then the directed line segment $\overrightarrow{P_1 P_2}$ is a vector P_2, the length of vector P_2 is $|P_2| = \sqrt{x_2^2 + y_2^2}$, called the magnitude of P_2.

In this section, experiments for points, line segments and plans are as follows:

1. Dot Product and Cross Product;
2. Line Segment Intersection;
3. Solving Polyhedron Problems by Euler's Formula.

443

8.1.1 Dot Product and Cross Product

First, dot product and cross product are introduced.

1. **Dot product.**

   Suppose coordinates for points are as follows: $A(x_1, y_1)$, $B(x_2, y_2)$, $C(x_3, y_3)$, $D(x_4, y_4)$. Vector $AB=(x_2-x_1, y_2-y_1)=(x_{AB}, y_{AB})$, where its magnitude $|AB| = \sqrt{x_{AB}^2 + y_{AB}^2}$. Vector $CD=(x_4-x_3, y_4-y_3)=(x_{CD}, y_{CD})$, where its magnitude $|CD| = \sqrt{x_{CD}^2 + y_{CD}^2}$. Vectors AB and CD are shown in Figure 8.1.

   The dot product of AB and CD is defined by $AB \bullet CD = x_{AB} \times x_{CD} + y_{AB} \times y_{CD} = |AB| \times |CD| \times cos(a)$, where a is the angle between vector AB and vector CD,

   $$a = acos\left(\frac{AB \bullet CD}{|AB| \times |CD|}\right), 0° \leq a \leq 180°.$$ Obviously, if the dot product $AB \bullet CD$ is negative, the angle a between vector AB and vector CD is an obtuse angle; if the dot product $AB \bullet CD$ is positive, the angle a between vector AB and vector CD is an acute angle; and if the dot product $AB \bullet CD$ is zero, vector AB and vector CD are vertical.

2. **Cross product.**

   In Figure 8.2, there are two vectors, P_1 and P_2.

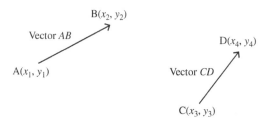

Figure 8.1

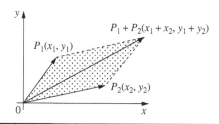

Figure 8.2

The cross product for vector P_1 and vector P_2 is defined by

$$P_1 \wedge P_2 = \begin{vmatrix} x_1 & y_1 \\ x_2 & y_2 \end{vmatrix} = x_1 \times y_2 - x_2 \times y_1 = -P_2 \wedge P_1.$$ The absolute value $|P_1 \wedge P_2|$ is

the area for the parallelogram whose points are $(0, 0)$, $P_1(x_1, y_1)$, $P_2(x_2, y_2)$ and $P_1 + P_2(x_1 + x_2, y_1 + y_2)$. And its positive or negative value is determined as follows:

If it goes clockwise from P_2 to P_1, then the cross product is $P_1 \wedge P_2 > 0$;
If it goes counterclockwise from P_2 to P_1, then the cross product is $P_1 \wedge P_2 < 0$;
If directions for vector P_2 and vector P_1 are same or opposite, then the cross product is $P_1 \wedge P_2 = 0$.

In Figure 8.3, by moving point P_0 horizontally and vertically to $(0, 0)$, we can determine whether it goes clockwise or counterclockwise from P_2 to P_1.

Suppose vectors $P_1' = P_1 - P_0$ and $P_2' = P_2 - P_0$, where $P_1' = (x_1', y_1') = (x_1 - x_0, y_1 - y_0)$; $P_2' = (x_2', y_2') = (x_2 - x_0, y_2 - y_0)$; $P_1' \wedge P_2' = (P_1 - P_0) \wedge (P_2 - P_0) = (x_1 - x_0)(y_2 - y_0) - (x_2 - x_0)(y_1 - y_0)$.

If the cross product is positive, it goes clockwise from $\overrightarrow{P_0 P_2}$ to $\overrightarrow{P_0 P_1}$, or the polar angle for P_2 is larger than the polar angle for P_1 with respect to point P_0. If the cross product is negative, it goes counterclockwise from $\overrightarrow{P_0 P_2}$ to $\overrightarrow{P_0 P_1}$, or the polar angle for P_1 is larger than the polar angle for P_2 with respect to point P_0. And if the cross product is zero, $\overrightarrow{P_0 P_1}$ and $\overrightarrow{P_0 P_2}$ are colinear, or the polar angle for P_1 is the same as the polar angle for P_2, with respect to point P_0.

Based on the cross product $P_1' \wedge P_2' = (P_1 - P_0) \wedge (P_2 - P_0) = (x_1 - x_0)(y_2 - y_0) - (x_2 - x_0)(y_1 - y_0)$, we can determine whether it goes clockwise or counterclockwise from $\overrightarrow{P_0 P_1}$ to $\overrightarrow{P_0 P_2}$.

If the cross product is positive, it goes counterclockwise from $\overrightarrow{P_0 P_1}$ to $\overrightarrow{P_0 P_2}$, that is, it turns left from P_1 to P_2 [Figure 8.4(a)].
If the cross product is negative, it goes clockwise from $\overrightarrow{P_0 P_1}$ to $\overrightarrow{P_0 P_2}$, that is, it turns right from P_1 to P_2 [Figure 8.4(b)].
If the cross product is 0, the P_0, P_1 and P_2 are colinear [Figure 8.4(c)].

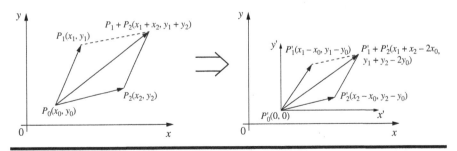

Figure 8.3

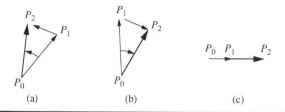

Figure 8.4

8.1.1.1 Transmitters

In a wireless network with multiple transmitters sending on the same frequencies, it is often a requirement that signals don't overlap, or at least that they don't conflict. One way of accomplishing this is to restrict a transmitter's coverage area. This problem uses a shielded transmitter that only broadcasts in a semicircle.

A transmitter T is located somewhere on a 1000-square-meter grid. It broadcasts in a semicircular area of radius r. The transmitter may be rotated any amount, but not moved. Given N points anywhere on the grid, compute the maximum number of points that can be simultaneously reached by the transmitter's signal. Figure 8.5 shows the same data points with two different transmitter rotations.

All input coordinates are integers (0–1000). The radius is a positive real number greater than 0. Points on the boundary of a semicircle are considered within that semicircle. There are 1–150 unique points to examine per transmitter. No points are at the same location as the transmitter.

Input

Input consists of information for one or more independent transmitter problems. Each problem begins with one line containing the (x, y) coordinates of the transmitter followed by the broadcast radius, r. The next line contains the number of points N on the grid, followed by N sets of (x, y) coordinates, one set per line. The end of the input is signaled by a line with a negative radius; the (x, y) values will be present but indeterminate. Figure 8.5 represents the data in the first two example data sets below, though they are on different scales. Figures 8.5(a) and 8.5(c) show transmitter rotations that result in maximal coverage.

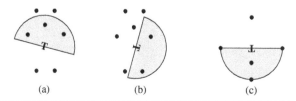

Figure 8.5

Output

For each transmitter, the output contains a single line with the maximum number of points that can be contained in some semicircle.

Sample Input	Sample Output
25 25 3.5	3
7	4
25 28	4
23 27	
27 27	
24 23	
26 23	
24 29	
26 29	
350 200 2.0	
5	
350 202	
350 199	
350 198	
348 200	
352 200	
995 995 10.0	
4	
1000 1000	
999 998	
990 992	
1000 999	
100 100 −2.5	

Source: ACM Mid-Central USA 2001

IDs for Online Judges: POJ 1106, ZOJ 1041, UVA 2290

 Analysis

Suppose the point for the transmitter is p_0. Because the transmitter may be rotated any amount, and broadcasts in a semicircular area of radius r, a straight line connecting any point and p_0 can be regarded as the lower boundary line for the semicircular. If the straight line containing $\overrightarrow{p_0 p_i}$ is the lower boundary line for the semicircular, point p_j in the semicircular area must meet the following conditions:

1. p_j must be in the semicircular area whose lower boundary line contains $\overrightarrow{p_0 p_i}$, that is, $\overrightarrow{p_0 p_i} \wedge \overrightarrow{p_0 p_j} \geq 0$;
2. The distance between p_j and p_0 must be less than the radius, that is, $\left|\overrightarrow{p_0 p_j}\right| \leq r$.

Each time point i is as a starting point. By using cross product, the number of points s_i in the semicircular area can be calculated. If the straight line containing $\overrightarrow{p_0 p_i}$ is as the lower boundary line for the semicircular, these points are in the semicircular area.

Obviously, the maximum number of points that can be contained in some semicircle are $S = \max_{1 \leq i \leq n}\{s_i\}$.

Program

```
#include <cstdio>
#include <cmath>
#include <cstring>
#include <algorithm>
using namespace std;
const double epsi = 1e-10;
const double pi = acos(-1.0);
const int maxn = 50005;
struct Point {     //Struct for point calculation
  double x, y;
  Point(double _x = 0, double _y = 0): x(_x), y(_y) { }
// Point vector
  Point operator -(const Point &op2) const {     //Vector
reduction
    return Point(x - op2.x, y - op2.y);
  }
  double operator ^(const Point &op2) const {     //Cross
product for 2 point vectors
    return x * op2.y - y * op2.x;
  }
};
inline int sign(const double &x) {     //return positive,
negative, or 0 mark for x
  if (x > epsi) return 1;
  if (x < -epsi) return -1;
  return 0;
}
```

```
inline double sqr(const double &x) {      //Calculating x²
   return x * x;
}
inline double mul(const Point &p0,const Point &p1,const Point
&p2) {//cross product for p₀p₁ and p₀p₂
   return (p1 - p0) ^ (p2 - p0);
}
inline double dis2(const Point &p0, const Point &p1) {
//Calculating |p₀p₁|²
   return sqr(p0.x - p1.x) + sqr(p0.y - p1.y);
}
inline double dis(const Point &p0, const Point &p1) {
//Calculating |p₀p₁|
   return sqrt(dis2(p0, p1));
}
int n ;
Point p[maxn], cp;      //p[ ]: the sequence of points, cp:
transmitter
double r;     // radius
int main() {
   while (scanf("%lf %lf %lf ", &cp.x, &cp.y, &r) && r >= 0 ) {
//Input coordinates of the transmitter and radius
      scanf("%d", &n);      //Number of points
      int ans = 0;
      for (int i=0;i<n;i++)scanf("%lf %lf",&p[i].x,&p[i].y);
// coordinates of points
      for (int i = 0 ; i < n ; i ++) {       //enumerating all
points
         int tmp = 0;      // the lower boundary line containing
point i and the transmitter, calculating number of points that
can be contained in some semicircle
         for (int j = 0 ; j < n ; j ++)
            if (sign( dis(p[j], cp)-r)!=1)
            if(sign( mul(cp,p[i],p[j]))!=-1)tmp++;      // number of
points +1
         ans = max( ans , tmp);      // adjust the maximum number
of points
      }
      printf("%d\n", ans);      //Output the result
   }
   return 0;
}
```

The absolute value of the cross product $P_1{}^{\wedge}P_2$ for vector P_1 and P_2 is the area of the parallelogram whose points are the origin (0, 0), P_1, P_2, and P_1+P_2 (Figure 8.6). And the area of the triangle whose points are origin, P_1 and P_2

$$S_{\Delta(0,0)P_1P_2} = \frac{|P_1 \wedge P_2|}{2}.$$

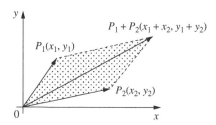

Figure 8.6

Therefore, the cross product can be used to calculate the area of a polygon. Points can be sorted clockwise or counterclockwise as $p_0 \ldots p_{n-1}$, and $p_n = p_0$. The area of the polygon is $S = \dfrac{\left| \displaystyle\sum_{i=1}^{n-2} P_i \wedge P_{i+1} \right|}{2}$, where vector P_i is $\overrightarrow{p_0 p_i}$, $1 \le i \le n-1$.

8.1.1.2 Area

You are going to compute the area of a special kind of polygon. One vertex of the polygon is the origin of the orthogonal coordinate system. From this vertex, you may go step by step to the following vertexes of the polygon until you go back to the initial vertex. For each step you may go North, West, South, or East with a step length of one unit, or go Northwest, Northeast, Southwest, or Southeast with a step length of the square root of two.

For example, Figure 8.7 shows a legal polygon to be computed and its area is 2.5.

Input

The first line of input is an integer t ($1 \le t \le 20$), the number of the test polygons. Each of the following lines contains a string composed of digits 1–9 describing how the polygon is formed by walking from the origin. Here 8, 2, 6, and 4 represent

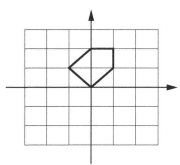

Figure 8.7

North, South, East and West, while 9, 7, 3, and 1 denote Northeast, Northwest, Southeast, and Southwest respectively. Number 5 only appears at the end of the sequence, indicating the end of walking. You may assume that the input polygon is valid, which means that the endpoint is always the start point and the sides of the polygon are not cross to each other. Each line may contain up to 1000000 digits.

Output

For each polygon, print its area on a single line.

Sample Input	Sample Output
4	0
5	0
825	0.5
6725	2
6244865	

Source: POJ Monthly, 2004.05.15 Liu Rujia@POJ

ID for Online Judge: POJ 1654

 Analysis

Suppose points for the polygon are $p_0, p_1, ..., p_{n-1}$, where p_0 is $(0, 0)$ and $p_n = p_0$. Based on the sequence of sides for the polygon, $\overrightarrow{p_i p_{i+1}}$ is the $i+1$-th side in the polygon, $0 \le i \le n-1$; and the n-th side is $\overrightarrow{p_{n-1} p_0}$. From $(0, 0)$, the vectors for points in the polygon $P_0, P_1, ..., P_{n-1}$ with respect to point p_0 are calculated

Calculate the cross product for two vectors for the front and the rear of each side $P_i \wedge P_{i+1}$ ($0 \le i \le n-1$). The area for the polygon is $S = \dfrac{\left| \sum\limits_{i=0}^{n-1} P_i \wedge P_{i+1} \right|}{2}$.

 Program

```
#include <cstdio>
#include <cmath>
#include <cstring>
#include <algorithm>
```

```
#include <iostream>
#include <string>
using namespace std;
const double epsi = 1e-10;
const double pi = acos(-1.0);
const int maxn = 100005;
inline int sign(const double &x) {     //positive or negative
  if (x > epsi) return 1;
  if (x < -epsi) return -1;
  return 0;
}
struct Point {     //Calculation for points
  long long x, y;
  Point(double _x = 0, double _y = 0): x(_x), y(_y) { }
//construct points
  Point operator +(const Point &op2) const {     // Vector
addition
    return Point(x + op2.x, y + op2.y);
  }
  long long operator ^(const Point &op2) const {     // Cross
product
    return x * op2.y - y * op2.x;
  }
};
int main() {
  int test = 0;
  string s;
  long long ans;
    scanf ("%d\n", &test );     // the number of the test
polygons
    for (; test; test --) {     //every polygon is dealt with
      cin >> s;     //polygon string
      ans = 0;
      Point p = Point( 0 , 0) , p1;     // the origin
      for (int i = 0 ; i < s.size() ; i ++) {
        if ( s[i] == '1') p1 = p+Point(-1, -1);  // Southwest
        if ( s[i] == '2') p1 = p+Point(0, -1);    // South
        if ( s[i] == '3') p1 = p+Point(1, -1);
// Southeast
        if ( s[i] == '4') p1 = p + Point(-1,0);    // West
        if ( s[i] == '5') p1 = Point(0, 0);    // the end of
walking
        if ( s[i] == '6') p1 = p + Point(1, 0);    // East
        if ( s[i] == '7') p1 = p+Point(-1, 1);
// Northwest
        if ( s[i] == '8') p1 = p + Point(0, 1);    // North
        if ( s[i] == '9') p1 = p + Point(1, 1);
// Northeast
        ans += p ^ p1;     //Accumulation for Cross product
        p = p1;     //continue to walk
```

```
    }
    if (ans<0 ) ans = -ans;      //absolute value for area
    cout<<ans/2;     //output area
    if (ans % 2 ) cout << ".5";      //odd
    cout << endl;
  }
  return 0;
}
```

8.1.2 Line Segment Intersection

In this section we focus on the following three problems:

1. Determining whether two line segments intersect or cross;
2. Calculating the intersection point when two line segments intersect;
3. Calculating the circumcenter of a triangle.

1. **Determining whether two line segments intersect or cross.**

    Crossing means that two points of a line segment are respectively on both sides of the straight line containing another line segment, or one point of the line segment is on the straight line. Obviously, if we need to determine whether line segment $\overline{p_1 p_2}$ and line segment $\overline{p_3 p_4}$ cross or not, we only need to determine whether the following two conditions are held or not:

    a. Line segment $\overline{p_1 p_2}$ crosses the straight line containing line segment $\overline{p_3 p_4}$;

    b. Line segment $\overline{p_3 p_4}$ crosses the straight line containing line segment $\overline{p_1 p_2}$.

    Two crossings are used to determine whether the above two conditions hold or not. Cross product is used to determine this. If we need to determine whether line segment $\overline{p_3 p_4}$ crosses the straight line containing line segment $\overline{p_1 p_2}$, we add two auxiliary lines $\overrightarrow{p_1 p_3}$ and $\overrightarrow{p_1 p_4}$, and then calculate two cross products: $(P_3-P_1)\wedge(P_2-P_1)$ and $(P_4-P_1)\wedge(P_2-P_1)$:

    If one cross product is positive, and the other is negative, then $\overline{p_3 p_4}$ can't cross the straight line containing line segment $\overline{p_1 p_2}$ [Figure 8.8(a)];

    If two cross products are all positive or negative, then $\overline{p_3 p_4}$ can't cross the straight line containing line segment $\overline{p_1 p_2}$ [Figure 8.8(b)];

    If one cross product is 0, p_3 or p_4 is on the straight line containing line segment $\overline{p_1 p_2}$ [Figure 8.8(c)].

8.1.2.1 Pick-up Sticks

Stan has n sticks of various lengths. He throws them on the floor, one at a time and in a random way. After he has finished throwing the sticks, Stan tries to find the top-most sticks, that is those with no sticks on top of them. Stan has noticed that the

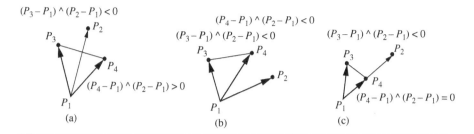

Figure 8.8

last thrown stick is always on top, but he wants to find all the sticks that are on top. Stan's sticks are quite thin, so thin that their thickness can be neglected (Figure 8.9).

Input

Input consists of a number of cases. The data for each case start with $1 \leq n \leq 100000$, the number of sticks for this case. The following n lines contain four numbers each; these numbers are the planar coordinates of the endpoints of one stick. The sticks are listed in the order in which Stan has thrown them. You may assume that there are not more than 1000 top sticks. The input is ended by the case with $n=0$. This case should not be processed.

Output

For each input case, print one line of output listing the top sticks in the format given in the sample. The top sticks should be listed in order in which they were thrown.

Figure 8.9 illustrates the first case from input.

Sample Input	Sample Output
5 1 1 4 2 2 3 3 1 1 –2.0 8 4 1 4 8 2 3 3 6 –2.0 3 0 0 1 1 1 0 2 1 2 0 3 1 0	Top sticks: 2, 4, 5. Top sticks: 1, 2, 3.

Source: Waterloo local 2005.09.17

IDs for Online Judges: POJ 2653, ZOJ 2551

Huge input, scanf is recommended.

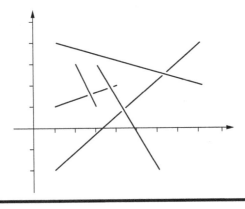

Figure 8.9

Analysis

The sticks are listed in the order in which Stan has thrown them. Each stick i is enumerated in ascending order of numbers (i.e., bottom-up), $1 \le i \le n$:

Each stick j which is over stick i is enumerated, $i+1 \le j \le n$. If there is a stick that stick i intersects with, then there is a stick on top of stick i, and stick $i+1$ is enumerated. If there is no stick on top of stick i, stick i is a stick on top.

Two crossings are used to determine whether two sticks intersect or not. Suppose stick i is $\overrightarrow{p_1^i p_2^i}$, and stick j is $\overrightarrow{p_1^j p_2^j}$. If $\overrightarrow{p_1^i p_2^i}$ and $\overrightarrow{p_1^j p_2^j}$ intersect, the two following conditions must hold:

1. $\overrightarrow{p_1^j p_2^j}$ crosses $\overrightarrow{p_1^i p_2^i}$, that is, positive and negative signs for $\overrightarrow{p_1^i p_2^i} \wedge \overrightarrow{p_1^i p_1^j}$ and $\overrightarrow{p_1^i p_2^i} \wedge \overrightarrow{p_1^i p_2^j}$ are different, or one of the two cross products is 0;

2. $\overrightarrow{p_1^i p_2^i}$ crosses $\overrightarrow{p_1^j p_2^j}$, that is, positive and negative signs for $\overrightarrow{p_1^j p_2^j} \wedge \overrightarrow{p_1^j p_1^i}$ and $\overrightarrow{p_1^j p_2^j} \wedge \overrightarrow{p_1^j p_2^i}$ are different, or one of the two cross products is 0.

The time complexity is $O(n^2)$.

Program

```
#include <cstdio>
#include <cmath>
#include <cstring>
#include <algorithm>
```

```cpp
#include <iostream>
using namespace std;
const double epsi = 1e-10;    // Infinitesimal
const double pi = acos(-1.0);
const int maxn = 100005;    //the upper limit of the number of
sticks
inline int sign(const double &x) {
  if (x > epsi) return 1;
  if (x < -epsi) return -1;
  return 0;
}
//structure and calculation for point
struct Point {
  double x, y;
  Point(double _x = 0, double _y = 0): x(_x), y(_y) { }
//construct a point
  Point operator +(const Point &op2) const {
    return Point(x + op2.x, y + op2.y);
  }
  Point operator -(const Point &op2) const {
    return Point(x - op2.x, y - op2.y);
  }
  double operator *(const Point &op2) const {
    return x * op2.x + y * op2.y;
  }
  Point operator *(const double &d) const {
    return Point(x * d, y * d);
  }
  Point operator /(const double &d) const {
    return Point(x / d, y / d);
  }
  double operator ^(const Point &op2) const {    // cross
product for vectors
    return x * op2.y - y * op2.x;
  }
  bool operator !=(const Point &op2) const {
    return sign (op2.x - x) != 0 || sign( op2.y - y) != 0;
  }
};
inline double sqr(const double &x) {    //x²
  return x * x;
}
inline double mul(const Point &p0, const Point &p1,
const Point &p2) {
  // cross product for $\overrightarrow{p_0p_1}$ and $\overrightarrow{p_1p_2}$
  return (p1 - p0) ^ (p2 - p0);
}
inline double dis2(const Point &p0, const Point &p1) {
  return sqr(p0.x - p1.x) + sqr(p0.y - p1.y);
}
```

```
inline double dis(const Point &p0, const Point &p1) {    //|p₀p₁|
  return sqrt(dis2(p0, p1));
}
inline int cross( const Point &p1 , const Point &p2 , const
Point &p3 , const Point &p4 , Point &p) {    //determine
whether p₁p₂ crosses p₃p₄
    double a1 = mul( p1, p2 , p3), a2 = mul( p1, p2 , p4 ) ;
    if (sign ( a1 ) ==0 && sign ( a2 ) == 0) return 2;    //if
p₁p₂ and p₃p₄ coincide, return 2
    if (sign ( a1 ) == sign ( a2 )) return 0;    //if p₁p₂
doesn't cross p₃p₄, return 0
    return 1;    // p₁p₂ crosses p₃p₄
}
int n;
Point p1[maxn] , p2[maxn] , tp;    //a sequence of coordinates
for sticks p1[] and p2[]
int main() {
  int test = 0;    //number of test cases
  while ( scanf ("%d", &n ) && n ) {    //number of sticks
    printf("Top sticks:");
    bool fl = false ;
    for ( int i = 1 ; i <= n ; i ++)    // a sequence of
coordinates for n sticks
      scanf("%lf %lf %lf %lf" , &p1[i].x , & p1[i].y , &
p2[i].x ,& p2[i].y);
    for ( int i = 1 ; i <= n ; i ++) {    // Each stick i is
enumerated bottom-up, 1≤i≤n
      bool flag = false ;
      for (int j = i+1 ; j <= n ; j ++)    // Each stick j
which is over stick i is enumerated
        if ( cross ( p1[i] , p2[i] , p1[j] , p2[j] , tp ) == 1
&& cross ( p1[j] , p2[j] , p1[i] , p2[i] , tp ) == 1) { flag =
true; break; }
      if (flag == false && fl == true ) printf(",");
      if (flag == false ) printf(" %d", i ), fl = true;
    }
    printf(".\n");
  }
  return 0;
}
```

2. **Calculating the intersection point when two line segments intersect.**

The formula for cross product can be used to calculate the intersection point when two line segments intersect. Suppose $mul(p_0, p_1, p_2)$ is the cross product for $\overrightarrow{p_0 p_1}$ and $\overrightarrow{p_0 p_2}$, thst is, $mul(p_0, p_1, p_2)=(p_1-p_0)^\wedge(p_2-p_0)$. The cross product can be calculated as the triangle area (the shadow area) in the parallelogram, whose points are p_0, p_1, p_2, and p_1+p_2 respectively, that is,

$$S_{\Delta p_0 p_1 p_2} = \frac{1}{2} \times mul(p_0, p_1, p_2) \text{ (Figure 8.10).}$$

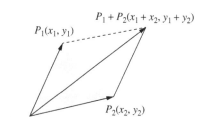

Figure 8.10

Based on this information, the intersection point when two line segments intersect can be calculated. For example, in Figure 8.11, point P is the intersection point for line segment AB and line segment CD.

DD' is a vertical line for segment line AB from point D, and CC' is a vertical line for line segment AB from point C. Because $\triangle DD'P \sim \triangle CC'P$, $\dfrac{|DD'|}{|CC'|} = \dfrac{|DP|}{|PC|}$. Because $S_{\triangle ABD} = \dfrac{|DD'| \times |AB|}{2}$, and $S_{\triangle ABC} = \dfrac{|CC'| \times |AB|}{2}$,

$$\frac{|DP|}{|PC|} = \frac{S_{\triangle ABD}}{S_{\triangle ACB}} = \frac{|\overrightarrow{AD} \wedge \overrightarrow{AB}|}{|\overrightarrow{AC} \wedge \overrightarrow{AB}|} = \frac{|mul(D,B,A)|}{|mul(C,B,A)|}.$$

Because $\dfrac{|DP|}{|PC|} = \dfrac{x_D - x_P}{x_P - x_C} = \dfrac{y_D - y_P}{y_P - y_C}$,

$$x_P = \frac{S_{\triangle ABD} \times x_c + S_{\triangle ABC} \times x_D}{S_{\triangle ABD} + S_{\triangle ABC}} = \frac{mul(D,B,A) \times x_c - mul(C,B,A) \times x_D}{mul(D,B,A) - mul(C,B,A)}, \text{ and}$$

$$y_P = \frac{S_{\triangle ABD} \times y_c + S_{\triangle ABC} \times y_D}{S_{\triangle ABD} + S_{\triangle ABC}} = \frac{mul(D,B,A) \times y_c - mul(C,B,A) \times y_D}{mul(D,B,A) - mul(C,B,A)}.$$

8.1.2.2 Intersecting Lines

We all know that a pair of distinct points on a plane defines a line and that a pair of lines on a plane will intersect in one of three ways: 1) no intersection because they

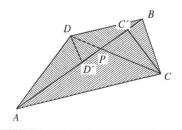

Figure 8.11

are parallel, 2) intersect in a line because they are on top of one another (i.e., they are the same line), 3) intersect in a point. In this problem you will use your algebraic knowledge to create a program that determines how and where two lines intersect.

Your program will repeatedly read in four points that define two lines in the *x-y* plane and determine how and where the lines intersect. All numbers required by this problem will be reasonable, say between -1000 and 1000.

Input

The first line contains an integer N between 1 and 10 describing how many pairs of lines are represented. The next N lines will each contain eight integers. These integers represent the coordinates of four points on the plane in the order $x_1 y_1 x_2 y_2 x_3 y_3 x_4 y_4$. Thus, each of these input lines represents two lines on the plane: the line through (x_1, y_1) and (x_2, y_2) and the line through (x_3, y_3) and (x_4, y_4). The point (x_1, y_1) is always distinct from (x_2, y_2). Likewise with (x_3, y_3) and (x_4, y_4).

Output

There should be $N+2$ lines of output. The first line of output should read "INTERSECTING LINES OUTPUT". There will then be one line of output for each pair of planar lines represented by a line of input, describing how the lines intersect: "NONE", "LINE", or "POINT". If the intersection is a point, then your program should output the x and y coordinates of the point, correct to two decimal places. The final line of output should read "END OF OUTPUT:".

Sample Input	Sample Output
5	INTERSECTING LINES OUTPUT
0 0 4 4 0 4 4 0	POINT 2.00 2.00
5 0 7 6 1 0 2 3	NONE
5 0 7 6 3 −6 4 −3	LINE
2 0 2 27 1 5 18 5	POINT 2.00 5.00
0 3 4 0 1 2 2 5	POINT 1.07 2.20
	END OF OUTPUT

Source: ACM Mid-Atlantic 1996

IDs for Online Judges: POJ 1269, ZOJ 1280, UVA 378

Analysis

The problem requires you to determine the relation between a pair of lines: they are parallel, they are the same line, or they intersect; and if they intersect, you need to output the x and y coordinates of the intersection point. One crossing is used to determine whether the two lines are parallel or the same line (whether $\overrightarrow{p_3 p_4}$ crosses $\overrightarrow{p_1 p_2}$ or not):

Suppose `a1=mul(`p_1`, `p_2`, `p_3`)`, and `a2=mul(`p_1`, `p_2`, `p_4`)`.
If `(a1==0)&&(a2==0)`, then line segments $\overrightarrow{p_1 p_2}$ and $\overrightarrow{p_3 p_4}$ are the same lines;
If positive and negative signs for `a1` and `a2` are same, then line segments $\overrightarrow{p_1 p_2}$ and $\overrightarrow{p_3 p_4}$ are parallel;
If positive and negative signs for `a1` and `a2` are different, the `x` and `y` coordinates of the intersection point is calculated directly: $p = \left(\dfrac{a2 \times p_3.x - a1 \times p_4.x}{a2 - a1}, \dfrac{a2 \times p_3.y - a1 \times p_4.y}{a2 - a1} \right)$.

 Program

```
#include <cstdio>
#include <cmath>
#include <cstring>
#include <algorithm>
#include <iostream>
using namespace std;
const double epsi = 1e-10;    // Infinitesimal
inline int sign(const double &x) {    // positive and negative
signs for x
  if (x > epsi) return 1;
  if (x < -epsi) return -1;
  return 0;
}
struct Point {    // structure and calculation for point
  double x, y;
  Point(double _x = 0, double _y = 0): x(_x), y(_y) { }
//Construct point
  Point operator -(const Point &op2) const {    // subtraction
for vectors
    return Point(x - op2.x, y - op2.y);
  }
  double operator ^(const Point &op2) const {    //cross
product
    return x * op2.y - y * op2.x;
  }
};
inline double sqr(const double &x) {    // x²
  return x * x;
}
inline double mul(const Point &p0,const Point &p1,const Point
&p2){    //cross product for p̄₀p̄₁ and p̄₀p̄₂
```

```
    return (p1 - p0) ^ (p2 - p0);
}
inline double dis2(const Point &p0, const Point &p1) {
    return sqr(p0.x - p1.x) + sqr(p0.y - p1.y);
}
inline double dis(const Point &p0, const Point &p1) {
//|p0p1|
    return sqrt(dis2(p0, p1));
}
inline int cross( const Point &p1 , const Point &p2 , const
Point &p3 , const Point &p4 , Point &p) {    //if p1p2 and p3p4
are the same lines, return 2; if parallel return 0; else
return 1 and the intersection point p
    double a1 = mul( p1, p2 , p3), a2 = mul( p1, p2 , p4 ) ;
    if (sign ( a1 ) ==0 && sign ( a2 ) == 0) return 2;
    if (sign ( a1 - a2 ) == 0) return 0;
    p.x = ( a2 * p3.x - a1 * p4.x) / ( a2 - a1 );
    p.y = ( a2 * p3.y - a1 * p4.y) / ( a2 -a1 );
    return 1;
}
Point p1 , p2 , p3 , p4 , p;
int main() {
    int test = 0;
    printf("INTERSECTING LINES OUTPUT\n");
    scanf("%d" , & test);    //number of test cases
    for ( ; test ; test --) {    //test cases are dealt with one
by one
        scanf( "%lf %lf %lf %lf %lf %lf %lf %lf" , &p1.x ,
&p1.y , &p2.x , & p2.y , &p3.x , &p3.y , &p4.x , &p4.y);
// coordinate for p1p2 and p3p4
        int m=cross(p1,p2,p3,p4,p);    //relationship between p1p2
and p3p4
        if (m == 0 ) printf("NONE\n");    //parallel
        else if (m==2)printf("LINE\n");    //same lines
        else printf("POINT %.2lf %.2lf\n", p.x , p.y);    // the
intersection point p
    }
    printf("END OF OUTPUT");
    return 0;
}
```

3. Calculating the circumcenter of a triangle.

In a triangle, the intersection point of perpendicular bisectors for three sides is the circumcenter of a triangle of the triangle. The distance between a point and the circumcenter is the radius of the circumcircle.

Suppose the three points for a triangle are $p_1=(x_1, y_1)$, $p_2=(x_2, y_2)$, and $p_3=(x_3, y_3)$, respectively; and the center of the circumcircle is $p=(x, y)$.

For edge vector $\overrightarrow{p_1 p_2}$, suppose $A_{\overline{p_1 p_2}} = x_2 - x_1$, $B_{\overline{p_1 p_2}}$ $B_{\overline{p_1 p_2}} = y_2 - y_1$, and $C_{\overline{p_1 p_2}} = -\dfrac{\left| \overline{p_1 p_2} \right|}{2}$; and for edge vector $\overrightarrow{p_1 p_3}$, suppose $A_{\overline{p_1 p_3}} = x_3 - x_1$, $B_{\overline{p_1 p_3}} = y_3 - y_1$, and $C_{\overline{p_1 p_3}} = -\dfrac{\left| \overline{p_1 p_3} \right|}{2}$; p_1 is as the origin. The intersection point of perpendicular bisectors for side $\overline{p_1 p_2}$ and side $\overline{p_1 p_3}$ in the triangle is $p_1^* = (x_1^*, y_1^*)$, where

$$x_1^* = -\frac{C_{\overline{p_1 p_3}} \times B_{\overline{p_1 p_2}} - C_{\overline{p_1 p_2}} \times B_{\overline{p_1 p_3}}}{A_{\overline{p_1 p_3}} \times B_{\overline{p_1 p_2}} - B_{\overline{p_1 p_3}} \times A_{\overline{p_1 p_2}}}, \text{ and } y_1^* = -\frac{C_{\overline{p_1 p_3}} \times A_{\overline{p_1 p_2}} - C_{\overline{p_1 p_2}} \times A_{\overline{p_1 p_3}}}{B_{\overline{p_1 p_3}} \times A_{\overline{p_1 p_2}} - B_{\overline{p_1 p_2}} \times A_{\overline{p_1 p_3}}}.$$

Therefore, the center of the circumcircle is $p = p_1 + p_1^*$, and the Cartesian coordinates of point p are $(x_1 + x_1^*, y_1 + y_1^*)$.

8.1.2.3 Circle Through Three Points

Your team is to write a program that, given the Cartesian coordinates of three points on a plane, will find the equation of the circle through them all. The three points will not be on a straight line. The solution is to be printed as an equation of the form

$$(x - h)^2 + (y - k)^2 = r^2 \tag{1}$$

and an equation of the form

$$x^2 + y^2 + cx + dy - e = 0 \tag{2}$$

Input

Each line of input to your program will contain the x and y coordinates of three points, in the order Ax, Ay, Bx, By, Cx, Cy. These coordinates will be real numbers separated from each other by one or more spaces.

Output

Your program must print the required equations on two lines using the format given in the sample below. Your computed values for h, k, r, c, d, and e in Equations 1 and 2 above are to be printed with three digits after the decimal point. Plus and minus signs in the equations should be changed as needed to avoid multiple signs before a number. Plus, minus, and equal signs must be separated from the adjacent characters by a single space on each side. No other spaces are to appear in the equations. Print a single blank line after each equation pair.

Sample Input	Sample Output
7.0 −5.0 −1.0 1.0 0.0 −6.0 1.0 7.0 8.0 6.0 7.0 −2.0	(x − 3.000)^2 + (y + 2.000)^2 = 5.000^2 x^2 + y^2 − 6.000x + 4.000y − 12.000 = 0 (x − 3.921)^2 + (y − 2.447)^2 = 5.409^2 x^2 + y^2 − 7.842x − 4.895y − 7.895 = 0

Source: ACM Southern California 1989

IDs for Online Judges: POJ 1329, UVA 190

Analysis

On a plane, if three points aren't on a straight line, the three points are points of a triangle, and the circle through the three points is a circumcircle.

For equation 1, $(x-h)^2+(y-k)^2=r^2$, (h, k) is the Cartesian coordinate for the center of the circumcircle, and r is the radius of the circumcircle.

For equation 2, $x^2+y^2+cx+dy-e=0$, $c=-2\times h$, $d=-2\times k$, $e=h^2+k^2-r^2$.

The key to the problem is to calculate the center (h, k) of the circumcircle for $\triangle ABC$. Distances between the center and the points of the triangle are same. The radius r is the distance between the center and any point of the triangle.

Program

```cpp
#include <cstdio>
#include <cmath>
#include <cstring>
#include <algorithm>
#include <iostream>
using namespace std;
const double epsi = 1e-10;     //precision
inline int sign(const double &x) {     //positive or negative
sign for x
   if (x > epsi) return 1;
   if (x < -epsi) return -1;
   return 0;
}
struct Point {     // structure and calculation for point
   double x, y;
   Point(double _x = 0, double _y = 0): x(_x), y(_y) { }
//point (x, y)
```

```
   Point operator +(const Point &op2) const {     //Addition for
vectors
      return Point(x + op2.x, y + op2.y);
   }
   Point operator -(const Point &op2) const {     //Subtraction
for vectors
      return Point(x - op2.x, y - op2.y);
   }
   Point operator *(const double &d) const {     //vector times
real
      return Point(x * d, y * d);
   }
   Point operator /(const double &d) const {     //vector is
divided by real
      return Point(x / d, y / d);
   }
   double operator ^(const Point &op2) const {     //cross
product for two vectors
      return x * op2.y - y * op2.x;
   }
};
inline double mul(const Point &p0,const Point &p1,const Point
&p2) {//Cross product for p̄₀p̄₁ and p̄₀p̄₂
   return (p1-p0) ^ (p2 - p0);
}
struct StraightLine {     // perpendicular bisector's structure
   double A, B, C;     // Perpendicular Bisector, where for
edge-vector pᵢpⱼ in a triangle, A=(xⱼ-xᵢ), B=(yⱼ-yᵢ),
```

$$C=\frac{\left|\overrightarrow{p_ip_j}\right|}{2}(1\le i,j\le 3)$$

```
   StraightLine(double _a=0, double _b=0, double _c=0): A(_a),
B(_b), C(_c){ }  // perpendicular bisector is constructed
      Point cross(const StraightLine &a) const {
// calculating the intersection point for the perpendicular
bisector and perpendicular bisector a
      double xx = - (C * a.B - a.C * B) / (A * a.B - B * a.A);
      double yy = - (C * a.A - a.C * A) / (B * a.A - a.B * A );
      return Point(xx, yy);
   }
};
inline double sqr(const double &x) {     // x²
   return x * x;
}
inline double dis2(const Point &p0, const Point &p1) { //|p̄₀p̄₁|²
   return sqr(p0.x - p1.x) + sqr(p0.y - p1.y);
}
inline double dis(const Point &p0, const Point &p1) {     //|p̄₀p̄₁|
   return sqrt(dis2(p0, p1));
}
```

```
inline double circumcenter(const Point &p1,const Point
&p2,const Point &p3,Point &p)      //calculating the center p
and radius for the circumcircle, p is the intersection point
for p₁p₃ and p₁p₂
{
   p=p1+StraightLine(p3.x-p1.x,p3.y-p1.y,-dis2(p3,p1)/2.0).
cross(StraightLine(p2.x-p1.x, p2.y-p1.y,-dis2(p2, p1)/2.0));
//center of circle p
   return dis( p , p1 );     //return radius
}
Point p1, p2, p3, p;
inline int print(double x) {      //output value x
   if (x > 0) printf(" + %.3lf", x);
   else printf(" - %.3lf", -x);
   return 0;
}
int main() {
    while (cin>>p1.x>>p1.y>>p2.x>>p2.y>>p3.x>>p3.y){
// coordinates of three points
      double r=circumcenter(p1,p2,p3,p);     // the center and
radius for circumcircle p and r
      printf(" (x");     //equation 1
      print(-p.x);
      printf(")^2 + (y");
      print(-p.y);
      printf(")^2 =");
      printf(" %.3lf", r);
      printf("^2\n");
      printf("x^2 + y^2");     //equation 2
      print(-2 * p.x);
      printf("x");
      print(-2 * p.y);
      printf("y");
      print(sqr(p.x) + sqr(p.y) - sqr(r));
      printf(" = 0\n\n");
   }
   return 0;
}
```

8.1.3 *Solving Polyhedron Problems by Euler's Polyhedron Formula*

Euler's Formula: If a finite, connected, planar graph is drawn in the plane without any edge intersections; and v is the number of vertices, e is the number of edges, and f is the number of faces (regions bounded by edges, including the outer, infinitely large region), then $v-e+f=2$.

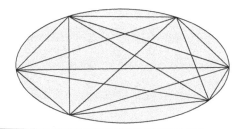

Figure 8.12

Euler's Polyhedron Formula: For a polyhedron, the number of vertices (corner points) v, plus the number of faces f, and minus the number of edges e, equals 2. Symbolically $v-e+f=2$.

8.1.3.1 How Many Pieces of Land?

You are given an elliptical-shaped piece of land (see Figure 8.12) and you are asked to choose n arbitrary points on its boundary. Then you connect all these points with each other with straight lines (that's $n \times (n-1)/2$ connections for n points). What is the maximum number of pieces of land you will get by choosing the points on the boundary carefully?

Input

The first line of the input file contains one integer S ($0 < S < 3500$), which indicates how many sets of input there are. The next S lines contain S sets of input. Each input contains one integer N ($0 \le N < 2^{31}$).

Output

For each set of input, you should output in a single line the maximum number of pieces of land possible to get for the value of N.

Sample Input	Sample Output
4	1
1	2
2	4
3	8
4	

Source: Math & Number Theory Lovers' Contest

ID for Online Judge: UVA 10213

Analysis

The number of pieces of land is the number of faces. Euler's Formula $v-e+f=2$ is used to solve the problem, where v is the number of vertices, e is the number of edges, and f is the number of faces.

First, the number of vertices v is calculated. There are n points on the ellipse's boundary. For a point x on the boundary, there are $n-1$ straight lines connecting point x and other points. For a straight line l, there are i points on the left, and there are $n-2-i$ points on the right. Because all these points are connected with one another with straight lines, there are at most $i \times (n-i-2)$ points on a straight line.

Each point is repeatedly counted four times. Therefore, $v = n + \dfrac{n}{4} \sum\limits_{i=1}^{n-3} i \times (n-i-2)$.

Second, the number of edges e is calculated. There are n points on the elliptical shaped land's boundary. There are n edges on the boundary. There are n straight lines connecting adjacent points. There are no intersection points on these edges. For other straight lines connecting points, there are $i \times (n-i-2)$ intersection points on a straight line. On a straight line, there are $i \times (n-i-2)+1$ edges. Therefore,

$$e = 2 \times n + \frac{n}{2} \sum_{i=1}^{n-3} i \times (n-i-2) + 1.$$

Euler's Formula $v-e+f=2$ is used to solve the problem. The maximum number of pieces of land $f = \dfrac{n^4 - 6n^3 + 23n^2 - 18n}{24} + 1$.

Because the upper limit for n is 2^{31}, the high-precision method is also used.

Program

```
# include <cstdio>
# include <cstring>
# include <cstdlib>
# include <iostream>
# include <string>
# include <cmath>
# include <algorithm>
using namespace std;
typedef long long int64;
int64 m=1e8;    //High-precision number: a decimal number with
8-digit
struct Bigint{    // High-precision number
```

```
    int64 s[50];int l;    // High-precision number: array s[],
  length l
    void print(){    //output the integer for s[]
      printf("%lld",s[l]);
      for(int i=l-1;i>=0;i--) printf("%08lld",s[i]);
    }
    void read(int64 x){    //integer x is stored in s[]
      l=-1; memset(s,0,sizeof(s))
      do{
        s[++l]=x%m;
        x/=m;
      }while(x);
    }
  } ans,tmp,t2;
  Bigint operator +(Bigint a,Bigint b){    // a[]+b[]
    int64 d=0;
    a.l=max(a.l,b.l);
    for(int i=0;i<=a.l;i++){    //addition bitwise
      a.s[i]+=d+b.s[i];
      d=a.s[i]/m;a.s[i]%=m;
    }
    if(d)  a.s[++a.l]=d;    //carry
    return a;
  }
  Bigint operator -(Bigint a,Bigint b){    // a[]-b[]
    int64 d=0;
    for(int i=0;i<=a.l;i++){    //subtraction bitwise
      a.s[i]-=d;
      if(a.s[i]<b.s[i])a.s[i]+=m,d=1;
      else d=0;
      a.s[i]-=b.s[i];
    }
    while(a.l&&!a.s[a.l]) a.l--;    //borrow
    return a;
  }
  Bigint operator *(int b,Bigint a){    // a[]*b
    int64 d=0;
    for(int i=0;i<=a.l;i++) {    //times bitwise
      d+=a.s[i]*b;a.s[i]=d%m;
      d/=m;
    }
    while(d){    //carry
      a.s[++a.l]=d%m;
      d/=m;
    }
    return a;
  }
  Bigint operator /(Bigint a,int b){    // a[]/b
    int64 d=0;
    for(int i=a.l;i>=0;i--){
      d*=m;d+=a.s[i];
```

```
         a.s[i]=d/b;d%=b;
      }
      while(a.l&&!a.s[a.l])    a.l--;    //omit 0
      return a;
}
Bigint operator *(Bigint a,Bigint b){    // a[]*b[]
   Bigint c; memset(c.s,0,sizeof(c.s))
   for(int i=0;i<=a.l;i++){
      for(int j=0;j<=b.l;j++){
         c.s[i+j]+=a.s[i]*b.s[j];
         if(c.s[i+j]>m){    //carry
            c.s[i+j+1]+=c.s[i+j]/m;
            c.s[i+j]%=m;
         }
      }
   }
   c.l=a.l+b.l+10;
   while(!c.s[c.l]&&c.l)c.l--;
   while(c.s[c.l]>m){
      c.s[c.l+1]+=c.s[c.l]/m;
      c.s[c.l++]%=m;
   }
   return c;
}
int v;
void work(){
   ans.read(v);tmp.read(24);    //ans: number of points
   ans=ans*ans*ans*ans+23*(ans*ans)+tmp-6*(ans*ans*ans)-18*ans;
//formula
   ans=ans/24;    //calculate and output the number of faces
   ans.print();printf("\n");
}
int main(){
   int casen;scanf("%d",&casen);    //number of test cases
   while(casen--){    //test cases are dealt with one by one
      scanf("%d",&v);    //number of points
      work();    //calculate and output the number of faces
   }
   return 0;
}
```

8.2 Calculating the Area for Union of Rectangles by Sweep Line Algorithms

Sweep line algorithms can be used to calculate the area for the union of rectangles.

Suppose there are n rectangles R_1, ..., R_n in a plane. $R_1 \cup R_2 \cup ... \cup R_n$ is the union of n rectangles. The area for the union of n rectangles is the area of coverage by these n rectangles. For example, in Figure 8.13, the shadow area is the area of $R_1 \cup R_2 \cup R_3$, that is, the area of coverage by the three rectangles.

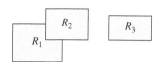

Figure 8.13 The area of $R_1 \cup R_2 \cup R_3$.

The steps for calculating the area for the union of rectangles are as follows:

1. **Discretization:** The plane is divided into several strips.
2. **Sweep:** A sweep algorithm is used to sweep strips. Strips are stored in a segment tree.
3. **Segment tree:** Calculating the area for union of rectangles is implemented by insertions and deletions in the segment tree.

Sweep line algorithms are introduced through two kinds of experiments:

1. Calculating the area for union of rectangles in the vertical direction;
2. Calculating the area for union of rectangles in the horizontal direction.

Sweep line algorithms can also be extended to the three-dimensional space to calculate the volume for union of cuboids.

8.2.1 Sweeping in the Vertical Direction

Calculating the area for union of rectangles in the vertical direction is as follows. Discretization is on the Y-axis. The plane is divided into several vertical strips by sweeping on the X-axis. A segment tree is used to accumulate areas of these vertical strips.

Discretization: Discrete points are intersection points for sides of rectangles (or their extended lines) and the Y-axis. In Figure 8.14, discrete points are A, B, C, and D.

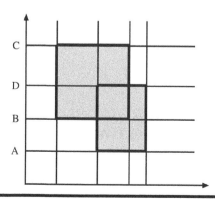

Figure 8.14 Discrete points A, B, C, and D are intersection points for sides of rectangles and Y-axis.

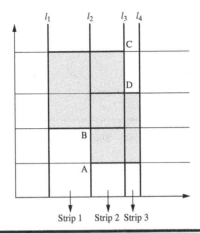

Figure 8.15   The plane is divided into three vertical strips by straight lines l_1, l_2, l_3, and l_4.

Segments of discrete units are distances between two adjacent discrete points in the ordered sequence of discrete points. For example, in Figure 8.14, the Y-axis for A is 1, the Y-axis for B is 2, the Y-axis for C is 3, and the Y-axis for D is 4. After the discretization, lengths for segment AB, BC, and CD are 1.

Sweep: First, the plane is divided into vertical strips, and each vertical strip is one-dimensional. In Figure 8.15, the plane is divided into three vertical strips by straight lines l_1, l_2, l_3, and l_4.

Each vertical strip's section can be regarded as a little modification for two adjacent vertical strips' sections. In Figure 8.16, the section for vertical strip 2 = the section for vertical strip 1 + segment AB = the section for vertical strip 3 + segment CD.

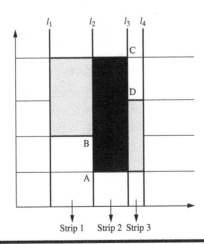

Figure 8.16

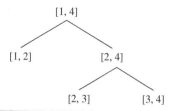

Figure 8.17   A segment tree representing interval [1, 4].

Segment tree: A segment tree is a rooted binary tree, where each vertex represents an interval $[a, b]$. For each vertex, if $(b-a)>1$, suppose $c = \left\lfloor \dfrac{a+b}{2} \right\rfloor$, and roots for its left subtree and right subtree represent intervals $[a, c]$ and $[c, b]$ respectively. In Figure 8.17, interval $[1, 4]$ can be divided into intervals $[1, 2]$ and $[2, 4]$. And interval $[2, 4]$ can be divided into intervals $[2, 3]$ and $[3, 4]$.

Because vertical strips can be represented as segments, a segment tree can be used to store vertical strips. Calculating the area for union of n rectangles can be implemented by insertion and deletion in a segment tree.

8.2.1.1 Mobile Phone Coverage

A mobile phone company ACMICPC (Advanced Cellular, Mobile, and Internet-Connected Phone Corporation) is planning to set up a collection of antennas for mobile phones in a city called Maxnorm. The company ACMICPC has several collections for locations of antennas as their candidate plans, and now they want to know which collection is the best choice.

For this purpose, they want to develop a computer program to find the coverage of a collection of antenna locations. Each antenna A_i has power r_i, corresponding to "radius". Usually, the coverage region of the antenna may be modeled as a disk centered at the location of the antenna (x_i, y_i) with radius r_i. However, in this city, Maxnorm, such a coverage region becomes the square $[x_i-r_i, x_i+r_i] \times [y_i-r_i, y_i+r_i]$. In other words, the distance between two points (x_p, y_p) and (x_q, y_q) is measured by the max norm $\max\{|x_p-x_q|, |y_p-y_q|\}$, or, the norm L_∞, in this city Maxnorm instead of the ordinary Euclidean norm $\sqrt{(x_p - x_q)^2 + (y_p - y_q)^2}$.

As an example, consider the following collection of three antennas as depicted in Figure 8.18:

4.0	4.0	3.0
5.0	6.0	3.0
5.5	4.5	1.0

where the i-th row represents x_i, y_i, r_i such that (x_i, y_i) is the position of the i-th antenna and r_i is its power. The area of regions of points covered by at least one antenna is 52.00 in this case.

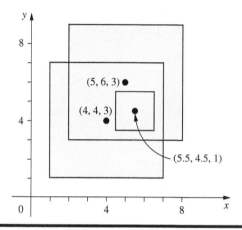

Figure 8.18

Write a program that finds the area of coverage by a given collection of antenna locations.

Input

The input contains multiple data sets, each representing a collection of antenna locations. A data set is given in the following format.

```
n
x₁ y₁ r₁
x₂ y₂ r₂
..........
xₙ yₙ rₙ
```

The first integer n is the number of antennas, such that $2 \le n \le 100$. The coordinate of the i-th antenna is given by (x_i, y_i), and its power is r_i. x_i, y_i and r_i are fractional numbers between 0 and 200 inclusive.

The end of the input is indicated by a data set with 0 as the value of n.

Output

For each data set, your program should output its sequence number (1 for the first data set, 2 for the second, etc.) and the area of the coverage region. The area should be printed with two digits to the right of the decimal point, after rounding it to two decimal places.

The sequence number and the area should be printed on the same line with no spaces at the beginning and end of the line. The two numbers should be separated by a space.

Sample Input	Sample Output
3	1 52.00
4.0 4.0 3.0	2 36.00
5.0 6.0 3.0	
5.5 4.5 1.0	
2	
3.0 3.0 3.0	
1.5 1.5 1.0	
0	

Source: ACM Asia Regional Contest Tokyo 1998

IDs for Online Judges: ZOJ 1659, UVA 688

 Analysis

Each antenna's coverage is a square whose center is (x_i, y_i) and the length of its side is $2 \times r_i$. The area of coverage by n antennas' locations is the area for the union of n corresponding squares. That is, the problem requires you to calculate the area for the union of n corresponding squares.

The area for the union of n corresponding squares is calculated by sweeping in the vertical direction. Discretization is on the Y-axis. The plane is divided into several vertical strips by sweeping on the X-axis. A segment tree is used to accumulate areas of vertical strips.

 Program

```
#include <cstdio>
#include <cmath>
#include <algorithm>
using namespace std;
const double epsi = 1e-10;
const int maxn = 100 + 10;
struct Line {      //coverage area
    double x, y1, y2;      // x-coordinate for the left or right,
y-coordinates for the above side and below sides,
```
$$s = \begin{cases} 1 & \text{x-coordinate for the left node} \\ -1 & \text{x-coordinate for the right node} \end{cases}$$

```
    int s;
    Line(double _a=0, double _b=0, double _c=0, int _d=0):
x(_a),y1(_b),y2(_c),s(_d){ }
//construct a segment
    bool operator <(const Line &op2) const {     //Sorting in
ascending order for x-coordinates
        return x < op2.x;
    }
};
extern double ly[maxn << 1];     //ly[ ] stores y-coordinates
for the above side and below sides of a square covered by
antennas, capacity is 2^{maxn}
class SegmentTree {     // Segment tree
    int cover;     // the flag for an open interval
    SegmentTree *child[2];     //left, right children pointers

    void deliver() {     // the length for covered interval
        if (cover)
            len = ly[r]-ly[l];
        else
            len = child[0]->len + child[1]->len;
    }
public:
    int l, r;     //the interval for a segment tree
    double len;     //the length for the current strip
    void setup(int _l, int _r) {     //set up a segment tree for
the interval [_l,_r]
        l = _l, r = _r;     //initialization
        cover = 0, len = 0;
        if (_l + 1 == _r) return;
        int mid = (_l + _r) >> 1;     //middle pointer
        child[0]=new SegmentTree(),child[1]=new SegmentTree();
//set up left and right subtrees
        child[0]->setup(_l, mid), child[1]->setup(mid, _r);
    }
    void paint(const int &_l, const int &_r, const int &v) {
//interval [_l,_r] is inserted into the segment tree for
interval [l,r]
        if (_l >= r || _r <= l) return;
        if (_l <= l && r <= _r) {
            if (cover += v) len = ly[r]-ly[l]; else {
                if (child[0]==NULL)len=0;else len = child[0]->len +
child[1]->len;
            }
            return;
        }
        child[0]->paint(_l, _r, v), child[1]->paint(_l, _r, v);
        deliver();
    }
    void die() {     //deletion
```

```
      if (child[0]) {
        child[0]->die();
        delete child[0];
        child[1]->die();
        delete child[1];
      }
   }
};
int cs(0);    //initialize the number of test cases
int n, tot, ty;    //n: number of antennas, tot: the length
of l[], ty: the length of ly[]
Line l[maxn << 1];    //l[] stores vertical strips
double ly[maxn << 1];    //ly[] stores y-coordinates
SegmentTree *seg_tr;    //Pointer for the segment tree
int main() {
   while (scanf("%d", &n), n) {    // number of antennas
     tot = ty = 0;
     for (int i = 0; i < n; ++i) {
        double x, y, r;
        scanf("%lf%lf%lf", &x, &y, &r);    //the i-th antenna
        l[tot++] = Line(x - r, y - r, y + r, 1);  //store strip
        l[tot++] = Line(x + r, y - r, y + r, -1);
        ly[ty++] = y-r, ly[ty++]=y + r;
//stores y-coordinates
     }
     sort(l, l + tot);    //sort strips from left to right
     sort(ly, ly + ty);    // sort y-coordinates top-down
     ty = unique(ly, ly + ty) - ly;    //eliminate duplicate
     double ans = 0;    //initialize the area of the coverage
region
     seg_tr = new SegmentTree();
     seg_tr->setup(0, ty - 1);    //set up a segment for
interval [0, ty-1]
     for (int i = 0, j; i < tot; i = j) {    // Enumerate
strips in l[]
        if (i) ans += seg_tr->len * (l[i].x-l[i-1].x);
//accumulate area of the coverage region
        j = i;    //Enumerate strips, [l, r, k] is inserted
into the segment tree
        while (j < tot && fabs(l[i].x - l[j].x) <= epsi) {
          seg_tr->paint(lower_bound(ly,ly+ty,l[j].y1)-
ly,lower_bound(ly,ly+ty,l[j].y2) -ly,l[j].s);
           ++j;
        }
     }
     seg_tr->die(); delete seg_tr;    //delete a segment
     printf("%d %.2lf\n", ++cs, ans);    // the area of the
coverage region
   }
   return 0;
}
```

8.2.2 Sweeping in the Horizontal Direction

Calculating the area for the union of rectangles in the horizontal direction is similar to calculating the area for the union of rectangles in the vertical direction. Discretization is on the X-axis. The plane is divided into horizontal strips by sweeping on the Y-axis. A segment tree is used to accumulate areas of horizontal strips. The method is as follows:

Discretization: Calculate intersection points for sides of rectangles (or their extended lines) and X-axis, sort intersection points in ascending order of x-coordinates, and calculate distances between two adjacent intersection points.

Sweep: The plane is divided into horizontal strips, and a segment tree is used to store these horizontal strips' cross sections.

Segment tree: Calculating the union of n rectangles' areas can be implemented by insertion and deletion in a segment tree.

8.2.2.1 Atlantis

There are several ancient Greek texts that contain descriptions of the fabled island Atlantis. Some of these texts even include maps of parts of the island. But unfortunately, these maps describe different regions of Atlantis. Your friend Bill has to know the total area for which maps exist. You (unwisely) volunteered to write a program that calculates this quantity.

Input

The input consists of several test cases. Each test case starts with a line containing a single integer n ($1 \leq n \leq 100$) of available maps. The n following lines describe one map each. Each of these lines contains four numbers x_1; y_1; x_2; y_2 ($0 \leq x_1 < x_2 \leq 100000$; $0 \leq y_1 < y_2 \leq 100000$), not necessarily integers. The values (x_1; y_1) and (x_2; y_2) are the coordinates of the top-left and bottom-right corners of the mapped area, respectively.

The input file is terminated by a line containing a single 0. Don't process it.

Output

For each test case, your program should output one section. The first line of each section must be "Test case #k", where k is the number of the test case (starting with 1). The second one must be "Total explored area: a", where a is the total explored area (i.e., the area of the union of all rectangles in this test case), printed exact to two digits to the right of the decimal point.

Output a blank line after each test case.

Sample Input	Sample Output
2 10 10 20 20 15 15 25 25.5 0	Test case #1 Total explored area: 180.00

Source: ACM Mid-Central European Regional Contest 2000

IDs for Online Judges: POJ 1151, ZOJ 1128, UVA 2184

 Analysis

An available map in the problem is represented as a rectangle. The total explored area is the area for the union of these rectangles.

The plane is divided into several rectangles (Figure 8.19). Areas of these rectangles are calculated respectively. The sum of areas is the total explored area (the area of the union of all rectangles). This is shown in Figure 8.20.

The algorithm is as follows:

For each map, its left boundary's x-coordinate and right boundary's x-coordinate are stored in a sequence q in ascending order.

For each map, its bottom edge's y-coordinate, top edge's y-coordinate, and x-coordinates for endpoints of edges are stored in a sequence f. The flag for bottom edges is 1, and the flag for top edges is −1. And f is sorted in ascending order of y-coordinates, to make f store horizontal strips from bottom to top.

Then, each horizontal strip is taken out from f and the segment $[x_l, x_r]$ (x-coordinates for endpoints of the edge) is inserted into the segment tree. There are two fields for vertices for the segment tree:

len, the length of the union of intervals;
mark: the mark of the union of intervals;

When a horizontal strip is added, the area covered by a horizontal strip is accumulated into the total explored area.

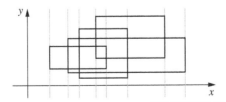

Figure 8.19

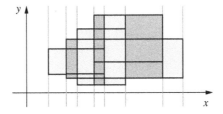

Figure 8.20

Program

```cpp
#include <cstdio>
#include <cmath>
#include <cstring>
#include <algorithm>
#include <iostream>
using namespace std;
const int maxn = 500;    //The upper limit of the number of
maps*2
struct node {
   double x;    //y-coordinate for horizontal strip
   int l, r, t;    //l, r: x-coordinates for two points in q, t:
the flag for bottom edge and top edge
} f[maxn];    // horizontal strips
int n;    // the number of maps
double q[maxn], x1[maxn], yy1[maxn], x2[maxn], yy2[maxn];
//q stores sorted x- coordinates, for the i-th map, the
coordinate for the top left corner (x1[i],yy1[i]), the
coordinate for the lower right corner (x2[i],yy2[i])
struct segment {
   int mark;
   double len;    //the length of the union for intervals
} tree[maxn * 20];    //segment tree
int cmp(node a, node b) {    //Comparison function for f[]
   return a.x < b.x;
}
int insert(const int k,const int l,const int r,const int
lc,const int rc,const int t) {    //horizontal strip[l, r]
is inserted into segment tree (k:root, interval [lc, rc]),
t: mark for bottom edge and top edge
   if (lc<=l && r<=rc) {    //[lc, rc] covers [l, r]
     tree[k].mark += t;
   } else  {
            if ((l+r)/2>=lc)insert(k*2,l,(l+r)/2, lc,rc,t);
```

```
                if((l+r)/2<rc) insert(k*2+1,(l+r)/ 2+1,r,lc, rc,t);
   }
   if (tree[k].mark == 0) tree[k].len=tree[k *2].
len+tree[k *2+1].len;
     else tree[k].len=q[r+1]-q[l];
   return 0;
}
int main() {
   int test = 0;      //number of test cases
   while (scanf("%d", &n) && n) {       //number of maps
     double ans = 0;      // the total explored area
     for (int i = 1; i <= n ; i ++) {       // the coordinates of
the top-left and bottom-right corner
        scanf("%lf %lf %lf %lf" , &x1[i], &yy1[i], &x2[i],
&yy2[i]);
        if (x1[i] > x2[i]) swap(x1[i], x2[i]);
        if (yy1[i] > yy2[i]) swap(yy1[i], yy2[i]);
        q[i * 2 - 2] = x1[i];      // x-coordinate
        q[i * 2 - 1] = x2[i];
     }
     sort(q, q+n*2);      //sort in ascending order of
x-coordinates in q
     int m = unique(q, q+n*2)-q;      //remove duplication in q,
m: the length of q
     for ( int i=1;i<= n ; i ++) {      //the i-th map is stored
in f
        f[i*2-2].l=lower_bound(q,q+m,x1[i])-q;
        f[i*2-2].r=lower_bound(q,q+m,x2[i])-q;
        f[i*2-2].x=yy1[i];
        f[i*2-2].t=1;
        f[i*2-1].l=lower_bound(q, q + m, x1[i]) - q;
        f[i*2-1].r=lower_bound(q, q + m, x2[i]) - q;
        f[i * 2 - 1].x = yy2[i];
        f[i * 2 - 1].t = -1;
     }
     sort(f,f+n*2,cmp);      //f is sorted bottom up
     for ( int i = 0 ; i < n * 2; i ++) {      //horizontal
strips are analyzed bottom-up
        if (i) ans += tree[1].len*(f[i].x-f[i-1].x);
//accumulate the current strip's area
        insert(1,0,m ,f[i].l,f[i].r-1,f[i].t);      //insert the
strip into the segment tree
     }
     printf("Test case #%d\n", ++ test);      //output the total
explored area
     printf("Total explored area: %.2lf \n\n", ans);

   }
   return 0;
}
```

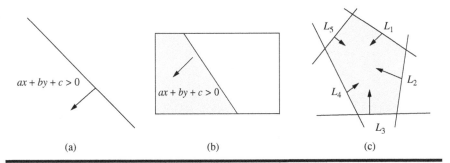

(a)   (b)   (c)

Figure 8.21

8.3 Intersection of Half-Planes

For a convex polygon, if its sides are represented by equations of lines or polar angles, the convex polygon can be represented by the intersection of half-planes.

A line $ax+by+c=0$, where a, b, and c are constants in a two-dimensional plane, divides the entire plane into two half-planes. A half-plane is thus defined by a line and one of its sides: either $ax+by+c \geq 0$ or $ax+by+c \leq 0$ [Figure 8.21(a)].

A half-plane in a bounded region, or an intersection of half-planes, can constitute a convex polygon [Figure 8.21(b) and (c)]. The intersection of n half-planes $H_1 \cap H_2 \cap \ldots \cap H_n$ is a convex polygon with at most n sides. For example, in Figure 8.21(c), there are five lines L_1, L_2, L_3, L_4, and L_5. One side of line L_i is the half-plane H_i, $1 \leq i \leq 5$. The intersection of five half-planes is a convex polygon with five sides.

Maybe an intersection of n half-planes is unbounded. Four half-planes, $x-c \leq 0$, $x+c \geq 0$, $y-c \leq 0$, and $y+c \geq 0$, can be added to make the intersection bounded (Figure 8.22).

An intersection of n half-planes can also be a line, a vertex, or an empty set.

The intersection of two convex polygons generates a new convex polygon [Figure 8.23(a)]. For the new convex polygon, its points are points of intersection of the two convex polygons' sides. The points are also boundary points that classify sides into outer sides and inner sides. Inner sides constitute the new convex polygon

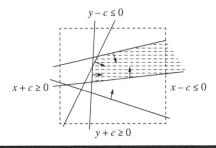

Figure 8.22

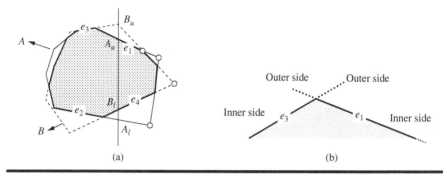

Figure 8.23

[Figure 8.23(b)]. Suppose there is a vertical sweep line sweeping from left to right. At any time, there are at most four points of intersection of the sweep line and the two convex polygons. For example, in Figure 8.23(a), the upper point and lower point of intersection of the sweep line and convex polygon A are A_u and A_l respectively; and the upper point and lower point of intersection of the sweep line and convex polygon A are B_u and B_l respectively. Obviously, the points of intersection of the sweep line and the intersection of two convex polygons are A_u and B_l. Sides containing A_u, A_l, B_u, and B_l are e_1, e_2, e_3, and e_4, respectively.

In this section, there are two kinds of experiments for the intersection of half-planes:

1. On-Line Algorithm for Intersection of Half-Planes;
2. Polar Angles.

8.3.1 On-Line Algorithm for Intersection of Half-Planes

Suppose the intersection of n half-planes $H_1 \cap H_2 \cap ... \cap H_n$ is the convex polygon A. Originally A is the entire plane. Then, cutting lines $a_i x + b_i y + c_i = 0$ for H_i (the line dividing A to generate the half-plane H_i) are used to divide A one by one, and the part that $a_i x + b_i y + c_i \geq 0$ is retained in A, $1 \leq i \leq n$. Finally, A is $H_1 \cap H_2 \cap ... \cap H_n$.

The key to the problem is how the cutting line $a_i x + b_i y + c_i = 0$ divides the convex polygon A, and how the part that $a_i x + b_i y + c_i \geq 0$ is retained in A is calculated. Suppose there are k points in A listed anticlockwise in $a[]$, the current cutting line is $\overrightarrow{p_1 p_2}$, and points are listed anticlockwise in $b[]$ after A is divided by $\overrightarrow{p_1 p_2}$. And $b[]$ is calculated as follows.

```
b[] is initialized empty;
    for (int i = 0; i<k; ++i) {      //enumerate points in a[]
  { if (p₁a[i] ^ p₂a[i] ≥ 0) {a[i] is added into b[]; continue;}
//if a[i]p₁ and p₁p₂ are connected anticlockwise, or a[i] is
over p₁p₂, then a[i] is retained [Figure 8.24(a)]
```

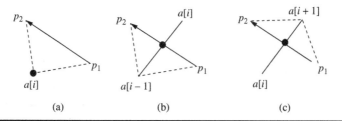

(a)   (b)   (c)

Figure 8.24

```
Calculate the left adjacent point j for point i;
   if (p₁a[j]∧ p₂a[j]>0)    //if a[j]p₁ and p₁p₂ are connected
anticlockwise, then the intersection point for p₁p₂ and a[j]a[i]
is retained [Figure 8.24(b)].
   { the intersection point for p₁p₂ and a[j]a[i] is added into
b[]; }
   Calculate the right adjacent point j for point i;
   if (p₁a[j] ∧ p₂a[j] > 0) {the intersection point for p₁p₂ and
a[j]a[i] is added into b[];}    //if a[j]p₁ and p₁p₂ are connected
anticlockwise, then the intersection point for p₁p₂ and a[i]a[j]
is retained [Figure 8.24(c)].
   }
```

The time complexity using a cutting line to divide the plane A to generate a half-plane is $O(n)$. The intersection of n half-planes can be calculated by using the following method n times.

Suppose the plane A is a square whose points' coordinates are $(-10^3, -10^3)$, $(10^3, -10^3)$, $(10^3, 10^3)$ and $(-10^3, 10^3)$, and the four points are stored in $a[]$. The cutting line for H_1 is used to divide A to generate a convex polygon whose points are stored in $b[]$, $b[]$ is assigned to $a[]$, and $b[]$ is cleared out. Then the cutting line for H_2 is used to divide A to generate a new convex polygon whose points are stored in $b[]$, ..., and so on. After the cutting line for H_n is used to divide A to generate a convex polygon, its points are stored in $b[]$. The time complexity is $O(n^2)$. This algorithm is called the On-Line Algorithm for Intersection of Half-Planes.

8.3.1.1 Feng Shui

Feng shui is the ancient Chinese practice of placement and arrangement of space to achieve harmony with the environment. George has recently become interested in feng shui, and now wants to apply it to his home and bring harmony to it.

There is a practice which says that bare floor is bad for living areas since spiritual energy drains through it, so George purchased two similar round-shaped carpets (feng shui says that straight lines and sharp corners must be avoided). Unfortunately, he is unable to cover the floor entirely since the room has the shape of a convex

Figure 8.25

polygon. But he still wants to minimize the uncovered area by selecting the best placing for his carpets, and he asks you to help.

You need to place two carpets in the room so that the total area covered by both carpets is the maximum possible. The carpets may overlap, but they may not be cut or folded (including cutting or folding along the floor border)—feng shui tells you to avoid straight lines. See Figure 8.25.

Input

The first line of the input file contains two integer numbers n and r—the number of corners in George's room ($3 \le n \le 100$) and the radius of the carpets ($1 \le r \le 1000$, both carpets have the same radius). The following n lines contain two integers x_i and y_i each—coordinates of the i-th corner ($-1000 \le x_i, y_i \le 1000$). Coordinates of all corners are different, and adjacent walls of the room are not colinear. The corners are listed in clockwise order.

Output

Write four numbers x_1, y_1, x_2, y_2 to the output file, where (x_1, y_1) and (x_2, y_2) denote the spots where carpet centers should be placed. Coordinates must be precise up to four digits after the decimal point.

If there are multiple optimal placements available, return any of them. The input data guarantees that at least one solution exists.

Sample Input	Sample Output
5 2	−2 3 3 2.5
−2 0	
−5 3	
0 8	
7 3	
5 0	
4 3	3 5 7 3
0 0	

Sample Input	Sample Output
0 8	
10 8	
10 0	

Source: ACM Northeastern Europe 2006, Northern Subregion

ID for Online Judge: POJ 3384

Analysis

Two circles are placed in a convex polygon so that the total area covered by the two circles is the maximum possible. The idea for solving this problem is to push sides for the convex polygon inward, and a new convex polygon is generated. Obviously, the two circles can be placed in the convex polygon, but they may not be cut or folded. The on-line algorithm for the intersection of half-planes is used to solve the problem.

Initially, the covered area *plan* is an infinitely great square. Then each side $\overrightarrow{p_i p_{i+1}}$ is enumerated anticlockwise, $0 \le i \le n-1$, $p_n = p_0$. $\overrightarrow{p_i p_{i+1}}$ is pushed r inward, and the side $\overrightarrow{q_i q_{i+1}}$ is a side for the new convex polygon ($\overrightarrow{p_i p_{i+1}}$ is rotated $90° \times \dfrac{r}{\left| \overrightarrow{p_i p_{i+1}} \right|}$, and $\overrightarrow{q_i q_{i+1}}$ is generated). $\overrightarrow{q_i q_{i+1}}$ is used to divide *plan*. Repeat the step until n sides are dealt with. Finally, *plan* is the new convex polygon.

Then distances between all pairs of points for the convex polygon are enumerated. Suppose the distance between a pair of points q_1 and q_2 is the longest. Points q_1 and q_2 are the spots where carpet centers should be placed. Obviouly, the total area covered by the two circles is maximal.

Program

```
#include <cstdio>
#include <iostream>
#include <cstdlib>
#include <cmath>
#include <cstring>
#include <ctime>
#include <climits>
#include <utility>
```

```
#include <algorithm>
using namespace std;
const double epsi = 1e-10;     // infinitesimal
const double pi = acos(-1.0);    //180°
const int maxn = 100 + 10;     //the upper limit of the number
of points
inline int sign(const double &x) {    // x is positive,
negative, or zero
  if (x > epsi) return 1;
  if (x < -epsi) return -1;
  return 0;
}
inline double sqr(const double &x) {    //x²
  return x * x;
}
struct Point {    //Structure for points
  double x, y;
  Point(double _x = 0, double _y = 0): x(_x), y(_y) { }
  Point operator +(const Point &op2) const {
    return Point(x + op2.x, y + op2.y);
  }
  Point operator -(const Point &op2) const {
    return Point(x - op2.x, y - op2.y);
  }
  double operator *(const Point &op2) const {    //Dot Product
    return x* op2.x + y*op2.y;
  }
  Point operator *(const double &d) const {
    return Point(x * d, y * d);
  }
  Point operator /(const double &d) const {
    return Point(x / d, y / d);
  }
  double operator ^(const Point &op2) const {    // vector
product
    return x * op2.y - y * op2.x;
  }
  bool operator ==(const Point &op2) const {    //coincidence
or not
    return sign(x - op2.x) == 0 && sign(y - op2.y) == 0;
  }
};
inline double mul(const Point &p0, const Point &p1, const
Point &p2)    // vector product for $\overrightarrow{p_1 p_0}$ and $\overrightarrow{p_2 p_0}$
{
  return (p1 - p0) ^ (p2 - p0);
}
inline double dot(const Point &p0, const Point &p1, const
Point &p2)    // Dot Product for $\overrightarrow{p_1 p_0}$ and $\overrightarrow{p_2 p_0}$
```

```
{
  return (p1 - p0) * (p2 - p0);
}
inline double dis2(const Point &p0, const Point &p1) { // |p1p0|²
  return sqr(p0.x - p1.x) + sqr(p0.y - p1.y);
}
inline double dis(const Point &p0, const Point &p1) {
// |p1p0|
  return sqrt(dis2(p0, p1));
}
inline double dis(const Point &p0, const Point &p1, const
Point &p2) {
  if(sign(dot(p1, p0, p2))<0) return dis(p0, p1);    // if the
included angle for p1p0 and p1p2 is larger than 90°, then
return |p1p0|
  if (sign(dot(p2,p0, p1))<0) return dis(p0, p2);    // if the
included angle for p2p0 and p2p1 is larger than 90°, then
return |p2p0|
  return fabs(mul(p0, p1, p2) / dis(p1, p2));    //the length
of the vertical line from p0 to p1p2
}
inline Point rotate(const Point &p, const double &ang) {
//return the point that point p is rotated degree ang
  return Point(p.x * cos(ang) - p.y * sin(ang), p.x *
sin(ang) + p.y * cos(ang));
}
inline void translation(const Point &p1, const Point &p2,
const double &d, Point &q1, Point &q2)
{    //p2p1 is pushed d inward and q2q1 is formed
  q1 = p1 + rotate(p2 - p1, pi / 2) * d / dis(p1, p2);
  q2 = q1 + p2 - p1;
}
inline void cross(const Point &p1, const Point &p2, const
Point &p3, const Point &p4, Point &q)
{    //the intersection point q for p1p2 and p3p4
  double s1 = mul(p1, p3, p4), s2 = mul(p2, p3, p4);
  q.x = (s1 * p2.x - s2 * p1.x) / (s1 - s2);
  q.y = (s1 * p2.y - s2 * p1.y) / (s1 - s2);
}
inline int half_plane_cross(Point*a, int n,Point *b, const
Point &p1, const Point &p2) {    // points for A are listed
anticlockwise in a[], the current cutting line is p1p2, and
points are listed anticlockwise in b[] after A is divided
by p1p2.
  int newn = 0;
  for (int i = 0, j; i < n; ++i) {
    if (sign(mul(a[i], p1, p2)) >= 0) {    // a[i] is added
into b[]
      b[newn++] = a[i];
```

```
        continue;
    }
    j = i-1; if (j == -1) j = n-1;      //point j is the left
adjacent point for point i
        if (sign(mul(a[j], p1, p2))>0)      //the intersection
point for p₁p₂ and a[j]a[i] is added into b[]
            cross(p1, p2, a[j], a[i], b[newn++]);
        j = i + 1; if (j == n) j = 0;      // point j is the right
adjacent point for point i
      if (sign(mul(a[j], p1, p2)) > 0)      // the intersection
point for p₁p₂ and a[j]a[i] is added into b[]
            cross(p1, p2, a[j], a[i], b[newn++]);
    }
    return newn;
}
int n;      // number of points
double r;      //radius
Point p[maxn];      //the sequence of points for a convex
polygon
int t[2];
Point plane[2][maxn], q1, q2;
int main() {
    scanf("%d%lf", &n, &r);      // the number of corners and the
radius of the carpets
    for (int i = 0; i < n; ++i)      // coordinates of corners
        scanf("%lf%lf", &p[i].x, &p[i].y);
    p[n] = p[0];
    int o1 = 0, o2;
    t[0] = 4;      //Initially the covered area plan is a square
initial
    t[0] = 4;
    plane[0][0] = Point(-1e3, -1e3);
    plane[0][1] = Point(1e3, -1e3);
    plane[0][2] = Point(1e3, 1e3);
    plane[0][3] = Point(-1e3, 1e3);
    for (int i = 0; i < n; ++i) {
        o2 = o1 ^ 1;
        translation(p[i + 1], p[i], r, q1, q2);      //p₁pᵢ₊₁ is
pushed r inward and forms q₂q₁
      t[o2] = half_plane_cross(plane[o1], t[o1], plane[o2],
q1, q2);
        o1 = o2;
    }
    double maxd = -1, curd;
    for (int i=0; i<t[o1];++i)      //distances between all pairs
of points for the convex polygon are enumerated, the distance
between a pair of points q₁ and q₂ is the longest
        for (int j = i; j < t[o1]; ++j) {
            curd = dis2(plane[o1][i], plane[o1][j]);
            if (sign(curd - maxd) > 0) {
```

```
        maxd = curd;
        q1 = plane[o1][i], q2 = plane[o1][j];
    }
  }
  printf("%.10lf %.10lf %.10lf %.10lf\n", q1.x, q1.y, q2.x,
q2.y);    //q₁ and q₂ are the spots where carpet centers
  return 0;
}
```

8.3.2 Polar Angles

In the plane, the polar angle θ is the counterclockwise angle from the x-axis to a line at which a point in the xy plane lies. See Figure 8.26.

For a half-plane $ax+by\le(\ge)c$, where $a=1$, $b\in\{1,-1\}$, its polar angle is as follows:

The polar angle for the half-plane $x-y\ge c$ is $\dfrac{1}{4}\pi$ [Figure 8.27(a)];

The polar angle for the half-plane $x-y\le c$ is $-\dfrac{3}{4}\pi$ [Figure 8.27(b)];

The polar angle for the half-plane $x+y\ge c$ is $-\dfrac{1}{4}\pi$ [Figure 8.27(c)];

The polar angle for the half-plane $x+y\le c$ is $\dfrac{3}{4}\pi$ [Figure 8.27(d)].

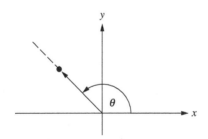

Figure 8.26

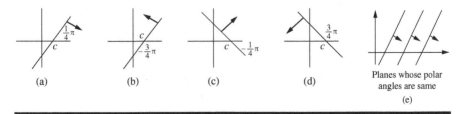

(a)   (b)   (c)   (d)   Planes whose polar angles are same
(e)

Figure 8.27

For the half-plane $ax+by \leq (\geq) c$, where a, b, and c are constants, its polar angle is $atan2(b, a)$. If there are several half-planes whose polar angles are the same, one half-plane is selected based on c. For example, in Figure 8.27(e), the plane whose c is the least is selected.

The insection of half-planes is a convex polygon, where lines whose polar angles are in $\left(-\dfrac{1}{2}\pi, \dfrac{1}{2}\pi \right]$ constitute the upper convex hull, and lines whose polar angles are in $\left(-\pi, -\dfrac{1}{2}\pi \right] \cup \left(\dfrac{1}{2}\pi, \pi \right]$ constitute the lower convex hull (Figure 8.28).

We can calculate a convex polygon in ascending order of polar angles (i.e., counterclockwise). The algorithm is as follows:

Suppose array a stores boundaries (A_i, B_i, and C_i for $A_ix+B_iy+C_i=0$, $1 \leq i \leq n$) for n half-planes H_1, H_2, ..., H_n. The convex polygon for the intersection of n half-planes $H_1 \cap H_2 \cap ... \cap H_n$ is stored by $b[\ ]$ and $c[\ ]$, where $b[\ ]$ is a deque that stores straight line equations for boundaries, $c[\ ]$ stores vertices, and h and t are the front and rear for the deque $b[\ ]$ respectively.

Step 1: Pretreatment for $a[]$: Sort $a[]$ using polars as the first key, and distances from the origin to boundaries as the second key. If there are more than one boundary with the same polar angle, the boundary with the shortest distance from the origin to the boundary is selected. If $A_i=B_i=0$, and if $C_i>0$, the line $A_ix+B_iy+C_i=0$ is removed; and if $C_i \leq 0$, the program exits.

Step 1: Step 1 is to determine the sequence for intersections of half-planes, and eliminate cases that intersections of half-planes don't exist and coincide.

Step 2: The first two boundaries are added into queue $b[\ ]$ as $b[0]$ and $b[1]$, and the insection point for the two boundaries is stored into $c[1]$, $h=0$, and $t=1$.

Step 3: Boundaries $a[3]...a[n]$ (half-planes) are dealt with one by one:

1. While the deque isn't empty, and when $c[t]$ is substituted in boundary $a[i]$, the equation is negative; then the rear for the deque is removed (t −−);
2. While the deque isn't empty, and when $c[h+1]$ is substituted in boundary $a[i]$, the equation is negative; then the front for the deque is removed (h++);

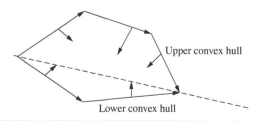

Upper convex hull

Lower convex hull

Figure 8.28

3. Boundary $a[i]$ is added into deque $b[]$ ($b[++t]=a[i]$), and the insection point for $b[t]$ and $b[t-1]$ is put into $c[t]$. It is to guarantee $A_i x + B_i y + C_i \geq 0$ when insection points in $c[]$ are substituted in each line.

Step 4: The front and the rear for queues $b[]$ and $c[]$ are joined. Redundant half-planes are removed.

1. While the deque isn't empty, and when $c[t]$ is substituted in boundary $b[h]$, the equation is negative, and the rear for the deque is removed ($t--$);
2. While the deque isn't empty, and when $c[h+1]$ is substituted in boundary $b[t]$, the equation is negative, and the front for the queue is removed from the queue ($h++$);
3. If the deque is empty ($h+1 \geq t$), then the program exits; else for the convex polygon, p_0 is the insection point for $b[h]$ and $b[t]$, and insection points $p_1 \ldots p_{i-h}$ are $c[h+1]\ldots c[t]$, and $p_{i-h+1} = p_0$.

The time complexity for the algorithm is $O(n\log_2 n)$.

8.3.2.1 Art Gallery

The art galleries of the new and very futuristic building of the Center for Balkan Cooperation have the form of polygons (not necessarily convex). When a big exhibition is organized, watching over all the pictures is a big security concern. Your task for a given gallery is to write a program that finds the surface of the area of the floor, from which each point on the walls of the gallery is visible. In Figure 8.29, a map of a gallery is given in some coordinate system. The area wanted is shaded on in the second half of the figure.

Input

The number of tasks T that your program needs to solve will be on the first row of the input file. Input data for each task start with an integer N, $5 \leq N \leq 1500$. Each of the next N rows of the input will contain the coordinates of a vertex of the polygon—two integers that fit in 16-bit integer type, separated by a single space.

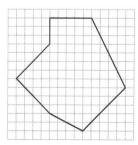 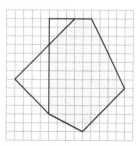

Figure 8.29

Following the row with the coordinates of the last vertex for the task comes the line with the number of vertices for the next test, and so on.

Output

For each test, you must write on one line the required surface—a number with exactly two digits after the decimal point (the number should be rounded to the second digit after the decimal point).

Sample Input	Sample Output
1	80.00
7	
0 0	
4 4	
4 7	
9 7	
13 −1	
8 −6	
4 −4	

Source: ACM Southeastern Europe 2002

IDs for Online Judges: POJ 1279, ZOJ 1369, UVA 2512

 Analysis

Based on the problem description "for a given gallery, to write a program which finds the surface of the area of the floor, from which each point on the walls of the gallery is visible", the surface of the area of the floor consists of a set of points with the following property:

Suppose *s* is a point on the boundary of the gallery, and *v* is a point in the set. A line segment connecting *s* and *v* must be in the gallery.

The gallery is a polygon. And the surface of the area of the floor is the insection of left half-planes divided by sides for the polygon. If the polygon is a convex polygon, the surface of the area of the floor is the polygon itself. And if the polygon isn't a convex polygon, the surface of the area of the floor is a subset for the polygon, and may be an empty set.

Based on the above discussion, the algorithm is as follows:

First, the coordinates of *n* vertices are input, and are transferred as equations of *n* lines. Second, the plane is divided by sides for polygons anticlockwise. The insection of *n* half-planes is the surface of the area of the floor. Finally, the area of the insection of *n* half-planes is calculated by the formula of cross product.

 Program

```
#include <iostream>
#include <cstdlib>
#include <cstdio>
#include <string>
#include <cmath>
#include <algorithm>
using namespace std;
const int maxn=2100;
const double eps=1e-10;
struct Point {      //Structure for points
   double x, y;
   Point(double _x = 0, double _y = 0): x(_x), y(_y) { }
//point
   double operator ^(const Point &op2) const {   //cross product
      return x * op2.y - y * op2.x;
   }
};
struct StraightLine{     //Intersection of half-planes
   double A, B, C;     //equation of line Ax+By+C=0
   StraightLine(double _a=0, double _b=0, double _c=0):A(_a),
B(_b), C(_c) { }     //line

   double f(const Point &p) const {     //point p is substituted
in an equation of line
      return A * p.x + B * p.y + C;
   }
   double rang() const{     // atan2(B, A), that is, the polar
angle for the line
      return atan2(B, A);
   }
   double d() const{     //the distance between the origin to
line Ax + By + C = 0: C/√(A² + B²)
      return C / (sqrt(A * A + B * B));
   }
```

```
    Point cross(const StraightLine &a) const {    //intersection
point for Ax+By+C=0 and line a
        double xx = - (C * a.B - a.C * B) / (A * a.B - B * a.A);
        double yy = - (C * a.A - a.C * A) / (B * a.A - a.B * A );
        return Point(xx, yy);
    }
};
StraightLine b[maxn], SL[maxn];    // SL[]: the sequence of
lines for half-planes, b[]: the sequence of lines for the
current intersection of half-planes
Point c[maxn], d[maxn];    // d[]: the sequence of points for
the intersection of half-planes,  c[]: the sequence of points
for the current intersection of half-planes
int n;    //number of points for a polygon
inline int sign(const double &x){    // x is positive or
negative
    if (x > eps) return 1;
    if (x < -eps) return -1;
    return 0;
}
int cmp(StraightLine a, StraightLine b){    // comparing lines
a and b, polar angles are as the first key, distances from the
origin to lines are as the second key
    if (sign( a.rang() - b.rang() ) != 0) return a.rang()
< b.rang();
    else return a.d() < b.d();
}
int half_plane_cross(StraightLine *a,  int n, Point *pt) {
// input a sequence of lines for a polygon a, whose length
is n. By polar angles, return a sequence of lines pt and its
length for the inner convex polygon a
    sort(a+1,a+n+1,cmp);    // a is sorted, where polar angles
are as the first key, distances from the origin to lines are
as the second key
    int tn = 1;    // initialize the length of a
    for (int i = 2; i <= n; i ++){    // enumerating adjacent
sides for the polygon, sides whose polar angles are same, or
A=B=0 and C>0, are deleted (C≤0, exit)
        if (sign( a[i].rang() - a[i-1].rang() )!=0) a[++tn]=a[i];
// polar angles for adjacent sides are different
        if (sign(a[tn].A )==0 && sign( a[tn].B )==0)    // A=B=0
            if (sign( a[tn].C )==1)    tn --;
            else return - 1;
    }
    n=tn;    // the length for a
    int h=0, t=1;
    b[0] = a[1];    // line 1 and line 2 are stored in b[]
    b[1] = a[2];
    c[1] = b[1].cross(b[0]);    //the intersection point for
line 1 and line 2 is stored in c[]
```

```
     for (int i = 3; i <= n; i ++){      // enumerate line 3…line n
        while (h < t && sign( a[i].f(c[t] ) )<0) t -- ;
        while (h<t && sign(a[i].f( c[h+1] ))<0) h++ ;
        b[ ++ t] = a[i];    // line i is added into the rear for b
        c[t] = b[t].cross( b[t-1] );      // the intersection for
the two lines at the rear of b is added into c
     }
     while (h < t && sign( b[h].f( c[t] ) )<0) t --;
     while (h < t && sign( b[t].f( c[h+1] ) )<0) h ++;
     if (h+1 >= t) return -1;      // the queue is empty
     pt[0] = b[h].cross( b[t] );      // the first point for the
convex polygon
     for(int i=h;i<t;i++) pt[i-h+1]=c[i+1];
     pt[t - h + 1] = pt[0];
     return t - h + 1;      // number of points for the convex
polygon
}
int main(){
   int x[maxn], y[maxn] ;    //the sequence of points for a
polygon
   double ans=0;    //initialize the area for the maximal
convex polygon
   int n, m;    //n: number of points for a polygon, m: number
of points for the maximal inner convex polygon
   int test;    //the number of test cases
   scanf("%d", & test );    //input the number of test cases
   for (; test ; test --){
      scanf("%d", & n);    // input number of vertices and the
coordinates of vertices of the polygon
      for (int i = 1; i <= n; i ++) scanf("%d %d", & x[i],
& y[i]);
      x[n+1]=x[1];y[n+1]=y[1];
      for(int i=1; i<=n;i++)    //calculate n equations of lines,
where SL[i] stores A, B, C for p_{i+1}p_i
         SL[i]=StraightLine(-(y[i]-y[i+1]),-(x[i+1]-x[i]),-(x[i]
*y[i+1]-x[i+1]*y[i])));
      m=half_plane_cross(SL,n,d);    //calculation by polar
angles
      ans = 0;
      if (m == -1) printf("0.00\n");    //if there is no convex
polygon
      else {
         for (int i = 0; i < m; i ++) ans += d[i] ^ d[i+1];
         printf("%.2lf\n", ans / 2);    //the area for the
maximal convex polygon
      }
   }
   return 0;
}
```

8.3.2.2 Hotter Colder Game

The children's game Hotter Colder is played as follows. Player A leaves the room while player B hides an object somewhere in the room. Player A re-enters at position (0,0) and then visits various other positions about the room. When player A visits a new position, player B announces "Hotter" if this position is closer to the object than the previous position; player B announces "Colder" if it is farther; and "Same" if it is the same distance.

Input

Input consists of up to 50 lines, each containing an x, y coordinate pair followed by "Hotter", "Colder", or "Same". Each pair represents a position within the room, which may be assumed to be a square with opposite corners at (0,0) and (10,10).

Output

For each line of input, print a line giving the total area of the region in which the object may have been placed, to two decimal places. If there is no such region, output 0.00.

Sample Input	Sample Output
10.0 10.0 Colder	50.00
10.0 0.0 Hotter	37.50
0.0 0.0 Colder	12.50
10.0 10.0 Hotter	0.00

Source: Waterloo local 2001.01.27

IDs for Online Judges: POJ 2540, ZOJ 1886

 Analysis

Suppose the position of the placed object is P, and player A moves into $D(x_2, y_2)$ from $C(x_1, y_1)$. The equation for the perpendicular bisector for the line segment CD is substituted by $P(x, y)$.

In the current round, if player B announces "Hotter", then for the position of $P(x, y)$ $|CP| > |DP|$ holds, that is,

$$2 \times (x_2 - x_1) \times x + 2 \times (y_2 - y_1) \times y + x_1^2 + y_1^2 - x_2^2 - y_2^2 > 0;$$

if player *B* announces "Colder", then for the position of *P*(*x, y*), $|CP|<|DP|$ holds, that is,

$$2\times(x_2 - x_1)\times x + 2\times(y_2 - y_1)\times y + x_1^2 + y_1^2 - x_2^2 - y_2^2 < 0;$$

if player *B* announces "Same", then for the position of *P*(x, y), $|CP|=|DP|$ holds, that is,

$$2\times(x_2 - x_1)\times x + 2\times(y_2 - y_1)\times y + x_1^2 + y_1^2 - x_2^2 - y_2^2 = 0.$$

For each time player *B* announces, a corresponding half-plane is added.

Initially, the total area of the region in which the object may have been placed is [0, 10]×[0, 10]. In each round, the intersection of the current half-plane and the new added half-plane is calculated. If the intersection doesn't exist, output 0.00; else output the area of the intersection.

In the program, the intersection of half-planes is calculated by polar angles.

 Program

```cpp
#include <iostream>
#include <cstdlib>
#include <cstdio>
#include <string>
#include <cmath>
#include <algorithm>
using namespace std;
const int maxn=21000;
const double eps=1e-10;
struct Point {      //structure for points
   double x, y;
   Point(double _x = 0, double _y = 0): x(_x), y(_y) { }
   double operator ^(const Point &op2) const {      //cross
product
      return x * op2.y - y * op2.x;
   }
};
struct StraightLine{      // the intersection of half-planes
   double A, B, C;      //equation of line Ax+By+C
   StraightLine(double _a=0, double _b=0, double _c=0): A(_a),
B(_b), C(_c) { }//construct equation of line
   double f(const Point &p) const {      // the equation of line
is substituted by point p
      return A * p.x + B * p.y + C;
   }
```

```
double rang() const{    //return the polar angle for the line
    return atan2(B, A);
}
double d() const{    //the distance from origin to line
Ax+By+C=0
    return C / (sqrt(A * A + B * B));
}
  Point cross(const StraightLine &a) const {    // intersection
point
    double xx = - (C * a.B - a.C * B) / (A * a.B - B * a.A);
    double yy = - (C * a.A - a.C * A) / (B * a.A - a.B * A );
    return Point(xx, yy);
}
};
StraightLine b[maxn], SL[maxn],S[maxn];    // SL[]: the
sequence of lines for half-planes, S[]: stores half-planes,
b[]: the sequence of lines for the current intersection of
half-planes
Point c[maxn], d[maxn];    // d[]: the sequence of points for
the intersection of half-planes,  c[]: the sequence of points
for the current intersection of half-planes
int n;    //number of points for a polygon
inline int sign(const double &x){    //x is positive or
negative
  if (x > eps) return 1;
  if (x < -eps) return -1;
  return 0;
}
int cmp(StraightLine a, StraightLine b){    //comparing lines
a and b, polar angles are as the first key, distances from the
origin to lines are as the second key
  if (sign( a.rang() - b.rang() ) != 0) return a.rang()
< b.rang();
  else return a.d() < b.d();
}
int half_plane_cross(StraightLine *a,  int n, Point *pt) {
//input a sequence of lines for a polygon a, whose length
is n. By polar angles, return a sequence of lines pt and its
length for the inner convex polygon a
  sort(a+1,a+n+1,cmp);    // a is sorted, where polar angles
are as the first key, distances from the origin to lines are
as the second key
  int tn = 1;    //initialize the length of a
  for (int i = 2; i <= n; i ++){    //enumerating adjacent
sides for the polygon, sides whose polar angles are same, or
A=B=0 and C>0, are deleted (C≤0, exit)
    if (sign( a[i].rang() - a[i-1].rang() )!=0) a[++tn]=a[i];
// polar angles for adjacent sides are different
    if (sign(a[tn].A)==0 && sign(a[tn].B)==0)    // A=B=0
      if (sign( a[tn].C )==1)    tn --;
```

```
        else return  - 1;
  }
  n=tn;     //the length for a
  int h = 0 , t = 1;
  b[0] = a[1];      //line 1 and line 2 are stored in b[]
  b[1] = a[2];
  c[1] = b[1].cross(b[0]);     //the intersection point for
line 1 and line 2 is stored in c[]
  for (int i = 3; i <= n; i ++){     //enumerate line 3…line n
    while (h < t && sign( a[i].f( c[t] ) )<0) t -- ;
    while (h<t && sign( a[i].f(c[h+1] ))<0) h++ ;
    b[ ++ t] = a[i];     //line i is added into the rear for b
  c[t] = b[t].cross( b[t-1] );     //the intersection for the
two lines at the rear of b is added into c
  }
  while (h < t && sign( b[h].f( c[t] ) )<0) t --;
  while (h < t && sign( b[t].f( c[h+1] ) )<0) h ++;
  if (h+1 >= t) return -1;     //the queue is empty
  pt[0] = b[h].cross( b[t] );     //the first point for the
convex polygon
  for(int i=h;i<t;i++) pt[i-h+1]=c[i+1];
  pt[t - h + 1] = pt[0];
  return t - h + 1;     //number of points for the convex
polygon
}
int main(){
  ios::sync_with_stdio(false);
  double x1, x2, y2, y1, ans=0;
  int n, m;     //n: number of half-planes, m: number of points
for the intersection of half-planes
  n=0;     //initially 4 half-planes for [0, 10]*[0, 10]
  SL[++n] = StraightLine(0, 1, 0);
  SL[++n] = StraightLine(1, 0, 0);
  SL[++n] = StraightLine(0, -1, 10);
  SL[++n] = StraightLine(-1, 0, 10);
  double px=0,py=0,nx, ny;     //position for before moving
(px,py) and after moving (nx,ny)
  string c;     // player B announces
  char s;
  while (cin >> nx >> ny){     // a position where play A enters
   cin >> c ;     // player B announces
   if (c[0] == 'C' )     //a corresponding plane is added based
on player B announces
     SL[++n]=StraightLine(-2*(nx-px),-2*(ny-py),-(px*px+py*py-
nx*nx-ny*ny));
    else if (c[0]=='H' )
     SL[++n]=StraightLine(2*(nx-px), 2*(ny-py) ,(px*px+py*py-
nx*nx -ny*ny));
      else SL[++n]=StraightLine(-2*(nx-px),-2*(ny-py),-
(px*px+py*py-nx* nx-ny*ny)),
```

```
        SL[++n]=StraightLine(2*(nx-px),2*(ny-py),(px*px+py*py-nx
*nx-ny*ny));
        px = nx ; py = ny ;    //(nx, ny) will be the next
position where player A enters
        ans=0;     //initialize the area for the intersection of
half-planes
        for (int i = 1 ; i <= n ; i ++) S[i] = SL[i];
        m = half_plane_cross(S, n, d);    //intersection of
half-planes
        if (m==-1) printf("0.00\n");    //the intersection doesn't
exist
        else {
            for (int i = 0; i < m; i ++) ans += d[i] ^ d[i+1];
            printf("%.2lf\n", ans / 2);
        }
    }
    return 0;
}
```

8.4 Convex Hull and Finding the Farthest Pair of Points

In this section, there are two kinds of experiments as follows.

1. Convex hull: Finding the smallest convex hull containing all given points.
2. Finding the farthest pair of points in a convex hull.

8.4.1 Convex Hull

Suppose Q is a set of n points, $Q = \{p_0, \ldots, p_{n-1}\}$. Its convex hull $CH(Q)$ is the smallest polygon P, in which each point in Q is either on the boundary of P or in its interior. A convex hull can be regarded as a shape formed by an elastic rubber band that surrounds all points. An example is shown in Figure 8.30.

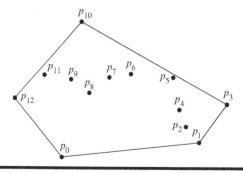

Figure 8.30

An algorithm, Graham's scan, computes the convex hull of a set of n points. Graham's scan inputs a set Q of n points, and outputs vertices of the convex hull $CH(Q)$ in counterclockwise order.

1. First, the point in Q with the minimum y-coordinate is selected. If there are more than one point with the minimum y-coordinate, the leftmost point is selected. The selected point is denoted as p_0, and p_0 is as the first vertice for the convex hull $CH(Q)$.
2. Second, other points in Q are sorted by polar angle in counterclockwise order. By calculating the cross product $(p_i-p_0)^\wedge(p_j-p_0)$ (i.e., $\text{Mul}(p_i, p_j, p_0)$), we can determine whose polar angle is larger.

   If $(p_i-p_0)^\wedge(p_j-p_0)>0$, then the polar angle for p_j is larger than the polar angle for p_i, with respect to p_0; and p_i is scanned before p_j.
   If $(p_i-p_0)^\wedge(p_j-p_0)<0$, then the polar angle for p_j is less than the polar angle for p_i, with respect to p_0; and p_j is scanned before p_i.
   If $(p_i-p_0)^\wedge(p_j-p_0)==0$, then the polar angle for p_j is the same as the polar angle for p_i, with respect to p_0. The point which is farther from p_0 is scanned. And the other points are removed.

Suppose pointers are sorted as a sequence $\{p_1, ..., p_{n-1}\}$. If $n \leq 2$, the convex hull is empty; else the sequence $\{p_1, ..., p_{n-1}\}$ is scanned.

A stack S is used to store candidate vertices in computing the convex hull. Initially, points p_0, p_1, and p_2 are pushed into stack S one by one. Then points $\{p_3, ..., p_{n-1}\}$ are scanned one by one. Suppose p_i is the current scanned point, and p_{top} is the point at the top of the stack S. Because vertices are traversed counterclockwise, if p_{top} is a vertice of the convex hull $CH(Q)$, a left turn should be made from p_{top} to p_i. If it is a nonleft turn, p_{top} isn't a vertice of the convex hull $CH(Q)$, and should be popped from S. After vertices making nonleft turns are popped, p_i is pushed into S. Then the next point p_{i+1} is scanned. Finally, points in S are vertices of the convex hull $CH(Q)$. The sequence from the bottom to the top in S are vertices of the convex hull $CH(Q)$ in counterclockwise order.

By calculating the vector product $(p_i - p_{top-1})^\wedge(p_{top} - p_{top-1})$ (i.e., $\text{Mul}(p_i, p_{top}, p_{top-1})$), where p_{top-1} is the point next to the top of S. If $(p_i - p_{top-1})^\wedge(p_{top} - p_{top-1}) \geq 0$, p_{top} makes a nonleft turn.

8.4.1.1 Wall

Once upon a time there was a greedy king who ordered his chief architect to build a wall around the king's castle. The king was so greedy that he would not listen to his architect's proposals to build a beautiful brick wall with a perfect shape and nice tall towers. Instead, he ordered the architect to build the wall around the whole castle using the least amount of stone and labor, but demanded that the wall should not come closer to the castle than a certain distance, as shown in Figure 8.31. If the

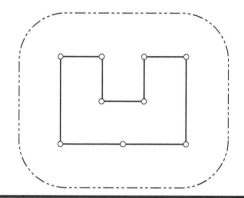

Figure 8.31

king finds that the architect has used more resources to build the wall than was absolutely necessary to satisfy those requirements, then the architect will lose his head. Moreover, he demanded that the architect introduce at once a plan of the wall listing the exact amount of resources that are needed to build the wall.

Your task is to help the poor architect to save his head, by writing a program that will find the minimum possible length of the wall that he could build around the castle to satisfy the king's requirements.

The task is somewhat simplified by the fact that the king's castle has a polygonal shape and is situated on flat ground. The architect has already established a Cartesian coordinate system and has precisely measured the coordinates of all castle's vertices in feet.

Input

The first line of the input file contains two integer numbers N and L separated by a space. N ($3 \leq N \leq 1000$) is the number of vertices in the king's castle, and L ($1 \leq L \leq 1000$) is the minimal number of feet that the king allows for the wall to come close to the castle.

The next N lines describe the coordinates of the castle's vertices in a clockwise order. Each line contains two integer numbers X_i and Y_i separated by a space ($-10000 \leq X_i, Y_i \leq 10000$) that represents the coordinates of the i-th vertex. All vertices are different, and the sides of the castle do not intersect anywhere except for vertices.

Output

Write to the output file the single number that represents the minimal possible length of the wall in feet that could be built around the castle to satisfy the king's requirements. You must present the integer number of feet to the king, because the floating numbers are not invented yet. However, you must round the result in such a way that it is accurate to 8 inches (1 foot is equal to 12 inches), since the king will not tolerate any larger error in the estimates.

Sample Input	Sample Output
9 100	1628
200 400	
300 400	
300 300	
400 300	
400 400	
500 400	
500 200	
350 200	
200 200	

Source: ACM Northeastern Europe 2001

IDs for Online Judges: POJ 1113, ZOJ 1465, UVA 2453

 Analysis

The shape for the king's castle is a polygon. The architect is required to build a wall around the king's castle. And the minimum number of feet that the king allows for the wall to come close to the castle is L.

First, a convex hull is computed by Graham's scan. The inputs for the algorithm are the castle's vertices. The built wall is a polygon with rounded corners around the convex hull. Edges for the polygon are parallel to the edges of the convex hull. Lengths of two parallel edges are the same. And the distance for two parallel edges is L. For the wall, each round corner is an arc connecting two adjacent edges, whose radius is L, and the center of the circle is a vertex for the convex hull. For a round corner, the sum of the radius angle and its corresponding interior angle for the convex hull is 180°. Because the sum of degrees of interior angles in a convex polygon with n edges is $(n-2)\times180°$, the sum of degrees of radius angles is 360°. Therefore, the sum of the lengths of arcs is the circumference of a circle whose radius is L.

The minimal possible length of the wall is the girth of the convex hull + the circumference of a circle whose radius is L.

Program

```
#include <cstdio>
#include <cmath>
#include <algorithm>
using namespace std;
const double epsi = 1e-8;    //infinitesimal
const double pi = acos(-1.0);     //Radian value for π
const int maxn = 1000 + 10;
struct Point {     //Calculation for point
  double x, y;     // coordinate
    Point(double _x = 0, double _y = 0): x(_x), y(_y) { }
  double operator ^(const Point &op2) const {    // vector
product for two point vectors
    return x * op2.y - y * op2.x;
  }
};
inline int sign(const double &x) {
  if (x > epsi) return 1;
  if (x < -epsi) return -1;
  return 0;
}
inline double sqr(const double &x) {    //calculate x²
  return x * x;
}
inline double mul(const Point &p0, const Point &p1,const Point
&p2){// vector product for p0p1 and p0p2
  return (p1.x-p0.x)*(p2.y-p0.y)-(p1.y-p0.y)*(p2.x-p0.x);
//(p1 - p0) ^ (p2 - p0);
}
inline double dis2(const Point &p0, const Point &p1) {
// |p0p1|²
  return sqr(p0.x - p1.x) + sqr(p0.y - p1.y);
}
inline double dis(const Point &p0, const Point &p1) {
// |p0p1|
  return sqrt(dis2(p0, p1));
}
int n, l;    //n: the number of vertices in the king's castle,
l: the minimal number of feet that king allows for the wall to
come close to the castle
Point p[maxn], convex_hull_p0;     // p[]: a sequence for
vertices for the polygon, convex_hull_p0: the point with the
minimum y-coordinate
inline bool convex_hull_cmp(const Point &a, const Point &b) {
```

```
    return sign(mul(convex_hull_p0, a, b))>0||sign(mul(convex_
hull_p0, a, b))==0 && dis2(convex_hull_p0, a)<dis2(convex_
hull_p0, b);
}
int convex_hull(Point *a, int n, Point *b){    // the convex
hull b[] is computed based on a set of points a[] (number of
points is n)
    if (n < 3) printf("Wrong in Line %d\n", __LINE__);    //the
number of points <3
    for (int i = 1; i < n; ++i)    //calculating convex_hull_p0
        if(sign(a[i].x-a[0].x)<0||sign(a[i].x-a[0].x)==0 &&
sign(a[i].y-a[0].y)<0)swap(a[0], a[i]);
    convex_hull_p0 = a[0];
    sort(a, a + n, convex_hull_cmp);    //with respect to
convex_hull_p0, a[] is sorted, polar angle is the first key,
distance is the second key
    int newn = 2;    // a[0], a[1] is pushed into the stack
    b[0] = a[0], b[1] = a[1];
    for (int i = 2; i < n; ++i) {    //points are dealt with one
by one
    while(newn>1 && sign(mul(b[newn-1],b[newn-2], a[i]))>=0)--
newn;    //pop nonleft turn points
        b[newn++] = a[i];    //point i is pushed into the stack
    }
    return newn;
}
int main() {
    scanf("%d%d", &n, &l);    // n: the number of vertices in
the king's castle, l: the minimal number of feet that king
allows for the wall to come close to the castle
    for (int i = 0; i < n; ++i)    // coordinates of castle's
vertices
        scanf("%lf%lf", &p[i].x, &p[i].y);
    n = convex_hull(p, n, p);    //calculating the convex
hull
    p[n] = p[0];
    double ans = 0;    // ans: length of the wall
    for (int i = 0; i < n; ++i)    // the girth of the convex
hull
    ans += dis(p[i], p[i + 1]);
    ans += 2 * pi * l;    // the circumference of a circle
    printf("%.0lf\n", ans);
    return 0;
}
```

8.4.2 *Finding the Farthest Pair of Points*

Given a set of n points in a plane, how can we find the farthest pair of points? The problem can be solved by finding the convex hull of the set of n points. The farthest

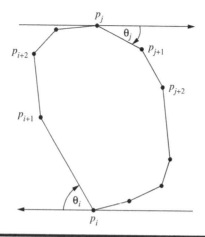

Figure 8.32

pair of points must be two vertices for the convex hull. The distance between the farthest pair of vertices for a convex hull is called the diameter of a convex hull.

For a convex hull, each pair of vertices can be enumerated to find the farthest pair of vertices. The method of rotating calipers is the optimal algorithm for finding the farthest pair of vertices of the convex hull.

Suppose P is a convex polygon with n vertices, and L is a line. If L intersects P, and the interior of P lies completely on one side of L, L is a line of support for P. It is shown in Figure 8.32. If L intersects P at a vertex v, or an edge e; v or e admits L. A pair of vertices p_i, $p_j \in P$ is an antipodal pair if it admits parallel lines of support for P. In Figure 8.32, an antipodal pair admits parallel lines of support. Lines of support can be rotated to generate the next antipodal pair. Suppose angles that the lines of support at p_i and p_j make with edges $p_i p_{i+1}$ and $p_j p_{j+1}$ are θ_i and θ_j respectively. If $\theta_j < \theta_i$, and the lines of supports are rotated by angle θ_j, then p_{j+1} and p_i become the next antipodal pair. If $\theta_j = \theta_i$, then three new antipodal pairs are generated. Therefore, there are three cases that lines of support interest P, shown in Figure 8.33, Figure 8.34, and Figure 8.35, respectively. In the case shown in Figure 8.33, there is

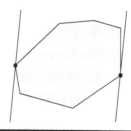

Figure 8.33

Figure 8.34

an antipodal pair; in the case shown in Figure 8.34, there are two antipodal pairs; and in the case shown in Figure 8.35, there are four antipodal pairs.

The diameter of a convex polygon P is the greatest distance between parallel lines of support of P. The diameter of a convex polygon P is the greatest distance between an antipodal pair of P. Therefore we need to check each antipodal pair. Initially q_a is the vertices for P with the minimum y-coordinate, and q_b is the vertices for P with with the maximum y-coordinate. Obviously, q_a and q_b are an antipodal pair. Suppose d_{ab} is the distance between q_a and q_b; C_a is the circle of radius d_{ab} centered at q_a, and C_b is the circle of radius d_{ab} centered at q_b; L_a is the tangent to C_a at q_a, and L_b is the tangent to C_b at q_b; L is the line through q_a and q_b. By the definition of tangent line, $L_a \perp L$ and $L_b \perp L$. Therefore, L_a and L_b are lines of support of P. L_a and L_b rotate to generate new antipodal pairs. The process that L_a and L_b rotate is continued until we come full circle to the starting position. Suppose q_a and q_b are the farthest pair of points. L_a and L_b are parallel lines of support that intersect q_a and q_b respectively. L_a and L_b can be rotated to generate each antipodal pair. This is shown in Figure 8.36.

For a convex polygon, suppose $u[0]$ is the lowest point, and if there are more than one lowest point, $u[0]$ is the rightmost point for the lowest points; and $u[2]$ is the highest point, and if there are more than one highest point, $u[0]$ is the leftmost point for the highest points. Obviously $u[0]$ and $u[2]$ is an antipodal pair. The algorithm calculating the distance for the farthest pair of points *ret* is as follows:

```
Calculate the sequence of vertices for convex hull;
Calculate u[0] and u[2], and initialize ret as |p_u[0]p_u[2]|;

Rotation degree sumang=0;
```

Figure 8.35

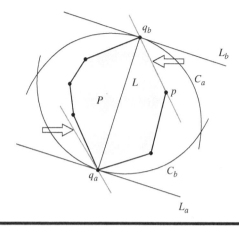

Figure 8.36

```
while (sumang≤2π) {
  calculate the current rotation degree curang to generate a
new antipodal pair u[0] and u[2];
  sumang+=curang;    //accumulation

  ret=max(ret,|p_{u[0]}p_{u[2]}|);    //adjust the distance for the
farthest pair of points ret

  }
  Output the distance for the farthest pair of points ret;
```

8.4.2.1 Beauty Contest

Bessie, Farmer John's prize cow, has just won first place in a bovine beauty contest, earning the title "Miss Cow World". As a result, Bessie will make a tour of N ($2 \le N \le 50{,}000$) farms around the world in order to spread goodwill between farmers and their cows. For simplicity, the world will be represented as a two-dimensional plane, where each farm is located at a pair of integer coordinates (x, y), each having a value in the range $-10{,}000...10{,}000$. No two farms share the same pair of coordinates.

Even though Bessie travels directly in a straight line between pairs of farms, the distance between some farms can be quite large, so she wants to bring a suitcase full of hay with her so she has enough food to eat on each leg of her journey. Since Bessie refills her suitcase at every farm she visits, she wants to determine the maximum possible distance she might need to travel so she knows the size of suitcase she must bring. Help Bessie by computing the maximum distance among all pairs of farms.

Input

Line 1: A single integer, N;
Line 2: $N+1$: Two space-separated integers x and y specifying the coordinate of each farm.

Output

Line 1: A single integer that is the squared distance between the pair of farms that are farthest apart from each other.

Sample Input	Sample Output
4 0 0 0 1 1 1 1 0	2

Source: USACO 2003 Fall

ID for Online Judge: POJ 2187

Farm 1 (0, 0) and farm 3 (1, 1) have the longest distance (square root of 2).

 Analysis

In this problem, there are N (2≤N≤50,000) points. The problem requires you to compute the maximum distance among all pairs of points. Obviously, the pair of points with the maximum distance must be two points for the convex hull. First, the convex hull of the set of N points is computed. Then the maximum distance among all pairs of points for the convex hull is computed. If each pair of points is enumerated, it will take more time. For this problem, rotating calipers is suitable to compute the maximum distance among all pairs of points for the convex hull. The program is shown as follows.

 Program

```
#include <cstdio>
#include <cstring>
#include <algorithm>
#include <cmath>
#include <queue>
#include <cstdlib>
using namespace std;
#define N 50005     // the upper limit for the number of points
struct point{    // p[]: a sequence of coordinates
    int x,y;
}p[N];
```

```
int n;      // the number of points
int stack[N],top = -1;      //stack stack[], pointer pointing to
the top of stack top
int multi(struct point a,struct point b,struct point c){
//cross product (b-a)^(c-a)
    return (b.x-a.x)*(c.y-a.y)-(b.y-a.y)*(c.x-a.x);
}
int dis(struct point a,struct point b){      //the distance
between points a and b |ab⃗|
    return (b.x-a.x)*(b.x-a.x)+(b.y-a.y)*(b.y-a.y);
}

int cmp(struct point a,struct point b){      // Comparison
function in sorting: If three collinear(points a, b, and
p[1]), the distance between b and p[1] is larger than the
distance between a and p[1], ab is sorted; else ba is sorted.
If no three collinear(points a, b, and p[1]), if the polar
angle for ap[1]⃗ is less than the polar angle for bp[1]⃗, ab is
sorted; else ba is sorted.
    int tmp = multi(p[1],a,b);
    if(tmp == 0)
        return dis(p[1],a) < dis(p[1],b);
    return tmp>0;
}

int main(){
    int i,j,res=0;      // res: the maximum distance
    struct point begin;      // the point with the minimum
y-coordinate
    scanf("%d",&n);      // number of farms
    begin.x = begin.y = 10005;      //Initialization for the
point with the minimum y-coordinate in convex hull
    for(i = 1;i<=n;i++){      //Input every farm's coordinate
        scanf("%d %d",&p[i].x,&p[i].y);
        if(p[i].y < begin.y){      //adjust begin, note down the
sequence number j
            begin = p[i];
            j = i;
        }else if(p[i].y==begin.y && p[i].x<begin.x){
            begin = p[i];
            j = i;
        }
    }
    if(n==2){      //output the distance between two points
        printf("%d\n",dis(p[1],p[2]));
        return 0;
    }
    p[j] = p[1];
    p[1] = begin;
    sort(p+2,p+n+1,cmp);      //Sorting point 2 … point n
```

```
        stack[++top] = 1;      //point 1 and point 2 are pushed into
    stack, and graham is used to calculate the convex hull stack[]
        stack[++top] = 2;
        for(i = 3;i<=n;i++){
            while(top>0 && multi(p[stack[top-1]], p[stack[top]],
    p[i])<=0) top--;
            stack[++top] = i;
        }
        // Rotating calipers are used to find the farthest pair of
    points
        j = 1;
        stack[++top] = 1;
        for(i = 0;i<top;i++){      //Enumerate point i
        //enumerate the farthest point j for the line segment
    p[stack[i]]p[stack[i+1]] anticlockwise
            while(multi(p[stack[i]],p[stack[i+1]],p[stack[j+1]])>mul
    ti(p[stack[i]],p[stack[i+1]], p[stack[j]])) j=(j+1)%top;
            res=max(res,dis(p[stack[i]],p[stack[j]]));      //
    calculate |p[stack[i]]p[stack[j]]|, and adjust the maximum distance
    res
        }
        printf("%d\n",res);      //output the maximum distance

    }
```

8.5 Problems

8.5.1 Segments

Given n segments in the two-dimensional space, write a program that determines if there exists a line such that after projecting these segments on it, all projected segments have at least one point in common.

Input

Input begins with a number T showing the number of test cases and then, T test cases follow. Each test case begins with a line containing a positive integer $n \le 100$ showing the number of segments. After that, n lines containing four real numbers $x_1\ y_1\ x_2\ y_2$ follow, in which (x_1, y_1) and (x_2, y_2) are the coordinates of the two endpoints for one of the segments.

Output

For each test case, your program must output "Yes!", if a line with the desired property exists and must output "No!" otherwise. You must assume that two floating-point numbers a and b are equal if $|a-b| < 10^{-8}$.

Sample Input	Sample Output
3	Yes!
2	Yes!
1.0 2.0 3.0 4.0	No!
4.0 5.0 6.0 7.0	
3	
0.0 0.0 0.0 1.0	
0.0 1.0 0.0 2.0	
1.0 1.0 2.0 1.0	
3	
0.0 0.0 0.0 1.0	
0.0 2.0 0.0 3.0	
1.0 1.0 2.0 1.0	

Source: Amirkabir University of Technology Local Contest 2006

ID for Online Judge: POJ 3304

 Hint

The problem description is equivalent to determining whether there is a line l intersecting with n segments or not. If there is a line l intersecting with n segments, let line m be perpendicular to line l, and line m is the line that the problem requires you to find.

For segment i, its endpoints are $p_{2 \times i}$ and $p_{2 \times i+1}$, $0 \leq i \leq n-1$. Each pair of endpoints p_i and p_j ($0 \leq i < j \leq 2n-1$) is enumerated: if the line through p_i and p_j intersects with or coincides with n segments, then all projected segments have at least one point in common, and "Yes!" is output; else the next pair of points is enumerated. If there is no line with the desired property, "No!" is output.

8.5.2 Titanic

It is a historical fact that during the legendary voyage of "Titanic", the wireless telegraph machine delivered six warnings about the danger of icebergs. Each of the telegraph messages described the point where an iceberg had been noticed. The first five warnings were transferred to the captain of the ship. The sixth one came late at night, and the telegraph operator did not notice that the coordinates mentioned were very close to the current ship's position.

Write a program that will warn the operator about the danger of icebergs!

Input

The input messages are of the following format:

Message #<n>.
Received at <HH>:<MM>:<SS>.
Current ship's coordinates are
<X1>^<X2>'<X3>" <NL/SL>
and <Y1>^<Y2>'<Y3>" <EL/WL>.
An iceberg was noticed at
<A1>^<A2>'<A3>" <NL/SL>
and <B1>^<B2>'<B3>" <EL/WL>.
===

Here <n> is a positive integer, <HH>:<MM>:<SS> is the time of the message reception, <X1>^<X2>'<X3>"<NL/SL> and <Y1>^<Y2>'<Y3>"<EL/WL> means "X1 degrees X2 minutes X3 seconds of North (South) latitude and Y1 degrees Y2 minutes Y3 seconds of East (West) longitude."

Output

Your program should print to the output file message in the following format:

The distance to the iceberg: <s> miles,

where <s> should be the distance between the ship and the iceberg (that is the length of the shortest path on the sphere between the ship and the iceberg). This distance should be printed up to (and correct to) two decimal digits. If this distance is less than (but not equal to!) 100 miles, the program should print one more line with the text: "DANGER!"

Sample Input	Sample Output
Message #513. Received at 22:30:11. Current ship's coordinates are 41^46'00" NL and 50^14'00" WL. An iceberg was noticed at 41^14'11" NL and 51^09'00" WL. ===	The distance to the iceberg: 52.04 miles. DANGER!

Source: Ural Collegiate Programming Contest 1999

IDs for Online Judges: POJ 2354, Ural 1030

Hint

For simplicity of calculations, assume that the Earth is an ideal sphere with the diameter of 6875 miles completely covered with water. Also, you can be sure that lines in the input file break exactly as shown in the input samples. The ranges of the ship and the iceberg coordinates are the same as the usual range for geographical coordinates, i.e., from 0 to 90 degrees inclusively for NL/SL and from 0 to 180 degrees inclusively for EL/WL.

Hint

The problem requires you to calculate the distance between two points on a sphere. The formula calculating spherical distance is used to solve the problem directly. If the distance is less than 100 miles, the program should print one more line with the text: "DANGER!"

8.5.3 *Intervals*

In the ceiling in the basement of a newly open developers' building, a light source has been installed. Unfortunately, the material used to cover the floor is very sensitive to light. It turns out that its expected lifetime is decreasing dramatically. To avoid this, authorities have decided to protect light-sensitive areas from strong light by covering them. The solution was not very easy because, as is common, in the basement there are different pipelines under the ceiling, and the authorities want to install the covers just on those parts of the floor that are not shielded from the light by pipes. To cope with the situation, the first decision was to simplify the real situation and, instead of solving the problem in 3-D space, to construct a 2-D model first (see Figure 8.37).

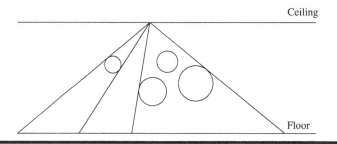

Figure 8.37

Within this model, the *x*-axis has been aligned with the level of the floor. The light is considered to be a point light source with integer coordinates $[b_x, b_y]$. The pipes are represented by circles. The center of the circle i has the integer coordinates $[c_{xi}, c_{yi}]$ and an integer radius r_i. As pipes are made from solid material, circles cannot overlap. Pipes cannot reflect the light and the light cannot go through the pipes. You have to write a program that will determine the non-overlapping intervals on the *x*-axis where there is, due to the pipes, no light from the light source.

Input

The input consists of blocks of lines, each of which except the last describes one situation in the basement. The first line of each block contains a positive integer number $N<500$ expressing the number of pipes. The second line of the block contains two integers b_x and b_y separated by one space. Each of the next N lines of the block contains integers c_{xi}, c_{yi} and r_i, where $c_{yi}+r_i<b_y$. Integers in individual lines are separated by one space. The last block consists of one line containing $n=0$.

Output

The output consists of blocks of lines, corresponding to the blocks in the input (except the last one). One empty line must be put after each block in the output. Each of the individual lines of the blocks in the output will contain two real numbers, the endpoints of the interval where there is no light from the given point light source. The reals are exact to two decimal places and separated by one space. The intervals are sorted according to increasing *x*-coordinate.

Sample Input	Sample Output
6	0.72 78.86
300 450	88.50 133.94
70 50 30	181.04 549.93
120 20 20	
270 40 10	75.00 525.00
250 85 20	
220 30 30	300.00 862.50
380 100 100	
1	
300 300	
300 150 90	
1	
300 300	
390 150 90	
0	

Source: ACM Central Europe 1996

IDs for Online Judges: POJ 1375, ZOJ 1309, UVA 313

Hint

Suppose the point light source is node b, and the center and the radius of the circle i are p_i and r_i respectively. There are two tangent lines from node b for the circle i, where the x-coordinates for the intersection points for two tangent lines and X-axis are L_i and R_i respectively.

First, for circle i, $1 \leq i \leq n$, L_i and R_i are calculated. Then circles are sorted in ascending order for L_i: $order[0..n-1]$. Finally, each circle in $order[0..n-1]$ is analyzed one by one to determine the interval where there is no light from the given point light source.

8.5.4 Treasure Hunt

Archeologists from the Antiquities and Curios Museum (ACM) have flown to Egypt to examine the great pyramid of Key-Ops. Using state-of-the-art technology, they are able to determine that the lower floor of the pyramid is constructed from a series of straightline walls, which intersect to form numerous enclosed chambers. Currently, no doors exist to allow access to any chamber. This state-of-the-art technology has also pinpointed the location of the treasure room. What these dedicated (and greedy) archeologists want to do is to blast doors through the walls to get to the treasure room. However, to minimize the damage to the artwork in the intervening chambers (and stay under their government grant for dynamite), they want to blast through the minimum number of doors. For structural integrity purposes, doors should only be blasted at the midpoint of the wall of the room being entered. You are to write a program which determines this minimum number of doors. An example is shown in Figure 8.38.

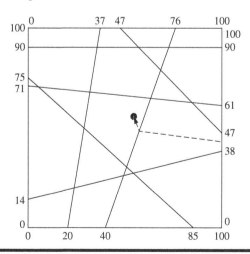

Figure 8.38

Input

The input will consist of one case. The first line will be an integer n ($0 \le n \le 30$) specifying the number of interior walls, followed by n lines containing the integer endpoints of each wall $x_1 \ y_1 \ x_2 \ y_2$. The four enclosing walls of the pyramid have fixed endpoints at (0,0); (0,100); (100,100), and (100,0) and are not included in the list of walls. The interior walls always span from one exterior wall to another exterior wall and are arranged such that no more than two walls intersect at any point. You may assume that no two given walls coincide. After the listing of the interior walls, there will be one final line containing the floating-point coordinates of the treasure in the treasure room (guaranteed not to lie on a wall).

Output

Print a single line listing the minimum number of doors that need to be created, in the format shown below.

Sample Input	Sample Output
7	Number of doors = 2
20 0 37 100	
40 0 76 100	
85 0 0 75	
100 90 0 90	
0 71 100 61	
0 14 100 38	
100 47 47 100	
54.5 55.4	

Source: ACM East Central North America 1999

IDs for Online Judges: POJ 1066, ZOJ 1158, UVA 754

 Hint

For each interior wall, the line segment connecting the treasure and its endpoint can be regarded as a route for archeologists entering the treasure room. The number of intersection points for the line segment and interior walls is the number of doors which archeologists need to create for interior walls.

The i-th interior wall is represented as an edge vector $\overrightarrow{p_{1i} p_{2i}}$, $0 \le i \le n-1$, where p_{1i} and p_{2i} are endpoints for the i-th interior wall; and the floating-point coordinate of the treasure in the treasure room is p.

The line segment connecting the treasure and the starting point for the i-th interior wall is represented as an edge vector $\overrightarrow{p p_{1i}}$, $0 \le i \le n-1$. Suppose A_i is the number of intersection points for the line segment and interior walls. $A = \min\{A_1, A_2, \dots, A_n\}$.

The line segment connecting the treasure and the terminal point for the i-th interior wall is represented as an edge vector $\overrightarrow{pp_{2i}}$, $0 \leq i \leq n-1$. Suppose B_i is the number of intersection points for the line segment and interior walls. $B=\min\{B_1, B_2, ..., B_n\}$.

The minimum number of doors which need to be created is $\min\{A, B\}+1$.

8.5.5 Intersection

You are to write a program that has to decide whether a given line segment intersects a given rectangle.

Here's an example:

line: start point: (4,9)
end point: (11,2)
rectangle: left-top: (1,5)
right-bottom: (7,1)

The line is said to intersect the rectangle if the line and the rectangle have at least one point in common. The rectangle consists of four straight lines and the area in between, as shown in Figure 8.39. Although all input values are integer numbers, valid intersection points do not have to lay on the integer grid.

Input

The input consists of n test cases. The first line of the input file contains the number n. Each following line contains one test case of the format:

xstart ystart xend yend xleft ytop xright ybottom

where (*xstart, ystart*) is the start and (*xend, yend*) the end point of the line and (*xleft, ytop*) the top-left and (*xright, ybottom*) the bottom-right corner of the rectangle. The

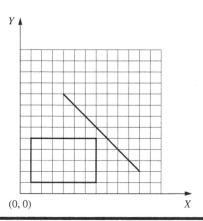

Figure 8.39   Line segment does not intersect rectangle.

eight numbers are separated by a blank. The terms "top left" and "bottom right" do not imply any ordering of coordinates.

Output

For each test case in the input file, the output file should contain a line consisting either of the letter "T" if the line segment intersects the rectangle, or the letter "F" if the line segment does not intersect the rectangle.

Sample Input	Sample Output
1 4 9 11 2 1 5 7 1	F

Source: ACM Southwestern European Regional Contest 1995

IDs for Online Judges: POJ 1410, UVA 191

Hint

Suppose the given line segment is $\overline{pt_1 pt_2}$, where the start point for the line segment is pt_1 and the end point is pt_2. Based on the left-top and the right-bottom corners for the given rectangle, the left-bottom corner, the right-top corner, and four edges for the rectangle are calculated.

The rectangle consists of four straight lines and the area in between. If pt_1 and pt_2 are in the area for the rectangle, the line intersects the rectangle. And if the line segment intersects any edge for the rectangle, the line segment intersects the rectangle. Otherwise, the line segment doesn't intersect the rectangle.

8.5.6 Space Ant

The most exciting space discovery occurred at the end of the 20th century. In 1999, scientists traced down an ant-like creature in the planet Y1999 and called it M11. It has only one eye on the left side of its head and has just three feet all on the right side of its body. It suffers from three walking limitations:

1. It cannot turn right due to its special body structure.
2. It leaves a red path while walking.
3. It hates to pass over a previously red colored path, and never does that.

The pictures transmitted by the Discovery space ship depict that plants in the Y1999 planet grow in special points on the planet. Analysis of several thousands of the pictures have resulted in discovering a magic coordinate system governing the grow points of the plants. In this coordinate system with x and y axes, **no two plants share the same x or y.**

An M11 needs to eat exactly one plant in each day to stay alive. When it eats one plant, it remains there for the rest of the day with no move. The next day, it looks for another plant to eat. If it cannot reach any other plant, it dies by the end of the day. Notice that it can reach a plant in any distance.

The problem is to find a path for an M11 to let it live as long as possible.

Input is a set of (x, y) coordinates of plants. Suppose A with the coordinates (x_A, y_A) is the plant with the least y-coordinate. M11 starts from point $(0, y_A)$ heading towards plant A. Notice that the solution path should not cross itself, and all of the turns should be counterclockwise. Also note that the solution may visit more than two plants located on a same straight line. See Figure 8.40.

Input

The first line of the input is M, the number of test cases to be solved ($1 \le M \le 10$). For each test case, the first line is N, the number of plants in that test case ($1 \le N \le 50$), followed by N lines for each plant data. Each plant data consists of three integers: the first number is the unique plant index $(1..N)$, followed by two positive integers x and y representing the coordinates of the plant. Plants are sorted by the increasing order on their indices in the input file. Suppose that the values of coordinates are at most 100.

Output

Output should have one separate line for the solution of each test case. A solution is the number of plants on the solution path, followed by the indices of visiting plants in the path in the order of their visits.

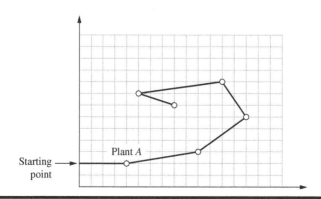

Figure 8.40

Sample Input	Sample Output
2	10 8 7 3 4 9 5 6 2 1 10
10	14 9 10 11 5 12 8 7 6 13 4 14 1 3 2
1 4 5	
2 9 8	
3 5 9	
4 1 7	
5 3 2	
6 6 3	
7 10 10	
8 8 1	
9 2 4	
10 7 6	
14	
1 6 11	
2 11 9	
3 8 7	
4 12 8	
5 9 20	
6 3 2	
7 1 6	
8 2 13	
9 15 1	
10 14 17	
11 13 19	
12 5 18	
13 7 3	
14 10 16	

Source: ACM Tehran 1999

IDs for Online Judges: POJ 1696, ZOJ 1429

Hint

Suppose N plants are a_0, a_1, ..., a_{N-1}. A with the coordinates (x_A, y_A) is the plant with the least y-coordinate. M11 starts from point $(0, y_A)$ heading towards plant A. Therefore A is as the first plant a_0. Then from a_i, $i \geq 0$, the next plant is analyzed one by one: based on a_i, the remaining plants are sorted as a_{i+1}, ..., a_{N-1}, where the first key is the direction, and the second key is the distance between a_i and the plant. The next plant is a_{i+1}.

8.5.7 Kadj Squares

In this problem, you are given a sequence S_1, S_2, ..., S_n of squares of different sizes. The sides of the squares are integer numbers. We locate the squares on the positive x-y

quarter of the plane, such that their sides make 45 degrees with x and y axes, and one of their vertices is on the $y=0$ line. Let b_i be the x coordinates of the bottom vertex of S_i. First, put S_l such that its left vertex lies on $x=0$. Then, put S_1, $(i>1)$ at minimum b_i such that

$b_i > b_{i-1}$ and

the interior of S_i does not have intersection with the interior of $S_1 \dots S_{i-1}$.

The goal is to find which squares are visible, either entirely or partially, when viewed from above. In Figure 8.41, the squares S_1, S_2, and S_4 have this property. More formally, S_i is visible from above if it contains a point p, such that no square other than S_i intersects the vertical half-line drawn from p upwards.

Input

The input consists of multiple test cases. The first line of each test case is n $(1\le n\le 50)$, the number of squares. The second line contains n integers between 1 to 30, where the i-th number is the length of the sides of S_i. The input is terminated by a line containing a zero number.

Output

For each test case, output a single line containing the index of the visible squares in the input sequence, in ascending order, separated by blank characters.

Sample Input	Sample Output
4	1 2 4
3 5 1 4	1 3
3	
2 1 2	
0	

Source: ACM Tehran 2006

IDs for Online Judges: POJ 3347, UVA 3799

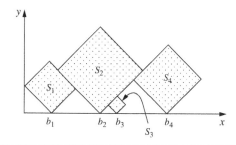

Figure 8.41

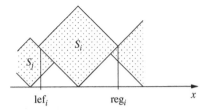

Figure 8.42

Hint

Suppose the length of one side for the *i*-th square is l_i, and projections on the *x*-axis for its left end and right end are lef_i and rig_i, respectively, $0 \le i \le n-1$. This is shown in Figure 8.42. If the *i*-th square is visible, the visible interval is $[le_i, ri_i]$.

Because of the precision of real numbers, the sides of squares are expanded $\sqrt{2}$ times. Obviously, $lef_0 = 0$, and $rig_0 = 2 \times l_0$.

First, for other squares, their lef_i and rig_i are calculated: $lef_i = \max_{0 \le j \le i-1} \{rig_j - |l_i - l_j|\}$, $rig_i = lef_i + 2 \times l_i$, $1 \le i \le n-1$.

Then, based on lef_i and rig_i, visible intervals for all squares are calculated: $le_i = \max_{0 \le j \le i-1} \{rig_j, lef_i\}$, $ri_i = \min_{i+1 \le j \le n-1} \{rig_i, lef_j\}$.

Finally, every square is analyzed. If $le_i < ri_i$, then the (*i*+1)th square is visible; else it is invisible.

8.5.8 Pipe

The GX Light Pipeline Company started to prepare bent pipes for the new transgalactic light pipeline. During the design phase of the new pipe shape, the company ran into the problem of determining how far the light can reach inside each component of the pipe. Note that the material which the pipe is made from is not transparent and is not light reflecting.

Each pipe component consists of many straight pipes connected tightly together. For programming purposes, the company developed the description of each component as a sequence of points $[x_1; y_1], [x_2; y_2], \ldots, [x_n; y_n]$, where $x_1 < x_2 < \ldots < x_n$. These are the upper points of the pipe contour. The bottom points of the pipe contour consist of points with *y*-coordinate decreased by 1. To each upper point $[x_i; y_i]$, there is a corresponding bottom point $[x_i; y_i-1]$ (see Figure 8.43). The company wants to find, for each pipe component, the point with maximal *x*-coordinate that the light will reach. The light is emitted by a segment source with endpoints $[x_1; y_1-1]$ and $[x_1; y_1]$ (endpoints are emitting light, too). Assume that the light is not bent at the pipe bent points and the bent points do not stop the light beam.

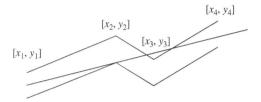

Figure 8.43

Input

The input file contains several blocks, each describing one pipe component. Each block starts with the number of bent points $2 \le n \le 20$ on a separate line. Each of the next n lines contains a pair of real values x_i, y_i separated by space. The last block is denoted with $n=0$.

Output

The output file contains lines corresponding to blocks in the input file. To each block in the input file, there is one line in the output file. Each such line contains either a real value, written with precision of two decimal places, or the message "Through all the pipe.". The real value is the desired maximal x-coordinate of the point where the light can reach from the source for the corresponding pipe component. If this value equals to x_n, then the message "Through all the pipe." will appear in the output file.

Sample Input	Sample Output
4	4.67
0 1	Through all the pipe.
2 2	
4 1	
6 4	
6	
0 1	
2 −0.6	
5 −4.45	
7 −5.57	
12 −10.8	
17 −16.55	
0	

Source: ACM Central Europe 1995

IDs for Online Judges: POJ 1039, UVA 303

Hint

Given a pipe component, the problem requires you to find the point with maximal *x*-coordinate that the light will reach, or the light can be through all the pipe.

There are *n* pairs of points: the upper point $[x_i, y_i]$, and its corresponding bottom point $[x_i, y_i-1]$, $1 \leq i \leq n$. Such a light must be through an upper point and a bottom point. Therefore, enumeration is used to solve the problem. Lines through an upper point and a bottom point are enumerated.

8.5.9 Geometric Shapes

While creating a customer logo, ACM uses graphical utilities to draw a picture that can later be cut into special fluorescent materials. To ensure proper processing, the shapes in the picture cannot intersect. However, some logos contain such intersecting shapes. It is necessary to detect them and decide how to change the picture.

Given a set of geometric shapes, you are to determine all of their intersections. Only outlines are considered; if a shape is completely inside another one, it is not counted as an intersection. See Figure 8.44.

Input

The input contains several pictures. Each picture describes at most 26 shapes, each specified on a separate line. The line begins with an uppercase letter that uniquely identifies the shape inside the corresponding picture. Then there is a kind of the shape and two or more points, everything separated by at least one space. Possible shape kinds are as follows:

Square: Followed by two distinct points giving the opposite corners of the square.

Rectangle: Three points are given; there will always be a right angle between the lines connecting the first point with the second and the second with the third.

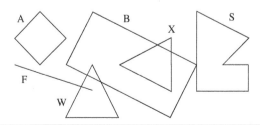

Figure 8.44

Line: Specifies a line segment; two distinct end points are given.

Triangle: Three points are given; they are guaranteed not to be colinear.

Polygon: Followed by an integer number N ($3 \leq N \leq 20$) and N points specifying vertices of the polygon in either clockwise or anticlockwise order. The polygon will never intersect itself and its sides will have non-zero length.

All points are always given as two integer coordinates X and Y separated with a comma and enclosed in parentheses. You may assume that $|X|,|Y| \leq 10000$.

The picture description is terminated by a line containing a single dash ("–"). After the last picture, there is a line with one dot (".").

Output

For each picture, output one line for each of the shapes, sorted alphabetically by its identifier (X). The line must be one of the following:

"X has no intersections", if X does not intersect with any other shapes.

"X intersects with A", if X intersects with exactly one other shape.

"X intersects with A and B", if X intersects with exactly two other shapes.

"X intersects with A, B, . . ., and Z", if X intersects with more than two other shapes.

Please note that there is an additional comma for more than two intersections. A, B, etc. are all intersecting shapes, sorted alphabetically.

Print one empty line after each picture, including the last one.

Sample Input	Sample Output
A square (1,2) (3,2)	A has no intersections
F line (1,3) (4,4)	B intersects with S, W, and x
W triangle (3,5) (5,5) (4,3)	F intersects with W
x triangle (7,2) (7,4) (5,3)	S intersects with B
S polygon 6 (9,3) (10,3) (10,4) (8,4) (8,1) (10,2)	W intersects with B and F
B rectangle (3,3) (7,5) (8,3)	x intersects with B
-	
B square (1,1) (2,2)	A has no intersections
A square (3,3) (4,4)	B has no intersections
-	
.	

Source: CTU Open 2007

ID for Online Judge: POJ 3449

Hint

Enumeration is used to solve the problem. All pairs of different shapes are enumerated. A square, a triangle, or a polygon is represented as a set of lines. That is, two shapes are intersected if and only if their lines are intersected.

8.5.10 *A Round Peg in a Ground Hole*

The DIY Furniture Company specializes in assemble-it-yourself furniture kits. Typically, the pieces of wood are attached to one another using a wooden peg that fits into pre-cut holes in each piece to be attached. The pegs have a circular cross section and so are intended to fit inside a round hole.

A recent factory run of computer desks were flawed when an automatic grinding machine was mis programmed. The result is an irregularly shaped hole in one piece that, instead of the expected circular shape, is actually an irregular polygon. You need to figure out whether the desks need to be scrapped or if they can be salvaged by filling a part of the hole with a mixture of wood shavings and glue.

There are two concerns. First, if the hole contains any protrusions (i.e., if there exist any two interior points in the hole that, if connected by a line segment, that segment would cross one or more edges of the hole), then the filled-in hole would not be structurally sound enough to support the peg under normal stress as the furniture is used. Second, assuming the hole is appropriately shaped, it must be big enough to allow for insertion of the peg. Since the hole in this piece of wood must match up with a corresponding hole in other pieces, the precise location where the peg must fit is known.

Write a program to accept descriptions of pegs and polygonal holes and to determine if the hole is ill-formed and, if not, whether the peg will fit at the desired location. Each hole is described as a polygon with vertices (x_1, y_1), $(x_2, y_2), \ldots, (x_n, y_n)$. The edges of the polygon are (x_i, y_i) to (x_{i+1}, y_{i+1}) for $i=1 \ldots n-1$ and (x_n, y_n) to (x_1, y_1).

Input

Input consists of a series of piece descriptions. Each piece description consists of the following data:

Line 1 *<nVertices> <pegRadius> <pegX> <pegY>*
number of vertices in polygon, *n* (integer)
radius of peg (real)
X and *Y* position of peg (real)
n Lines *<vertexX> <vertexY>*

On a line for each vertex, listed in order, the *X* and *Y* position of vertex. The end of input is indicated by a number of polygon vertices less than three.

Output

For each piece description, print a single line containing the string:

"HOLE IS ILL-FORMED" if the hole contains protrusions.

"PEG WILL FIT" if the hole contains no protrusions and the peg fits in the hole at the indicated position.

"PEG WILL NOT FIT" if the hole contains no protrusions but the peg will not fit in the hole at the indicated position.

Sample Input	Sample Output
5 1.5 1.5 2.0	HOLE IS ILL-FORMED
1.0 1.0	PEG WILL NOT FIT
2.0 2.0	
1.75 2.0	
1.0 3.0	
0.0 2.0	
5 1.5 1.5 2.0	
1.0 1.0	
2.0 2.0	
1.75 2.5	
1.0 3.0	
0.0 2.0	
1	

Source: ACM Mid-Atlantic 2003

IDs for Online Judges: POJ 1584, ZOJ 1767, UVA 2835

Hint

Suppose the position for the peg is *peg*, its radius is *r*; and the sequence for vertices of the polygon are $p_0, \ldots, p_n$, where $p_n = p_0$.

1. Determine whether the hole contains protrusions or not.
    Initially $c0$ and $c1$ are 0. Vertices are enumerated ($0 \le i \le n-1$) as follows.

```
If  p̄ᵢ₊₁pᵢ ∧ p̄ᵢ₊₂pᵢ < 0 , then pᵢ₊₁ is a concave vertex, the
number of concave vertices is c0++;
If  p̄ᵢ₊₁pᵢ ∧ p̄ᵢ₊₂pᵢ > 0 , then pᵢ₊₁ is a convex vertex, the number
of convex vertices is c1++;
If (c0!=0&&c1!=0), then the hole contains protrusions.
```

2. Determine whether the position for the peg is in the polygon or not.

The necessary condition that the peg fits in the hole is that the position for the peg is in the polygon. Suppose $\overrightarrow{peg}$ is a line segment from the origin to vertex *peg*. Initially $c0$ and $c1$ are 0. Each side $\overrightarrow{p_{i+1}p_i}$ ($0 \le i \le n-1$) of the polygon is enumerated. If $\overrightarrow{p_{i+1}p_i} \wedge \overrightarrow{peg} < 0$, $c0$++; and if $\overrightarrow{p_{i+1}p_i} \wedge \overrightarrow{peg} > 0$, $c1$++. After all sides are enumerated, if ($c0!=0$&&$c1!=0$), then the position for the peg isn't in the polygon.

3. Determine whether the peg fits in the hole or not.

If the position for the peg is in the polygon, then the distance from *peg* to

sides for the polygon $m = \min\limits_{0 \le i \le n-1} \left\{ \dfrac{\overrightarrow{p_i peg} \wedge \overrightarrow{p_i p_{i+1}}}{\left| \overrightarrow{p_i p_{i+1}} \right|} \right\}$ is calculated. If $r > m$, then

the peg can't fit in the hole; else the peg can fit in the hole.

8.5.11 Triangle

A *lattice point* is an ordered pair (x, y) where x and y are both integers. Given the coordinates of the vertices of a triangle (which happen to be lattice points), you are to count the number of lattice points which lie completely inside of the triangle (points on the edges or vertices of the triangle do not count).

Input

The input test file will contain multiple test cases. Each input test case consists of six integers x_1, y_1, x_2, y_2, x_3, and y_3, where (x_1, y_1), (x_2, y_2), and (x_3, y_3) are the coordinates of vertices of the triangle. All triangles in the input will be non-degenerate (will have positive area), and $-15000 \le x_1$, y_1, x_2, y_2, x_3, $y_3 \le 15000$. The end-of-file is marked by a test case with $x_1 = y_1 = x_2 = y_2 = x_3 = y_3 = 0$ and should not be processed.

Output

For each input case, the program should print the number of internal lattice points on a single line.

Sample Input	Sample Output
0 0 1 0 0 1	0
0 0 5 0 0 5	6
0 0 0 0 0 0	

Source: Stanford Local 2004

ID for Online Judge: POJ 2954

Hint

Suppose p_1, p_2, and p_3 are vertices of a triangle. Then the area for the triangle is
$S_\Delta = \dfrac{\left|p_1 \wedge p_2\right| + \left|p_2 \wedge p_3\right| + \left|p_3 \wedge p_1\right|}{2}$. Suppose $g(\overrightarrow{p_i p_j})$ is the number of points on p_i,
p_j, and $\overrightarrow{p_i p_j}$.

$$g(\overrightarrow{p_i p_j}) = \begin{cases} \left|p_i.y - p_j.y\right| & \left|p_i.x - p_j.x\right| = 0 \\ \left|p_i.x - p_j.x\right| & \left|p_i.y - p_j.y\right| = 0 \\ \gcd\left(\left|p_i.y - p_j.y\right|, \left|p_i.x - p_j.x\right|\right) & \text{otherwise} \end{cases}$$

Based on Pick's theorem, the area for a triangle S_Δ= the number of internal
lattice points $+\dfrac{g(\overrightarrow{p_1 p_2}) + g(\overrightarrow{p_2 p_3}) + g(\overrightarrow{p_3 p_1})}{2} - 1$. Therefore, the number of internal
lattice points $= S_\Delta - \dfrac{g(\overrightarrow{p_1 p_2}) + g(\overrightarrow{p_2 p_3}) + g(\overrightarrow{p_3 p_1})}{2} + 1$.

8.5.12 Ants

Young naturalist Bill studies ants in school. His ants feed on plant-louses that live on apple trees. Each ant colony needs its own apple tree to feed itself.

Bill has a map with coordinates of n ant colonies and n apple trees. He knows that ants travel from their colony to their feeding places and back using chemically tagged routes. The routes cannot intersect each other, or ants will get confused and get to the wrong colony or tree, thus spurring a war between colonies.

Bill would like to connect each ant colony to a single apple tree so that all n routes are non-intersecting straight lines. In this problem, such a connection is always possible. Your task is to write a program that finds such a connection.

In Figure 8.45, ant colonies are denoted by empty circles and apple trees are denoted by filled circles. One possible connection is denoted by lines.

Input

The first line of the input file contains a single integer number n ($1 \leq n \leq 100$)—the number of ant colonies and apple trees. It is followed by n lines describing n ant colonies, followed by n lines describing n apple trees. Each ant colony and apple tree is described by a pair of integer coordinates x and y ($-10{,}000 \leq x, y \leq 10{,}000$) on a Cartesian plane. All ant colonies and apple trees occupy distinct points on a plane. No three points are on the same line.

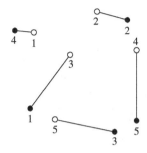

Figure 8.45

Output

Write to the output file n lines with one integer number on each line. The number written on the i-th line denotes the number (from 1 to n) of the apple tree that is connected to the i-th ant colony.

Sample Input	Sample Output
5	4
−42 58	2
44 86	1
7 28	5
99 34	3
−13 −59	
−47 −44	
86 74	
68 −75	
−68 60	
99 −60	

Source: ACM Northeastern Europe 2007

IDs for Online Judges: POJ 3565, UVA 4043

Hint

Suppose a_i is the position for ant colony i, b_j is the position for apple tree j, and $1 \leq i$, $j \leq n$. Set x consisits of n ant colonies, and set y consisits of n apple trees. The weight of edge (a_i, b_j) is $-\left|\overline{a_i b_j}\right|$. Then a KM algorithm is used to calculate the maximum matching.

8.5.13 The Doors

You are to find the length of the shortest path through a chamber containing obstructing walls. The chamber will always have sides at $x=0$, $x=10$, $y=0$, and $y=10$. The initial and final points of the path are always (0, 5) and (10, 5). There will also be from 0 to 18 vertical walls inside the chamber, each with two doorways. Figure 8.46 illustrates such a chamber and also shows the path of minimal length.

Input

The input data for the illustrated chamber would appear as follows:

```
2
4 2 7 8 9
7 3 4.5 6 7
```

The first line contains the number of interior walls. Then there is a line for each such wall, containing five real numbers. The first number is the x coordinate of the wall ($0 < x < 10$), and the remaining four are the y coordinates of the ends of the doorways in that wall. The x coordinates of the walls are in increasing order, and within each line, the y coordinates are in increasing order. The input file will contain at least one such set of data. The end of the data comes when the number of walls is −1.

Output

The output should contain one line of output for each chamber. The line should contain the minimal path length rounded to two decimal places past the decimal point, and always showing two decimal places past the decimal point. The line should contain no blanks.

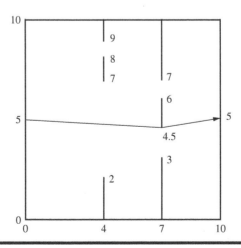

Figure 8.46

Sample Input	Sample Output
1	10.00
5 4 6 7 8	10.06
2	
4 2 7 8 9	
7 3 4.5 6 7	
−1	

Source: ACM Mid-Central USA 1996

IDs for Online Judges: POJ 1556, ZOJ 1721, UVA 393

Hint

There are $4n+2$ vertices, where $p_0=(0, 5)$, $p_{4 \times n+1}=(10, 5)$, $\overrightarrow{p_{4 \times i-3} p_{4 \times i-2}}$ is the first door for the i-th wall, and $\overrightarrow{p_{4 \times i-1} p_{4 \times i}}$ is the second door for the i-th wall ($1 \le i \le n$).

The distance between each pair of vertices p_i and p_j is calculated ($0 \le i < 4 \times n+1$, $i < j \le 4 \times n+1$), $d_{ij} = \left| \overrightarrow{p_i p_j} \right|$. If the intersection point for $\overrightarrow{p_i p_j}$ and the k-th wall ($p_i.x \le p_{4 \times k}.x \le p_j.x$) isn't at $\overrightarrow{p_{4 \times k-3} p_{4 \times k-2}}$ or $\overrightarrow{p_{4 \times k-1} p_{4 \times k}}$, then $d_{ij} = \infty$.

Then the Floyd algorithm is used to calculate the shortest distance between each pair of vertices d_{ij}.

Finally, $d_{0,4n+1}$ is the minimal path length.

8.5.14 *Toys*

Calculate the number of toys that land in each bin of a partitioned toy box.

Mom and dad have a problem—their child John never puts his toys away when he is finished playing with them. They gave John a rectangular box to put his toys in, but John is rebellious and obeys his parents by simply throwing his toys into the box. All the toys get mixed up, and it is impossible for John to find his favorite toys.

John's parents came up with the following idea. They put cardboard partitions into the box. Even if John keeps throwing his toys into the box, at least toys that get thrown into different bins stay separated. Figure 8.47 shows a top view of an example toy box.

Figure 8.47

For this problem, you are asked to determine how many toys fall into each partition as John throws them into the toy box.

Input

The input file contains one or more problems. The first line of a problem consists of six integers, n m x_1 y_1 x_2 y_2. The number of cardboard partitions is n ($0 < n \leq 5000$) and the number of toys is m ($0 < m \leq 5000$). The coordinates of the upper-left corner and the lower-right corner of the box are (x_1, y_1) and (x_2, y_2), respectively. The following n lines contain two integers per line, U_i L_i, indicating that the ends of the i-th cardboard partition is at the coordinates (U_i, y_1) and (L_i, y_2). You may assume that the cardboard partitions do not intersect each other and that they are specified in sorted order from left to right. The next m lines contain two integers per line, x_j y_j specifying where the j-th toy has landed in the box. The order of the toy locations is random. You may assume that no toy will land exactly on a cardboard partition or outside the boundary of the box. The input is terminated by a line consisting of a single 0.

Output

The output for each problem will be one line for each separate bin in the toy box. For each bin, print its bin number, followed by a colon and one space, followed by the number of toys thrown into that bin. Bins are numbered from 0 (the leftmost bin) to n (the rightmost bin). Separate the output of different problems by a single blank line.

Sample Input	Sample Output
5 6 0 10 60 0	0: 2
3 1	1: 1
4 3	2: 1
6 8	3: 1
10 10	4: 0
15 30	5: 1
1 5	
2 1	0: 2
2 8	1: 2
5 5	2: 2
40 10	3: 2
7 9	4: 2
4 10 0 10 100 0	
20 20	
40 40	
60 60	

Sample Input	Sample Output
80 80	
5 10	
15 10	
25 10	
35 10	
45 10	
55 10	
65 10	
75 10	
85 10	
95 10	
0	

Source: ACM Rocky Mountain 2003

IDs for Online Judges: POJ 2318, UVA 2910

Hint

Suppose the coordinates of the ends of the i-th cardboard partition are p_i' and p_i'' respectively, $0 \le i \le n-1$; and the coordinates of the upper-right corner and the lower-right corner of the box are p_n' and p_n'' respectively.

For each input toy, dichotomy is used to calculate the bin *ret* in which John throws the toy. After m toys are input, the numbers of toys thrown into each bin are calculated.

8.5.15 Area

Being well known for its highly innovative products, Merck would definitely be a good target for industrial espionage. To protect its brand new research and development facility, the company has installed the latest system of surveillance robots patrolling the area. These robots move along the walls of the facility and report suspicious observations to the central security office. The only flaw in the system a competitor agent could find is the fact that the robots radio their movements unencrypted. Not being able to find out more, the agent wants to use that information to calculate the exact size of the area occupied by the new facility. It is public knowledge that all the corners of the building are situated on a rectangular grid and that only straight walls are used. Figure 8.48 shows the course of a robot around an example area.

You are hired to write a program that calculates the area occupied by the new facility from the movements of a robot along its walls. You can assume that this

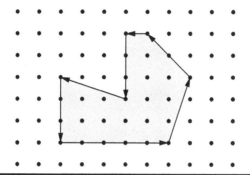

Figure 8.48

area is a polygon with corners on a rectangular grid. However, your boss insists that you use a formula that he is proud to have found somewhere. The formula relates the number I of grid points inside the polygon, the number E of grid points on the edges, and the total area A of the polygon. Unfortunately, you have lost the sheet on which he had written down that simple formula for you, so your first task is to find the formula yourself.

Input

The first line contains the number of scenarios.

For each scenario, you are given the number m, $3 \leq m < 100$, of movements of the robot in the first line. The following m lines contain pair "*dx dy*" of integers, separated by a single blank, satisfying $-100 \leq dx$, $dy \leq 100$ and $(dx, dy) != (0, 0)$. Such a pair means that the robot moves on to a grid point dx units to the right and dy units upwards on the grid (with respect to the current position). You can assume that the curve along which the robot moves is closed and that it does not intersect or even touch itself except for the start and end points. The robot moves anticlockwise around the building, so the area to be calculated lies to the left of the curve. It is known in advance that the whole polygon would fit into a square on the grid with a side length of 100 units.

Output

The output for every scenario begins with a line containing "Scenario #i:" where i is the number of the scenario starting at 1. Then print a single line containing I, E, and A, the area A rounded to one digit after the decimal point. Separate the three numbers by two single blanks. Terminate the output for the scenario with a blank line.

Sample Input	Sample Output
2	Scenario #1:
4	0 4 1.0
1 0	
0 1	Scenario #2:
−1 0	12 16 19.0
0 −1	
7	
5 0	
1 3	
−2 2	
−1 0	
0 −3	
−3 1	
0 −3	

Source: ACM Northwestern Europe 2001

IDs for Online Judges: POJ 1265, ZOJ 1032, UVA 2329

Hint

Given a polygon on a rectangular grid, the number I of grid points inside the polygon, the number E of grid points on the edges, and the total area A of the polygon are required to be calculated.

Based on Pick's theorem, the total area A of the polygon is $I+E/2-1$. The total area A of the polygon is the sum of all cross products for vectors from the origin to each pair of adjacent vertices. $E=gcd(abs(x_2-x_1), abs(y_2-y_1))$. Finally I is calculated.

8.5.16 Line of Sight

An architect is very proud of his new home and wants to be sure that it can be seen by people passing by his property line along the street. The property contains various trees, shrubs, hedges, and other obstructions that may block the view. For the purpose of this problem, model the house, property line, and obstructions as straight lines parallel to the x axis, as shown in Figure 8.49.

To satisfy the architect's need to know how visible the house is, you must write a program that accepts as input the locations of the house, property line, and surrounding obstructions and calculates the longest continuous portion of the property line from which the entire house can be seen, with no part blocked by any obstruction.

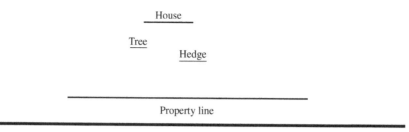

Figure 8.49

Input

Because each object is a line, it is represented in the input file with a left and right x coordinate followed by a single y coordinate:

$$<x_1><x_2><y>$$

where x_1, x_2, and y are non-negative real numbers. $x_1<x_2$.

An input file can describe the architecture and landscape of multiple houses. For each house, the first line will have the coordinates of the house. The second line will contain the coordinates of the property line. The third line will have a single integer that represents the number of obstructions, and the following lines will have the coordinates of the obstructions, one per line.

Following the final house, a line "0 0 0" will end the file.

For each house, the house will be above the property line (house y>property line y). No obstruction will overlap with the house or property line, e.g., if obstacle y = house y, you are guaranteed that the entire range obstacle $[x_1, x_2]$ does not intersect with house $[x_1, x_2]$.

Output

For each house, your program should print a line containing the length of the longest continuous segment of the property line from which the entire house can be seen to a precision of two decimal places. If there is no section of the property line where the entire house can be seen, print "No View".

Sample Input	Sample Output
2 6 6	8.80
0 15 0	No View
3	
1 2 1	
3 4 1	
12 13 1	
1 5 5	

Sample Input	Sample Output
0 10 0 1 0 15 1 0 0 0	

Source: ACM Mid-Atlantic 2004

IDs for Online Judges: POJ 2074, ZOJ 2325, UVA 3112

Hint

The key to the problem is to calculate the line equation through two points and the intersection point for two lines.

On the property line, each obstruction corresponds to a line segment where the house can't be seen, as dotted lines in Figure 8.50. All corresponding dotted lines are sorted and scanned to calculate the result.

8.5.17 An Easy Problem?!

It's raining outside. Farmer Johnson's bull, Ben, wants some rain to water his flowers. Ben nails two wooden boards on the wall of his barn. Shown in Figure 8.51, the two boards on the wall just look like two segments on the plane, as they have the same width.

Your mission is to calculate how much rain these two boards can collect.

Input

The first line contains the number of test cases.

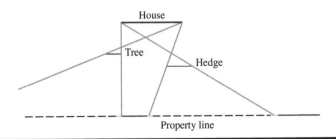

Property line

Figure 8.50

Figure 8.51

Each test case consists of eight integers not exceeding 10,000 by absolute value, $x_1, y_1, x_2, y_2, x_3, y_3, x_4, y_4$. (x_1, y_1), (x_2, y_2) are the endpoints of one board, and (x_3, y_3), (x_4, y_4) are the endpoints of the other one.

Output

For each test case, output a single line containing a real number with precision up to two decimal places—the amount of rain collected.

Sample Input	Sample Output
2	1.00
0 1 1 0	0.00
1 0 2 1	
0 1 2 1	
1 0 1 2	

Source: POJ Monthly, 2006.04.28, Dagger@PKU_RPWT

ID for Online Judge: POJ 2826

 Hint

Two wooden boards (line segments) constitute a container. The problem requires you to calculate the amount of rain that the container can collect.

Cases are shown in Figure 8.52.

The shaded area is collected rain.

Figure 8.52

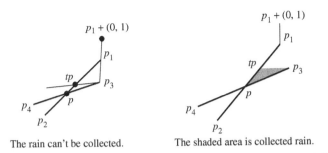

The rain can't be collected.   The shaded area is collected rain.

Figure 8.53

Suppose two wooden boards (line segments) are $\overline{p_1 p_2}$ and $\overline{p_3 p_4}$ respectively, where $p_1.y \geq p_2.y$, $p_3.y \geq p_4.y$, and $p_1.y \geq p_3.y$.

If there is no intersection point for $\overline{p_1 p_2}$ and $\overline{p_3 p_4}$, the rain can't be collected;

Suppose the intersection point for $\overline{p_1 p_2}$ and $\overline{p_3 p_4}$ is p, and the intersection point for $\overline{p_1 p_2}$ and the horizontal line is tp (as shown in Figure 8.53). If signs for $\overrightarrow{pp_1} \wedge \overrightarrow{pp_3}$ and $\overrightarrow{p_1(p_1 + (0,1))} \wedge \overrightarrow{p_1 p_3}$ are the same, or $\overrightarrow{p_1(p_1 + (0,1))}$ and $\overrightarrow{p_1 p_3}$ coincide, then the rain can't be collected; else the amount of collected rain is $S_{\triangle tpp_3 p}$.

8.5.18 Road Accident

Two cars crashed on the road, receiving some damage each, and raising the usual question: "Who to blame?" To answer this question, it is essential to thoroughly reconstruct the sequence of events. By gathering witness testimonies and analyzing tire tracks, positions and speeds of cars just before the impact were determined. From these positions until the crash the cars moved straight forward.

Your program must, given the available data, calculate for each car what part of it first came into contact with the other car. Parts are numbered as shown in Figure 8.54.

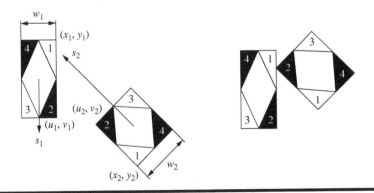

Figure 8.54

Input

The input file contains twelve floating-point numbers: x_1 y_1 u_1 v_1 w_1 s_1 x_2 y_2 u_2 v_2 w_2 s_2, where (x, y) and (u, v)—coordinates of back-left and forward-left corners of the car, w—width of the car, s—speed of the car.

Constraints

$1 \leq x_i, y_i, u_i, v_i, w_i \leq 10^6, 0 \leq s_i \leq 10^6$. Input data is such that a crash certainly happens. Initially cars don't have common points.

Output

The output file must contain two integers: p_1 p_2, where p is the number of the part which first contacted the other one (if two parts came into contact simultaneously, output the lesser of the part numbers).

Sample Input 1	Sample Output 1
1.0 2.0 10.0 2.0 1.0 10.0 50.0 1.0 40.0 1.0 1.0 20.0	2 2
Sample Input 2	Sample Output 2
1 1 10 1 1 20 40 1 50 1 1 10	2 1

Source: ACM Northeastern Europe 2005, Far-Eastern Subregion

ID for Online Judge: POJ 3433

 Hint

Two cars go at their speeds. Suppose one car stops, and the other car goes at a relative speed (see Figure 8.55).

Then the moving trail is determined. There are three cases: The first car's point comes into contact with the second car's side; the second car's point comes into contact with the first car's side; or two cars' points crash together.

8.5.19 Wild West

Once upon a time in the west... The quiet life of the villages on the western frontier are often stirred up by the appearance of mysterious strangers. A stranger might

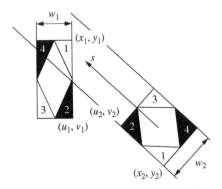

Figure 8.55

be a bounty hunter looking for a notorious villain, or he might be a dangerous criminal escaping the hand of justice. The number of strangers has become so large that they formed the Mysterious Strangers' Union. If you want to be a mysterious stranger, then you have to apply to the Union, and you have to pass three exams that test the three most important skills: shooting, fist-fighting, and harmonica playing. For each skill, the Admission Committee gives a score between 1 (worst) and m (best). Interestingly enough, there are no two members in the Union having exactly the same skills: for every two members,s there is always at least one skill for which they have different scores. Furthermore, it turns out that for every possible combination of scores, there is exactly one member having these scores. This means that there are exactly m^3 strangers in the union.

Recently, some members left the Union and they formed the Society of Evil Mysterious Strangers. The aim of this group is to commit as many evil crimes as possible, and they are quite successful at it. Therefore, the Steering Committee of the Union decided that a Hero is needed who will destroy this evil society. A Hero is a mysterious stranger who can defeat every member of the Society of Evil Mysterious Strangers. A Hero can defeat a member if the Hero has a higher score in at least one skill. For example, if the evil society has two members:

- Colonel Bill, with a score of 7 for shooting, 5 for knife throwing and 3 for harmonica playing, and
- Rabid Jack, with a score 10 for shooting, 6 for knife throwing and 8 for harmonica playing

Then a Hero with score 8 for shooting, 7 for knife throwing, and 3 for harmonica playing can defeat both of them. However, someone with a score of 8 for shooting, 6 for knife throwing, and 8 for harmonica playing cannot be the Hero. Moreover, the Hero cannot be a member of the evil society.

Your task is to determine whether there is a member in the Union who can be the Hero. If so, then you have to count how many members are potential heroes.

Input

The input contains several blocks of test cases. Each block begins with a line containing two integers: the number $1 \leq n \leq 100000$ of members in the Society of Evil Mysterious Strangers and the maximum value $2 \leq m \leq 100000$ of the scores. The next n lines describe these members. Each line contains three integers between 1 and m: the scores for the three skills.

The input is terminated by a block with $n=m=0$.

Output

For each test case, you have to output a single line containing the number of members in the Union who satisfy the requirements for becoming a Hero. If there is no such member, then output '0'. It can be assumed that the output is always at most 10^{18}.

Sample Input	Sample Output
3 10	848
2 8 5	19
6 3 5	999999999992
1 3 9	
1 3	
2 2 2	
1 10000	
2 2 2	
0 0	

Source: ACM Central Europe 2005

IDs for Online Judges: POJ 2944, UVA 3525

 ## Hint

There are M^3 gunmen characterized by three different skills, each ranging from 1 to M. A subset containing N of these gunmen are The Bad Guys. We want to select one of the other gunmen to be a Hero. The Hero must beat each of the Bad Guys in at least one skill (not necessarily the same skill for all Bad Guys).

The task is to compute the number of gunmen that can be selected to be the Hero.

Consider a Bad Guy with skills [a, b, c]. The set of gunmen that can't beat him is a cuboid with opposite corners [1,1,1] and [a, b, c]. The union U of all these cuboids is exactly the set of gunmen that can't be heroes. Thus the answer is M^3 minus the volume of U.

We can compute the volume of U in $O(M\log M)$ by sweeping in one direction and maintaining the intersection of the sweeping plane and U in some tree-like structure.

In C++, STL sets can be used, so there is no need to implement the tree-like structure.

8.5.20 *The Skyline Problem*

You are to design a program to assist an architect in drawing the skyline of a city, given the locations of the buildings in the city. To make the problem tractable, all buildings are rectangular in shape and they share a common bottom (the city they are built in is very flat). The city is also viewed as two-dimensional. A building is specified by an ordered triple (L_i, H_i, R_i) where L_i and R_i are left and right coordinates, respectively, of building i and H_i is the height of the building. In Figure 8.56, buildings are shown on the left with triples (1,11,5), (2,6,7), (3,13,9), (12,7,16), (14,3,25), (19,18,22), (23,13,29), (24,4,28), and the skyline, shown on the right, is represented by the sequence: (1, 11, 3, 13, 9, 0, 12, 7, 16, 3, 19, 18, 22, 3, 23, 13, 29, 0).

Input

The input is a sequence of building triples. All coordinates of buildings are positive integers less than 10,000, and there will be at least one and at most 5,000 buildings in the input file. Each building triple is on a line by itself in the input file. All integers in a triple are separated by one or more spaces. The triples will be sorted by L_i, the left x-coordinate of the building, so the building with the smallest left x-coordinate is first in the input file.

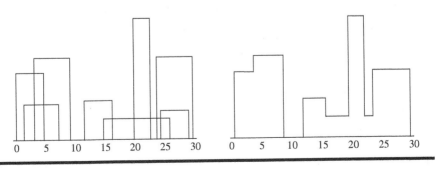

Figure 8.56

Output

The output should consist of the vector that describes the skyline as shown in Figure 8.56. In the skyline vector $(v_1, v_2, v_3, ..., v_{n-2}, v_{n-1}, v_n)$, the v_i such that i is an even number represents a horizontal line (height). The v_i such that i is an odd number represents a vertical line (x-coordinate). The skyline vector should represent the "path" taken, for example, by a bug starting at the minimum x-coordinate and traveling horizontally and vertically over all the lines that define the skyline. Thus the last entry in the skyline vector will be a 0. The coordinates must be separated by a blank space.

Sample Input	Sample Output
1 11 5 2 6 7 3 13 9 12 7 16 14 3 25 19 18 22 23 13 29 24 4 28	1 11 3 13 9 0 12 7 16 3 19 18 22 3 23 13 29 0

Source: Internet Programming Contest 1990

ID for Online Judge: UVA 105

Hint

The sweep line method is used to solve the problem.

Suppose the current building is represented by an ordered triple (*left, height, right*), the right border for the skyline is *rightmost*, and for each point i on the bottom, $a[i]$ is its height for the current skyline, $1 \leq i \leq rightmost$.

After a building triple is input, $a[i]$ in the interval (*left* $\leq i \leq$ *right*) and *rightmost* are adjusted: $a[i] = max\{a[i], height\}$ (*left* $\leq i \leq$ *right*); *rightmost* = *max*{*rightmost, right*}.

Then the sweep line method is used to enumerate points on the bottom. If $a[i] \neq a[i-1]$, then output i and $a[i]$. Finally, *rightmost*+1 and 0 are output.

8.5.21 Lining Up

"How am I ever going to solve this problem?" said the pilot.

Indeed, the pilot was not facing an easy task. She had to drop packages at specific points scattered in a dangerous area. Furthermore, the pilot could only fly over the area once in a straight line, and she had to fly over as many points as possible. All points were given by means of integer coordinates in a two-dimensional space.

The pilot wanted to know the largest number of points from the given set that all lie on one line. Can you write a program that calculates this number?

Your program has to be efficient!

Input

The input consists of several cases. The first line of each case is an integer N ($1 < N < 700$), and then follow N pairs of integers. Each pair of integers is separated by one blank and ended by a newline character. The input ended by $N=0$.

Output

Output one integer for each input case, representing the largest number of points that all lie on one line.

Sample Input	Sample Output
1 1	3
2 2	
3 3	
9 10	
10 11	

Source: ACM 1994 East-Central Regionals

IDs for Online Judges: POJ 1118, UVA 270

 Hint

Given n distinct points in a two-dimensional plane, find the maximum number of points which lie in an arbitrary line.

An algorithm whose time complexity is $O(n^2 \lg(n))$ is as follows:

The initial step to solve this problem is by sorting the points based on their y-coordinates in ascending order. In case of ties, sort the points based on their x-coordinates in ascending order.

The next step is to give each point a turn to become a pivot point. In each turn, create a new set based on the remaining points (the ones which have greater indices than the pivot point index) and sort these points angularly with respect to the pivot point. Using the fact that this new set is sorted angularly, a simple $O(n)$ algorithm can be devised to find the maximum number of points on a line whose bottom-left endpoint is the pivot.

One may ask why the points whose indices are less than pivot point are omitted during each pivot turn. Those points can be safely ignored without worrying that the final result won't be optimal. When an earlier point is a part of the optimal line, then the optimal result should have already been computed in an earlier pivot turn (remember the fact that the points were initially sorted based on their Cartesian coordinates).

8.5.22 Triathlon

A triathlon is an athletic contest consisting of three consecutive sections that should be completed as fast as possible as a whole. The first section is swimming, the second section is riding a bicycle, and the third one is running.

The speed of each contestant in all three sections is known. The judge can choose the length of each section arbitrarily provided that no section has zero length. As a result, sometimes she could choose their lengths in such a way that some particular contestant would win the competition.

Input

The first line of the input file contains integer number N ($1 \leq N \leq 100$), denoting the number of contestants. Then N lines follow, and each line contains three integers V_i, U_i, and W_i ($1 \leq V_i, U_i, W_i \leq 10000$), separated by spaces, denoting the speed of i-th contestant in each section.

Output

For every contestant, write to the output file one line, that contains the word "Yes" if the judge could choose the lengths of the sections in such a way that this particular contestant would win (i.e., she is the only one who would come first), or the word "No" if this is impossible.

Sample Input	Sample Output
9	Yes
10 2 6	Yes
10 7 3	Yes
5 6 7	No
3 2 7	No
6 2 6	No
3 5 7	Yes
8 4 6	No
10 4 2	Yes
1 8 7	

Source: ACM Northeastern Europe 2000

IDs for Online Judges: POJ 1755, ZOJ 2052, UVA 2218, URAL 1062

Hint

Suppose lengths of swimming, riding bicycle, and running are A, B and C (A, B, $C>0$) respectively. If the i-th contestant can win, for any other contestant j ($i \neq j$),

$$\frac{A}{v_i} + \frac{B}{u_i} + \frac{C}{w_i} < \frac{A}{v_j} + \frac{B}{u_j} + \frac{C}{w_j}.$$

$$\frac{A}{v_i} + \frac{B}{u_i} + \frac{C}{w_i} < \frac{A}{v_j} + \frac{B}{u_j} + \frac{C}{w_j}$$

$$\Rightarrow \frac{1}{v_i} \times \frac{A}{C} + \frac{1}{u_i} \times \frac{B}{C} + \frac{1}{w_i} < \frac{1}{v_j} \times \frac{A}{C} + \frac{1}{u_j} \times \frac{B}{C} + \frac{1}{w_j}$$

$$\Rightarrow \left(\frac{1}{v_j} - \frac{1}{v_i}\right) \times \frac{A}{C} + \left(\frac{1}{u_j} - \frac{1}{u_i}\right) \times \frac{B}{C} + \left(\frac{1}{w_j} - \frac{1}{w_i}\right) < 0.$$

If $\frac{A}{C}$ and $\frac{B}{C}$ are regarded as variables x and y respectively, the above formula is an inequality representing a half-plane. If the intersection of half-planes is a convex polygon, then the i-th contestant can win. The algorithm is as follows:

For the i-th contestant ($1 \leq i \leq n$), there are $n+2$ line equations ($A_k x + B_k y + C_k = 0$) representing half-planes H_1, H_2, ..., H_{n+2}, where the first $n-1$ line equations represent that the i-th contestant defeats the j-th contestant: $A_k = \frac{1}{u_j} - \frac{1}{u_i} - \left(\frac{1}{w_j} - \frac{1}{w_i}\right)$, $B_k = \frac{1}{v_j} - \frac{1}{v_i} - \left(\frac{1}{w_j} - \frac{1}{w_i}\right)$, $C_k = \frac{1}{w_j} - \frac{1}{w_i}$, $1 \leq j \leq n$, $j \neq i$, and $1 \leq k \leq n-1$; and the last three line equations represent $x=0$ ($A_n=1$, $B_n=0$, $C_n=0$), $y=0$ ($A_{n+1}=0$, $B_{n+1}=1$, $C_{n+1}=0$), and $x+y=1$ ($A_{n+2}=-1$, $B_{n+2}=-1$, $C_{n+2}=1$).

If the intersection of $n+2$ half-planes $H_1 \cap H_2 \cap \ldots \cap H_{n+2}$ is a convex polygon, then the i-th contestant can win; else the i-th contestant can't win.

8.5.23 Rotating Scoreboard

This year, ACM/ICPC World finals will be held in a hall in the form of a simple polygon. The coaches and spectators are seated along the edges of the polygon. We want to place a rotating scoreboard somewhere in the hall such that a spectator sitting anywhere on the boundary of the hall can view the scoreboard (i.e., his line of sight is not blocked by a wall). Note that if the line of sight of a spectator is tangent to the polygon boundary (either in a vertex or in an edge), he can still view the scoreboard. You may view spectator's seats as points along the boundary of the simple polygon, and consider the scoreboard as a point as well. Your program is given the corners of the hall (the vertices of the polygon), and must check if there is a location for the scoreboard (a point inside the polygon) such that the scoreboard can be viewed from any point on the edges of the polygon.

Input

The first number in the input line, T, is the number of test cases. Each test case is specified on a single line of input in the form n x_1 y_1 x_2 y_2 ... x_n y_n where n ($3 \leq n \leq 100$) is the number of vertices in the polygon, and the pair of integers x_i y_i sequence specifies the vertices of the polygon sorted in order.

Output

The output contains T lines, each corresponding to an input test case in that order. The output line contains either "YES" or "NO" depending on whether the scoreboard can be placed inside the hall conforming to the problem conditions.

Sample Input	Sample Output
2	YES
4 0 0 0 1 1 1 1 0	NO
8 0 0 0 2 1 2 1 1 2 1 2 2 3 2 3 0	

Source: ACM Tehran 2006 Preliminary

ID for Online Judge: POJ 3335

Hint

The hall is in the form of a simple polygon. The rotating scoreboard is placed somewhere in the hall such that a spectator sitting anywhere on the boundary of the hall can view the scoreboard. That is, the rotating scoreboard is placed at the core for the simple polygon. This problem is similar to **8.3.2.1 Art Gallery.**

8.5.24 How I Mathematician Wonder What You Are!

After counting so many stars in the sky in his childhood, Isaac, now an astronomer and a mathematician, uses a big astronomical telescope and lets his image processing program count stars. The hardest part of the program is to judge if a shining object in the sky is really a star. As a mathematician, the only way he knows is to apply a mathematical definition of *stars*.

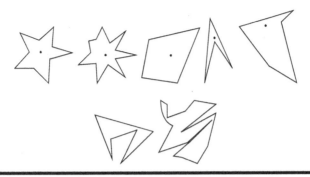

Figure 8.57

The mathematical definition of a star shape is as follows: A planar shape *F* is *star-shaped* if and only if there is a point *C*∈*F* such that, for any point *P*∈*F*, the line segment *CP* is contained in *F*. Such a point *C* is called a *center* of *F*. To get accustomed to the definition, let's see some examples in Figure 8.57.

The first two are what you would normally call stars. According to the above definition, however, all shapes in the first row are star-shaped. The two in the second row are not. For each star shape, a center is indicated with a dot. Note that a star shape in general has infinitely many centers. For example, for the third quadrangular shape, all points in it are centers.

Your job is to write a program that tells whether a given polygonal shape is star-shaped or not.

Input

The input is a sequence of datasets followed by a line containing a single zero. Each dataset specifies a polygon, and is formatted as follows:

> *n*
> x_1 y_1
> x_2 y_2
> ...
> x_n y_n

The first line is the number of vertices, *n*, which satisfies $4 \leq n \leq 50$. Subsequent *n* lines are the *x*- and *y*-coordinates of the *n* vertices. They are integers and satisfy $0 \leq x_i \leq 10000$ and $0 \leq y_i \leq 10000$ ($i = 1, ..., n$). Line segments $(x_i, y_i)-(x_{i+1}, y_{i+1})$ ($i = 1, ..., n-1$) and the line segment $(x_n, y_n)-(x_1, y_1)$ form the border of the polygon in the counterclockwise order. That is, these line segments see the inside of the polygon in the left of their directions.

You may assume that the polygon is *simple*, that is, its border never crosses or touches itself. You may also assume that no three edges of the polygon meet at a single point even when they are infinitely extended.

Output

For each dataset, output "1" if the polygon is star-shaped and "0" otherwise. Each number must be in a separate line, and the line should not contain any other characters.

Sample Input	Sample Output
6	1
66 13	0
96 61	
76 98	
13 94	
4 0	
45 68	
8	
27 21	
55 14	
93 12	
56 95	
15 48	
38 46	
51 65	
64 31	
0	

Source: ACM Japan 2006

IDs for Online Judges: POJ 3130, ZOJ 2820, UVA 3617

 Hint

Based on the definition of a star shape (a planar shape F is *star-shaped* if and only if there exists a point $C \in F$ such that, for any point $P \in F$, the line segment CP is contained in F. Such a point C is called a *center* of F.). If a planar shape F is *star-shaped*, then there must exist a core for F, and *centers* of F constitute the core; and if there is no core, then F isn't *star-shaped*. The intersection of half-planes is used to calculate the core for a planar shape F. The solution is the same as **8.5.23 Rotating Scoreboard**.

8.5.25 Video Surveillance

A friend of yours has taken the job of security officer at the Star-Buy Company, a famous department store. One of his tasks is to install a video surveillance system to guarantee the security of the customers (and the security of the merchandise, of course) on all of the store's countless floors. As the company has only a limited budget, there will be only one camera on every floor. But these cameras may turn around to look in every direction.

The first problem is to choose where to install the camera for every floor. The only requirement is that every part of the room must be visible from there. In Figure 8.58, the left floor can be completely surveyed from the position indicated by a dot, while for the right floor, there is no such position, the given position failing to see the lower-left part of the floor.

Before trying to install the cameras, your friend first wants to know whether there is indeed a suitable position for them. He therefore asks you to write a program that, given a ground plan, determines whether there is a position from which the whole floor is visible. All floor ground plans form rectangular polygons, whose edges do not intersect each other and touch each other only at the corners.

Input

The input contains several floor descriptions. Every description starts with the number n of vertices that bound the floor ($4 \leq n \leq 100$). The next n lines contain two integers each, the x and y coordinates for the n vertices, given in clockwise order. All vertices will be distinct and at corners of the polygon. Thus the edges alternate between horizontal and vertical.

A zero value for n indicates the end of the input.

Output

For every test case, first output a line with the number of the floor, as shown in the sample output. Then print a line stating "Surveillance is possible." if there exists

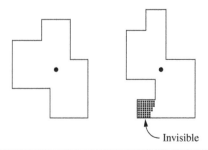

→ Invisible

Figure 8.58

a position from which the entire floor can be observed, or print "Surveillance is impossible." if there is no such position.

Print a blank line after each test case.

Sample Input	Sample Output
4	Floor #1
0 0	Surveillance is possible.
0 1	
1 1	Floor #2
1 0	Surveillance is impossible.
8	
0 0	
0 2	
1 2	
1 1	
2 1	
2 2	
3 2	
3 0	
0	

Source: ACM Southwestern European Regional Contest 1997

IDs for Online Judges: POJ 1474, ZOJ 1248, UVA 588

Hint

The only requirement for installing the camera is that every part of the room must be visible from the camera. A ground plan is a polygon. If there exists a core for the polygon, the camera can be installed in the core. The intersection of half-planes is used to determine whether there exists a core for the polygon. The solution is the same as **8.5.23 Rotating Scoreboard**.

8.5.26 Most Distant Point from the Sea

The main land of Japan, called Honshu, is an island surrounded by the sea. In such an island, it is natural to ask a question: "Where is the most distant point from the sea?" The answer to this question for Honshu was found in 1996. The most distant point is located in former Usuda Town, Nagano Prefecture, whose distance from the sea is 114.86 km.

In this problem, you are asked to write a program which, given a map of an island, finds the most distant point from the sea in the island, and reports its distance from the sea. In order to simplify the problem, we only consider maps representable by convex polygons.

Input

The input consists of multiple datasets. Each dataset represents a map of an island, which is a convex polygon. The format of a dataset is as follows:

n

x_1 y_1

$\vdots$

x_n y_n

Every input item in a dataset is a non-negative integer. Two input items in a line are separated by a space. n in the first line is the number of vertices of the polygon, satisfying $3 \leq n \leq 100$. Subsequent n lines are the x- and y-coordinates of the n vertices. Line segments $(x_i, y_i)-(x_{i+1}, y_{i+1})$ ($1 \leq i \leq n-1$) and the line segment $(x_n, y_n)-(x_1, y_1)$ form the border of the polygon in counterclockwise order. That is, these line segments see the inside of the polygon in the left of their directions. All coordinate values are between 0 and 10000, inclusive.

You can assume that the polygon is simple, that is, its border never crosses or touches itself. As stated above, the given polygon is always a convex one.

The last dataset is followed by a line containing a single zero.

Output

For each dataset in the input, one line containing the distance of the most distant point from the sea should be output. An output line should not contain extra characters such as spaces. The answer should not have an error greater than 0.00001 (10^{-5}). You may output any number of digits after the decimal point, provided that the above accuracy condition is satisfied.

Sample Input	Sample Output
4	5000.000000
0 0	494.233641
10000 0	34.542948
10000 10000	0.353553
0 10000	
3	
0 0	
10000 0	

(*Continued*)

Sample Input	Sample Output
7000 1000	
6	
0 40	
100 20	
250 40	
250 70	
100 90	
0 70	
3	
0 0	
10000 10000	
5000 5001	
0	

Source: ACM Japan 2007

IDs for Online Judges: POJ 3525, UVA 3890

Hint

Given a convex polygon (a map of an island), the problem requires you to find the most distant point from the sides of the convex polygon, and report its distance from sides, that is, the radius for the largest circle in the convex polygon.

The problem can be solved by the intersection of half-planes and dichotomy. The algorithm is as follows:

Suppose line equations for n sides for the convex polygon are $A_i x + B_i y + C_i = 0$, where $A_i = y_{i+1} - y_i$, $B_i = x_{i+1} - x_i$, and $C_i = x_i \times y_{i+1} - x_{i+1} \times y_i$, $(1 \le i \le n-1)$; and $A_n = y_1 - y_n$, $B_n = x_1 - x_n$, and $C_n = x_n \times y_1 - x_1 \times y_n$.

Suppose the interval for the distance is $[l, r]$. Initially the interval is $[0, 20000]$. Dichotomy is used to calculate the most distance. Suppose $mid = \frac{l+r}{2}$. Sides for the convex polygon are pushed inward mid: For each line equation for a side for the convex polygon, $A_i x + B_i y + C_i = 0$, A_i and B_i aren't changed, and C_i is decreased $mid \times \sqrt{A_i^2 + B_i^2}$.

If there is an intersection of n half-planes, then the circle with radius mid can be in the convex polygon, and the right subinterval is searched ($l=mid$); else the left subinterval is searched ($r=mid$). Repeat the above process until $l=r$. And l is the distance.

8.5.27 Uyuw's Concert

Prince Remmarguts solved the chess puzzle successfully. As a reward, Uyuw planned to hold a concert in a huge piazza named after its great designer Ihsnayish.

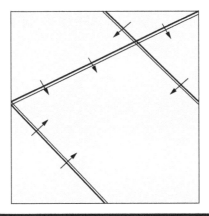

Figure 8.59

The piazza in United Delta of Freedom's (UDF) downtown was a square of [0, 10000]×[0, 10000]. Some basket chairs had been standing there for years, but in a terrible mess. Look at the graph in Figure 8.59.

In this case we have three chairs, and the audiences face the direction as the arrows in Figure 8.59 have pointed out. The chairs were old and too heavy to be moved. Princess Remmarguts told the piazza's current owner, Mr. Uw, to build a large stage inside it. The stage must be as large as possible, but he should also make sure the audience in every position of every chair would be able to see the stage without turning aside (that means the stage is in the forward direction of their own).

To make it simple, the stage could be set highly enough to make sure that even if thousands of chairs were in front of you, as long as you were facing the stage, you would be able to see the singer/pianist—Uyuw.

Being a mad idolater, can you tell them the maximal size of the stage?

Input

In the first line, there's a single non-negative integer N ($N \leq 20000$), denoting the number of basket chairs. Each of the following lines contains four floating numbers x_1, y_1, x_2, y_2, which means there's a basket chair on the line segment of (x_1, y_1)–(x_2, y_2), and facing to its LEFT (that a point (x, y) is at the LEFT side of this segment means that $(x-x_1) \times (y-y_2) - (x-x_2) \times (y-y_1) \geq 0$).

Output

Output a single floating number, rounded to one digit after the decimal point. This is the maximal area of the stage.

Sample Input	Sample Output
3 10000 10000 0 5000 10000 5000 5000 10000 0 5000 5000 0	54166666.7

Source: POJ Monthly, Zeyuan Zhu

ID for Online Judge: POJ 2451

Hint

Lines on which *n* basket chairs are can be regarded as *n* half-planes. There are four additional half-planes: $x=0$, $x=10000$, $y=0$, and $y=10000$. The intersection of $n+4$ half-planes constitute a polygon. The polygon is the stage. And the audience in every position of the sides of the polygon can see the stage.

8.5.28 Moth Eradication

Entomologists in the Northeast have set out traps to determine the influx of Jolliet moths into the area. They plan to study eradication programs that have some potential to control the spread of the moth population.

The study calls for organizing the traps in which moths have been caught into compact regions, which will then be used to test each eradication program. A region is defined as the polygon with the minimum length perimeter that can enclose all traps within that region. For example, the traps (represented by dots) of a particular region and its associated polygon are shown in Figure 8.60.

You must write a program that can take as input the locations of traps in a region and output the locations of traps that lie on the perimeter of the region as well as the length of the perimeter.

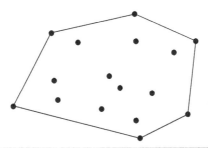

Figure 8.60

Input

The input file will contain records of data for several regions. The first line of each record contains the number (an integer) of traps for that region. Subsequent lines of the record contain two real numbers that are the *x*- and *y*-coordinates of the trap locations. Data within a single record will not be duplicated. End of input is indicated by a region with 0 traps.

Output

Output for a single region is displayed on at least three lines:

First line: The number of the region. (The first record corresponds to Region #1, the second to Region #2, etc.)

Next line(s): A listing of all the points that appear on the perimeter of the region. The points must be identified in the standard form "(*x*-coordinate, *y*-coordinate)" rounded to a single decimal place. The starting point for this listing is irrelevant, but the listing must be oriented *clockwise* and *begin and end with the same point*. For collinear points, any order which describes the minimum length perimeter is acceptable.

Last line: The length of the perimeter of the region rounded to 2 decimal places.

One blank line must separate the output from consecutive input records.

Sample Input	Sample Output
3	Region #1:
1 2	(1.0,2.0)–(4.0,10.0)–(5.0,12.3)–(1.0,2.0)
4 10	Perimeter length = 22.10
5 12.3	
6	Region #2:
0 0	(0.0,0.0)–(3.0,4.5)–(6.0,2.1)–(2.0,–3.2)–(0.0,0.0)
1 1	Perimeter length = 19.66
3.1 1.3	
3 4.5	Region #3:
6 2.1	(0.0,0.0)–(2.0,2.0)–(4.0,1.5)–(5.0,0.0)–(2.5,–1.5)–(0.0,0.0)
2 –3.2	Perimeter length = 12.52
7	
1 0.5	
5 0	
4 1.5	
3 –0.2	

(Continued)

Sample Input	Sample Output
2.5 −1.5 0 0 2 2 0	

Source: ACM World Finals 1992

ID for Online Judge: UVA 218

Hint

Given *n* points in the plane, the smallest perimeter polygon containing all of the given points is required to be found. It is a straightforward planar Convex Hull problem. An O(*n*log*n*) solution can solve it.

8.5.29 Bridge Across Islands

Thousands and thousands of years ago, there was a small kingdom located in the middle of the Pacific Ocean. The territory of the kingdom consists of two separated islands. Due to the impact of the ocean current, the shapes of both the islands became convex polygons. The king of the kingdom wanted to establish a bridge to connect the two islands. To minimize the cost, the king asked you, the bishop, to find the minimal distance between the boundaries of the two islands, as shown in Figure 8.61.

Input

The input consists of several test cases.

Each test case begins with two integers *N*, *M*. (3≤*N*, *M*≤10000)

Each of the next *N* lines contains a pair of coordinates, which describes the position of a vertex in one convex polygon.

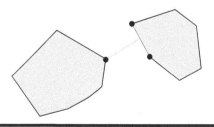

Figure 8.61

Each of the next M lines contains a pair of coordinates, which describes the position of a vertex in the other convex polygon.

A line with $N = M = 0$ indicates the end of input.

The coordinates are within the range [−10000, 10000].

Output

For each test case, output the minimal distance. An error within 0.001 is acceptable.

Sample Input	Sample Output
4 4	1.00000
0.00000 0.00000	
0.00000 1.00000	
1.00000 1.00000	
1.00000 0.00000	
2.00000 0.00000	
2.00000 1.00000	
3.00000 1.00000	
3.00000 0.00000	
0 0	

Source: POJ Founder Monthly Contest, 2008.06.29, Lei Tao

ID for Online Judge: POJ 3608

Hint

Suppose the first convex polygon is p_1, and the second convex polygon is p_2. The problem requires you to calculate the minimal distance between the two convex polygons.

Because p_1 and p_2 are separated, the algorithm for rotating calipers is used to solve the problem.

8.5.30 Useless Tile Packers

Yes, as you have guessed, the *Useless Tile Packers* (UTP) pack tiles. The tiles are of uniform thickness and have a simple polygonal shape. For each tile, a container is custom-built. The floor of the container is a convex polygon, and under this constraint, it has the minimum possible space inside to hold the tile it is built for. But this strategy leads to wasted space inside the container, as shown in Figure 8.62.

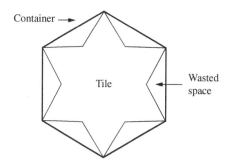

Figure 8.62

The UTP authorities are interested to know the percentage of wasted space for a given tile.

Input

The input file consists of several data blocks. Each data block describes one tile.

The first line of a data block contains an integer N ($3 \leq N \leq 100$) indicating the number of corner points of the tile. Each of the next N lines contains two integers giving the (x, y) coordinates of a corner point (determined using a suitable origin and orientation of the axes) where $0 \leq x, y \leq 1000$. Starting from the first point given in the input, the corner points occur in the same order on the boundary of the tile as they appear in the input. No three consecutive points are colinear.

The input file terminates with a value of 0 for N.

Output

For each tile in the input, output the percentage of wasted space rounded to two digits after the decimal point. Each output must be on a separate line. Print a blank line after each output block.

Sample Input	Sample Output
5	Tile #1
0 0	Wasted Space = 25.00 %
2 0	
2 2	Tile #2
1 1	Wasted Space = 0.00 %
0 2	
5	
0 0	
0 2	

Sample Input	Sample Output
1 3 2 2 2 0 0	

Source: BUET/UVA Occidental (WF Warmup) Contest 1, 2001

ID for Online Judge: UVA 10065

Hint

The floor of the container is a convex polygon, and under this constraint it has the minimum possible space inside to hold the tile it is built for. It is a straightforward Convex Hull problem.

The convex hull algorithm (choose one of them) is applied to calculate the area of the convex hull. Then the area that the points cover is calculated. Finally, the percentage of wasted space is calculated.

8.5.31 Nails

Arash is tired of working hard, so he wants to surround some nails on the wall of his room by a rubber ribbon to make fun of it! Now, he wants to know what will be the final length of the rubber ribbon after surrounding the nails. You must assume that the radius of nails and rubber ribbon is negligible.

Input

The first line of input gives the number of cases, *N*. *N* test cases will follow. Each test case starts with a line containing two integers, the initial length of rubber ribbon and the number of nails $0 < n \leq 100$, respectively. Each of the next *n* lines contains two integers denoting the location of a nail. There will be a blank line after each test case.

Output

Your program must output the final length of rubber ribbon precise to five decimal digits.

Sample Input	Sample Output
2	4.00000
2 4	5.00000
0 0	
0 1	
1 0	
1 1	
5 4	
0 0	
0 1	
1 0	
1 1	

Source: Annual Contest 2006 Qualification Round

ID for Online Judge: UVA 11096

Hint

The problem requires you to find the length of an elastic band around a set of nails on a 2-D surface. You need to find the convex hull and calculate its perimeter.

The initial length of the elastic is given; remember that it might be longer than the convex perimeter. Also remember to output in the correct format (to five decimal places).

8.5.32 Scrambled Polygon

A closed polygon is a figure bounded by a finite number of line segments. The intersections of the bounding line segments are called the vertices of the polygon. When one starts at any vertex of a closed polygon and traverses each bounding line segment exactly once, one comes back to the starting vertex.

A closed polygon is called convex if the line segment joining any two points of the polygon lies in the polygon. Figure 8.63 shows a closed polygon which is convex and one which is not convex. (Informally, a closed polygon is convex if its border doesn't have any "dents".)

The subject of this problem is a closed convex polygon in the coordinate plane, one of whose vertices is the origin ($x=0$, $y=0$). Figure 8.64 shows an example. Such a polygon will have two properties significant for this problem.

The first property is that the vertices of the polygon will be confined to three or fewer of the four quadrants of the coordinate plane. In the example shown in Figure 8.64, none of the vertices are in the second quadrant (where $x<0$, $y>0$).

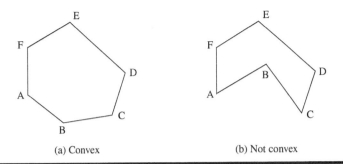

(a) Convex   (b) Not convex

Figure 8.63

To describe the second property, suppose you "take a trip" around the polygon: start at (0, 0), visit all other vertices exactly once, and arrive at (0, 0). As you visit each vertex (other than (0, 0)), draw the diagonal that connects the current vertex with (0, 0), and calculate the slope of this diagonal. Then, within each quadrant, the slopes of these diagonals will form a decreasing or increasing sequence of numbers, i.e., they will be sorted. Figure 8.65 illustrates this point.

Input

The input lists the vertices of a closed convex polygon in the plane. The number of lines in the input will be at least three but no more than 50. Each line contains the *x* and *y* coordinates of one vertex. Each *x* and *y* coordinate is an integer in the range −999..999. The vertex on the first line of the input file will be the origin, i.e.,

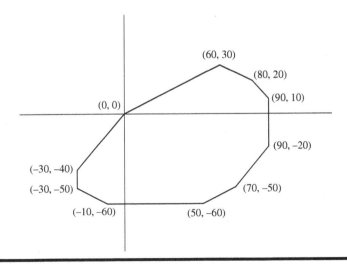

Figure 8.64

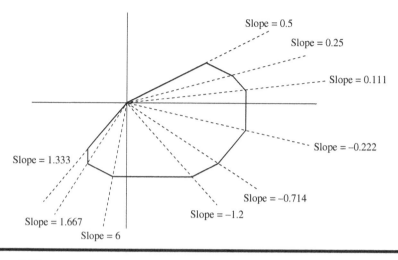

Figure 8.65

$x=0$ and $y=0$. Otherwise, the vertices may be in a scrambled order. Except for the origin, no vertex will be on the x-axis or the y-axis. No three vertices are colinear.

Output

The output lists the vertices of the given polygon, one vertex per line. Each vertex from the input appears exactly once in the output. The origin (0,0) is the vertex on the first line of the output. The order of vertices in the output will determine a trip taken along the polygon's border, in the counterclockwise direction. The output format for each vertex is (x, y) as shown below.

Sample Input	Sample Output
0 0	(0,0)
70 −50	(−30,−40)
60 30	(−30,−50)
−30, −50	(−10,−60)
80 20	(50,−60)
50 −60	(70,−50)
90 −20	(90,−20)
−30 −40	(90,10)
−10, −60	(80,20)
90 10	(60,30)

Source: ACM Rocky Mountain 2004

IDs for Online Judges: POJ 2007, ZOJ 2352, UVA 3052

Hint

The problem requires you to "take a trip" around the polygon: starting at (0, 0), visiting all other vertices exactly once in the counterclockwise direction, and finally arriving at (0, 0).

The problem is solved by sorting polar angles. A cross product is used to sort polar angles. The program segments are as follows.

```
double cross(point p0, point p1, point p2)
{
    return (p1.x-p0.x)×(p2.y-p0.y)-(p2.x-p0.x)×(p1.y-p0.y);
}
bool cmp(const point &a, const point &b)// sorting in the
counterclockwise direction
    {
        point origin;      // the origin
        origin.x = origin.y = 0;
        return cross(origin, b, a)<EPS;
    }
```

8.5.33 Grandpa's Estate

Being the only living descendant of his grandfather, Kamran the Believer inherited all of the grandpa's belongings. The most valuable one was a piece of convex polygon-shaped farm in the grandpa's birth village. The farm was originally separated from the neighboring farms by a thick rope hooked to some spikes (big nails) placed on the boundary of the polygon. But, when Kamran went to visit his farm, he noticed that the rope and some spikes are missing. Your task is to write a program to help Kamran decide whether the boundary of his farm can be exactly determined only by the remaining spikes.

Input

The first line of the input file contains a single integer t ($1 \le t \le 10$), the number of test cases, followed by the input data for each test case. The first line of each test case contains an integer n ($1 \le n \le 1000$), which is the number of remaining spikes. Next, there are n lines, one line per spike, each containing a pair of integers, which are the x and y coordinates of the spike.

Output

There should be one output line per test case containing "YES" or "NO" depending on whether the boundary of the farm can be uniquely determined from the input.

Sample Input	Sample Output
1	NO
6	
0 0	
1 2	
3 4	
2 0	
2 4	
5 0	

Source: ACM Tehran 2002 Preliminary

ID for Online Judge: POJ 1228, ZOJ 1377

 Hint

Given a set of points, these points are on the boundary of the convex polygon-shaped farm. The problem requires you to determine whether the convex hull is a stable convex hull. A convex hull isn't stable if a larger convex polygon can be gotten by adding some points, and the larger convex polygon' sides contains the given set of points. Therefore, if a convex hull is stable, then there are at least three points on each side. If there are only two points on a side, a larger convex polygon can be gotten by adding a point.

The algorithm is as follows. First, a convex hull is calculated for the set of spikes. If the number of spikes is less than six, the boundary of the farm can't be determined. Second, if there are at least three spikes on each side for the convex hull, the boundary of the farm can be determined; else the boundary of the farm can't be determined.

8.5.34 The Fortified Forest

Once upon a time, in a faraway land, there lived a king. This king owned a small collection of rare and valuable trees, which had been gathered by his ancestors on their travels. To protect his trees from thieves, the king ordered that a high fence be built around them. His wizard was put in charge of the operation.

Alas, the wizard quickly noticed that the only suitable material available to build the fence was the wood from the trees themselves. In other words, it was necessary to cut down some trees in order to build a fence around the remaining trees. Of course, to prevent his head from being chopped off, the wizard wanted to minimize the value of the trees that had to be cut. The wizard went to his tower and stayed there until he had found the best possible solution to the problem. The fence was then built and everyone lived happily ever after.

You are to write a program that solves the problem the wizard faced.

Input

The input contains several test cases, each of which describes a hypothetical forest. Each test case begins with a line containing a single integer n, $2 \leq n \leq 15$, the number of trees in the forest. The trees are identified by consecutive integers 1 to n. Each of the subsequent n lines contains four integers x_i, y_i, v_i, l_i that describe a single tree. (x_i, y_i) is the position of the tree in the plane, v_i is its value, and l_i is the length of fence that can be built using the wood of the tree. v_i and l_i are between 0 and 10,000.

The input ends with an empty test case ($n=0$).

Output

For each test case, compute a subset of the trees such that, using the wood from that subset, the remaining trees can be enclosed in a single fence. Find the subset with minimum value. If more than one such minimum-value subset exists, choose the one with the smallest number of trees. For simplicity, regard the trees as having zero diameter.

Display, as shown below, the test case numbers (1, 2, ...), the identity of each tree to be cut, and the length of the excess fencing (accurate to two fractional digits).

Display a blank line between test cases.

Sample Input	Sample Output
6	Forest 1
0 0 8 3	Cut these trees: 2 4 5
1 4 3 2	Extra wood: 3.16
2 1 7 1	
4 1 2 3	Forest 2
3 5 4 6	Cut these trees: 2
2 3 9 8	Extra wood: 15.00
3	
3 0 10 2	
5 5 20 25	
7 -3 30 32	
0	

Source: ACM World Finals 1999

IDs for Online Judges: POJ 1873, UVA 811

Hint

The problem requires you to compute such a subset of the trees that, using the wood from that subset, the remaining trees can be enclosed in a single fence; and the subset must be with minimum value.

The fence built around the remaining trees is the convex hull containing all remaining trees. The search with state compression is used to find which trees will be cut, and which trees will remain. A binary number i with n digits is used to represent n trees, $0 \leq i \leq 2^n - 1$, where a digit being 0 means that the corresponding tree will be cut, and a digit being 1 means that the corresponding tree will remain. Suppose P_k is the k-th tree's position, $1 \leq k \leq n$; $pt[]$ is used to store remaining trees (i.e., sequence numbers for digits being 1 in $i+1$), and the number of remaining trees is tt; sums of cut trees' values and lengths are *valu* and *len* respectively; the set of points of the convex hull for $pt[]$ is h, and the perimeter for the convex hull is ll; *ans* is the sum of the current cut trees' values, *anst* is the number of cut trees, *ansi* is the state for trees, and *lef* is the length of the excess fencing.

The algorithm is as follows:

All states i for trees are enumerated, $(0 \leq i \leq 2^n - 1)$:

1. In state i, remaining trees are stored in $pt[]$. Then the sums of cut trees' values and lengths as *valu* and *len* are calculated;
2. The convex hull for remaining trees $pt[]$ and the perimeter for the convex hull ll is calculated;
3. If the sum of the cut trees' lengths can enclose remaining trees ($ll \leq len$), the current best solution should be adjusted:

```
If (valu<ans), then ans=valu, anst=n-tt, ansi=i, and the
length of the excess fencing is calculated (lef=len-ll);
   If (valu==ans) and (n-tt<anst), then anst=n-tt, ansi=i, and
the length of the excess fencing is calculated (lef=len-ll).
```

Finally, output the result.

8.5.35 *The Picnic*

The annual picnic of the Zeron company will take place tomorrow. This year they have agreed on the Gloomwood Park as the place to be. The girl responsible for the arrangement, Lilith, thinks it would be nice if everyone is able to watch everyone else during the occasion. From geometry class, she remembers that a region in the plane with the property that a straight line between any two points in the region, lies entirely in the region, is called convex. So that is what she is looking for.

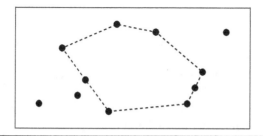

Figure 8.66

Unfortunately, this seems hard to fulfill, since Gloomwood Park has many opaque obstacles, such as large trees, rocks, and so on.

Because the staff of the Zeron company is large, Lilith has a rather intricate problem to solve: finding a location to hold them all. Therefore, some of her friends help her to draw a map of the whereabouts of the largest obstacles. To mark out the place, she will use a ribbon stretched around the obstacles on the circumference of the chosen region. The opaque obstacles should be thought of as points of zero extension.

Figure 8.66 shows the Gloomwood Park from above with black dots representing obstacles. The picnic area is the region whose circumference is dashed.

Input

The first line of the input contains a single positive integer n, specifying the number of test scenarios to follow. Each test scenario begins with a line containing an integer m, the number of obstacles in the park ($2<m<100$). The next line contains the coordinates of the m obstacles, in the order $x_1\, y_1\, x_2\, y_2\, x_3\, y_3 \ldots \ldots$ All coordinates are integers in the range [0, 1000]. Each scenario has at least three obstacles that are not on a straight line, and no two obstacles have the same coordinates.

Output

For each test scenario, one line of output should be generated, stating the area with one decimal of the largest convex polygon having obstacles as corners, but no enclosed obstacles.

Sample Input	Sample Output
1	129.0
11	
3 3 8 4 12 2 22 3 23 5 24 7 27 12 18 12 13 13 6 10 9 6	

Source: ACM Northwestern Europe 2002

IDs for Online Judges: POJ 1259, ZOJ 1562, UVA 2674

Hint

Given a set of vertices, the problem requires you to calculate the largest convex polygon whose vertices are a subset of the set of vertices, and in which there is no vertex. Suppose the given set of vertices is $\{p_i|0\leq i\leq n-1\}$; and $f[j][k]$ is the area of the largest convex polygon having vertex k and vertex j.

Each vertex p_i in the convex hull is enumerated:

For the convex hull, vertex p_i is as the bottom vertex, and $tp[0...m-1]$ is the sequence for vertices in counterclockwise direction.

In $tp[]$ all intervals $[k, j]$ $(0\leq k\leq j-1, 1\leq j\leq m)$ are enumerated:

1. If $p_{k+1}...p_{j-1}$ aren't on the inside of the convex hull having vertex p_i, vertex $tp[k]$ and $tp[j]$ $(((\text{mul}(tp[k], tp[j], tp[l])\leq 0)\|(\text{mul}(p[i], tp[k], tp[l])=0))$, $k+1\leq l\leq j-1)$, then the area $f[j][k]=\dfrac{\overline{p_i tp_k} \wedge \overline{p_i tp_j}}{2}$ is calculated.

2. If $p_{k+1}...p_{j-1}$ are on the right of $\overline{tp_k tp_j}$ and tp_l, tp_k, and tp_j are sorted in counterclockwise direction, then neither the convex polygon having vertices p_i, tp_k, and tp_j $\left(\text{its area is } s1=\dfrac{\overline{p_i tp_j} \wedge \overline{p_i tp_k}}{2}\right)$ nor the convex polygon having vertices p_i, tp_k, and tp_l (its area is $s2=f[k][l]$) contain $p_{k+1}... p_{j-1}$, then $f[j][k]=max\{f[j][k], s1+s2\}$.

Then the area *ans* is adjusted: $ans=max\{f[j][k], ans\}$.

Finally, *ans* is the area of the largest convex polygon in which there are no vertices.

8.5.36 Triangle

Given n distinct points on a plane, your task is to find the triangle that has the maximum area, whose vertices are from the given points.

Input

The input consists of several test cases. The first line of each test case contains an integer n, indicating the number of points on the plane. Each of the following n lines contains two integers x_i and y_i, indicating the i-th points. The last line of the input is an integer -1, indicating the end of input, which should not be processed. You may assume that $1\leq n\leq 50000$ and $-10^4\leq x_i,y_i\leq 10^4$ for all $i=1 \ldots n$.

Output

For each test case, print a line containing the maximum area, which contains two digits after the decimal point. You may assume that there is always an answer which is greater than zero.

Sample Input	Sample Output
3	0.50
3 4	27.00
2 6	
2 7	
5	
2 6	
3 9	
2 0	
8 0	
6 5	
−1	

Source: ACM Shanghai 2004 Preliminary

IDs for Online Judges: POJ 2079, ZOJ 2419

Hint

First, the set of n vertices for the convex hull based on the given set of points are calculated $\{p_0, p_1, \ldots, p_{n-1}\}$. Obviously, vertices for the triangle with the maximum area are vertices for the convex hull.

Each vertex p_i is enumerated, $0 \le i \le n-1$:

The other two vertices p_k and p_j for the triangle are calculated as follows:.

Initially k is $(i+1)\%n$;

The length $\_j$ of $i-j$ is enumerated, and j is calculated (for (int $\_j=1, j=(\_j+i)\%n$; $\_j<n-1$; $\_j++, j=(\_j+i)\% n$))), and p_k is calculated based on p_i and p_j: k is calculated by "rotating" in counterclockwise direction until $\overrightarrow{p_j p_i} \wedge \overrightarrow{p_{(k+1)\%n} p_k} \le 0$. The area for the triangle whose vertices p_i, p_j, and p_k $S\Delta_{p_i p_j p_k} = \dfrac{\left| \overrightarrow{p_i p_j} \wedge \overrightarrow{p_i p_k} \right|}{2}$. The maximum area for the triangle is adjusted $ans = \max\{ans, S\Delta_{p_i p_j p_k}\}$.

Finally, ans is the maximum area for the triangle.

8.5.37 Smallest Bounding Rectangle

Given the Cartesian coordinates of n (>0) two-dimensional points, write a program that computes the area of their smallest bounding rectangle (smallest rectangle containing all the given points).

Input

The input file may contain multiple test cases. Each test case begins with a line containing a positive integer n (<1001) indicating the number of points in this test case. Then follow n lines, each containing two real numbers giving respectively the x- and y-coordinates of a point. The input terminates with a test case containing a value 0 for n which must not be processed.

Output

For each test case in the input, print a line containing the area of the smallest bounding rectangle rounded to the fourth digit after the decimal point.

Sample Input	Sample Output
3	80.0000
−3.000 5.000	100.0000
7.000 9.000	
17.000 5.000	
4	
10.000 10.000	
10.000 20.000	
20.000 20.000	
20.000 10.000	
0	

Source: 2001 Regionals Warmup Contest

ID for Online Judge: UVA 10173

Hint

The problem requires you to calculate the area of the smallest rectangle containing all given points. First, the convex hull containing all given points is calculated. Then the method of rotating calipers is used to calculate the area of the smallest rectangle containing all given points:

The rightmost point and the leftmost point are calculated to guarantee the minimal width covering all points;

The lowest point and the highest point are calculated to guarantee the minimal height covering all points;

The area of the smallest rectangle containing all given points is calculated through adjustment in the procedure of rotating calipers.

8.5.38 Exocenter of a Triangle

Given a triangle ABC, the Extriangles of ABC are constructed as follows:

On each side of ABC, construct a square (ABDE, BCHJ, and ACFG in Figure 8.67).

Connect adjacent square corners to form the three Extriangles (AGD, BEJ, and CFH in Figure 8.67).

The Exomedians of ABC are the medians of the Extriangles, which pass through vertices of the original triangle, extended into the original triangle (LAO, MBO, and NCO in Figure 8.67). As the figure indicates, the three Exomedians intersect at a common point called the Exocenter (point O in Figure 8.67).

This problem is to write a program to compute the Exocenters of triangles.

Input

The first line of the input consists of a positive integer n, which is the number of datasets that follow. Each dataset consists of three lines; each line contains two floating-point values which represent the (two-dimensional) coordinate of one vertex of a triangle. So, there are a total of $(n \times 3)+1$ lines of input. Note: All input

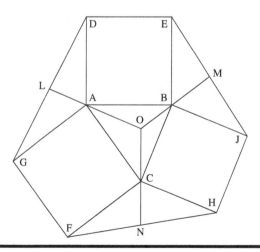

Figure 8.67

triangles will be strongly non-degenerate in that no vertex will be within one unit of the line through the other two vertices.

Output

For each dataset, you must print out the coordinates of the Exocenter of the input triangle correct to four decimal places.

Sample Input	Sample Output
2	9.0000 3.7500
0.0 0.0	−48.0400 23.3600
9.0 12.0	
14.0 0.0	
3.0 4.0	
13.0 19.0	
2.0 −10.0	

Source: ACM Greater New York 2003

IDs for Online Judges: POJ 1673, ZOJ 1821, UVA 2873

Hint

The problem is solved based on the definition of exocenters of triangles.

Let $\overrightarrow{p_1a}$ be a vertical line through p_1 for $\overrightarrow{p_2p_3}$, and the intersection point for $\overrightarrow{p_1a}$ and $\overrightarrow{p_2p_3}$ is a. And let $\overrightarrow{p_2b}$ be a vertical line through p_2 for $\overrightarrow{p_1p_3}$, and the intersection point for $\overrightarrow{p_2b}$ and $\overrightarrow{p_1p_3}$ is b. The intersection point o for $\overrightarrow{p_1p_a}$ and $\overrightarrow{p_2p_b}$ (the orthocenter of a triangle) is the exocenter of the triangle (Figure 8.68).

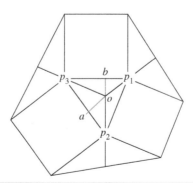

Figure 8.68

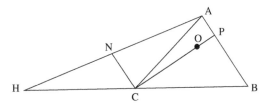

Figure 8.69

For a triangle, its exocenter is its orthocenter. The proof is as follows.

In Figure 8.67, ΔFCN is rotated clockwise 90°. Line segment AC and line segment CF coincide. Line segment OC is lengthened. And the intersection point of the lengthened line and AB is point P (Figure 8.69).

Because BC=CH and AN=NH, CN∥AB. Because ∠NCP=90°, ∠APC=90°.

Similarly for the other two sides, we can prove, for a triangle, that its exocenter is its orthocenter.

8.5.39 Picture

A number of rectangular posters, photographs, and other pictures of the same shape are pasted on a wall. Their sides are all vertical or horizontal. Each rectangle can be partially or totally covered by the others. The length of the boundary of the union of all rectangles is called the perimeter.

Write a program to calculate the perimeter. An example with seven rectangles is shown in Figure 8.70.

The corresponding boundary is the whole set of line segments drawn in Figure 8.71.

The vertices of all rectangles have integer coordinates.

Input

Your program is to read from standard input. The first line contains the number of rectangles pasted on the wall. In each of the subsequent lines, one can find the integer

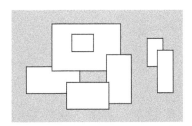

Figure 8.70

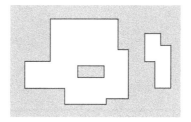

Figure 8.71

coordinates of the lower-left vertex and the upper-right vertex of each rectangle. The values of those coordinates are given as ordered pairs consisting of an *x*-coordinate followed by a *y*-coordinate. 0≤number of rectangles<5000. All coordinates are in the range [−10000,10000], and any existing rectangle has a positive area.

Output

Your program is to write to standard output. The output must contain a single line with a non-negative integer which corresponds to the perimeter for the input rectangles.

Sample Input	Sample Output
7	228
−15 0 5 10	
−5 8 20 25	
15 −4 24 14	
0 −6 16 4	
2 15 10 22	
30 10 36 20	
34 0 40 16	

Source: IOI 1998

ID for Online Judge: POJ 1177

 Hint

The problem is solved by the Sweep Line Algorithm.

First, discretization is on the *X*-axis. The plane is divided into several horizontal strips by sweeping on the *Y*-axis. A segment tree is used to accumulate lengths of these horizontal strips s_1.

Second, discretization is on the *Y*-axis. The plane is divided into several vertical strips by sweeping on the *X*-axis. A segment tree is used to accumulate lengths of these vertical strips s_2.

Obviously, the result is $s_1 + s_2$.

8.5.40 Fill the Cisterns! (Water Shortage)

During the next century, certain regions on earth will experience severe water shortages. The old town of Uqbar has already started to prepare itself for the worst. Recently they created a network of pipes connecting the cisterns that distribute water in each neighborhood, making it easier to fill them at once from a single source of water. But in case of a water shortage, the cisterns above a certain level will be empty since the water will go to the cisterns below, as shown in Figure 8.72.

You have been asked to write a program to compute the level to which cisterns will be filled with a certain volume of water, given the dimensions and position of each cistern. To simplify, we will neglect the volume of water in the pipes.

Write a program which for each data set:

reads the description of cisterns and the volume of water, computes the level to which the cisterns will be filled with the given amount of water, writes the result.

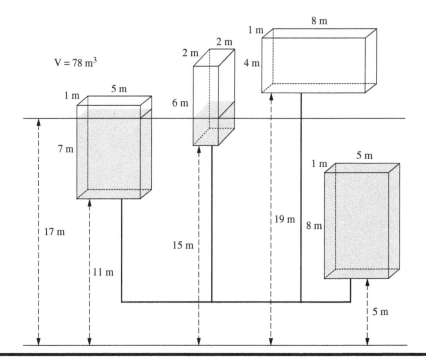

Figure 8.72

Input

The first line of the input contains the number of data sets k, $1 \le k \le 30$. The data sets follow.

The first line of each data set contains one integer n, the number of cisterns, $1 \le n \le 50000$. Each of the following n lines consists of four non-negative integers, separated by single spaces: b, h, w, d—the base level of the cistern, and its height, width, and depth in meters, respectively. The integers satisfy $0 \le b \le 10^6$ and $1 \le h \times w \times d \le 40000$. The last line of the data set contains an integer V—the volume of water in cubic meters to be injected into the network. Integer V satisfies $1 \le V \le 2 \times 10^9$.

Output

The output should consist of exactly d lines, one line for each data set.

Line i, $1 \le i \le d$, should contain the level that the water will reach, in meters, rounded up to two fractional digits, or the word "OVERFLOW", if the volume of water exceeds the total capacity of the cisterns.

Sample Input	Sample Output
3	1.00
2	OVERFLOW
0 1 1 1	17.00
2 1 1 1	
1	
4	
11 7 5 1	
15 6 2 2	
5 8 5 1	
19 4 8 1	
132	
4	
11 7 5 1	
15 6 2 2	
5 8 5 1	
19 4 8 1	
78	

Note: Descriptions for Problem F Fill the Cisterns! in ACM Central Europe 2001 and Problem D Water Shortage are similar.

Source: ACM Central Europe 2001, ACM Southwestern Europe 2001

IDs for Online Judges: POJ 1434, ZOJ 1389, UVA 2428

Hint

For the i-th cistern, its base level, height, width, and depth are b_i, h_i, w_i, and d_i, respectively, $0 \leq i \leq n-1$.

First, we need to calculate how much water is in the n cisterns if the height of the water level is m. Suppose the amount of water is v_m.

For the i-th cistern ($0 \leq i \leq n-1$), if its base level $b_i \leq$ the water level m, then the water level for the i-th cistern $tmp=min\{ m-b_i, h_i\}$, and $v_m += tmp \times w_i \times d_i$.

Dichotomy is used to calculate the level that the water will reach. Suppose the interval for the water level is [l, r]. Initially, the interval for the water level is [0, ∞]. Let $mid=(l+r)/2$, and calculate the amount of filled water v_{mid} that makes the water level to be mid. If $v_{mid} \geq V$ (the volume of water in cubic meters to be injected into the network), then the left subinterval is searched; else the right subinterval is searched. Repeat the procedure until the search ends and l is the level that the water will reach.

8.5.41 Area of Simple Polygons

There are N, $1 \leq N \leq 1,000$ rectangles in the 2-D xy plane. The four sides of a rectangle are horizontal or vertical line segments. Rectangles are defined by their lower-left and upper-right corner points. Each corner point is a pair of two non-negative integers in the range of 0 through 50,000 indicating its x and y coordinates.

Assume that the contour of their union is defined by a set S of segments. We can use a subset of S to construct simple polygon(s). Please report the total area of the polygon(s) constructed by the subset of S. The area should be as large as possible. In a 2-D xy plane, a polygon is defined by a finite set of segments such that every segment extreme (or endpoint) is shared by exactly two edges, and no subsets of edges have the same property. The segments are edges and their extremes are the vertices of the polygon. A polygon is simple if there is no pair of nonconsecutive edges sharing a point.

Example: Consider the following three rectangles:

rectangle 1: <(0, 0) (4, 4)>,
rectangle 2: <(1, 1) (5, 2)>,
rectangle 3: <(1, 1) (2, 5)>.

The total area of all simple polygons constructed by these rectangles is 18.

Input

The input consists of multiple test cases. A line of four −1's separates each test case. An extra line of four −1's marks the end of the input. In each test case, the rectangles are given one by one in a line. In each line for a rectangle, four non-negative integers are given. The first two are the x and y coordinates of the lower-left corner. The next two are the x and y coordinates of the upper-right corner.

Output

For each test case, output the total area of all simple polygons in a line.

Sample Input	Sample Output
0 0 4 4	18
1 1 5 2	10
1 1 2 5	
−1 −1 −1 −1	
0 0 2 2	
1 1 3 3	
2 2 4 4	
−1 −1 −1 −1	
−1 −1 −1 −1	

Source: ACM Taiwan 2001

IDs for Online Judges: POJ 1389, UVA 2447

 Hint

First, the convex hull containing all rectangles' points are calculated. Then the method of rotating calipers is used to calculate the total area of all simple polygons.

8.5.42 Squares

A square is a four-sided polygon whose sides have equal length and adjacent sides form 90-degree angles. It is also a polygon such that rotating about its center by 90 degrees gives the same polygon. It is not the only polygon with the latter property, however, as a regular octagon also has this property.

So we all know what a square looks like, but can we find all possible squares that can be formed from a set of stars in a night sky? To make the problem easier, we will assume that the night sky is a two-dimensional plane, and each star is specified by its x and y coordinates.

Input

The input consists of a number of test cases. Each test case starts with the integer *n* (1≤*n*≤1000) indicating the number of points to follow. Each of the next *n* lines specify the *x* and *y* coordinates (two integers) of each point. You may assume that the points are distinct and the magnitudes of the coordinates are less than 20000. The input is terminated when *n*=0.

Output

For each test case, print on a line the number of squares one can form from the given stars.

Sample Input	Sample Output
4	1
1 0	6
0 1	1
1 1	
0 0	
9	
0 0	
1 0	
2 0	
0 2	
1 2	
2 2	
0 1	
1 1	
2 1	
4	
−2 5	
3 7	
0 0	
5 2	
0	

Source: ACM Rocky Mountain 2004

IDs for Online Judges: POJ 2002, ZOJ 2347, UVA 3047

Hint

Suppose *m* is the container storing coordinates for all given stars; where the *i*-th star's coordinate is (a_i, b_i), 0≤*i*≤*n*−1. Initially *ans*=0.

After the i-th star's coordinate (a_i, b_i) is input, the first $i-1$ stars' coordinates (a_j, b_j), $0 \leq j \leq i-1$, are enumerated:

If $(a_i+b_i-b_j, b_i+a_j-a_i)$ and $(b_i+a_j-b_j, a_j+b_j-a_i)$ are in the container m, then ans++;
If $(a_i+b_j-b_i, b_i-a_j+a_i)$ and $(a_j+b_j-b_i, a_i+b_j-a_j)$ are in the container m, then ans++;

Finally, the number of squares one can form from the given stars is $\dfrac{ans}{2}$.

8.5.43 That Nice Euler Circuit

Little Joey invented a Scrabble machine that he called Euler, after the great mathematician. In his primary school, Joey heard about the nice story of how Euler started the study about graphs. The problem in that story was—let me remind you—to draw a graph on a paper without lifting your pen, and finally return to the original position. Euler proved that you could do this if and only if the (planar) graph you created has the following two properties: (1) The graph is connected; and (2) Every vertex in the graph has even degree.

Joey's Euler machine works exactly like this. The device consists of a pencil touching the paper, and a control center issuing a sequence of instructions. The paper can be viewed as the infinite two-dimensional plane; that means you do not need to worry whether the pencil will ever go off the boundary.

In the beginning, the Euler machine will issue an instruction of the form (X_0, Y_0) which moves the pencil to some starting position (X_0, Y_0). Each subsequent instruction is also of the form (X', Y'), which means to move the pencil from the previous position to the new position (X', Y'), thus drawing a line segment on the paper. You can be sure that the new position is different from the previous position for each instruction. At last, the Euler machine will always issue an instruction that moves the pencil back to the starting position (X_0, Y_0). In addition, the Euler machine will definitely not draw any lines that overlay other lines already drawn. However, the lines may intersect.

After all the instructions are issued, there will be a nice picture on Joey's paper. You see, since the pencil is never lifted from the paper, the picture can be viewed as an Euler circuit.

Your job is to count how many pieces (connected areas) are created on the paper by those lines drawn by Euler.

Input

There are no more than 25 test cases. Each case starts with a line containing an integer $N \geq 4$, which is the number of instructions in the test case. The following N pairs of integers give the instructions and appear on a single line separated by single spaces. The first pair is the first instruction that gives the coordinates of the starting position. You may assume there are no more than 300 instructions in each test case, and all the integer coordinates are in the range (−300, 300). The input is terminated when *N* is 0.

Output

For each test case, there will be one output line in the format:

Case *x*: There are *w* pieces.
where *x* is the serial number starting from 1.

Note: Figure 8.73 illustrates the two sample input cases.

Sample Input	Sample Output
5 0 0 0 1 1 1 1 0 0 0 7 1 1 1 5 2 1 2 5 5 1 3 5 1 1 0	Case 1: There are 2 pieces. Case 2: There are 5 pieces.

Source: ACM Shanghai 2004

IDs for Online Judges: POJ 2284, ZOJ 2394, UVA 3263

Hint

The problem is solved by Euler's formula: for any convex polyhedron, the number of vertices and faces together is exactly two more than the number of edges. That is, $V-E+F=2$.

First, the number of vertices V is calculated. We calculate coordinates for intersection points, sort all points, and eliminate recurring points.

Then the number of edges E is calculated. Initially, E is the number of input edges ($N-1$). Then, for each vertex v, if v is on a line segment and isn't an endpoint for the line segment, E++.

Finally, the number of faces F is calculated and output.

Figure 8.73

8.5.44 Can't Cut Down the Forest for the Trees

Once upon a time, in a country far away, there was a king who owned a forest of valuable trees. One day, to deal with a cash flow problem, the king decided to cut down and sell some of his trees. He asked his wizard to find the largest number of trees that could be safely cut down.

All the king's trees stood within a rectangular fence, to protect them from thieves and vandals. Cutting down the trees was difficult, since each tree needed room to fall without hitting and damaging other trees or the fence. Each tree could be trimmed of branches before it was cut. For simplicity, the wizard assumed that when each tree was cut down, it would occupy a rectangular space on the ground, as shown in Figure 8.74. One of the sides of the rectangle is a diameter of the original base of the tree. The other dimension of the rectangle is equal to the height of the tree.

Many of the king's trees were located near other trees (that being one of the tell-tale signs of a forest.) The wizard needed to find the maximum number of trees that could be cut down, one after another, in such a way that no fallen tree would touch any other tree or the fence. As soon as each tree falls, it is cut into pieces and carried away so it does not interfere with the next tree to be cut.

Input

The input consists of several test cases each describing a forest. The first line of each description contains five integers, *xmin, ymin, xmax, ymax,* and *n*. The first four numbers represent the minimal and maximal coordinates of the fence in the *x*- and *y*-directions (*xmin<xmax, ymin<ymax*). The fence is rectangular and its sides are parallel to the coordinate axes. The fifth number *n* represents the number of trees in the forest (1≤*n*≤100).

The next *n* lines describe the positions and dimensions of the *n* trees. Each line contains four integers, *xi, yi, di,* and *hi*, representing the position of the

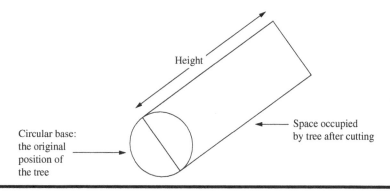

Figure 8.74

tree's center (*xi, yi*), its base diameter *di*, and its height *hi*. No tree bases touch each other, and all the trees are entirely inside the fence, not touching the fence at all.

The input is terminated by a test case with $x_{min}=y_{min}=x_{max}=y_{max}=n=0$. This test case should not be processed.

Output

For each test case, first print its number. Then print the maximum number of trees that can be cut down, one after another, such that no fallen tree touches any other tree or the fence. Follow the format in the sample output given below. Print a blank line after each test case.

Sample Input	Sample Output
0 0 10 10 3 3 3 2 10 5 5 3 1 2 8 3 9 0 0 0 0 0	Forest 1 2 tree(s) can be cut

Source: ACM World Finals 2001

ID for Online Judge: UVA 2235

Hint

The polar angle for the central axis of a rectangle is used to represent the state that a tree is cut down, as shown in Figure 8.75.

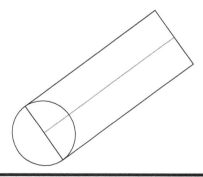

Figure 8.75

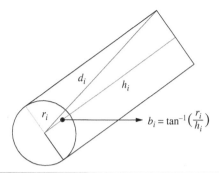

Figure 8.76

Other trees or the fence may prevent a tree from falling. Because the number of trees in the forest $n \leq 100$, each tree is enumerated. We calculate the range for the polar angle that the tree can't be cut down.

In Figure 8.76, for the i-th tree, its radius of the original base of the tree is r_i, the length of its central axis is h_i, the length for the line segment from the centre of the circle to the another endpoint for the rectangle is d_i, and the included angle for the line segment and its central axis is b_i. States that trees or the fence prevent a tree from falling are as follows.

There are two states that the fence can prevent a tree from falling.

Case 1: The distance between the center of a circle and the fence is in $[0, h_i]$.

The range for the included angle for two dotted lines (the polar angle for the central axis of a rectangle) is $\left[mid - \left(\cos^{-1}\left(\frac{d_i}{dist} \right) + b_i \right), mid + \left(\cos^{-1}\left(\frac{d_i}{dist} \right) + b_i \right) \right]$ (as shown in Figure 8.77), where the distance between the center of a circle and the fence is $dist$, and the polar angle for the vertical line is mid.

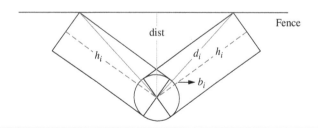

Figure 8.77

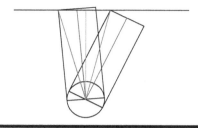

Figure 8.78

Case 2: The distance between the center of a circle and the fence is $[h_i, d_i]$.
The rectangle (tree) can also be perpendicular to the fence. When the tree is rotated, it can touch the fence (Figure 8.78). Then the polar angle for the central axis of a rectangle is $\left[mid - \cos^{-1}\left(\frac{d_i}{dist}\right) + b_i, mid + \cos^{-1}\left(\frac{d_i}{dist}\right) + b_i \right]$.

For tree j, there are two cases that tree j prevents tree i fall.
Case 1: It is similar to the above Case 1 (Figure 8.79):

Suppose the distance between centers of two circles is *dist*, and the polar angle for j with respect to i is *mid*. If the height of the tree exceeds $\sqrt{dist^2 + (r_i + r_j)^2}$, then $h_i = \sqrt{dist^2 + (r_i + r_j)^2}$. The range for the polar angle for the central axis of a rectangle that will prevent the tree from falling is $\left[mid - \cos^{-1}\left(\frac{d_i}{dist}\right) + b_i, mid + \cos^{-1}\left(\frac{d_i}{dist}\right) + b_i \right]$.

Case 2: It is similar to the above Case 2.

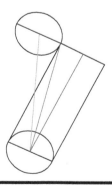

Figure 8.79

Chapter 9

Practice for State Space Search

Search technologies are fundamental technologies in computer science and technology.

In data structure, search spaces are static, and a search algorithm is used to find items with specified properties among a collection of items. There are three kinds of static search methods for data structure: Sequential Search, Binary Search, and Binary Search Trees (BST).

Sometimes search spaces are dynamic. Searched objects (also called states) are generated during the search.

Classical algorithms for trees and graphs are based on explicit graph models and tree models. But sometimes graph models and tree models are implicit.

In this chapter, we're back to the starting point: how to represent a search problem? And practices for state space search are shown.

A search problem can be represented as a state space. And a state space can be represented as an implicit tree or an implicit graph. States are represented as vertices. There are operations that lead from one state to other states. The goal for a search problem is to find a path from an initial state to a set of goal states. During the search, a search tree or a search graph is generated in the state space, called a search space. That is, a search space is a part of a state space.

DFS and BFS are the most widely used dynamic search algorithms.

A search problem can be analyzed from different viewpoints:

1. **The state space.**

   Is the state space limited or unlimited?

   Is the state space static or generated dynamically? For example, in AI, searched objects (states) are generated during the search.

2. **The search goal.**

Is the search goal clear or unclear in the state space? For example, in a game of chess, the search goal is unclear.

Does the problem require you to calculate the goal and/or the paths to the goal?

3. **Search.**

Are there any constraints or not for the search? For example, in the Eight Queens' problem, there are constraints among queens, and backtracking is used to find solutions.

Is the search data-driven or goal-driven? Data-driven search is also called forward search. Goal states are searched from current states. Conversely, the search is a goal-driven search, also called backward search.

Is the search unidirectional search or bidirectional search? If only data-driven search or only goal-driven search is used, the search is unidirectional search. And if data-driven search and goal-driven search are used together, the search is bidirectional search.

Is the search a game search (i.e., there is an opponent) or not? A two-person zero-sum game is a game search, such as Weiqi, Chinese Chess, Chess, and so on.

Is the search a blind search or a heuristic search? Heuristic search is using problem-specific knowledge to find solutions.

9.1 Constructing a State Space Tree

A state space consists of a set of states and a set of operations. A state is a situation for a problem. A state can be a situation in a game of chess, or a situation that cars move, stop, or turn on a road, and so on. For a problem, there is one initial state or more than one initial state. An operation is applied to a state of the problem to get a new state. The relationship between states can be discrete, such as a game of chess; or continuous, such as cars on a road. In a game of chess, a chesspiece can be moved into another square to change the current state. On the road, cars can move, stop, or turn to change the current state. Operations applied to states can be represented as a successor function. If there is only one initial state, the state space is represented as a tree, called a state space tree. And if there are more than one initial state, the state space is represented as a graph, called a state space graph.

If there is only one initial state, the state space search is to find a path from an initial state to a set of goal states.

There are costs for transformations from a state to other states.

Figure 9.1 shows a state space tree.

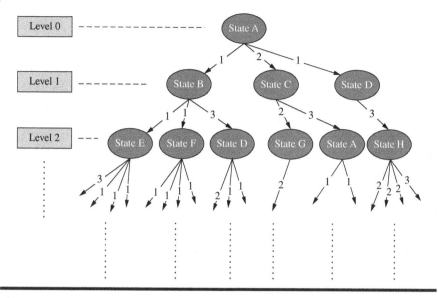

Figure 9.1

Therefore, to solve a problem of state space search, we need to define states, a successor function, costs, and a state space. For example, in a game of chess:

States: Chessboards according to rules;
A successor function: Rules moving a chesspiece;
Costs: The cost for a state transformation is 1, and represents moving a chesspiece one time;
A State Space: A set of chessboards according to rules.

For the problem of calculating single-source shortest paths in a graph, states, a successor function, costs, and a state space are as follows:

States: all vertices in the graph;
A successor function: all edges in the graph;
Costs: Weights of edges;
A State Space: A set of reachable vertices.

A state space tree is used to represent transformations from an initial state to a set of goal states, and calculate costs for transformations. There are two kinds of cost calculations:

1. Evaluating Function $g(x)$: The cost from the initial state to the current state x;
2. Heuristic Function $h(x)$: The estimated cost from the current state x to goal states.

Therefore, if a state space is regarded as a graph, a state space tree can be regarded as a problem for the shortest path.

9.1.1 Robot

The Robot Moving Institute is using a robot in their local store to transport different items. Of course, the robot should spend only the minimum time necessary when traveling from one place in the store to another. The robot can move only along a straight line (track). All tracks form a rectangular grid, as shown in Figure 9.2. Neighboring tracks are one meter apart. The store is a rectangle $N \times M$ meters and it is entirely covered by this grid. The distance of the track closest to the side of the store is exactly one meter. The robot has a circular shape with diameter equal to 1.6 meters. The track goes through the center of the robot. The robot always faces north, south, west, or east. The tracks are in the south-north and in the west-east directions. The robot can move only in the direction it faces. The direction in which it faces can be changed at each track crossing. Initially, the robot stands at a track crossing. The obstacles in the store are formed from pieces occupying 1m×1m on the ground. Each obstacle is within a 1×1 square formed by the tracks. The movement of the robot is controlled by two commands—GO and TURN.

The GO command has one integer parameter n in {1, 2, 3}. After receiving this command, the robot moves n meters in the direction it faces.

The TURN command has one parameter, which is either left or right. After receiving this command, the robot changes its orientation by 90° in the direction indicated by the parameter.

The execution of each command lasts one second.

Help researchers of RMI to write a program which will determine the minimal time in which the robot can move from a given starting point to a given destination.

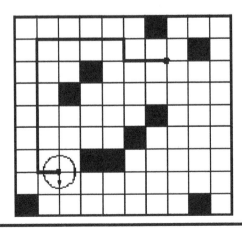

Figure 9.2 The circle is the Robot, the black squares are obstacles, and the heavy lines are the path that the Robot moves through.

Input

The input consists of blocks of lines. The first line of each block contains two integers $M \leq 50$ and $N \leq 50$, separated by one space. In each of the next M lines there are N numbers one or zero separated by one space. One represents obstacles and zero represents empty squares. (The tracks are between the squares.) The block is terminated by a line containing four positive integers $B1$ $B2$ $E1$ $E2$, each followed by one space and the word indicating the orientation of the robot at the starting point. $B1$ and $B2$ are the coordinates of the square in the north-west corner of which the robot is placed (starting point). $E1$ and $E2$ are the coordinates of square to the north-west corner of which the robot should move (destination point). The orientation of the robot when it has reached the destination point is not prescribed. We use (row, column)-type coordinates, i.e., the coordinates of the upper left (the most north-west) square in the store are 0,0 and the lower right (the most south-east) square are $M-1$, $N-1$. The orientation is given by the words north or west or south or east. The last block contains only one line with $N=0$ and $M=0$.

Output

The output contains one line for each block except the last block in the input. The lines are in the order corresponding to the blocks in the input. The line contains a minimal number of seconds in which the robot can reach the destination point from the starting point. If there does not exist any path from the starting point to the destination point, the line will contain -1.

Sample Input	Sample Output
9 10	12
0 0 0 0 0 0 1 0 0 0	
0 0 0 0 0 0 0 0 1 0	
0 0 0 1 0 0 0 0 0 0	
0 0 1 0 0 0 0 0 0 0	
0 0 0 0 0 0 1 0 0 0	
0 0 0 0 0 1 0 0 0 0	
0 0 0 1 1 0 0 0 0 0	
0 0 0 0 0 0 0 0 0 0	
1 0 0 0 0 0 0 0 1 0	
7 2 2 7 south	
0 0	

Source: ACM Central Europe 1996

IDs for Online Judges: POJ 1376, ZOJ 1310, UVA 314

Analysis

First, for a test case, in the area whose coordinate for the upper-left corner is (0, 0), and the coordinate for the lower-right corner is (M–1, N–1), we find grids that the Robot can't move through. The Robot has a circular shape with diameter equal to 1.6 meter, and row 0, row M–1, column 0, and column N–1 are boundaries for the Robot. Therefore, in the area whose coordinate for the upper-left corner is (1, 1), and the coordinate for the lower-right corner is (M–2, N–2), if there is an obstacle at (i, j), the Robot can't move through (i–1, j), (i. j–1), and (i–1, j–1). That is, (i–1, j), (i. j–1), and (i–1, j–1) should also be set as obstacles. See Figure 9.3.

State (x, y, s, step): The current coordinate (x, y) at which the Robot is, the current orientation s that the Robot faces; and the number of commands that has been executed is *step*.

A successor function *move*[ ][ ][ ]: After the Robot moves j meters in direction i, the horizontal increment is *move*[i][j][0] meters, the vertical increment is *move*[i][j][1] meters, and the orientation is *move*[i][j][2]. That is, the Robot moves from (x, y), and moves j meters in direction i, then the coordinate for the Robot is (x+*move*[i][j][0], y+*move*[i][j][1]), and the orientation that the Robot faces is *move*[i][j][2]. In order to avoid repeated searches, if the coordinate and the orientation hasn't appeared before, then the command is valid, a new state is generated, and the number of commands for the new state equals the number of commands for the previous state + 1; else the state is omitted.

Obviously, *move*[ ][ ][ ] are contestant arrays.

```
Byte move[4][5][4] = {    // the Robot moves j meters in
direction i the horizontal increment is move[i][j][0] meters,
the vertical increment is move[i][j][1] meters, and the
orientation is move[i][j][2]

    {{0, 0, 1}, {0, 0, 2}, {1, 0, 0}, {2, 0, 0}, {3, 0, 0}},
    {{0, 0, 0}, {0, 0, 3}, {0, 1, 1}, {0, 2, 1}, {0, 3, 1}},
```

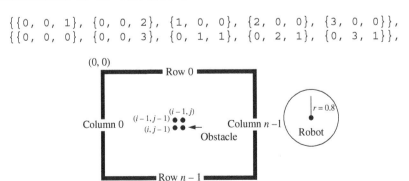

Figure 9.3

```
     {{0, 0, 0}, {0, 0, 3}, {0, -1, 2}, {0, -2, 2}, {0, -3, 2}},
     {{0, 0, 1}, {0, 0, 2}, {-1, 0, 3}, {-2, 0, 3}, {-3, 0, 3}},
};
```

State Space: A set of state graphs generated by legal commands.

Costs: The cost that the Robot executes one command is 1, represented as an edge in the graph. The number of edges in the path is the number of commands that the Robot executes from the starting point to the destination point, and is also the minimal number of seconds in which the Robot can reach the destination point from the starting point.

Obviously, BFS is suitable to calculate the best path in such a state space. The algorithm is as follows.

First, the coordinate for the Robot's starting point, the current orientation that the Robot faces; and the number 0 are as the first state. The first state is added into the queue. Then the front for the queue is removed until the queue is empty or the destination point is reached.

Each time the front is removed from the queue, the numbers of meters i ($0 \leq i \leq 4$) that the Robot moves are enumerated to calculate the reached coordinate (x', y') and the orientation s':

If (x', y') is an obstacle, then the new state is invalid;

If (x', y') is the destination point, then the minimal number of seconds in which the Robot can reach the destination point from the starting point is the number of commands for the previous state +1, and return successfully;

Otherwise, if (x', y') and s' haven't been visited, the visited mark is set, the number of executed commands $step' =$ the number of executed commands for the previous state $step+1$, and the new state containing (x', y'), s' and $step'$ is added into the queue.

The cost for the execution of one command is 1 second. BFS is done layer by layer. If the destination point is reached, the number of executed commands is the minimal number of seconds in which the robot reaches the destination point from the starting point.

 Program

```cpp
#include <iostream>
#include <queue>
using namespace std;
typedef int Byte;
struct Node {     //State
```

```
  Byte x, y, s, step;      // current coordinate (x, y), the
current orientation s, and the number of commands has been
executed step
};
Node Qt[300000], start, end;      //Queue Qt[ ], starting point
start, destination point end
bool used[51][51][4];      //memorized state list, used[x][y]
[d]: the Robot has visited (x, y) in direction d
bool map[51][51];      //the matrix representing the store
Byte move[4][5][4] = {      // the Robot moves j meters in
direction i the horizontal increment is move[i][j][0] meters,
the vertical increment is move[i][j][1] meters, and the
orientation is move[i][j][2]
  {{0, 0, 1}, {0, 0, 2}, {1, 0, 0}, {2, 0, 0}, {3, 0, 0}},
  {{0, 0, 0}, {0, 0, 3}, {0, 1, 1}, {0, 2, 1}, {0, 3, 1}},
  {{0, 0, 0}, {0, 0, 3}, {0, -1, 2}, {0, -2, 2}, {0, -3, 2}},
  {{0, 0, 1}, {0, 0, 2}, {-1, 0, 3}, {-2, 0, 3}, {-3, 0, 3}},
};
int n, m;      // The size of the matrix is n*m
int SearchAns() {      //BFS is used to calculate and return the
minimal number of seconds in which the Robot can reach the
destination point from the starting point
  if (start.x == end.x && start.y == end.y) return 0;
// starting point and destination point are same, return 0
  Node *cur = Qt, *next = Qt;      // Initialize pointers
pointing to the front and rear for the queue
  int i;
  memset(used, 0, sizeof(used));      //all states havn't been
visited
  start.step = 0;
  used[start.x][start.y][start.s] = 1;      // starting point
has been visited
  *next++ = start;      // starting point is added into the
queue
  while (cur!=next) {      // while the queue isn't empty
    for (i = 0; i < 5; i++) {      //enumerating the number of
meters
      next->x=cur->x+move[cur->s][i][0];      // coordinate
that the Robot moves i meters is (next->x, next->y), and the
orientation is next->s
      next->y = cur->y + move[cur->s][i][1];
      next->s = move[cur->s][i][2];
      if (map[next->x][next->y]) break;      // Obstacle
      if (next->x == end.x && next->y == end.y) return cur-
>step + 1;      // destination point
      if (!used[next->x][next->y][next->s])      // a new state
is generated
      {
        used[next->x][next->y][next->s] = 1;
        next->step = cur->step + 1;
```

```
            next++;      // the new state is added into the queue
        }
    }
    cur++;      // the front is removed from the queue
    }
    return -1;      //there is no path reaching the destination
point from the starting point, return -1
}
int main() {
    int i, j, t, t1, t2, t3, t4;      // starting point (t1, t2),
destination point (t3, t4)
    char buf[10];      //orientation string
    memset(map[0], 1, sizeof(map[0]));
    while(scanf("%d%d", &n, &m) != EOF){      //input the size of
the store until 0 0
        if (n == 0 && m == 0) break;
        for (i = 1; i <= n; i++) {      // input each row
            memset(map[i], 0, sizeof(map[i]));
            map[i][0]=map[i][m]=1;      // for row i, column 0 and
column m are obstacles
            for(j=1;j<=m;j++){      // input row i
                scanf("%d", &t);
                if (t == 1)      //if (i, j) is an obstacle, then (i-1, j),
(i. j-1), and (i-1, j-1) are obstacles
                    map[i][j]=map[i-1][j]=map[i][j-1]=map[i-1][j-1]=1;
            }
        }
        memset(map[n], 1, sizeof(map[n]));      //column n are
obstacles
            scanf("%d%d%d%d%s",&t1,&t2,&t3,&t4,buf);      // starting
point (t1, t2), destination point (t3, t4), the orientation
that the Robot faces at starting point buf
        start.x=t1; start.y=t2; end.x=t3; end.y=t4;
        if (buf[0] == 's') start.s = 0;      // the orientation
numbers
        else if (buf[0] == 'e') start.s = 1;
        else if (buf[0] == 'w') start.s = 2;
        else if (buf[0] == 'n') start.s = 3;
        printf("%d\n", SearchAns());      //calculation and output
the result
    }
    return 0;
}
```

In state space search, generated states need to be stored in a queue. Sometimes states should be compressed to store. In **9.1.2 The New Villa**, states are stored as binary numbers.

9.1.2 The New Villa

Mr. Black recently bought a villa in the countryside. Only one thing bothers him: although there are light switches in most rooms, the lights they control are often in other rooms than the switches themselves. While his estate agent saw this as a feature, Mr. Black has come to believe that the electricians were a bit absent-minded (to put it mildly) when they connected the switches to the outlets.

One night, Mr. Black came home late. While standing in the hallway, he noted that the lights in all other rooms were switched off. Unfortunately, Mr. Black was afraid of the dark, so he never dared to enter a room that had its lights out and would never switch off the lights of the room he was in.

After some thought, Mr. Black was able to use the incorrectly wired light switches to his advantage. He managed to get to his bedroom and to switch off all lights except for the one in the bedroom.

You are to write a program that, given a description of a villa, determines how to get from the hallway to the bedroom if only the hallway light is initially switched on. You may never enter a dark room, and after the last move, all lights except for the one in the bedroom must be switched off. If there are several paths to the bedroom, you have to find the one which uses the smallest number of steps, where "move from one room to another", "switch on a light" and "switch off a light" each count as one step.

Input

The input file contains several villa descriptions. Each villa starts with a line containing three integers r, d, and s. r is the number of rooms in the villa, which will be at most 10. d is the number of doors/connections between the rooms, and s is the number of light switches in the villa. The rooms are numbered from 1 to r; room number 1 is the hallway, and room number r is the bedroom.

This line is followed by d lines containing two integers i and j each, specifying that room i is connected to room j by a door. Then follow s lines containing two integers k and l each, indicating that there is a light switch in room k that controls the light in room l.

A blank line separates the villa description from the next one. The input file ends with a villa having $r=d=s=0$, which should not be processed.

Output

For each villa, first output the number of the test case ('Villa #1', 'Villa #2', etc.) in a line of its own.

If there is a solution to Mr. Black's problem, output the shortest possible sequence of steps that leads him to his bedroom and only leaves the bedroom light switched on. (Output only one shortest sequence if you find more than one.) Adhere to the output format shown in the sample below.

If there is no solution, output a line containing the statement "The problem cannot be solved."

Output a blank line after each test case.

Sample Input	Sample Output
3 3 4 1 2 1 3 3 2 1 2 1 3 2 1 3 2 2 1 2 2 1 1 1 1 2 0 0 0	Villa #1 The problem can be solved in 6 steps: – Switch on light in room 2. – Switch on light in room 3. – Move to room 2. – Switch off light in room 1. – Move to room 3. – Switch off light in room 2. Villa #2 The problem cannot be solved.

Source: ACM Southwestern European Regional Contest 1996

IDs for Online Judges: POJ 1137, ZOJ 1301, UVA 321

 Analysis

Suppose the interval for the room numbers is $[0, r-1]$.

State u: A state u is represented as a $r+4$-digit binary number, where the last four digits for u ($u\%16$) represents the current room number, and the r–digit prefix for u ($u/16$) represents the current lights' status for all rooms: one binary digit represents one room's light: 1 represents that the light is on, and 0 represents that the light is off; for the upper limit for the number of rooms is 10. The initial state $u_0=2^4$. That is, the light in the hallway (room 0) is on, and lights in other rooms are off. The goal state $u_{target}=(1<<(r+4-1))+r-1$. That is, the light in in the bedroom (room $r-1$) is on, and lights in other rooms are off.

Successor Function (the rule generating a new state u_new): For state u, there are three operations:

1. **Operation 1—Moving.** If there is a door between the current room (room $u\%16$) and room i whose light is on, then Mr. Black enters room i, and the new state $u\_new=u- u\%16+i$ is generated. That is, room i becomes the current room.

2. **Operation 2—Switching off a light.** If there is a light switch in the current room (room $u\%16$) that controls the light in room i and the light in room i is on (the binary digit corresponding to room i in $u/16$ $u_{4+i}==1$), then a new state is generated $u\_new=u-2^{4+i}$. That is, the light in room i is switched off.

3. **Operation 3—Switching on a light.** If there is a light switch in room $u\%16$ that controls the light in room i and the light in room i is off (the binary digit corresponding to room i in $u/16$ $u_{4+i}==0$), then a new state is generated $u\_new=u+2^{4+i}$. That is, the light in room i is switched on.

The generated state u_new is valid if the operation meets two following conditions.

> In order to avoid repeated searches, u_new hasn't been visited before; In u_new, the light in the current room must be on $(((u\_new/16)\&(2^{u\_new\%16}))==1)$.

State Space: From the initial state, new states are generated to construct a state space tree.

Cost: In the state space tree, the cost for each edge is 1.

The problem requires you to calculate the shortest possible sequence of steps. The upper limit for the number of states is 1024×10. For each state, the upper limit of the number of operations is 30 (10 moving methods + 10 switching on lights + 10 switching off lights). Therefore, BFS is suitable to solve the problem.

 Program

```
#include <cstdio>
#include <cstdlib>
#include <cstring>
#include <string>
#include <queue>
#include <algorithm>
using namespace std;
#define maxn 15      // The upper limit for the size of a matrix
#define maxs 20010    // The upper limit for states
#define MOVETO 20     // moving
#define SWITCHON 10    // switching on lights
#define SWITCHOFF 0    // switching off lights
int r;     // the number of rooms
int control[maxs];     //state transformation: state u is
generated by control[u]
int op[maxs];     //op[i] stores control information for state i:
the number of room that Mr. Black will enter or control its light
```

```
bool visited[maxs];    // visited[i]: visited mark for state i
bool g[maxn][maxn], light[maxn][maxn];    //adjacency matrix
g[ ][ ] for rooms, where g[i][j]==true represents there is a
door between room i and room j; adjacency matrix light[ ][ ]
represents controlling lights, where light[i][j]==true
represents there is a switch in room i controling the light in
room j
bool init()    //If the input is a test case, input and return
true; else return false
{
    int t1, t2, d, s;
    scanf("%d%d%d", &r, &d, &s);    // the number of rooms r,
the number of doors d, and the number of switches s
    if (r==0) return false;    //Input end mark
    memset(g, false, sizeof(g));    //Initialize adjacency
matrix g[ ][ ] for rooms
    for (int i=0; i<d; i++)    // constructing g[ ][ ]
    {
        scanf("%d%d", &t1, &t2);    //a door between room t1
and room t2
        t1--; t2--;
        g[t1][t2]=g[t2][t1]=true;
    }
    memset(light, false, sizeof(light));    // Initialize
adjacency matrix light
    for (int i=0; i<s; i++)    // constructing light[ ][ ]
    {
        scanf("%d%d", &t1, &t2);    //a switch in room t1
controls the light in room t2
        t1--; t2--;
        light[t1][t2]=true;
    }
    return true;
}
bool checkstay(int u)    //In state u, the light in the
current room is on, return true; else return false
{
    int pos=u%16; int tmp=u/16;    //In state u, the current
room number pos, and its light tmp
    int j=1<<pos;
    if (tmp&j) return true;    // the light in room pos is on,
return true; else return false
    return false;
}
int bfs()    //BFS is used to calculate and return the
shortest possible sequence of steps that leads Mr. Black to
his bedroom and only leaves the bedroom light switched on; if
there is no solution, return -1
{
    queue<int> q;    // state queue q
```

```
    queue<int> step;      // the queue storing the number of
step step
    int target =(1<<(r+4-1))+r-1;      //goal state 2^{r-1+4}+r-1
    int u=(1<<4), k=0;      //initial state u=2^4, the number of
steps is 0
    int u_new, uu, pos;
    memset(visited, 0, sizeof(visited));      //all states are
unvisited
    memset(control, 255, sizeof(control));      // all states'
fathers are empty
    visited[u]=true; q.push(u);      // initial state u is added
into queue q
    step.push(k);      //the number of steps is k is added into
queue step
    while (!q.empty())
    {
        u=q.front(); q.pop();      //the front u is popped from
queue q
        k=step.front(); step.pop();      // the number of steps
is k, is popped from queue step
        pos=u%16; uu=u>>4;      //current room pos and lights
for all rooms uu
        if (u==target){return k;}      //if u is a goal state
        for (int i=0; i<r; i++)      //move operation:
enumerating each room i
        {
         if (g[pos][i])      //a door between room pos and
room i
         {
           u_new=u-pos+i;      //calculate new state u_new
           if (!visited[u_new] && checkstay(u_new)) //if u_new
hasn't been and is valid
           {
             q.push(u_new); step.push(k+1);      // state u_new
is added into q, the number of steps is added into step
             visited[u_new]=true;      //Set visited mark for
u_new
             control[u_new]=u;      //u_new is generated from u
             op[u_new]=MOVETO+i;      //state u_new, Mr. Black
enters room i
           }
         }
        }
        for (int i=0, j=(1<<4); i<r; i++, j=j<<1, uu=uu>>1) //
switch off lights: enumerate each room i
         if (light[pos][i])      //in room pos a switch controls
the light in room i
         {
          if (uu&1)      //If the light is on, turn it off
```

```
            {
             u_new=u-j;      //calculate state u_new
             if (!visited[u_new] && checkstay(u_new)    //u_new
hasn't been visited, and is valid
               {
                q.push(u_new);    //u_new is added into q
step.push(k+1);    //the number of steps is added into step
visited[u_new]=true;     //set u_new visited mark
                control[u_new]=u;
                op[u_new]=SWITCHOFF+i;    //state u_new turns the
light in room i off
               }
             }
             else    //turn on the light in room i
               {
                u_new=u+j;    //new state u_new
                if (!visited[u_new] && checkstay(u_new))
// u_new hasn't been visited, and is valid
                  {
                   q.push(u_new);    // u_new is added into q
step.push(k+1);     // the number of steps is added into step
visited[u_new]=true;    // set u_new visited mark
                   control[u_new]=u;
                   op[u_new]=SWITCHON+i;    // state u_new turns
the light in room i on
                  }
                }
              }
     }
     return -1;    //return no solution
}
void dfsprint(int u)    //from goal state u, output the
shortest possible sequence of steps
{
     int u_new;
     if (u==(1<<4)) return;
u_new=control[u];
dfsprint(u_new);
//Backtracking: if op[u]>=20, then Move to room op[u]-20+1; if
op[u] is in [19,10], then switch on light in room op[u]-10+1;
if op[u] is in [9,0], Switch off light in room op[u] +1
     if (op[u]>=MOVETO) printf("- Move to room %d.\n",
op[u]-MOVETO+1);
      else if (op[u]>=SWITCHON) printf("- Switch on light in
room %d.\n", op[u]-SWITCHON+1);
      else if (op[u]>=SWITCHOFF) printf("- Switch off light in
room %d.\n", op[u]-SWITCHOFF+1);
}
void print(int cs, int steps)    //calculate and output the
cs-th test case
```

```
{
   printf("Villa #%d\n", cs);      // the number of test cases
   if (steps==-1)      // no solution
       printf("The problem cannot be solved.\n");
   else
   {
      printf("The problem can be solved in %d steps:\n",
steps);
      dfsprint((1<<(r+4-1))+r-1);      // from goal state output
the shortest possible sequence of steps
   }
   printf("\n");
}
int main()
{
   int steps;      //the number of the shortest possible
sequence of steps
   for (int cs=1; ;cs++)      // deal with test cases
   {
       if (!init()) break;      // Input a villa
       steps=bfs();      //BFS is used to calculate the number
of the shortest possible sequence of steps
       print(cs, steps);      //calculate and output the cs-th
test case
   }
   return 0;
}
```

9.2 Optimizing State Space Search

For **9.1.1 Robot** and **9.1.2 The New Villa**, searches aren't blind. In constructing a state space tree and finding the best path in the state space tree, some optimization strategies can be taken to improve the algorithm efficiency. In state space search, there are six kinds of optimization strategies, as follows:

1. Branching;
2. Memorization;
3. Indexing;
4. Pruning;
5. Bounding;
6. A* algorithm;

Strategy 1: Branching
A state space can be very large. And there is no need to construct a state space before the state space search begins. Branching means that a state space is searched as its state space tree is constructed. For **9.1.1 Robot** and **9.1.2**

The New Villa, branching is used to solve problems. Branching is used in almost all state space searches.

Strategy 2: Memorization

In state space searches, searched states need to be memorized to avoid being repeatedly searched. For **9.1.1 Robot** and **9.1.2 The New Villa**, *used*[ ][ ][ ] and *visited*[ ] are used to store searched states respectively.

Strategy 3: Indexing

In state space searches, indexing means searched states are numbered. For **9.1.2 The New Villa**, binary numbers are used to represent states.

The goal of using memorization and indexing is to improve efficiencies for state space searches. Memorization and indexing are always combined with pruning, bounding, and A* algorithm.

Strategy 4: Pruning

Pruning means removing some branches (subtrees) in state space searches to improve search efficiencies. For **9.1.1 Robot** and **9.1.2 The New Villa**, if the current state has been visited before, it is pruned. Pruning can be combined with memorization and indexing. And for **9.1.2 The New Villa**, the adjacent room whose light is off is also pruned.

9.2.1 Be Wary of Rose

You've always been proud of your prize rose garden. However, some jealous fellow gardeners will stop at nothing to gain an edge over you. They have kidnapped, blindfolded, and handcuffed you, and dumped you right in the middle of your treasured roses! You need to get out, but you're not sure how you can do it without trampling any precious flowers.

Fortunately, you have the layout of your garden committed to memory. It is an $N \times N$ collection of square plots (N odd), some containing roses. You are standing on a square marble plinth in the exact center. Unfortunately, you are quite disoriented, and have no idea which direction you're facing! Thanks to the plinth, you can orient yourself facing one of **North**, **East**, **South**, or **West**, but you have no way to know which.

You must come up with an escape path that tramples the fewest possible roses, no matter which direction you're initially facing. Your path must start in the center, consist only of horizontal and vertical moves, and end by leaving the grid.

Input

Every case begins with N, the size of the grid ($1 \le N \le 21$), on a line. N lines with N characters each follow, describing the garden: "." indicates a plot without any roses, "R" indicates the location of a rose, and "P" stands for the plinth in the center.

Input will end on a case where $N=0$. This case should not be processed.

Output

For each case, output a line containing the minimum guaranteed number of roses you can step on while escaping.

Sample Input	Sample Output
5 .RRR. R.R.R R.P.R R.R.R .RRR. 0	At most 2 rose(s) trampled.

IDs for Online Judges: UVA 10798

Analysis

According to the problem description, you are blindfolded; you have the layout of your garden committed to memory; you are standing on a square marble plinth in the exact center; and you are quite disoriented: you can orient yourself facing one of **North**, **East**, **South**, or **West**, but you have no way to know which.

Your garden is an $N{\times}N$ collection of square plots (N is odd). Because of symmetry, the current square plot on which you are standing is a square plot (x, y), or $(n-1-y, x)$, or $(y, n-1-x)$ or $(n-1-x, n-1-y)$. And (x, y), $(n-1-y, x)$, $(y, n-1-x)$ and $(n-1-x, n-1-y)$ constitute a square whose center is the square marble plinth in the exact center. Suppose you move into an adjacent square plot (x', y') from (x, y) in direction 1, (x, y) and (x', y') are adjacent, either $| x - x'|{==}1$ or $| y - y'|{==}1$. Also because of symmetry, moves from three square plots in other three directions are similar. In Figure 9.4, you move into (x', y') from (x, y), or move from $(n-1-y, x)$, $(y, n-1-x)$ and $(n-1-x, n-1-y)$ into $(n-1-y', x')$, $(y', n-1-x')$ and $(n-1-x', n-1-y')$ respectively. Obviously, $(n-1-y, x)$ and $(n-1-y', x')$ are adjacent, $(y, n-1-x)$ and $(y', n-1-x')$ are adjacent, and $(n-1-x, n-1-y)$ and $(n-1-x', n-1-y')$ are adjacent. And (x', y'), $(n-1-y', x')$, $(y', n-1-x')$ and $(n-1-x', n-1-y')$ constitute a new square whose center is the square marble plinth in the exact center. The four adjacent square plots represent your four moving directions.

Based on the above information, the successor function is as follows:

```
Suppose you move into an adjacent square plot (x', y') from
(x, y) in direction 1. Because of symmetry, if

   (x', y') contains a rose, then the number of roses you step
      on in direction one into (x', y') = the number of roses
      you step on when you move into (x, y) + 1;
```

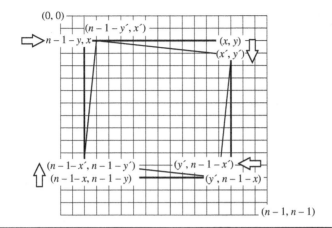

Figure 9.4

(y', n-1-x') contains a rose, then the number of roses you
   step on in direction two into (y', n-1-x') = the number
   of roses you step on when you move into (y, n-1-x) +1;
(n-1-x', n-1-y') contains a rose, then the number of roses
   you step on in direction three into (n-1-x', n-1-y') =
   the number of roses you step on when you move into (n-1-x,
   n-1-y) + 1;
(n-1-y', x') contains a rose, then the number of roses you
   step on in direction four into (n-1-y', x') = the number
   of roses you step on when you move into (n-1-y, x) + 1;

Obviously, the maximum number of roses *val* you step on when you move into
(x', y') in four directions (*up, left, down, right*) is the upper limit of the number of
roses you step on when you move into (x', y').

Memorization BFS is used to calculate the minimum guaranteed number of
roses you step on while escaping. The greedy method is used. Each time, the state
with minimum *val* is removed from the priority queue. Therefore, the priority
queue is a min heap in which *val* is the priority.

Memorization is used in the search. In states there are the current position (x, y),
the number of roses you can step on in four directions (*up, left, down, right*) and
its maximum value *val*; and a Boolean array $vis[x][y][d_1][d_2][d_3][d_4]$ is used to mark
whether the state that (x, y) is the position and the number of roses you step on are
$d_1, d_2, d_3,$ and d_4 respectively has been searched before. If a generated state has been
searched before, the state is pruned.

Obviously, if $((x==0)||(x==n-1)||(y==0)||(y==n-1))$, then you have escaped
from the garden, and $f[x, y]$ is the minimum guaranteed number of roses you can
step on while escaping.

Program

```
#include <cstdio>
#include <cstring>
#include <algorithm>
#include <queue>
using namespace std;
const int N = 21;      //the upper limit of your garden size
const int d[4][2]={{1, 0}, {-1, 0}, {0, -1}, {0, 1}};      //four
directions: horiztontal and vertical
int n, vis[N][N][11][11][11][11];      //memorization list,
where vis[x][y][d₁][d₂][d₃][d₄] is the mark that when you move
into (x, y), the number of roses you step on are d₁, d₂, d₃,
and d₄ respectively in directions (up, left, down, right)
char g[N][N];      //the graph for the garden
struct State {     // definition for State
    int x, y, val;      //the current square plot(x, y), the
maximum number of roses val you step on when you move into
(x', y') in four directions
    int up, left, down, right;      // the number of roses you
step on in 4 directions
    State() {x= y=up=left=down=right=0;}      //Initial state
(starting position (0, 0), numbers of roses you step on in
4 directions are 0)
    State(int x, int y, int up, int left, int down, int right)
{    //current state
        this->x = x;
        this->y = y;
        this->up = up;
        this->left = left;
        this->down = down;
        this->right = right;
        val = max(max(max(up,left), down), right);
    }
    bool operator<(const State& c)const {      //priority for
states: val
        return val > c.val;
    }
} s;      //state s
void init() {      //Input the garden
    for (int i = 0; i < n; i++) {
        scanf("%s", g[i]);
        for (int j = 0; j < n; j++)      // the plinth in the
center
            if (g[i][j] == 'P') s.x = i, s.y = j;
    }
```

```
}
int bfs() {      //memorization BFS: calculate the minimum
guaranteed number of roses you step on while escaping
    memset(vis, 0, sizeof(vis));    //Initialization
    priority_queue<State> Q;     //priority queue Q storing
states: number of roses you step on val is priority
    Q.push(s);     //initial state is added into the queue
    vis[s.x][s.y][0][0][0][0]=1;     //initialize memorization
list
    while (!Q.empty()) {     // remove the front u from the
queue
        State u = Q.top();
        Q.pop();
        if (u.x==0||u.x==n-1||u.y==0||u.y==n-1)return u.val;
//if escaping, return the minimum number of roses you step on
        for (int i = 0; i < 4; i++) {     //enumerating 4
directions
            int xx = u.x + d[i][0];     //calculate the
adjacent square plot (xx, yy) in direction i
            int yy = u.y + d[i][1];
            int up = u.up;     // the number of roses you step
on in the original 4 directions
            int left = u.left;
            int down = u.down;
            int right = u.right;
            if (g[xx][yy] == 'R') up++;     //accumulation for
4 directions
            if (g[n - 1 - yy][xx] == 'R') left++;
            if (g[n - 1 - xx][n - 1 - yy] == 'R') down++;
            if (g[yy][n - 1 - xx] == 'R') right++;
            if (!vis[xx][yy][up][left][down][right]) {     //if
the new state hasn't been visited, add it into the
memorization list and queue
                vis[xx][yy][up][left][down][right] = 1;
                Q.push(State(xx, yy, up, left, down, right));
            }
        }
    }
}
int main() {
    while (~scanf("%d", &n) && n) {     //Input the size of
garden N
        init();     // Input garden and the plinth in the
center
        printf("At most %d rose(s) trampled.\n",bfs());
//Calculating and output the result
    }
    return 0;
}
```

Strategy 5: Bounding
Before enumerating the candidate solutions of a branch, the branch is checked against upper and lower estimated bounds on the optimal solution, and it is discarded if it cannot produce a better solution than the best one found so far by the algorithm.

9.2.2 Fill

There are three jugs with a volume of a, b, and c liters. (a, b, and c are positive integers not greater than 200). The first and the second jug are initially empty, while the third is completely filled with water. It is allowed to pour water from one jug into another until either the first one is empty or the second one is full. This operation can be performed zero, one, or more times.

You are to write a program that computes the least total amount of water that needs to be poured; so that at least one of the jugs contains exactly d liters of water (d is a positive integer not greater than 200). If it is not possible to measure d liters this way, your program should find a smaller amount of water $d' < d$ which is closest to d and for which d' liters could be produced. When d' is found, your program should compute the least total amount of poured water needed to produce d' liters in at least one of the jugs.

Input

The first line of input contains the number of test cases. In the next T lines, T test cases follow. Each test case is given in one line of input containing four space-separated integers—a, b, c, and d.

Output

The output consists of two integers separated by a single space. The first integer equals the least total amount (the sum of all waters you pour from one jug to another) of poured water. The second integer equals d, if d liters of water could be produced by such transformations, or it equals the closest smaller value d' that your program has found.

Sample Input	Sample Output
2	2 2
2 3 4 2	9859 62
96 97 199 62	

Source: Bulgarian National Olympiad in Informatics 2003

IDs for Online Judges: UVA 10603

Analysis

Suppose that volumes for three jugs are *A*, *B*, and *C* liters respectively. Finally, one of the jugs contains exactly *D* liters of water.

State (*a*, *b*, *c*, *tot*): In the current three jugs, there are *a*, *b*, and *c* liters of water respectively. And the current total amount of poured water is *tot* liters.

Successor function: There are six cases that you pour water from one jug into another.

Case 1: If all water in jug 1 can be poured into jug 2 ($a<B-b$), then all water in jug 1 is poured into jug 2, and a new state (0, $b+a$, *c*, $tot+a$) is generated; else water in jug 1 is poured into jug 2 until jug 2 is full, and a new state ($a-(B-b)$, *B*, *c*, $tot+(B-b)$) is generated.

Case 2: If all water in jug 1 can be poured into jug 3 ($a<C-c$), then all water in jug 1 is poured into jug 3, and a new state (0, *b*, $c+a$, $tot+a$) is generated; else water in jug 1 is poured into jug 3 until jug 3 is full, and a new state ($a-(C-c)$, *b*, *C*, $tot+(C-c)$) is generated.

Case 3: If all water in jug 2 can be poured into jug 1 ($b<A-a$), then all water in jug 2 is poured into jug 1, and a new state ($a+b$, 0, *c*, $tot+b$) is generated; else water in jug 2 is poured into jug 1 until jug 1 is full, and a new state (*A*, $b-(A-a)$, *c*, $tot+(A-a)$) is generated.

Case 4: If all water in jug 2 can be poured into jug 3 ($b<C-c$), then all water in jug 2 is poured into jug 3, and a new state (*a*, 0, $c+b$, $tot+b$) is generated; else water in jug 2 is poured into jug 3 until jug 3 is full, and a new state (*a*, $b-(C-c)$, *C*, $tot+(C-c)$) is generated.

Case 5: If all water in jug 3 can be poured into jug 1 ($c<A-a$), then all water in jug 3 is poured into jug 1, and a new state ($a+c$, *b*, 0, $tot+c$) is generated; else water in jug 3 is poured into jug 1 until jug 1 is full, and a new state (*A*, *b*, $c-(A-a)$, $tot+(A-a)$) is generated.

Case 6: If all water in jug 3 can be poured into jug 2 ($c<B-b$), then all water in jug 3 is poured into jug 2, and a new state (*a*, $b+c$, 0, $tot+c$) is generated; else water in jug 3 is poured into jug 2 until jug 2 is full, and a new state (*a*, *B*, $c-(B-b)$, $tot+(B-b)$) is generated.

State Space: Generated states constitute the state space.

Cost: The amount of poured water each time is as a weight of the corresponding edge in the graph. The problem requires you to calculate the path with the minimum sum of weights from the initial state (0, 0, *C*, 0) to the goal state (One of the jugs contains exactly *d* liters of water or *d'* liters of water. If it is not possible to measure *d* liters this way, we should find a smaller amount of water $d'<d$, which is

closest to *d* and for which *d'* liters could be produced.) Obviously the problem is a problem of finding a shortest path. BFS is suitable to solve the problem.

> Suppose *QA*, *QB*, *QC* are queues storing the current amount of water in the three jugs respectively; *QTOT* is the queue storing the current total amount of poured water.
> *dp*[ ][ ][ ] is the matrix storing upper limits for the total amount of poured water, where *dp*[*a*][*b*][*c*] is the upper limit for the total amount of poured water when the amount of water in the three jugs are *a*, *b*, and *c* respectively. Initially *dp*[ ][ ][ ] is ∞.
> *res*[ ] is the goal matrix, where *res*[*D*] is the least total amount of poured water when at least one of the jugs contains exactly *D* liters of water. Initially *res*[ ] is ∞.
> *dp*[ ][ ][ ] and *res*[ ] are used in bounding. There are two boundings: If (*tot*≥*res*[*D*]) or (*tot*≥*dp*[*a*][*b*][*c*]), then the case needn't be considered.

BFS is as follows:

An initial state (0, 0, *C*, 0) is added into queues *QA*, *QB*, *QC* and *QTOT* respectively;
> Repeat the following process until queue *QA* is empty:
>> Remove fronts for queues *QA*, *QB*, *QC* and *QTOT*, and constitute a new state (*a*, *b*, *c*, *tot*);
>>> if ((*tot*<*res*[*D*])&&(*tot*<*dp*[*a*][*b*][*c*]))
>>> {
>>>> *dp*[*a*][*b*][*c*]=*tot*;   //adjust *dp*[*a*][*b*][*c*]
>>>> *res*[*a*]=min(*res*[*a*], *tot*); *res*[*b*]=min(*res*[*b*], *tot*); *res*[*c*]=min(*res*[*c*], *tot*);
>>>> Simulating the above six cases;
>>>> if a state satisfies the condition, then the state is added into queues *QA*, *QB*, *QC* and *QTOT*;
>>> }

When BFS ends, from *D*, the first *res*[*D'*]≠∞ is searched in descending order. That is, *res*[*D'*] is the least total amount of poured water.

 Program

```c
#include <stdio.h>
#include <queue>
using namespace std;
#define min(x, y) ((x) < (y) ? (x) : (y))
```

```
#define oo 0xffffffff    //Define ∞
int A, B, C, D, JUG[3];    // A, B, C: three jugs' volumes,
one of the jugs contains exactly D liters of water finally
int dp[201][201][201], res[201];    // in the current three
jugs there are a, b, and c liters of water; the upper limit
for poured water is dp[a][b][c]; if one of the jugs contains
exactly D liters of water finally, res[D] is minimal poured
water
queue<int> QA, QB, QC, QTOT;    // QA, QB, QC are queues
storing the current amount of water in the three jugs
respectively; QTOT is the queue storing the current total
amount of poured water
void pushNode(int a, int b, int c, int tot) {    // the
current amount of water in the three jugs a, b, c is added
into queues OA, OB, OC, total amount of poured water tot is
added into queue OTOT
    QA.push(a), QB.push(b), QC.push(c), QTOT.push(tot);
}
void update(int a,int b,int c,int tot){    //6 cases pouring
water from one jug into another is simulated
    if(tot >= res[D]) return;    // Bounding: If (tot≥res[D])
or (tot≥dp[a][b][c]), then the case needn't be considered.
    if(tot >= dp[a][b][c]) return;
    dp[a][b][c]=tot;    //the upper limit for the total amount
of poured water tot when three jugs have a, b, and c liters of
water
    res[a] = min(res[a], tot);    // the total amount of
poured water can't exceed tot when three jugs have a, b, and c
liters of water
    res[b] = min(res[b], tot);
res[c] = min(res[c], tot);
//case 1
    if(a< B-b) pushNode(0, b+a, c, tot+a);
      else pushNode(a-(B-b), B, c, tot+(B-b));
//case 2
    if(a < C-c) pushNode(0, b, c+a, tot+a);
else pushNode(a-(C-c), b, C, tot+(C-c));
//case 3
    if(b < A-a) pushNode(a+b, 0, c, tot+b);
      else pushNode(A, b-(A-a), c, tot+(A-a));
//case 4
    if(b < C-c) pushNode(a, 0, c+b, tot+b);
       else pushNode(a, b-(C-c), C, tot+(C-c));
//case 5
    if(c < A-a) pushNode(a+c, b, 0, tot+c);
      else pushNode(A, b, c-(A-a), tot+(A-a));
//case 6
    if(c < B-b) pushNode(a, b+c, 0, tot+c);
      else pushNode(a, B, c-(B-b), tot+(B-b));
}
```

```
void bfs(int a,int b,int c,int tot){    //BFS calculate the
result
    QA.push(a), QB.push(b),QC.push(c), QTOT.push(tot);  //initial
state is added into the queue
    while (!QA.empty()) {
        a = QA.front(), QA.pop();    //fronts of queues are
removed
        b = QB.front(), QB.pop();
        c = QC.front(), QC.pop();
        tot = QTOT.front(), QTOT.pop();
        update(a, b, c, tot);    //6 cases pouring water from
one jug into another is simulated
    }
}
int main() {
    int t;
    scanf("%d", &t);    //number of test cases
    while(t--) {    //each test case is dealt with
        int i, j, k;
        scanf("%d %d %d %d",&A,&B,&C,&D);    //input a test
case
        for(i = 0; i <= A; i++)    //initialization
            for(j = 0; j <= B; j++)
                for(k = 0; k <= C; k++) dp[i][j][k] = oo;
        JUG[0]=A, JUG[1]=B, JUG[2]=C;
        for(i=0;i<=D;i++)res[i]=oo;
        bfs(0, 0, C, 0);    //BFS from initial state
        while(res[D] == oo) D--;    //from D, find the first
res[D]≠∞ in descending order
        printf("%d %d\n", res[D], D);    //output the result
    }
    return 0;
}
```

9.2.3 Package Pricing

The Green Earth Trading Company sells four different sizes of energy-efficient fluorescent light bulbs for use in home lighting fixtures. The light bulbs are expensive, but last much longer than ordinary incandescent light bulbs and require much less energy. To encourage customers to buy and use the energy-efficient light bulbs, the company catalog lists special packages which contain a variety of sizes and numbers of the light bulbs. The price of a package is always substantially less than the total price of the individual bulbs in the package. Customers typically want to buy several different sizes and numbers of bulbs. You are asked to write a program to determine the least expensive collection of packages that satisfy any customer's request.

Input

The input file is divided into two parts. The first one describes the packages which are listed in the catalogue. The second part describes individual customer requests. The four sizes of light bulbs are identified in the input file by the characters "a", "b", "c", and "d".

The first part of the input file begins with an integer n ($1 \leq n \leq 50$) indicating the number of packages described in the catalog. Each of the n lines that follows is a single package description. A package description begins with a catalog number (a positive integer) followed by a price (a real number), and then the sizes and corresponding numbers of the light bulbs in the package. Between one and four different sizes of light bulbs will be listed in each description. The listing format for these size-number pairs is a blank, a character ("a", "b", "c", or "d") representing a size, another blank, and then an integer representing the number of light bulbs of that size in the package. These size-number pairs will not appear in any particular order, and there will be no duplicate sizes listed in any package. The following line describes a package with catalog number 210 and price $76.95 which contains three size "a" bulbs, one size "c" bulb, and four size "d" bulbs.

210 76.95 a 3 c 1 d 4

The second part of the input file begins with a line containing a single positive integer m representing the number of customer requests. Each of the remaining m lines is a customer request. A listing of sizes and corresponding numbers of light bulbs constitutes a request. Each list contains only the size-number pairs, formatted the same way that the size-number pairs are formatted in the catalogue descriptions. Unlike the catalogue descriptions, however, a customer request may contain duplicate sizes. The following line represents a customer request for one size "a" bulb, two size "b" bulbs, two size "c" bulbs, and five size "d" bulbs.

a 1 d 5 b 1 c 2 b 1

Output

For each request, print the customer number (1 through m, 1 for the first customer request, 2 for the second, ..., m for the m-th customer), a colon, the total price of the packages which constitute the least expensive way to fill the request, and then the combination of packages that the customer should order to fill that request.

Prices should be shown with exactly two significant digits to the right of the decimal. The combination of packages must be written in ascending order of catalog numbers. If more than one of the same type package is to be ordered, then the number ordered should follow the catalog number in parentheses. You may assume that each customer request can be filled. In some cases, the least expensive way to fill a customer request may contain more light bulbs of some sizes than necessary to fill the actual request. This is acceptable. What matters is that the customers receive *at least* what they request.

Sample Input	Sample Output
5	Input set #1:
10 25.00 b 2	1: 27.50 55
502 17.95 a 1	2: 50.00 10(2)
3 13.00 c 1	3: 65.50 3 10 55
55 27.50 b 1 d 2 c 1	4: 52.87 6
6 52.87 a 2 b 1 d 1 c 3	5: 90.87 3 6 10
6	6: 100.45 55(3) 502
d 1	
b 3	
b 3 c 2	
b 1 a 1 c 1 d 1 a 1	
b 1 b 2 c 3 c 1 a 1 d 1	
b 3 c 2 d 1 c 1 d 2 a 1	
0	

Source: ACM World Finals 1994

IDs for Online Judges: POJ 1889, UVA 233

DFS is used to calculate the least expensive way to fill the request and the combination of packages that the customer should order to fill that request. The following strategies are used in the search.

1. **Memorization:** Current states are memorized. States constitute the current number of stored package *st*, the total price *now*, the combination of packages *nowmet*[ ], and the remainder requirement for four sizes of light bulbs *need*[ ]. In order to avoid overflow, *st* and *now* are as parameters for DFS, and *nowmet*[ ] and *need*[ ] are as global variables. Initially, *nowmet*[ ] is set 0, *need*[ ] is a customer's request, *st*=0, and *now*=0.

2. **Bounding:** The current total price is checked. The key to the problem is to determine whether the current price is better. If the current total price isn't better, then it backtracks.

   The problem shows *n* packages' descriptions: their prices and numbers of four sizes of light bulbs. Obviously, the interval of the price for the *i*-th light bulb in package *j* is in $\left[0, \dfrac{\text{the price for package j}}{\text{the number of the i-th light bulbs}} \right]$,

   $0 \le j \le n-1, 0 \le i \le 3$. That is, the upper limit of the price for the *i*-th light bulb in package *j* is $\dfrac{\text{the price for package j}}{\text{the number of the i-th light bulbs}}$.

For each size of light bulb, packages are sorted in ascending order of the upper limits of its prices. Suppose *rankby*[*i*][*j*] stores the number of package storing the light bulb with size *i* whose price is ranked *j*; *minave*[*i*][*j*] stores the least upper limit of prices for light bulb with size *i* from package *j* to package *n*. And *rankby*[ ][ ] and *minave*[ ][ ] are calculated when a test case is input.

There are two boundings:

Bounding 1: If the current total price *now* is higher than the current cheapest price *ans*, it backtracks directly;

Bounding 2: Four sizes of light bulbs are searched. If *now+minave*[*i*] [*st*]×*need*[*i*] > *ans* (0≤*i*≤3) for four sizes of light bulbs, then it backtracks.

3. **Optimized search strategy:** A greedy algorithm is used. The size of bulb which is demanded most is searched, that is, $need[br] = \max_{0 \leq i \leq 3}\{need[i]\}$. In order to buy light bulbs with the cheapest price, based on the ascending order for the upper limit of prices, in *st* the first package *p* which meets the customer's requirement is searched and is put into the current combination, that is, (*p*=*rankby*[*br*][*i*])&&(*p*>*st*)&&(*need*[*j*]>0)&&(there exist light bulbs with size *j* in package *p*), 0≤*i*≤*n*−1.

DFS is implemented by *search*(*st*, *now*):

```
If now > ans, it backtracks directly (bounding 1);
If need[0]<=0 && need[1]<=0 && need[2]<=0 && need[3]<= 0,
then ans=now; memcpy(met, nowmet, sizeof(met)), and it
backtracks;
If now+minave[i][st]*need[i] > ans (0≤i≤3) for four sizes of
light bulbs, then it backtracks ( bounding 2);
Calculate such size br for a light bulb that
need[br] = max {need[i]} ;
         0≤i≤3
Select the suitable package p (p=rankby[br][i])&&(p>st)
&&(need[j]>0)&&(there exist light bulbs with size j in
package p), 0≤i≤n-1;
Add a package p: ++nowmet[p]; need[j] - the number of light
bulbs with size j in package p (0≤j≤3);
search(p, now + the price for package p);
recover need[ ] and nowmet[ ] before the recursion;
```

Program

```
#include <cstdio>
#include <cstring>
#include <iostream>
```

```cpp
#include <vector>
#include <algorithm>
#include <utility>
#include <sstream>
#include <map>
using namespace std;
struct pacnode      //package struct
{
  int q[4];        //number of bulbs q[i] with size i
  double price;     //price
  int id;          // catalog number
}pac[60];         //packages
int n, met[60], nowmet[60], need[4], rankby[4][60];
double ans, ave[4][60], minave[4][60];
void init();       // packages' information
void work();       //current customer's request
void search(int st, double now);    //DFS is used to calculate
the result, the st-th package, the current price now
int main()
{
  int testno = 0;
  while (true)     //number of packages n
  {
    if (scanf("%d", &n) == EOF) break;
    if (n == 0) break;
    init();       //n packages' information
    ++testno;     //number of the test case
    printf("Input set #%d:\n", testno);
    int m;
    scanf("%d\n", &m);   //number of customers
    for (int i = 0; i < m; ++i)   //customers' requests
    {
      printf("%d:", i + 1);   //number of the customer
      work();    //deal with the request for customer i
    }
  }
  return 0;
}
void init()      // n packages' information
{
  for (int i = 0; i < n; ++i)     // n packages' information
  {
    scanf("%d%lf", &pac[i].id, &pac[i].price);    // catalog
number and price for package i
      memset(pac[i].q, 0, sizeof(pac[i].q));
      char tmp[1000];
      gets(tmp);    // numbers and sizes for bulbs in package i
      istringstream in(tmp);
      while (true)   //size kind and number x
```

```
          {
            char kind;
            int x;
            if (in >> kind >> x == NULL) break;
            pac[i].q[kind - 97] += x;      //the number of bulbs
with size kind in package i
          }
     }
     for (int i = 0; i < 4; ++i)          //calculate ave[ ][ ] by
enumerating sizes and packages
        for (int j = 0; j < n; ++j)
           if (pac[j].q[i] == 0) ave[i][j] = 1e100;
           else ave[i][j] = pac[j].price / pac[j].q[i];
     for (int i = 0; i < 4; ++i)     //enumerate size i, sorting
packages x[ ], the upper limit for the price is as the first
key, number of packages is the second key
     {
        pair<double, int> x[60];      // x[ ]: a pair of elements
        for (int j = 0; j < n; ++j)
        {
          x[j].first = ave[i][j]; x[j].second = j;
        }
        sort(x, x + n);
//calculate rankby[i][j] and minave[i][j]
        for (int j = 0; j < n; ++j) rankby[i][j] = x[j].second;
        minave[i][n - 1] = ave[i][n - 1];
        for (int j = n - 2; j >= 0; --j) minave[i][j] =
min(minave[i][j + 1], ave[i][j]);
     }
}
void work()      // current customer's request
{
   memset(need, 0, sizeof(need));
   char tmp[1000];
   gets(tmp);
   istringstream in(tmp);
   while (true)      //size kind, number x
   {
      char kind;
      int x;
      if (in >> kind >> x == NULL) break;
      if (kind == 'a') need[0] += x;      //4 bulbs' numbers
      else if (kind == 'b') need[1] += x;
      else if (kind == 'c') need[2] += x;
      else if (kind == 'd') need[3] += x;
   }
   memset(nowmet, 0, sizeof(nowmet)); ans = 1e100; //Initialize:
ans=∞
   search(0, 0.0);      // DFS is used to calculate the result,
the 0-th package, the current price 0
```

```
    printf("%8.2lf", ans);     // the least expensive way to fill
the request
    vector<pair<int, int> > oa;
    for (int i = 0; i < n; ++i)  // the combination of packages
is stored in oa
        if (met[i] != 0) oa.push_back(make_pair(pac[i].id,
met[i]));
    sort(oa.begin(), oa.end());     //sorting oa
    for (int i = 0; i < oa.size(); ++i)  //output the
combination of packages
        if (oa[i].second != 1) printf(" %d(%d)", oa[i].first,
oa[i].second);
        else printf(" %d", oa[i].first);
    printf("\n");
}
void search(int st, double now)    // DFS is used to calculate
the result, the st-th package, the current price now
{
    if (now > ans) return;     // the current price is higher
than the current cheapest price (bounding 1),
    if (need[0] <= 0 && need[1] <= 0 && need[2] <= 0 && need[3]
<= 0) //fill the request, adjust the current price, and
current package into the combination
    {
        ans = now; memcpy(met, nowmet, sizeof(met)); return;
    }
    for (int i = 0; i < 4; ++i)  //search all sizes, after
adding it, the price is higher than ans
        if (now + minave[i][st] * need[i] > ans) return;
    int br = 0;      // The size br of bulb which is demanded most
    for (int i = 1; i < 4; ++i)
        if (need[i] > need[br]) br = i;
    for (int i = 0; i < n; ++i) //search the bulb with size br
in n packages in ascending order of price
    {
        int p = rankby[br][i];      //package p
        if (p < st) continue;      // package p has been searched
        bool use = false;      //determine whether package p fills
the request
        for (int j = 0; j < 4; ++j)
            if (need[j] > 0 && pac[p].q[j] > 0)
            {
                use = true; break;
            }
        if (!use) continue;      // package p can't fill the
request
        ++nowmet[p];      // package p is added into the
combination, and adjust remainder request
        for (int j = 0; j < 4; ++j) need[j] -= pac[p].q[j];
```

```
        search(p, now + pac[p].price);
        --nowmet[p];
        for (int j = 0; j < 4; ++j) need[j] += pac[p].q[j];
    }
}
```

Strategy 6: Heuristic Search (A* Search)

Heuristic search is suitable to find the best path. The most widely known heuristic search is A* search. A* search evaluates vertices by combining $g(v)$, the minimal cost to reach vertex v from the initial vertex; $h(v)$, the estimated minimal cost to get from vertex v to the goal vertex; and $f(v)=g(v)+h(v)$, $f(v)$ is the estimated minimal cost through vertex v.

Obviously, for initial state (vertex) s, $f(s)=0+h(s)=h(s)$.

That is, if we try to find the best solution, it is reasonable to find the vertex v with the lowest value $f(v)$. Therefore, A* search is also called best-first search.

Heuristic search is implemented by BFS. Two lists are constructed:

OPEN is used to store states to be extended. And *OPEN* is a priority queue. For an element v in *OPEN*, and $f(v)$ is the key for the priority queue.
CLOSED is used to store visited states. That is, states that have been deleted from *OPEN*.

The reason why two lists are constructed is that we need to determine whether the current state is unvisited, visited, or generated. Based on that, the methods calculating $f[v]$ are different: if state v is unvisited, $f[v]$ is calculated by $f(v)=g(v)+h(v)$; else $f[v]$ is adjusted. The process for A* search is as follows:.

```
    For the initial state s, f(s)=h(s), and state s is
added into OPEN;
    Each time the state u with the minimal f is removed
from OPEN, and its successor state v is extended: if v is
in the queue and the g(v) from u to v is better than the
previous g(v), then the f(v) should be adjusted, and v is
set as the successor state for u; if v isn't in the
queue, then f(v)=g(v)+h(v), v is set as the successor
state for u and added into OPEN.
```

The above process is repeated until the goal state is reached or *OPEN* is empty.

If *OPEN* is empty, the search fails; else the best path is found.

Two aspects should be noted:

1. The selection for the estimated cost $h(v)$ is the key to find the best path.
2. There is a balance between the computation of $h(v)$ and the efficiency solving a problem.

1	2	3	4
5	6	7	8
9	10	11	12
13	14	15	*x*

Figure 9.5

9.2.4 Eight

The 15-puzzle has been around for over 100 years; even if you don't know it by that name, you've seen it. It is constructed with 15 sliding tiles, each with a number from 1 to 15 on it, and all packed into a 4×4 frame with one tile missing. Let's call the missing tile "x"; the object of the puzzle is to arrange the tiles so that they are ordered as shown in Figure 9.5 where the only legal operation is to exchange "x" with one of the tiles with which it shares an edge. As an example, the following sequence of moves shown in Figure 9.6 solves a slightly scrambled puzzle.

The letters in the previous row indicate which neighbor of the "x" tile is swapped with the "x" tile at each step; legal values are "r", "l", "u", and "d", for right, left, up, and down, respectively.

Not all puzzles can be solved; in 1870, a man named Sam Loyd was famous for distributing an unsolvable version of the puzzle, and frustrating many people. In fact, all you have to do to make a regular puzzle into an unsolvable one is to swap two tiles (not counting the missing "x" tile, of course).

In this problem, you will write a program for solving the less well-known 8-puzzle, composed of tiles on a three by three arrangement.

Input

You will receive, in "eight.inp", a description of a configuration of the 8-puzzle. The description is just a list of the tiles in their initial positions, with the rows listed from top to bottom, and the tiles listed from left to right within a row, where the tiles are

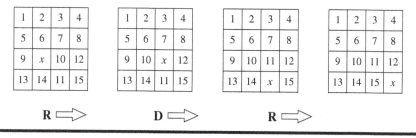

Figure 9.6

1	2	3
x	4	6
7	5	8

Figure 9.7

represented by numbers 1 to 8, plus "x". For example, see the puzzle shown in Figure 9.7 is described by this list:

$$1\ 2\ 3\ \times\ 4\ 6\ 7\ 5\ 8$$

Output

You will print to standard output either the word "unsolvable", if the puzzle has no solution, or a string consisting entirely of the letters "r", "l", "u", and "d" that describes a series of moves that produce a solution. The string should include no spaces and start at the beginning of the line.

Sample Input	Sample Output
2 3 4 1 5 × 7 6 8	ullddrurdllurdruldr

Source: ACM South Central USA 1998

IDs for Online Judges: POJ 1077, ZOJ 1217, UVA 652

 Analysis

An 8-puzzle is a 3×3 matrix, where its elements are 1 to 9 respectively, and 9 represents "x". An 8-puzzle can be represented as a permutation for 1...9, where the k-th element in the permutation is the element whose position is $\left(\left\lfloor \dfrac{k}{3} \right\rfloor, k\%3\right)$ in the matrix, $0{\le}k{\le}8$.

States: A permutation for 1...9 can be regarded as a state. The number of states is 9!=362880. In order to save memory, the alphabet order for permutations is used as the index for states:

1,2,3.4,5,6,7,8,9: 0

......

9,8,7.6,5,4,3,2,1: 362879

Obviously, 0 represents the goal state.

Heuristic Function $f(u)$: The cost to the goal state, from the initial state and through state u, $f(u)=d(u)+h(u)$, where $d(u)$ is the minimal number of moves from the initial state to state u, and $h(u)$ is the estimated cost from state u to the goal state: $h(u) = \sum_{i=1}^{8} |x'_i - x_i| + |y'_i - y_i|$ (the position for number i in state u is (x_i, y_i), and in the goal state is (x'_i, y'_i)).

Obviously, for the initial state s, $f(s)=h(s)$.

Successor Function: If the position for "x" in state u is (x, y), that is, the number in the position is 9, and k $(1 \le k \le 8)$ is the number in one adjacent position (x', y'), then numbers in (x, y) and (x', y') are exchanged and a new state v is generated.

State Space: Generated states constitute the state space.

Cost: In the state space tree, the cost for generating a new state is 1.

The problem requires you to calculate the path with the minimal cost from the initial state to the goal state. Obviously, BFS is suitable to solve the problem. A priority queue in which the heuristic function $f(\ )$ is used as the key is used to store states, and each time the state with the minimal $f(\ )$ is selected.

Suppose $d(u)$ is the length of the path from the initial state to state u. If state v is generated from state u, and $d(u)+1<d(v)$, then $d(v)$ should be adjusted: $d(v)= d(u)+1$, state u is set as the precursor state for state v, and state v is added into the queue. Therefore, the key is to calculate the estimated cost from the initial state through state v to the goal state $f(v)$. The heuristic function $f(\ )$ is as the key for the priority queue. There are three cases:

Case 1: State v hasn't been visited before: $f(v)=d(v)+h(v)$.

Case 2: State v has been in the queue: There is a $f(v)$ for v. If the estimated cost should be adjusted, that is, the path to state v is adjusted through state u, $f(v)=f(v)-d(v)+d(u)+1$.

Case 3: State v has been visited and removed from the queue: If the estimated cost should be adjusted, $f(v) = f(v)-d(v)+d(u)+1$. The reason is the same as case 2.

A* algorithm is as follows:

```
Suppose the precursor state for the initial state s is -1;
  d[s]=0; f[s]=h[s];
 State s is added into the priority queue q; Set s the flag
entering q, and set flags not entering q for other states;
   while ( the priority queue q isn't empty)
   {
        the state u with the minimal f[ ] in the priority
queue q is selected;
        if u is the goal state (the alphabet order is 0)
return the result;
        State u is removed from q;
```

```
        Calculate the position (x, y) for 'x' in state u;
        for(int i=0; i<4; ++i){     // four search directions
          calculate the adjacent cell (a, b) for (x, y) in
direction i;
            If ((a, b) is in the puzzle){
              9 in (x, y) and the number in (a, b) are
exchanged, and a new state v is generated;
                If (d[u] + 1) < d[v])
                { d[v]=d[u]+1;
                State u is set as the precursor state for state
v, and direction i is recorded;
                The heuristic function f[v] is adjusted based
on above three cases;
                State v is added into q; set v the flag
entering q;
                }
            }
        Set the search flags for the four directions for state u;
      }
      if (the goal state has been visited (the alphabet order
is 0))
        Output a series of moves that produce a solution;
        else output "unsolvable";
```

 Program

```cpp
#include<iostream>
#include<cstdio>
#include<cstring>
#include<cstdlib>
#include<queue>
using namespace std;
int fac[] = { 1, 1, 2, 6, 24, 120, 720, 5040, 40320, 362880 };
// fac[i]=i!   ..
int order(const char *s, int n) {     // the alphabet order for
permutations s
```

$$k_s = \sum_{i=0}^{n-1} (s[i] - 1)! * \text{(the number of elements after } s[i] \text{ and less than}$$

s[i]),
$0 \le k_s \le n!-1$

```cpp
    int i, j, temp, num;
    num = 0;     // the alphabet order 0
    for (i = 0; i < n-1; i++) {     //from right to left,
enumeration
```

```
            temp = 0;      //the number of elements after s[i] and
less than s[i] temp
        for (j = i + 1; j < n; j++) {
          if (s[j] < s[i]) temp++;
        }
        num += fac[s[i] -1] * temp;      //accumulating
(s[i]-1)!*temp
    }
    return num;      //return the alphabet order for
permutations for s
}
bool is_equal(const char *b1, const char *b2){      // permutations
b1 and b2 are equal or not
    for(int i=0; i<9; i++)   //scan 9 positions
        if(b1[i] != b2[i])
            return false;
    return true;       //b1 and b2 are equal
}
struct node{      //States
    char board[9];      //permutation
    char space;      //position for space
};
const int TABLE_SIZE = 362880;      //the upper limit for the
number of states 9!
int hash(const char *cur){      // the alphabet order for
permutation cur
    return order(cur, 9);
}
void get_node(int num, node &tmp) {      //map alphabet order
num to state tmp
    int n=9;      //the length for the permutation
    int a[9];
    for (int i = 2; i <= n; ++i) {
        a[i - 1] = num % i;
        num = num / i;
        tmp.board[i - 1] = 0;
    }
    tmp.board[0] = 0;
    int rn, i;
    for (int k = n; k >= 2; k--) {
        rn = 0;
        for (i = n - 1; i >= 0; --i) {
            if (tmp.board[i] != 0) continue;
            if (rn == a[k - 1]) break;
            ++rn;
        }
        tmp.board[i] = k;
    }
    for (i = 0; i < n; ++i)
        if (tmp.board[i] == 0) {
```

```
            tmp.board[i] = 1;
            break;
        }
    tmp.space = n - a[n-1] -1;
}
int goal_state[9][2] = {{0,0}, {0,1}, {0,2}, {1,0}, {1,1},
{1,2}, {2,0}, {2,1}, {2,2}};      //In the goal state, number i
is at (goal_state[i][0], goal_state[i][1])
int h(const char *board){     // calculate h(board): the
estimated cost from state board to the goal state
 int k;
 int hv = 0;     //sum of distances
 for(int i=0; i<3; ++i)      //search each position (i, j)
  for(int j=0; j<3; ++j){
   k = i*3+j;
   if(board[k] !=9){      //the number in position isn't 9
     hv += abs(i - goal_state[board[k]-1][0]) +abs(j - goal_
state[board[k] -1][1]);
   }
  }
return hv;     //return the sum of distances
}
int f[TABLE_SIZE], d[TABLE_SIZE];     // Heuristic Function
f[u]: the cost to the goal state, from the initial state and
through state u, d[u]: the minimal number of moves from the
initial state to state u,  (u is the alphabet order for
permutations)
struct cmp{     //Comparison function for the priority queue:
Heuristic Function f[u] is the key, sorting in ascending
order
    bool operator () (int u, int v){
        return f[u] > f[v];
    }
};
```

$$
// \; color[u] = \begin{cases} 0 & u \text{ hasn't been visited} \\ 1 & u \text{ is in the queue} \\ 2 & u \text{ has been removed from the queue} \end{cases} , \text{ in the}
$$

best path the precursor for u is *parent*[u], the direction
enter u is *move*[u], u is the alphabet order for states

```
char color[TABLE_SIZE];
int parent[TABLE_SIZE];
char move[TABLE_SIZE];
int step[4][2] = {{-1, 0},{1, 0}, {0, -1}, {0, 1}};
//direction i (step[i][0], step[i][1])
void A_star(const node & start){     //from the initial state
start, A* algorithm is used to calculate a series of moves
    int x, y, k, a, b;
    int u, v;
```

```
      priority_queue<int, vector<int>, cmp> open;        //priority
queue open(Heuristic Function f[u] is the key, sorting in
ascending order)
      memset(color, 0, sizeof(char) * TABLE_SIZE);       //there is
one state in the queue
      u = hash(start.board);
      parent[u] = -1;      //the precursor for u is empty
      d[u] = 0;       //number of moves for u is 0
      f[u] = h(start.board);      // Heuristic Function f[u]
      open.push(u); color[u] = 1;       //u is added into the
priority queue
      node tmp, cur;      //the front for the queue cur, generated
state tmp
      while(!open.empty()){       //while the queue isn't empty, u
is the state with the minimal f[u]
            u = open.top();
            if(u == 0) return;       //u is the goal state
            open.pop();
            get_node(u, cur);       // the alphabet order u
corresponds to state cur
            k = cur.space;       //in state cur, the space position
(x, y)
            x = k / 3; y = k % 3;
            for(int i=0; i<4; ++i){       //search 4 directions
                  a=x+step[i][0]; b=y+step[i][1];       //the adjacent
position (a, b) for (x, y) in direction i
                  if(0<=a && a<=2 && 0<=b && b<=2){       //(a, b) is
the puzzle
                        tmp = cur;       //the space at (x, y) and the
number at (a, b) are exchanged, and the state tmp is generated
                        tmp.space = a*3 + b;
                        swap(tmp.board[k], tmp.board[tmp.space]);
                        v = hash(tmp.board);       // the alphabet order
v for state tmp
                        if(color[v]==1 &&(d[u]+1)<d[v]){       //v is in
the queue, and d[u]+1<d[v], (u, v) is adjusted in the best
path
                              move[v] = i;       //the direction i entering v
                              f[v]=f[v]-d[v]+d[u]+1;
                              d[v] = d[u] + 1;
                              parent[v] = u;
                              open.push(v);       //v is added into the
queue
                        }
                        else if(color[v]==2 && (d[u]+1)<d[v]){       //4
directions for v have been searched, and d[u]+1<d[v], (u, v)
is adjusted in the best path
                              move[v] = i;      /the direction i entering v
                              f[v]=f[v]-d[v]+d[u]+1;
```

```
                        d[v] = d[u] + 1;
                        parent[v] = u;
                        open.push(v);     //v is added into the
queue again
                        color[v] = 1;     //v is in the queue
                    }
                else if(color[v] == 0){     // v hasn't been
visited
                        move[v] = i;     //direction i entering v
                        d[v] = d[u] + 1;
                        f[v]=d[v]+h(tmp.board);
                        parent[v] = u;
                        open.push(v);     //v is added into the
queue
                        color[v] = 1;     //v is in the queue
                }
            }
        }
        color[u] = 2;     //4 directions for u have been searched
    }
}
void print_path(){     // output a series of moves
    int n, u;
    char path[1000];     // a series of moves
    n = 1;     //number of moves
    path[0] = move[0];
    u = parent[0];     //the precursor for the goal state u
    while(parent[u] != -1){     //from the goal state to the
initial state
        path[n] = move[u];     //the direction entering state u
        ++n;     //number of moves
        u = parent[u];     //the precursor
    }
    for(int i=n-1; i>=0; --i){     //output a series of moves
from the initial state
        if(path[i] == 0) printf("u");
        else if(path[i] == 1) printf("d");
            else if(path[i] == 2) printf("l");
                else printf("r");
    }
}
int main(){
    //freopen("in", "r", stdin);
    node start;     //initial state
    char c;     //input character
for(int i=0; i<9; ++i){     //input 9 characters, the initial
state start is constructed
    cin>>c;
        if(c == 'x'){     //the i-th character "x" is 9, it's a
space
```

```
            start.board[i] = 9;
            start.space = i;
    }
    else  start.board[i]=c-'0';
}
A_star(start);     // A* algorithm is used to calculate a
series of moves
    if(color[0]!= 0)print_path();     // if the goal state has
been visited, output a series of moves; else output "unsolvable-"
        else printf("unsolvable");
    return 0;
}
```

9.2.5 Remmarguts' Date

"A good man never makes girls wait or breaks an appointment!" said the mandarin duck father. Softly touching his little ducks' heads, he told them a story.

"Prince Remmarguts lives in his kingdom UDF—United Delta of Freedom. One day their neighboring country sent them Princess Uyuw on a diplomatic mission."

"Erenow, the princess sent Remmarguts a letter, informing him that she would come to the hall and hold commercial talks with UDF if and only if the prince would go and meet her via the K-th shortest path. (In fact, Uyuw does not want to come at all)."

Being interested in the trade development and in such a lovely girl, Prince Remmarguts really became enamored. He needs you—the prime minister's—help!

Details: UDF's capital consists of N stations. The hall is numbered S, while the station numbered T denotes the prince's current place. M muddy directed sideways connect some of the stations. Remmarguts' path to welcome the princess might include the same station twice or more than twice, even it is the station with number S or T. Different paths with the same length will be considered disparate.

Input

The first line contains two integer numbers N and M ($1 \le N \le 1000$, $0 \le M \le 100000$). Stations are numbered from 1 to N. Each of the following M lines contains three integer numbers A, B and T ($1 \le A$, $B \le N$, $1 \le T \le 100$). It shows that there is a directed sideway from the A-th station to the B-th station with time T.

The last line consists of three integer numbers S, T, and K ($1 \le S$, $T \le N$, $1 \le K \le 1000$).

Output

A single line consisting of a single integer number: the length (time required) to welcome Princess Uyuw using the *K*-th shortest path. If *K*-th shortest path does not exist, you should output "−1" (without quotes) instead.

Sample Input	Sample Output
2 2	14
1 2 5	
2 1 4	
1 2 2	

Source: POJ Monthly, Zeyuan Zhu

ID for Online Judge: POJ 2449

 Analysis

The problem can be represented as a weighted directed graph *G*. Stations in UDF are represented as vertices, muddy directed sideways connecting some of the stations are represented as arcs, and the time cost on a sideway is represented as the weight for the corresponding arc.

The problem requires you to calculate the length of the *K*-th shortest path from the starting point to the terminal point in *G*.

The naive algorithm solving the problem is using BFS from the starting point. When the terminal point is searched *K* times, the length of the path is the time required to welcome Princess Uyuw using the *K*-th shortest path. If *K* is larger or the number of vertices is more, the solution will consume more memory than permitted.

The method to solve the problem is using algorithm calculating the single-source shortest paths and A* search.

Step 1: For *G*, its converse digraph *G'* is constructed. Then the terminal point *T* is as the single-source, and an SPFA algorithm is used to calculate the lengths of the shortest paths from the terminal point *T* to other vertices. It is used for the estimated cost.

$$
H[i] = \begin{cases} \text{the length of the shortest path from vertex } i \text{ to } T & T \text{ is reachable from } i \\ \infty & T \text{ isn't reachable from } i \end{cases}
$$

Step 2: A* search is used to calculate the length of the *K*-th shortest path.

A* search can be used to calculate the shortest path. Therefore the *K*-th shortest path can be calculated in the *K*-th times.

The key to A* search is to design the function for the estimated cost $F[i]$: $F[i]=G[i]+H[i]$, where $G[i]$ is the length of the shortest path from the starting point *S* to vertex *i*, and $H[i]$ is the length of the shortest path from vertex *i* to the terminal point *T*, calculated in **Step 1**. Therefore, the estimated cost $F[i]$ is the length of the shortest path from *S* to *T*, through vertex *i*.

Each time the minimal value is gotten out from $F[\,]$. Therefore $F[\,]$ is stored as a priority queue.

Initially the starting point *S* is added into the priority queue $F[\,]$. Then A* search is used. Each time, the vertex with the minimal value is obtained from $F[\,]$. For each adjacent vertex *i*, $G[i]$ is calculated, $F[i]$ is calculated based on $H[i]$, and the number of times that Prince Remmarguts goes through vertex *i* is accumulated.

1. For each vertex, the number of paths that Prince Remmarguts goes through it is at most *K*.
2. If *T* isn't reachable from some vertices, in $H[\,]$ the vertices' values are ∞.

 Program

```
#include<stdio.h>
#include<iostream>
#include<queue>
#include<vector>
using namespace std;
#define inf 99999999
#define N 1100
typedef struct nnn     //the struct for priority queue
{
   int F,G,s;     //vertex s, G: the length from the starting
point to s, F; the length of the path through s
   friend bool operator<(nnn a,nnn b)   //the priority for the
priority queue
   {
      return a.F>b.F;
   }
}PATH;
typedef struct nn                   // adjacency list
{
   int v,w;     //adjacent vertex v, the length of the arc w
}node;
```

```
vector<node>map[N],tmap[N];      // adjacency list map[ ],
associated adjacency list tmap[ ], where map[i] and tmap[i]
stores adjacent vertices for vertex i
int H[N];     // H[ ]: lengths of the shortest paths from the
terminal point T to other vertices
void findH(int s)     //SPFA algorithm to calculate H[ ]
{
  queue<int>q;
  int inq[N]={0};
  q.push(s); inq[s]=1; H[s]=0;     //s is added into queue q
  while(!q.empty())     //the front s for q is removed
  {
    s=q.front(); q.pop(); inq[s]=0;
    int m=tmap[s].size();     //out-degree for s
    for(int i=0;i<m;i++)     //enumerate arcs from s
    {
      int j=tmap[s][i].v;     //for s, the other vertex j for
the i-th arc
      if(H[j]>tmap[s][i].w+H[s])
      {
        H[j]=tmap[s][i].w+H[s];     //H[j]
        if(!inq[j]) inq[j]=1,q.push(j);     //j isn't an
element in the queue, is added into the queue
      }
    }
  }
}
int Astar(int st,int end,int K)     // calculate the length of
the K-th shortest path from the starting point st to the
terminal point end
{
  priority_queue<PATH>q;     //priority queue q, elements' type
is PATH
  PATH p,tp;
  int k[N]={0};     //k[ ]: the number of times through
vertices
  findH(end);     //calculate H[ ]
  if(H[st]==inf)return -1;     //end is reachable from st
  p.s=st; p.G=0; p.F=H[st];     // p is added into the priority
queue
  q.push(p);
  while(!q.empty())     //the priority queue isn't empty
  {
    p=q.top(); q.pop();
    k[p.s]++;     //the number of times through the vertex +1
    if(k[p.s]>K)continue;     // the number of times through
the vertex at most K
    if(p.s==end&&k[end]==K) return p.F;     // the number of
times arriving at the terminal point is K, return the length
of the path
```

```
    int m=map[p.s].size();    //degree for p.s
    for(int i=0;i<m;i++)      //arcs from p.s are enumerated
    {
      int j=map[p.s][i].v;    //vertex j for the i-th arc
      if(H[j]!=inf)    //the terminal point is reachable from j
      {
        tp.G=p.G+map[p.s][i].w;
        tp.F=H[j]+tp.G;
        tp.s=j;
        q.push(tp);
      }
    }
  }
  return -1;
}
int main()
{
  int n,m,S,T,K,a,b,t;
  node p;
  scanf("%d%d",&n,&m);    //numbers of vertices and edges
  for(int i=1;i<=n;i++)
    {
      map[i].clear(); tmap[i].clear(); H[i]=inf;  //Initialize
adjacency list and H[]
    }
    while(m--)    // m muddy directed sideways
    {
      scanf("%d%d%d",&a,&b,&t);    // arc (a,b) with length t
is stored in map[a]
      p.v=b; p.w=t; map[a].push_back(p);    //flip arc (b,a)
with length t is stored in map[b]
      p.v=a; tmap[b].push_back(p);
    }
    scanf("%d%d%d",&S,&T,&K);    // the starting point, the
terminal point, and K
    if(S==T)K++;    // the starting point and the terminal
point are same
    printf("%d\n",Astar(S,T,K));
}
```

Iterative deepening A* Algorithm (IDA* Algorithm)

Iterative deepening A* (IDA*) is a graph traversal and path search algorithm that can find the shortest path between a designated start node and any member of a set of goal nodes in a weighted graph. It is a variant of iterative deepening depth-first search that borrows the idea to use a heuristic function to evaluate the remaining cost to get to the goal from the A* search algorithm. Since it is a depth-first search algorithm, its memory usage is lower than in A*, but unlike ordinary iterative deepening search, it concentrates on exploring the most promising nodes and thus doesn't go to the same depth everywhere in the search tree. Unlike A*, IDA* doesn't utilize dynamic programming and therefore often ends up exploring the same nodes many times.

The keys to the IDA* algorithm are bounding and pruning.

9.2.6 Jaguar King

In a deep forest, a war is going to begin. Like other animals, the jaguars are preparing for this ultimate battle. Though they are mighty strong and lightning fast, they have an extra advantage over other animals. It's their wise and brave king, the Jaguar King.

The king knows that only speed and strength is not enough for winning the war. They have to make a perfect formation. The king has set up a nice formation and placed all the **Jaguar Warriors** according to that formation. There are N positions for N jaguar warriors (including the king). The king is marked by 1 and the other jaguar warriors are marked by a number from 2 to N. The warriors are placed according to their number.

But then the king realizes that to make the formation perfect and effective, some positions should have stronger jaguars and some should have faster jaguars. Since the strength and speed of all the jaguars are not equal, the king decided to change the positions of some jaguars. The wise king knows the ability of each and every jaguar, so his decision is perfect, but the problem is how to change the jaguars.

One of the wise jaguars has given an idea. The idea is simple. All the jaguars will wait for the king's signal, all eyes upon the king. Suppose the king is in the i-th position. The king jumps to the j-th position, and when the jaguar at the j-th position sees the king coming, he immediately jumps to the i-th position. The king repeats this procedure until they are formatted like the new formation. Now there is another problem. Collision can occur while jumping. So, some wise jaguars have made a jumping scheme so that no collision can occur. The scheme is noted below.

```
If the king is in the i-th position
  If (i % 4=1) The king can jump to position (i+1), (i+3),
(i+4), (i-4)
  If (i % 4=2) The king can jump to position (i+1), (i-1),
(i+4), (i-4)
  If (i % 4=3) The king can jump to position (i+1), (i-1),
(i+4), (i-4)
  If (i % 4=0) The king can jump to position (i-3), (i-1),
(i+4), (i-4)
```

Any position is valid if it lies in between 1 and N.

Actually the king can jump much higher between these positions so, no collision can occur. Now you are one of the prisoners (actually they kept you to eat after the war). You now have a chance to get out alive. You know all their ideas and the new formation.

If you can tell the king the minimum number of times the king has to jump to gain the new formation, they could be generous and release you.

Input

The input file contains several sets of inputs. The total number of sets will be less than 50. The description of each set is given below:

Each set starts with one integer N ($4 \leq N \leq 40$) which indicates the total number of jaguar warriors. You can assume that N is a multiple of 4. The next line will contain N numbers which indicates the final formation of the jaguars. Consecutive numbers will be separated by a single space.

The input will be terminated by the set where $N=0$. And this set should not be processed.

Output

For each set in the input, you should first print the set number starting from 1. And the next line should be the minimum number of times the king has to jump to gain the new formation.

Check the sample input-output for more details. Output should be formatted like the sample output.

Hope you get out alive.

Sample Input	Sample Output
4	Set 1:
1 2 3 4	0
4	Set 2:
4 2 3 1	1
8	Set 3:
5 2 3 4 8 6 7 1	2
8	Set 4:
5 2 8 3 6 7 1 4	7
0	

Source: Next Generation Contest III

ID for Online Judge: UVA 11163

 Analysis

Initially N jaguars are placed at N positions in a queue, $N\%4==0$. The king is placed at position 1. Other jaguars are marked by a number from 2 to N. The jaguars are placed according to their number. Only the king can exchange its position with another jaguar. Suppose the king's current position is i:

If (i % 4==1), then the king can jump to position (i+1), (i+3), (i+4), (i−4);
If (i % 4==2) , then the king can jump to position (i+1), (i−1), (i+4), (i−4);
If (i % 4==3), then the king can jump to position (i+1), (i−1), (i+4), (i−4);
If (i % 4 = 0), then the king can jump to position (i−3), (i−1), (i+4), (i−4).

Given the final formation of the jaguars, the minimum number of times the king jumps to gain the final formation is required to calculate.

An array $dx[i][j]$ is used to represent the above rule, where i is the remainder that the king's current position is divided by 4 ($0 \leq i \leq 3$), and each remainder is a kind of the king's jumping; and j is the sequence number for the kind of the king's jumping ($0 \leq j \leq 3$). That is, int $dx[4][4] = \{\{-3, -1, +4, -4\}, \{+1, +3, +4, -4\}, \{+1, -1, +4, -4\}, \{+1, -1, +4, -4\}\}$. Therefore, if the king's current position is k, the king can jump to four positions, where the j-th position is $k+dx[k \% 4][j]$, $0 \leq j \leq 3$.

The estimated cost for the IDA* algorithm is $F(v)=G(v)+H(v)$. $F(v)$ is the estimated number of times that the king jumps to gain the final formation through the current state v. $G(v)$ is the number of times that the king jumps to the current state v from the initial formation. Initially, the king's state is x, $G(x)=0$. And $H(v)$ is the estimated cost that the king jumps to gain the final formation from the state v. $H(v)$ is the sum of the numbers of times that $n-1$ jaguars (not the king) jump to their current positions from their final positions using rule $dx[i \% 4][0..3]$, for $n-1$ jaguars (not the king) jump in the opposite directions for the king's jumping. A Floyd algorithm is used to calculate $H(x)$ before the IDA* algorithm is used.

In the IDA* algorithm, states are represented as (x, *prev*, *dep*, *hv*); where x is the current position for the king; *prev* is the precursor position for the king, that is, the king jumps to position x from position *prev*, and initially *prev*=−1; *dep* is the number of times the king jumps to position x, that is, *dep* is $G(x)$, and initially, *dep*=0; and *hv* is the estimated number of times that the king jumps to gain the final formation from the position x, that is, *hv* is $H(v)$.

In the program, the function *IDA*(x, *prev*, *dep*, *hv*) is used to calculate the minimum number of times *mxdep* the king jumps to gain the final formation. In order to improve the efficiency, the following optimization strategies are used:

1. Bounding. The current minimum number of times *mxdep* that the king jumps to gain the final formation is as the bounding. Based on the estimated cost, if the king can jump to gain the final formation through the current state, the estimated cost is *dep+hv*. Obviously, if *dep+hv>mxdep*, it can't be a solution.
2. The king can't return to *prev* from x. That is, the endless loop needs to be avoided.

3. There are at most four positions to which the king can jump. Therefore, there are at most four generated states. The function *IDA*(*x*, *prev*, *dep*, *hv*) returns the minimal value *submxdep* that is *min* {numbers of times for the king's jumping to gain the final formation in four directions}.

The function *mxdep* =*IDA*(*x*, *prev*, *dep*, *hv*) is as follows:

```
if (hv==0) return the number of times the king jumps dep and
the successful mark;
if (dep+hv> mxdep) it can't be a solution;
submxdep=∞;
Four positions that the king can jump are enumerated tx(tx =
x+dx[x%4][i], 0≤i≤3):
  {
    if (tx is in the bound)&&(tx≠prev)
    { the number of times the king jumps to gain the final
formation is calculated;
      The generated state (tx, x, dep+1, shv) is recursively
calculated, and the number of times the king jumps is tmp;
      submxdep= min(submxdep, tmp);
    }
  };
Return submxdep;
IDA (x, -1, 0, E(s)) is called to calculate the number of
times the king jumps mxdep until the final formation is gained.
```

The size of the problem isn't large (4≤*N*≤40).

 Program

```c
#include <stdio.h>
#include <algorithm>
#include <string.h>
using namespace std;
int g[45][45];    // g[i][j]: the length of the shortest path
between position i and position j
int dx[4][4] = {{-3,-1,+4,-4},{+1,+3,+4,-4},{+1,-1,+4,-4},
{+1,-1,+4,-4}};    // if the king's current position is k, the
king can jump to 4 positions, where the j-th position is
k+dx[k %4][j]
int A[45], n;    // A[i] is the number of the jaguar at
position i in the final formation, the numder of jaguars n
void build() {    //using Floyd algorithm to calculate g[ ][ ]
    int i, j, k;
    for(i = 1; i <= 40; i++) {    //initialize g[ ][ ]
```

```
        for(j = 1; j <= 40; j++) g[i][j] = 0xffffffff;
        g[i][i] = 0;
    }
    for(i = 1; i <= 40; i++)      //each position i is
enumerated
        for(j = 0; j < 4; j++)      //4 positions can be jumped
            if(i+dx[i%4][j]>0 && i+dx[i%4][j]<=40)    // in the
bound
                g[i][i+dx[i%4][j]] = 1;
    for(k = 1; k <= 40; k++)
        for(i = 1; i <= 40; i++)
            for(j = 1; j <= 40; j++)
                g[i][j] = min(g[i][j], g[i][k]+g[k][j]);
//adjust the length of the shortest path from i to j
}
int H() {     //accumulate the minimal numbers of times that
n-1 jaguars (not the king) jump to initial positions from
final positions
    int i, sum = 0;     //initialize the sum of the minimal
numbers of times sum 0
    for(i=1; i <= n; i++){     //enumerate each jaguar (not the
king) position in the final formation, and accumulate
        if(A[i] == 1) continue;
        sum += g[i][A[i]];
    }
    return sum;
}
int singleH(int pos) {     // the minimal numbers of times that
jaguar at pos jumps to A[pos]
    return g[pos][A[pos]];
}
int mxdep, solved;     //mxdep, the minimum number of times the
king jumps to gain the final formation
int IDA(int x, int prev, int dep, int hv) {     // IDA(x, prev,
dep, hv) is defined in the analysis
    if(hv == 0){     //return the minimum number of times the
king jumps to gain the final formation
        solved = dep;
        return dep;
    }
    if(dep+hv>mxdep) return dep+hv;     //it isn't the solution
    int i, tx;
    int submxdep = 0xffffffff, shv, tmp;
    for(i = 0; i < 4; i++) {     // 4 positions to which the
king can jump
        tx = x+dx[x%4][i];     //the i-th position tx
        if(tx==prev || tx<=0 || tx>n) continue;     // (tx isn't
in the bound) or (tx=prev), the next position
        shv=hv;               /hv+dist[x][tx]-dist[x][A[tx]]
        shv -= singleH(tx);
```

```
        swap(A[x], A[tx]);
        shv += singleH(x);
        tmp = IDA(tx, x, dep+1, shv);    //the king jumps from tx
        if(solved)return solved;    //gain the final formation
        swap(A[x], A[tx]);
        submxdep=min(submxdep,tmp);    //adjust the current
minimal number of times
    }
    return submxdep;    //return the current minimal number of
times
}
int main() {
    int i, j, k, x;
    int cases = 0;    //the number of test cases
    build();    // Floyd algorithm is used to calculate the
matrix for the length of shortest path g[ ][ ]
    while(scanf("%d", &n) == 1 && n) {    // the total number
of jaguar warriors
        for(i=1; i<=n; i++) scanf("%d", &A[i]);    // the
final formation of the jaguars
        printf("Set %d:\n", ++cases);    //output the number
of test cases
        int initH = H();    // accumulate the minimal numbers
of times that n-1 jaguars (not the king) jump to initial
positions from final positions
        if(initH == 0) puts("0");    //initial formation is
the final formation, the king needn't jump
        else {
            solved = 0;    //initialize the minimal numbers
of times the king jumps
            mxdep = initH;    //initialize
            for(i = 1; i <= n; i++) if(A[i] == 1) x=i;
            while(solved==0) mxdep=IDA(x,-1,0,initH);
            printf("%d\n", solved);
        }
    }
    return 0;
}
```

9.3 A Game Tree Used to Solve a Game Problem

In holographic, zero-sum, turn-taking, and two-player games, the Game Tree and the Minimax Algorithm are used to find the best steps.

"Holographic" means the game's information is clear to two players. For example, Chinese Chess, Chess, and Weiqi are holographic games.

"Zero-sum" means the sum of interests of two players is 0: either A defeats B, B defeats A, or A and B end in a draw.

The state space for a game can be represented as a game tree, where nodes represent states, and edges represent moves between states. The root for a game tree is the initial state. For a node in a game tree, its children are states generated by next possible moves. Leaves are goal states for a game.

For two players, "Minimax" means that one player makes a move to maximize his or her utility, and minimize the utility for his or her opponent. The Minimax algorithm is to calculate the minimal or maximal value, that is, the best move, based on the current state. If the size for the game tree isn't large, DFS can be used to get all possible moves, and to calculate the best move for the current state. Otherwise, pruning should be used to eliminate some parts of the tree.

9.3.1 Find the Winning Move

4×4 tic-tac-toe is played on a board with four rows (numbered 0 to 3 from top to bottom) and four columns (numbered 0 to 3 from left to right). There are two players, *x* and *o*, who move alternately, with *x* always going first. The game is won by the first player to get four of his or her pieces on the same row, column, or diagonal. If the board is full and neither player has won, then the game is a draw.

Assuming that it is *x*'s turn to move, *x* is said to have a **forced win** if *x* can make a move such that no matter what moves *o* makes for the rest of the game, *x* can win. This does not necessarily mean that *x* will win on the very next move, although that is a possibility. It means that *x* has a winning strategy that will guarantee an eventual victory regardless of what *o* does.

Your job is to write a program that, given a partially completed game with *x* to move next, will determine whether *x* has a forced win. You can assume that each player has made at least two moves, that the game has not already been won by either player, and that the board is not full.

Input

The input file contains one or more test cases, followed by a line beginning with a dollar sign that signals the end of the file. Each test case begins with a line containing a question mark and is followed by four lines representing the board; formatting is exactly as shown in the example. The characters used in a board description are the period (representing an empty space), lowercase *x*, and lowercase *o*.

Output

For each test case, output a line containing the (*row, column*) position of the *first* forced win for *x*, or '#####' if there is no forced win. Format the output exactly as shown in the example.

For this problem, the *first* forced win is determined by board position, not the number of moves required for victory. Search for a forced win by examining positions (0, 0), (0, 1), (0, 2), (0, 3), (1, 0), (1, 1), ..., (3, 2), (3, 3), in that order, and output the first forced win you find. In the second test case below, note that *x* could win immediately by playing at (0, 3) or (2, 0), but playing at (0, 1) will still ensure victory (although it unnecessarily *delays* it), and position (0, 1) comes first.

Sample Input	Sample Output
?	#####
....	(0,1)
.xo.	
.ox.	
....	
?	
o...	
.ox.	
.xxx	
xooo	
$	

Source: University of Valladolid Local Contest

IDs for Online Judges: UVA 10111

 Analysis

There are 16 squares in a 4×4 tic-tac-toe. There are three possibilities: empty, occupied by *x*, and occupied by *o*. Therefore, state compression can be used to represent states. A 16-digit ternary number *state* is used to represent a state for a 4×4 tic-tac-toe. In *state*, the *i*×4+*j*-th digit represents square (*i*, *j*):

If square (*i*, *j*) is "*x*", then the *i*×4+*j*-th digit for *state* is 1. That is, *state* |= 1UL<<((*i*×4+*j*)×2);

If square (*i*, *j*) is "*o*", then the *i*×4+*j*-th digit for *state* is 2. That is, *state* |= 2UL<<((*i*×4+*j*)×2);

If square is ".", then the *i*×4+*j*-th digit for *state* is 0.

For two players, *x* and *o*, there are 10 situations for winning a game, respectively: four situations getting four pieces on the same row, four situations getting four pieces on the same column, and two situations getting four pieces on the same diagonal. For *x*, 10 situations winning a game are stored in *xw*[10]. For *o*, 10 situations winning a game are stored in *ow*[10].

Four pieces on the same row or column:

```
Four pieces on the same row (0≤i≤3):
    for(j = 0; j < 4; j++) {
        xw[n] |= 1UL<<((i*4+j)*2);      //For x, the
i-th row is 1, x wins the game.
        ow[n] |= 2UL<<((i*4+j)*2);      //For o, the
i-th row is 1, o wins the game.
    }
    n++;
```

Four pieces on the same column (0≤*j*≤3):

```
    for(j = 0; j < 4; j++) {
        xw[n] |= 1UL<<((j*4+i)*2);      //For x, the
i-th column is 1, x wins the game.
        ow[n] |= 2UL<<((j*4+i)*2);      //For o, the
i-th column is 1, o wins the game.
    }
    n++;
```

Four pieces on the left diagonal:

```
for(i = 0; i < 4; i++) {
        xw[n] |= 1UL<<((i*4+i)*2);      //For x, the left
diagonal is 1, x wins the game.
        ow[n] |= 2UL<<((i*4+i)*2);      // For o, the left
diagonal is 1, o wins the game.
    }
    n++;
```

Four pieces on the right diagonal:

```
for(i = 0; i < 4; i++) {
        xw[n] |= 1UL<<((i*4+3-i)*2);    // For x, the
right diagonal is 1, x wins the game.
        ow[n] |= 2UL<<((i*4+3-i)*2);    // For o, the
right diagonal is 1, o wins the game.
    }
    n++;
```

In order to avoid repeated searches, for each state, a mark *R*[ ] is set. If *turn* will win board *node*, then *R*[*node*]=1; else *R*[*node*]=0.

Because the size for the board is small, the search depth is limited. Assuming that it is *x*'s turn to move, *x* is said to have a ***forced win*** if *x* can make a move such that no matter what moves *o* makes for the rest of the game, *x* can win.

A recursive function *dfs(node, rx, ry, turn)* is used to calculate the result for *turn*; where *node* represents the board, initially *node* is a test case; *turn*: 1 represents *x*, and 2 represents *o*. Obviously, after *turn* makes a move, *3-turn* makes the next move; initially *turn* is 1; *(rx, ry)* is a position of the forced win; initially *(rx, ry)* is (−1,−1). *dfs(node, rx, ry, turn)* returns the result for *turn*. If the result is 0, then *turn* is defeated; and if the result is 1, then *turn* must win, and the position of the forced win *(rx, ry)* is also returned.

```
dfs(node, rx, ry, turn);
{
   if ( board node has appeared before) return R[node];
   if (board node is a situation winning a game) return 0 (turn
is defeated);
   for each empty square(i, j) (0≤i, j≤3, node>>((i*4+j)*2))&3==0)
     { dfs(node', rx, ry, 3-turn);     //turn selects (i, j), the
new board node'=node|(turn<<((i*4+j)*2)), then 3-turn makes
the next move
       if (dfs returns 0){     //3-turn is defeated, turn wins node'
         (i, j) is as the position of the first forced win
(rx, ry);
         R[node']=1;     //board node' wins
         return 1;     //turn wins
       }
     }
   return 0;     //turn is defeated
}
```

Program

```
#include <stdio.h>
#include <string.h>
#include <map>
using namespace std;
map<unsigned int, int> R;     // mark R[ ], where R[x] is for
board x
unsigned int ow[10] = {}, xw[10] = {};     //xw[10]: 10
situations that x wins the game, ow[10]: 10 situations that o wins
the game
int check(unsigned int node) {     //the result for board node,
```

$$
check(node) = \begin{cases} 0 & undetermined \\ 1 & x \ wins \\ 2 & o \ wins \end{cases}
$$

```
   int i;
   for(i = 0; i < 10; i++)     //x wins, return 1
```

```
        if((node&xw[i]) == xw[i]) return 1;
        for(i = 0; i < 10; i++)      // o wins, return 2
        if((node&ow[i]) == ow[i]) return 2;
        return 0;      //undetermined
}
int dfs(unsigned int node, int &rx, int &ry, unsigned int
turn) {      //node is the current board, returns the result for
turn. If the result is 0, then there is no forced win for
turn; else turn must win, and the position of the first forced
win (rx, ry) is also returned
        if (R.find(node)!=R.end()) return R[node];// board node
has appeared before
        int f=check(node);      //if board node can determine the
result, return 0
        if (f) return 0;
        int i, j;
        int &ret = R[node];
        for(i = 0; i < 4; i++) {      //From top to down, from left
to right, search empty square
          for(j = 0; j < 4; j++) {
          if((node>>((i*4+j)*2))&3) continue;//if (i, j) isn't
empty, continue to search; else turn makes a move at (i, j),
new board node|(turn<<((i*4+j)*2)), 3-turn makes the next move
          f = dfs(node|(turn<<((i*4+j)*2)), rx, ry, 3-turn);  // new
board
          if(f == 0)                  {   // if in new board turn
wins, (i, j) is the position of the first forced win
          rx = i, ry = j;
          ret = 1;
          return 1;      //return the mark that turn wins
          }
        }
      }
    return 0;      // return the mark that turn is defeated
}
int main() {
    char end[10], g[10][10];
    int i, j, n = 0;
    //x->1, o->2
    for(i = 0; i < 4; i++) {      //4 situations getting pieces
on the same row, 4 situations getting pieces on the same
column
          for(j = 0; j < 4; j++) {
              xw[n] |= 1UL<<((i*4+j)*2);      //For x, the i-th
row is 1, x wins the game.
              ow[n] |= 2UL<<((i*4+j)*2);      // For o, the i-th
row is 1, o wins the game.
          }
          n++;
          for(j = 0; j < 4; j++) {
```

```
            xw[n]  |= 1UL<<((j*4+i)*2);     // For x, the i-th
column is 1, x wins the game.
            ow[n]  |= 2UL<<((j*4+i)*2);     // For o, the i-th
column is 1, o wins the game.
    }
    n++;
    }
    for(i = 0; i < 4; i++) {     // left diagonal
        xw[n]  |= 1UL<<((i*4+i)*2);     // For x, the left
diagonal is 1, x wins the game.
        ow[n]  |= 2UL<<((i*4+i)*2);     // For o, the left
diagonal is 1, o wins the game.
    }
    n++;
    for(i = 0; i < 4; i++) {     // right diagonal
        xw[n]  |= 1UL<<((i*4+3-i)*2);     // For x, the right
diagonal is 1, x wins the game.
        ow[n]  |= 2UL<<((i*4+3-i)*2);     // For o, the right
diagonal is 1, o wins the game.
    }
    n++;
    while(scanf("%s", end)==1) {     //Input test cases until
end mark "$"
        if(end[0] == '$') break;
        for(i = 0; i < 4; i++) scanf("%s", g[i]);     //4 rows
        unsigned int state = 0;     // initial state 0
        for(i = 0; i < 4; i++) {     //From top to down, and
from left to right construct the initial state
            for(j = 0; j < 4; j++) {
                if(g[i][j] == '.') {}     //(i, j) is ".", then
omit
                else if(g[i][j] == 'x')     //if (i, j) is "x",
then the i*4+j-th digit for state is 1
                    state |= 1UL<<((i*4+j)*2);
                else     // if (i, j) is "o", then the i*4+j-th
digit for state is 2
                    state |= 2UL<<((i*4+j)*2);
            }
        }
        int rx = -1, ry = -1;     // initialize the position of
the first forced win for x
        int f = dfs(state, rx, ry, 1);
        if(f == 0)     //no forced win for x
            puts("#####");
        else
            printf("(%d,%d)\n", rx, ry);     // the position of
the first forced win for x (rx, ry)
    }
    return 0;
}
```

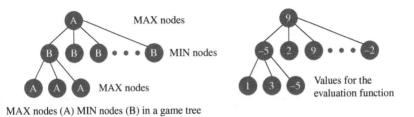

MAX nodes (A) MIN nodes (B) in a game tree

Figure 9.8

There are some games whose search depths are larger. Therefore, we need to restrict the search. For the current state, a score is evaluated. The method for evaluating such scores is called an evaluation function. Suppose there are two players A and B for a game. A makes the first move. In the game tree, nodes that specify A's moves are called "MAX nodes", for A always moves to a state with the maximum score. And nodes that specify B's moves are called "MIN nodes", for B always moves to a state with the minimum score. That is to say, both A and B play optimally (Figure 9.8).

In order to get the result more quickly, a kind of heuristic method, called α–β (alpha-beta) pruning, is used in DFS in a game tree.

For a MAX node in the game tree, the value of the best choice (i.e., the maximum score) is obtained from leaves through DFS, called the value of α.

For a MIN node in the game tree, the value of the best choice (i.e., the minimum score) is obtained from leaves through DFS, called the value of β.

The DFS updates values of α and value of β as it goes along and prunes remaining branches at a node as soon as the value of the current node is known to be worse than the current α or β value for MAX node or MIN node respectively.

α pruning: When DFS is used in a game tree, the score from leaves in the leftmost branch to a MAX node A is denoted as α. And the score α is as the lower bound for A's score.

Then DFS is used for other children for A. If after a round (two moves), a score is lower than α, the corresponding branch (the subtree whose root is the child for A) is pruned [Figure 9.9(a)]. In Figure 9.9(b), scores are in an ascending order from left to right, and no branch is pruned.

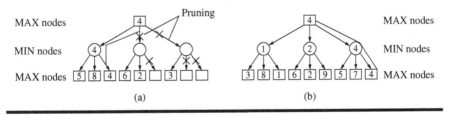

(a)   (b)

Figure 9.9

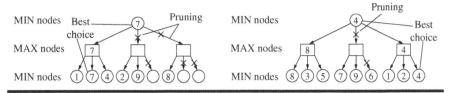

Figure 9.10

β pruning: By the same reason, the score from leaves in the leftmost branch to a MIN node B is denoted as β. Obviously, the score β is as the upper bound for B's score.

Then DFS is used for other children for B. If after a round (two moves), a score is higher than β, the corresponding branch (the subtree whose root is the child for B) is pruned. See Figure 9.10.

9.3.2 The Pawn Chess

Consider the following mini-version of chess: We have a 4×4 chessboard, with four white pawns on the first rank (bottom line in the input) and four black pawns on the last rank. The goal is for the player to get one of his pawns to the other end of the board (last rank for the white player, first rank for the black player), *or* to stalemate his opponent. A player is stalemated if he has no legal moves when it's his turn to move (this includes having no pawns left).

The pawns move as in ordinary chess, except that they can never move two steps. That is, a pawn may move one step forward (toward the opposite rank), assuming that this square is empty. A pawn can also capture a pawn of the opposite color if it's one step ahead and to the left or right. Captured pieces are removed from the game.

Given the position of the pawns on a chessboard, determine who will win the game, assuming both players play optimally. You should also determine the number of moves the game will last before it's decided (assuming the player who will win tries to win as fast as possible and the player to lose will lose as slowly as possible). It will always be white's turn to move from the given position.

Input

The first line in the input contains the number of test cases (at most **50**). Each case contains four lines describing the chessboard, preceded by a blank line. The first of the four lines will be the last rank of the chessboard (the starting point for the black pawns). Black pawns will be denoted with a '**p**', white pawns with a '**P**', and empty squares with **a** '.'. There will be between one and four pawns of each color. The initial position will not be a final game position, and white will always have at least one legal move. Note that the input position may not necessarily be one that could have arisen from legal play from the games starting position.

Output

For each test case, output a line containing the text white (*xx*) if white will win, or black (*xx*) if black will win. Replace *xx* with the number of moves (which will always be an odd number if white wins and an even number if black wins).

Sample Input	Sample Output
2	white (7)
.ppp	black (2)
....	
.PPP	
....	
...p	
...p	
pP.P	
...P	

Source: ACM ICPC World Finals Warmup 2 (2004–2005)

IDs for Online Judges: UVA 10838

 Analysis

The problem requires you to calculate the winner and its number of moves. It is a game problem. And α–β (alpha-beta) pruning is used to improve the search efficiency.

DFS is used to construct a game tree, where nodes represent states for the 4×4 chessboard, and the root is the input test case. Goal states, that is, leaves, are states that one player uses to get one of his pawns to the other end of the board (last rank for the white player, first rank for the black player), or to stalemate his opponent. Middle games are internal nodes.

The level number for the root of the game tree is *depth*=36. As it moves down one level, *−−depth*. At even levels, it is white's turn to move. And at odd levels, it is black's turn to move. The game tree is constructed top-down. Based on game theory, at even levels, states of the maximum score should be calculated; and at odd levels, states of the minimum score should be calculated. Suppose *alpha*(*v*) is the maximum number of moves for white from *v* to leaves. That is, *alpha*(*v*)=max$_{u∈v's\,children}${the number of moves from *u* to leafs}. The initial value for *alpha* is −9999.

beta(*v*) is the minimum number of moves for black from *v* to leaves. *beta*(*v*)=min$_{u∈v's\,children}${*the number of moves from u to leafs*}. The initial value for *beta* is 9999.

The value of *alpha* (or the value of *beta*) for each node is gotten from leaves. Therefore, when it is white's turn to move, return the value of *alpha* for the internal nodes; and return −*depth* for leaves. And when it is black's turn to move, return the value of *beta* for internal nodes; and return *depth* for leaves.

During the process calculating *alpha*(*v*) and *beta*(*v*), if for a parent node *v*, *alpha*(*v*) is calculated; for its children, the minimal *beta* is calculated. If *u* is a child for *v*, *beta*(*u*) is *p*; and *u'* is another child for *v*; *u"* is a child for *u'*, and return *q*<*p*, then the subtree whose root is *u'* can be pruned.

Repeat the process until a goal state is reached.

If the recursion result is a positive number, white wins, and the number of moves is 36 − the recursion result. And if the recursion result is a negative number, black wins, and the number of moves is 36 + the recursion result.

 Program

```c
#include <stdio.h>
#include <algorithm>
using namespace std;
struct state {      // the chessboard
  char g[4][5];     // the chessboard
  int isEnd() {     // if the chessboard is the final state,
return 1; else return 0
    int b = 0, w = 0;
    for (int i = 0; i < 4; i++) {      //the number of white w
and the number of black b
      for (int j = 0; j < 4; j++)  {
        if (g[i][j] == 'p') b++;
        if (g[i][j] == 'P') w++;
      }
    }
    if (b == 0 || w == 0) return 1;    // having no pawns
left, return 1
    for (int i = 0; i < 4; i++) {      // get one of his pawns
to the other end of the board, return 1
      if (g[0][i] == 'P') return 1;
```

```
        if (g[3][i] == 'p') return 1;
    }
    return 0;      //else return 0
  }
};
int alpha_beta(state board,int depth,int alpha,int beta) {  //the
current state for the chessboard board, level depth, alpha and
beta is as the definition
  if (board.isEnd())return depth%2==0 ?-depth:depth;    //if
the current state for the chessboard is a final state, if
depth is even (white) return -depth; else (black) return depth
   int movable = 0;    //initialize the moveable sign
   if (depth% 2 == 0) {    // it is white's turn to move
     for (int i = 1; i < 4; i++) {     // each pawn's position
is searched
       for (int j = 0; j < 4; j++) {
       if (board.g[i][j] == 'P') {    // at (i, j) there is a 'P'
         for (int k = j - 1; k <= j + 1; k++) {    //3 moveable
positions for (i, j) are searched
           if (k < 0 || k >= 4) continue;    // the position is
out of the chessboard
           if ((k != j && board.g[i-1][k] == 'p') || (k == j &&
board.g[i-1][k] == '.')) {
             state s=board;    // new state for chessboard s: (i, j)
is empty, 'p' is moved to (i-1, k)
             s.g[i][j] = '.', s.g[i-1][k] = 'P';
             alpha=max(alpha,alpha_beta(s,depth-1,alpha,beta));
// recursion, calculate alpha
             if (beta<=alpha) return alpha;
             movable = 1;    // the sign for move
           }
         }
       }
     }
    }
    }
    if (!movable)return -depth;    // leaf (white has no legal
moves), return -depth
return alpha;    //return alpha
   } else {    // it is black's turn to move
     for (int i = 0; i < 3; i++) {    //each pawn's
position is searched
       for (int j = 0; j < 4; j++) {
       if (board.g[i][j] == 'p') {    //at (i, j) there is a 'p'
         for (int k=j-1; k<=j+1; k++) {    //3 moveable
positions for (i, j) are searched
           if (k < 0 || k >= 4)continue;    //the position is
out of the chessboard
           if ((k != j && board.g[i+1][k] == 'P') || (k == j &&
board.g[i+1][k] == '.')) {
```

```
            state s=board;    //new state for chessboard s: (i, j)
is empty, 'p' is moved to (i+1, k)
            s.g[i][j] = '.', s.g[i+1][k] = 'p';
            beta=min(beta,alpha_beta(s,depth-1,alpha,beta));
//recursion, calculate beta
            if (beta <= alpha) return beta;
            movable = 1;     //the sign for move
          }
        }
      }
    }
  }
    if(!movable)return depth;     //leaf (black has no legal
moves), return depth
    return beta;     //return beta
  }
}
int main() {
  int testcase;
  scanf("%d", &testcase);     //number of test cases
  while (testcase--) {
    state init;    //initial chessboard
    for (int i = 0; i < 4; i++) scanf("%s", init.g[i]);
    int ret = alpha_beta(init, 36, -9999, 9999);
    if (ret >= 0) printf("white (%d)\n", 36-ret);     // the
number of moves (white wins)
      else printf("black (%d)\n", 36+ret);     // the number
of moves (black wins)
  }
  return 0;
}
```

9.4 Problems

9.4.1 The Most Distant State

The 8puzzle is a square tray in which eight square tiles are placed. The remaining ninth square is uncovered. Each tile has a number on it. A tile that is adjacent to the blank space can be slid into that space. A game consists of a starting state and a specified goal state. The starting state can be transformed into the goal state by sliding (moving) the tiles around. The 8puzzle problem asks you to do the transformation in a minimum number of moves, as shown in Figure 9.11.

However, our current problem is a bit different. In this problem, given an initial state of the puzzle, you are asked to discover a goal state which is the most distant (in terms of number of moves) of all the states reachable from the given state.

2	8	3
1	6	4
7		5

⇨

1	2	3
8		4
7	6	5

Start   Goal

Figure 9.11

Input

The first line of the input file contains an integer representing the number of test cases to follow. A blank line follows this line.

Each test case consists of three lines of three integers each, representing the initial state of the puzzle. The blank space is represented by a 0 (zero). A blank line follows each test case.

Output

For each test case, first output the puzzle number. The next three lines will contain three integers each representing one of the most distant states reachable from the given state. The next line will contain the shortest sequence of moves that will transform the given state to that state. The move is actually the movement of the blank space represented by four directions : U (Up), L (Left), D (Down), and R (Right). After each test case, output an empty line.

Sample Input	Sample Output
1 2 6 4 1 3 7 0 5 8	Puzzle #1 8 1 5 7 3 6 4 0 2 UURDDRULLURRDLLDRRULULDDRUULDDR

Source: BUET/UVA World Finals Warm-up

ID for Online Judge: UVA 10085

 Hint

The problem is a special 8-puzzle. Normally the 8-puzzle requires you to calculate the minimum number of moves from a starting state to a goal state. In this problem, there is a starting state, but there isn't a specified goal state. You need to discover a goal state which is the most distant (in terms of number of moves) of all the states reachable from the starting state.

Normally the 8-puzzle is solved by BFS. In this problem, a *while* repetition ends when the queue is empty, to guarantee that the most distant of all the states reachable from the starting state can be calculated.

A queue q is used to store states. The type for a state is *struct*, including: a state for the current nine squares $ch[3][3]$, represented as a nine-digit novenary number; the position for the square containing 0 (x, y); and the sequence of moves that transforms from the starting state to the current state *str*;

Suppose $m[]$ is used to show whether a state has been visited or not. That is,

$$m[p] = \begin{cases} 1 & \text{state } p \text{ has appeared before} \\ 0 & \text{state } p \text{ hasn't appeared before} \end{cases}. \text{ It can be defined as map<long}$$

long,int> m;

Function $hash(p)$ is used to determine whether a state has visited or not.

```
int hash(p) {
    long long cnt, k;
    cnt=k=0;
    for(int i=0; i<N; i++))      // calculate the value cnt for
state p
        for(int j=0; j<N; j++) cnt+=p.ch[i][j]*pow(9, k++);
    if (!m[cnt]) {      //if cnt hasn't appeared
        m[cnt]=1;
        return 1;
    }
    return 0;
}
```

BFS is used to solve the problem.

```
void bfs() {
  initial state st is pushed into the queue q;
  while (q isn't empty) {
    the front st is poped from the queue;
    Four directions are enumerated (0≤i≤3):
      {
    Calculate the position (x1, y1) which the blank space can
be slid into from (st.x, st.y) in direction i;
      if ((x1, y1) is out of the puzzle) continue;
      st1=st;    //new state st1
      st1.ch[st1.x][st1.y]=st1.ch[x1][y1]; st1.ch[x1][y1]=0;
      st1.x=x1; st1.y=y1;
      st1.str+= the character for direction i;
      if (hash(st1)) q.push(st1);    //if st1 hasn't been
visited, push st1 into the queue
      }
  }
}
```

Figure 9.12

The main program is as follows:

1. Initialization;
   m.clear();
   st.str is empty;
   Set *st* visited mark (*hash*(*st*));
2. *bfs*(): search all reachable states;
3. For the last state *st* removed from the queue, output *st.ch*[3][3] and *st.str*.

9.4.2 15-Puzzle Problem

The 15-puzzle is a very popular game; even if you don't know it by that name, you've seen it. It is constructed with 15 sliding tiles, each with a number from 1 to 15 on it, and all packed into a 4×4 frame with one tile missing. The object of the puzzle is to arrange the tiles so that they are ordered as shown in Figure 9.12.

Here the only legal operation is to exchange a missing tile with one of the tiles with which it shares an edge. As an example, in Figure 9.13, the following sequence of moves changes the status of a puzzle.

The letters in the previous row indicate which neighbor of the missing tile is swapped with it at each step; legal values are 'R', 'L', 'U', and 'D', for "RIGHT", "LEFT", "UP", and "DOWN", respectively.

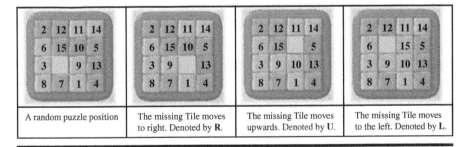

Figure 9.13

Given an initial configuration of a 15-puzzle, you will have to determine the steps that would make you reach the final stage. The input for 15-puzzles requires at most 45 steps to be solved with our judge solution. So you will not be allowed to use more than 50 steps to solve a puzzle. If the given initial configuration is not solvable, you just need to print the line "This puzzle is not solvable."

Input

The first line of the input contains one integer *N*, which indicates how many sets of puzzles will be given as input. The next 4*N* lines contain *N* sets of inputs. It means four lines make one set of input. Zero denotes the missing tile.

Output

For each set of input, you will have to give one line of output. If the input puzzle is not solvable, then print the line "This puzzle is not solvable." If the puzzle is solvable, then print the move sequence as described above to solve the puzzle.

Sample Input	Sample Output
2	LLLDRDRDR
2 3 4 0	This puzzle is not solvable.
1 5 7 8	
9 6 10 12	
13 14 11 15	
13 1 2 4	
5 0 3 7	
9 6 10 12	
15 8 11 14	

Source: 2001 Regionals Warmup Contest

ID for Online Judge: UVA 10181

Hint

The initial state is 16 sliding tiles, numbered from 0 to 15, are packed into a 4×4 frame. (The tile numbered 0 means there is no tile at the position). The goal state is as shown in Figure 9.12. Only the tile numbered 0 can be exchanged with the tile with which it shares an edge. It can be regarded as the tile moving horizontally or vertically. The problem requires you to calculate the steps from the initial state to the goal state.

Tiles' positions are also numbered from 0 to 15. A position $(x, y)=(p/4, p\%4)$, $0 \le p \le 15$.

First, we determine whether a puzzle is solvable or not.

If the title numbered 0 is moved horizontally, the sequence for the other 15 numbers can't be changed. And if the tile number 0 is moved vertically, the inverse number's parity for the sequence for other 15 numbers will be changed (+−3 or +−1). And the tile numbered 0 should be moved to the lower right corner. If a puzzle is solvable, the parity for the inverse number for the sequence for 15 numbers plus the number of rows at which the title numbered 0 is the same as the parity for the inverse number for the sequence for 15 numbers in the goal state. The inverse number for the sequence for 15 numbers in the goal state is an even.

Therefore, if the inverse number for the sequence for 15 numbers plus the number of rows at which the tile numbered 0 is an even, the puzzle is solvable; else the puzzle is not solvable.

For example, in Figure 9.14, the tile numbered 0 is at the end of the fourth row. The first tile is numbered 13, and the number of tiles whose number is less than 13 after the first title is 12. The first tile is denoted as 12(13). The second tile is numbered 10, and the number of tiles whose number is less than 10 after the second title is 9. The second title is denoted as 9(10). Similarly, other titles are denoted as 9(11), 5(6), 4(5), 4(7), 3(4), 3(8), 0(1), 3(12), 3(14), 2(9), 1(3), 1(15), and 0(2).

The inverse number for the sequence for 15 numbers is 12+9+9+5+4+4+3+3+ 0+3+3+2+1+1+0=59. Because 59+4=63 is an odd, this puzzle is not solvable.

The function to determine whether a puzzle is solvable or not is as follows:

```
bool solvable ( )      //Whether the puzzle is solvable or not
{
    int cnt = 0;
    for (int i = 0; i < 16; ++i){     //each position
      if (puz[i]==0) cnt+=3-i/4;      // the row at which the
title numbered 0 is (puz[i]; the number at poistion i)
        else{
              for (int j=0; j<i; ++j)  // inverse number
                  if (puz[j] && puz[j] > puz[i]) cnt++;
        }
    }
    return !(cnt&1);      //even, return true; else return false
}
```

There are three algorithms for solving the problem.

13	10	11	6
5	7	4	8
1	12	14	9
3	15	2	0

Figure 9.14

Solution 1. DFS Algorithm

In DFS, there may be many produced states. The maximal search depth should be fixed. The input 15-puzzles requires at most 45 steps to be solved with our judge solution. Therefore the maximal search depth can be set to 50.

The key to DFS is to determine efficiently whether a state has been visited or not. In this problem, a state is represented as a hexadecimal number. A tile is represented as a digit for the hexadecimal number. All tiles are sorted from left to right and from top to bottom. For example, Figure 9.14 is represented as a hexadecimal number:

$$13 \times 16^{15} + 10 \times 16^{14} + 11 \times 16^{13} + 6 \times 16^{12} + 5 \times 16^{11} + 7 \times 16^{10} + 4 \times 16^9 + 8 \times 16^8$$
$$+ 1 \times 16^7 + 12 \times 16^6 + 14 \times 16^5 + 9 \times 16^4 + 3 \times 16^3 + 15 \times 16^2 + 2 \times 16^1 + 0 \times 16^0$$
$$= (DAB657481CE93F20)_{16} = (15759879913263939360)_{10}.$$

If two states' values are same, the two states are same. Hash technology is used to determine whether two states are same or not.

For this problem, the DFS algorithm is as follows, where *path*[] is the sequence for steps moving the tile whose number is 0.

```
void dfs(p, d) // p is the position for the tile whose
number is 0, and the state with search depth d
{
    if (d>50) return fail;
    if (puz[] is the goal state) output path[] and return;
    (x, y) is the position for the title whose number is 0;
    4 directions are enumerated (0≤i≤3):
    {
        (x', y') is the new position for the title whose
number is 0 moving in direction i and p' is the position
value;
        if ((x', y') is in the puzzle){
        puz[p] and puz[p'] are exchanged;
        Calculate the state value s for puz[] and the hash
function h(s);
        if ( in hash table hash[h(s)] there is no state
whose value is s) {
            path[d]= the value for direction i;      //the
value for the dth step
            dfs(p', d+1,);      //recursion for the next
step
            puz[p] and puz[p'] are exchanged;      //the
state before the recursion
            path[d] = ` `;
        }
    }
}
}
```

The main program is as follows:

```
    Input the initial state for the puzzle puz[], and
calculate the position value p for the title whose number is 0;
    if (solvable()) dfs(p, 0);    //if the puzzle is
solvable
         else output "This puzzle is not solvable.";
```

Solution 2. BFS Algorithm

BFS is to try to find a shortest path. If there exists a path from the initial state to the goal state, BFS can find the shortest path from the initial state to the goal state. The difference between DFS and BFS is that DFS uses the system stack (the stack in a recursion) to store visited states, and BFS uses a queue to store visited states.

For this problem, the BFS algorithm is as follows, where *ch*[][] represents states for the puzzle, (x, y) is the position for the title whose number is 0, and *str* is the move sequence.

```
The initial state st: the input puzzle st.ch[][], the
initial position for the title whose number is 0 (st.x,
st.y), and st.str=' ';
if (the puzzle isn't solvable) output "This puzzle is not
solvable.";
  else {
            st is added into q;
         while(q isn't empty) {
         the front st for q is removed;
            if (st is the goal state){ output st.str;
break;};
            Four directions are enumerated (0≤i≤3):
            {
            the title whose number is 0 moves to (x1, y1)
from (st.x, st.y) in direction i;
            if ((x1, y1) is out of the puzzle) continue;
            st1=st;    //new state st1
            st1.ch[st1.x][st1.y]=st1.ch[x1][y1];
st1.ch[x1][y1]=0;
            st1.x=x1; st1.y=y1;
            st1.str+= the value representing direction i;
            if (there is no state st in the hash table)
st1 is added into the hash table and q;
            }
         }
      }
```

When the number of steps is less than 15, BFS and DFS can get the solution to the problem quickly. As the number of steps increases, the spent time and memory for BFS and DFS will increase.

Solution 3: IDA* Algorithm

1. Heuristic Function

   In this problem, the heuristic function is used to restrain the search depth. The heuristic function $f^*(n)=g^*(n)+h^*(n)$, where $g^*(n)$ is the minimal number of steps moving from the initial puzzle to the current puzzle n; $h^*(n)$ is the minimal number of steps moving from the current puzzle n to the goal puzzle; that is, the sum of Manhattan Distances from all numbers' current positions to their goal positions. The function $h^*(n)$ is as follows:

```
int h( )
{
    int s = 0;
    for (int i = 0; i < 16; ++i){    //each position is
enumerated
        x is the number at position i;
        if (x == 0) continue;
        s+= abs(i/4 - the row number for x in the goal
state)+abs(i%4 - the column number for x in the goal state);
    }
    return s;
}
```

2. Determine whether the current puzzle is solvable or not under the limited search depth.

   There is no function to determine repetitions in the IDA* algorithm. In IDA* algorithm, each search step can't move in the opposite direction.

   The Boolean function *dfs*(*p, pre, g, maxd*) is used to determine whether the current puzzle is solvable or not under the limited search depth, where p is the position for the title whose number is 0, *pre*: the last moving direction, g is the current search depth, and *maxd* is the limit for the search depth.

```
bool dfs(p, pre, g, maxd)
{
    if(g+h()>maxd)return false;    //the search depth
will exceed the limit
    if(g == maxd)    //the search depth reaches the
limit, return the comparison result for the current and
the goal puzzle
        return memcmp(the current puzzle, the goal
puzzle, sizeof(the goal puzzle))==0;
    calculate the position (x, y) for the tile whose
number is 0, that is x=p/4, y=p%4
        Four directions are enumerated (0≤j≤3):
        If (pre+j==3) continue;    //j is the opposite
direction for the previous move (x', y') is the new
position for the tile whose number is 0 in direction j
and its position value is p';
```

```
        if ((x', y')  in the puzzle){
            the number at position p and the number at
position p' are exchanged;
            path[g] = the character for direction j;  //the
character for the gth step
            if(dfs(p', j, g+1, maxd))return true;  //if the
puzzle is solvable
            the number at position p and the number at
position p' are exchanged;    //recovering the state
before the recursion
        }
    }
    return false;
}
```

3. The main program.

The above function *dfs(p, pre, g, maxd)* is the kernel program for the IDA* algorithm. Based on that, the main program is as follows:

```
A puzzle is input, and the position value p for the title
whose number is 0;
    if(solvable()){
            int maxd = 0;    //initialize the
search depth
            for(;!dfs(p,-1,0,maxd); ++maxd);  //find
the solution as the search depth increases
            path[maxd] =0,
            output the move sequence path[];
    }
    else  output "This puzzle is not solvable.";
```

The IDA* algorithm is more efficient than DFS or BFS.

9.4.3 Addition Chains

An addition chain for n is an integer sequence $<a_0, a_1, a_2, ..., a_m>$ with the following four properties:

$a_0=1$

$a_m=n$

$a_0<a_1<a_2<...<a_{m-1}<a_m$

For each k ($1 \leq k \leq m$) there exist two (not necessarily different) integers i and j ($0 \leq i, j \leq k-1$) with $a_k=a_i+a_j$

You are given an integer n. Your job is to construct an addition chain for n with minimal length. If there is more than one such sequence, any one is acceptable.

For example, <1,2,3,5> and <1,2,4,5> are both valid solutions when you are asked for an addition chain for 5.

Input

The input file will contain one or more test cases. Each test case consists of one line containing one integer n ($1 \leq n \leq 100$). Input is terminated by a value of zero (0) for n.

Output

For each test case, print one line containing the required integer sequence. Separate the numbers by one blank.

Hint

The problem is a little time-critical, so use proper break conditions where necessary to reduce the search space.

Sample Input	Sample Output
5	1 2 4 5
7	1 2 4 6 7
12	1 2 4 8 12
15	1 2 4 5 10 15
77	1 2 4 8 9 17 34 68 77
0	

Source: Ulm Local Contest 1997

IDs for Online Judges: POJ 2248, UVA 529

Hint

Given an integer n, the problem requires you to construct an addition chain for n with minimal length.

Obviously, it is inefficient that only DFS is used. IDA* (DFS + pruning) is used to solve the problem. Suppose *best* is the current length of the addition chain. Initially *best*=∞. In array $d[]$, $d[i]$ is the maximal number of integers that can be added into the addition chain after integer i.

The addition chain is constructed from a_0=1. In order to make the length of the addition chain minimal, each time the maximum that can be added into is considered first. That is, if the length of the addition chain is $k+1$, and the current maximum in the addition chain is a_k, then the possible maximum for a_{k+1} is $a_k+a_k=2a_k$. Therefore, the upper limit for the length of the addition chain after a_k is extended is $k+d[a_k]$. Therefore, $d[i]=\begin{cases} 0 & n \le i \le 2*n \\ 1+d[2*i] & 1 \le i \le n-1 \end{cases}$.

The IDA* process solving the problem is as follows:

```
void DFS(k)        //extend the addition chain from a[k]
{
    if (k+d[a[k]]>=best) return;      //The minimal addition
chain can't be generated from a[k], backtracking
    if (a[k]==n)     // The addition chain is generated
    {
      best = k;
      a[]is stored in b[]; ;
      return;
    }
  for (i=k; i>=0; i--)      //Enumerate each pair of a[i] and
a[j] from a[k] to a[0]
    for (j=k; j>=i; j--)
      {
      a[k+1]=a[i]+a[j];      // a[k+1] is the sum of a[i] and a[j]
      if (a[k+1]>a[k] && a[k+1]<=n)  DFS(k+1);      //if a[k+1]
meets requirements, the addition chain is extended from a[k+1]
      }
}
```

The main program is as follows:

```
d[] is constructed;
    best =∞; a[0] = 1;
    DFS(0);      // the addition chain is extended from a[0]
    Output b[0]...b[best];
```

9.4.4 Bombs! No, They Are Mines!!

It's the year **3002**. The robots of **"ROBOTS 'R US (R:US)"** have taken control over the world. You are one of the few people who remain alive only to act as their guinea pigs. From time to time, the robots use you to find if they have been able to become more intelligent. You, being the smart guy, have always been successful in proving to be more intelligent.

Today is your big day. If you can beat the fastest robot in the **IRQ2003** land, you'd be free. These robots are intelligent. However, they have not been able to overcome a major deficiency in their physical design—they can only move in four directions: Forward, Backward, Upward, and Downward. And they take one unit of time to travel one unit of distance. As you have only one chance, you're planning it thoroughly. The robots have left one of the fastest robots to guard you. You'd need to program another robot which would carry you through the rugged terrain. A crucial part of your plan requires you to find how much time the guard robot would need to reach your destination. If you can beat him, you're through. See Figure 9.15.

We must warn you that the **IRQ2003** land is not a pleasant place to roam. The R:US have dropped numerous bombs while they invaded the human land. Most of the bombs have exploded. Still some of the bombs remain, acting as land mines. We have managed to get a map that shows the unsafe regions of the **IRQ2003** land; unfortunately your guard has a copy of the map, too. We know that at most 40 percent of the area can be unsafe. If you are to beat your guard, you'd have to find his fastest route long before he finds it.

Input

Input consists of several test cases. Each test begins with two integers R ($1 \leq R \leq 1000$), C ($1 \leq C \leq 1000$)—they give you the total number of rows and columns in the grid map of the land. Then follow the grid locations of the bombs. It starts with the number of *rows*, ($0 \leq rows \leq R$) containing bombs. For each of the rows, you'd have one line of input. These lines start with the row number, followed by the number of bombs in that row. Then you'd have the column locations of that many bombs in that row. The test case ends with the starting location (row, column) followed by your destination (row, column). All the points in the region are in the range **(0,0)** to **(R–1, C–1)**. Input ends with a test case where $R=0$ and $C=0$, and you must not process this test case.

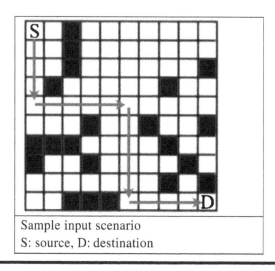

Sample input scenario
S: source, D: destination

Figure 9.15

Output

For each test case, print the time the guard robot would take to go from the starting location to the destination.

Sample Input	Sample Output
10 10	18
9	
0 1 2	
1 1 2	
2 2 2 9	
3 2 1 7	
5 3 3 6 9	
6 4 0 1 2 7	
7 3 0 3 8	
8 2 7 9	
9 3 2 3 4	
0 0	
9 9	
0 0	

Source: UVA Local and May Monthly Contest (2004)

ID for Online Judge: UVA 10653

Hint

Given an *R*×*C* grid map of the land, a robot can only move in four directions: <u>F</u>orward, <u>B</u>ackward, <u>U</u>pward, and <u>D</u>ownward. There are bombs in some grids and the robot can't move into the grids. The problem requires you to calculate the shortest path from the starting location to the destination.

BFS is used to calculate the shortest path. A state is represented as (x, y, s), where your current location is (x, y), and the distance from the starting location to (x, y) is s; and the visited sequence is represented as $Vist[][]$, where $Vist[x][y]$ is the mark that (x, y) is visited.

The process for BFS is as follows:

```
The initial state (the starting location, 0) is added into
the queue;
Vist[the starting location]=1;
while (the queue isn't empty) {
  The front for the queue p is removed from the queue;
  4 directions are enumerated (0≤i≤3):
```

```
    {    a new state q is generated: the new location (q.x,
q.y) and the distance q.s(=p.s+1);
if ((q.x, q.y) is in the grid map)&&(! Vist[q.x][q.y])&&( there
is no bomb at (q.x, q.y)){
        Vist[q.x][q.y]=1;
        The new state q is added into the queue;
        }
        if ((q.x,q.y) is the destination) return q.s;
    }
}
```

9.4.5 Jugs

In the movie "Die Hard 3", Bruce Willis and Samuel L. Jackson were confronted with the following puzzle. They were given a three-gallon jug and a five-gallon jug and were asked to fill the five-gallon jug with exactly four gallons. This problem generalizes that puzzle.

You have two jugs, *A* and *B*, and an infinite supply of water. There are three types of actions that you can use: (1) you can fill a jug, (2) you can empty a jug, and (3) you can pour from one jug to the other. Pouring from one jug to the other stops when the first jug is empty or the second jug is full, whichever comes first. For example, if *A* has five gallons and *B* has six gallons and a capacity of eight gallons, then pouring from *A* to *B* leaves *B* full and three gallons in *A*.

A problem is given by a triple (*Ca*, *Cb*, *N*), where *Ca* and *Cb* are the capacities of the jugs *A* and *B*, respectively, and *N* is the goal. A solution is a sequence of steps that leaves exactly *N* gallons in jug *B*. The possible steps are

```
fill A
fill B
empty A
empty B
pour A B
pour B A
success
```

where "pour *A B*" means "pour the contents of jug *A* into jug *B*", and "success" means that the goal has been accomplished.

You may assume that the input you are given does have a solution.

Input

Input to your program consists of a series of input lines each defining one puzzle. Input for each puzzle is a single line of three positive integers: *Ca*, *Cb*, and *N*. *Ca* and *Cb* are the capacities of jugs *A* and *B*, and *N* is the goal. You can assume $0 < Ca \le Cb$ and $N \le Cb \le 1000$ and that *A* and *B* are relatively prime to one another.

Output

Output from your program will consist of a series of instructions from the list of the potential output lines which will result in either of the jugs containing exactly *N* gallons of water. The last line of output for each puzzle should be the line "success". Output lines start in column 1 and there should be no empty lines nor any trailing spaces.

Sample Input	Sample Output
3 5 4 5 7 3	fill B pour B A empty A pour B A fill B pour B A success fill A pour A B fill A pour A B empty B pour A B success

Source: ACM South Central USA 1997

ID for Online Judges: POJ 1606, UVA 571

Hint

There are two methods to solve the problem.

Solution 1. Mathematic method.

An equation *ax−by=c* is used to represent the problem, where *a* and *b* are the capacities of the jugs *A* and *B* respectively; *x* and *y* are the numbers pouring water to jugs *A* and *B* respectively; and finally *c* gallons of water is left in jug *B*.

Each time, first water is poured into the jug with small capacity. Then water is poured into the jug with larger capacity. The solution with minimum integers is the solution to the problem. The equation can be solved by simulation.

```
Initially the two jugs A and B are empty;
 while (water in jug B isn't N gallons)
 {
    if (jug B is full){
    empty jug B;
```

```
      Output "empty B";
      }
  else if (jug A is empty)
      {
          fill jug A;
          output "fill A";
      }
      else
      {
        pour water from jug A to jug B;
          jug A is empty;
          if (water in jug B is more than its capacity)
          {
            extra water is poured into jug A;
            jug B is full;
          }
          Output "pour A B";
      }
  }
 Output "success";
}
```

Solution 2. BFS

1. The struct for a vertex p includes:

   A state (a, b, opr); where $p.a$ and $p.b$ are the current amount of water in jug A and jug B respectively, and $p.opr$ are six actions numbered 0~5, and represent "fill A", "fill B", "empty A", "empty B", "pour A B", and "pour B A" respectively;

   A precursor pointer $p.pre$ pointing to the state generating p. When the goal state (a, n, opr) is reached, a series of instructions can be output through $p.pre$:

   ```
   void Outpath(p);
       {
           if (p.pre != NULL) Outpath(*(p.pre));
           output the instruction whose index is p.opr;
       }
   ```

   ① The visited mark $vis[][]$, where $vis[a][b]$ shows the state that there are a gallons and b gallons water in jug A and jug B respectively has been visited;

2. States are added into the queue:
   Push $(\&t, h, a, b, opr)$ is used to add the state (a, b, opr) into the queue, where t is the pointer pointing to the rear, h is the pointer pointing to the front, and $vis[a][b]=1$, t++.

3. BFS is used to calculate a series of instructions

```
The initial state p (p.a=p.b=0, p.opr=-1) is added into the
queue, and the precursor pointer p.pre is set as empty, vis[0]
[0] = 1;
 while (the queue isn't empty)
 {
     p is the front for the queue;
     if (p.b==n)
   {
   Outpath (p)     //output the series of instructions
     output "success";
      return;
   }
    if (!vis[ca][p.b]) Push(t, h, ca, p.b, 0);    //if jug A
isn't full, fill A
    if (!vis[p.a][cb]) Push(t, h, p.a, cb, 1);    //if jug B
isn't full, fill B
    if (!vis[0][p.b]) Push(t, h, 0, p.b, 2);    // if jug A
isn't empty, empty A
    if (!vis[p.a][0]) Push(t, h, p.a, 0, 3);    // if jug B
isn't full, empty B
    ta=p.a; tb=p.b;
 // pour A B
  if (ta+tb<=cb) {tb+= ta; ta = 0; }
      else ta-=(cb - tb); tb=cb;
  if (!vis[ta][tb]) Push(t, h, ta, tb, 4);
// pour B A
  if (ta+tb<=ca){ta+=tb; tb=0; }
else{tb-=(ca-ta); ta=ca; };
  if (!vis[ta][tb]) Push(t, h, ta, tb, 5);
h ++;    //the front is removed from the queue
}
```

9.4.6 Knight's Tour Problem

You must have heard of the Knight's Tour problem. In that problem, a knight is placed on an empty chess board and you are to determine whether it can visit each square on the board exactly once.

Let's consider a variation of the Knight's Tour problem. In this problem, a knight is place on an infinite plane and it's restricted to make certain moves. For example, it may be placed at (0, 0) and can make two kinds of moves: Denote its current place as $(x; y)$, it can only move to $(x+1; y+2)$ or $(x+2; y+1)$. The goal of this problem is to make the knight reach a destination position as quickly as possible (i.e., make as few moves as possible).

Input

The first line contains an integer T ($T<20$) indicating the number of test cases.

Each test case begins with a line containing four integers: $f_x f_y t_x t_y$ ($-5000 \leq f_x$, f_y, t_x, $t_y \leq 5000$). The knight is originally placed at (f_x, f_y) and (t_x, t_y) is its destination.

The following line contains an integer m ($0 < m \leq 10$), indicating how many kinds of moves the knight can make.

Each of the following m lines contains two integers $m_x m_y$ ($-10 \leq m_x$, $m_y \leq 10$; $|m_x| + |m_y| > 0$), which means that if a knight stands at (x, y), it can move to $(x+m_x, y+m_y)$.

Output

Output one line for each test case. It contains an integer indicating the least number of moves needed for the knight to reach its destination. Output "IMPOSSIBLE" if the knight may never gets to its target position.

Sample Input	Sample Output
2	3
0 0 6 6	IMPOSSIBLE
5	
1 2	
2 1	
2 2	
1 3	
3 1	
0 0 5 5	
2	
1 2	
2 1	

Source: ACM 2010 Asia Fuzhou Regional Contest

IDs for Online Judges: POJ 3985, UVA 5098

 Hint

Given an infinite chess board, an initial position and a destination, and some kinds of moves the knight can make, you are required to calculate the least number of moves for the knight from the initial position to the destination.

The algorithm is achieved by combining BFS and pruning.

The structure for elements in the queue is *struct*, and contains the knight's current coordinate (x, y); and the distance s, from the knight's initial coordinate to knight's current coordinate.

Each time the front of the queue is removed, and m kinds of moves are enumerated. If the knight's new coordinate (x, y) is legal and isn't in the queue, then the new element $q(q.x=x', q.y=y', q.s=p.s+1)$ is added into the queue, else the branch is pruned.

Suppose the longest moving distance $d = \max\limits_{1\le i\le m}\left\{\left(m_{x_i}^2 + m_{y_i}^2\right)\right\}$; the initial position is (sx, sy) and the destination is (tx, ty); $a=ty-sy$; $b=sx-tx$; and $c=sy\times tx-sx\times ty$.

We analyze cases that the knight's new coordinate (x, y) are legal or not as follows:

Case 1: If the square of the Euclidean distance between (x, y) and the initial position (sx, sy) $((x-sx)^2+(y-sy)^2)$ isn't larger than d, then (x, y) is legal;

Case 2: If the square of the Euclidean distance between (x, y) and the destination (tx, ty) $((x-tx)^2+(y-ty)^2)$ isn't larger than d, then (x, y) is legal;

Case 3: If (x, y) deviates from the initial position (sx, sy), that is, $(tx-sx)\times(x-sx)+(ty-sy)\times(y-sy)<0$, then (x, y) isn't legal;

Case 4: If (x, y) deviates from the destination (tx, ty), that is, $(sx-tx)\times(x-tx)+(sy-ty)\times(y-ty)<0$, then (x, y) isn't legal;

Case 5: If the distance between (x, y) and the line segment $(sx, sy)\rightarrow(tx, ty)$ isn't more than d, that is, $(a\times x+b\times y+c)^2/(a^2+b^2)\le d$, then (x, y) is legal; else (x, y) isn't legal.

Therefore, the algorithm based on the above analysis is as follows:

```
if (case 1) then (x, y) is legal
else if (case 2) then (x, y) is legal
      else if (case 3) then (x, y) isn't legal
            else if (case 4) then (x, y) isn't legal
                  else if (case 5) then (x, y) is legal
                        else (x, y) isn't legal;
```

Hash technology is used to avoid repetitions. The hash function is $h(x,y)=((x<<15)\wedge y)\% \; 999997+999997)\% \; 999997$, where $(x<<15)\wedge y$ is a 32-digit binary number, the first 16 digits is x, the last 16 digits is y, and $h(x, y)$ is a positive integer. For each searched (x, y), we search whether there exists (x, y) in the hash table $head[h(x, y)]$.

9.4.7 Playing with Wheels

In this problem we will be considering a game played with four wheels. Digits ranging from 0 to 9 are printed consecutively (clockwise) on the periphery of each wheel. The topmost digits of the wheels form a four-digit integer. For example, in Figure 9.16, the wheels form the integer 8056. Each wheel has two buttons associated with it. Pressing the button marked with a *left arrow* rotates the wheel one digit

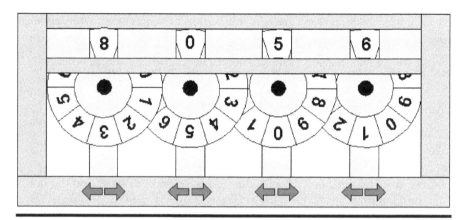

Figure 9.16

in the clockwise direction and pressing the one marked with the *right arrow* rotates it by one digit in the opposite direction.

The game starts with an initial configuration of the wheels. Suppose that in the initial configuration the topmost digits form the integer $S_1S_2S_3S_4$. You will be given some (say, n) forbidden configurations $F_{i1}F_{i2}F_{i3}F_{i4}$ ($1 \le i \le n$) and a target configuration $T_1T_2T_3T_4$. Your job will be to write a program that can calculate the minimum number of button presses required to transform the initial configuration to the target configuration by never passing through a forbidden one.

Input

The first line of the input contains an integer N giving the number of test cases to follow.

The first line of each test case contains the initial configuration of the wheels specified by four digits. Two consecutive digits are separated by a space. The next line contains the target configuration. The third line contains an integer n giving the number of forbidden configurations. Each of the following n lines contains a forbidden configuration.

There is a blank line between two consecutive input sets.

Output

For each test case in the input, print a line containing the minimum number of button presses required. If the target configuration is not reachable, then print −1.

Sample Input	Sample Output
2	14
8 0 5 6	−1
6 5 0 8	
5	
8 0 5 7	
8 0 4 7	
5 5 0 8	
7 5 0 8	
6 4 0 8	
0 0 0 0	
5 3 1 7	
8	
0 0 0 1	
0 0 0 9	
0 0 1 0	
0 0 9 0	
0 1 0 0	
0 9 0 0	
1 0 0 0	
9 0 0 0	

Source: BUET/UVA Occidental (WF Warmup) Contest 1

ID for Online Judge: UVA 10067

Hint

The problem can be represented as a connected graph, where each four-digit integer is represented as a vertex. Because each integer can become eight other integers through pressing buttons, the degree for each vertex is 8. In the graph, the weight of edges is 1. The initial configuration is as the initial state. And the target configuration is the goal state.

If we press a button marked with a *left arrow*, the corresponding digit becomes (the original digit +1)%10; and if we press a button marked with a *right arrow*, the corresponding digit becomes (the original digit +9)%10.

Because there are many test cases, the offline method can be used. First, the graph is constructed. Then, for each test case, vertices for forbidden configurations and their edges are deleted from the graph. Finally, the shortest path from the initial configuration's vertex to the target configuration's vertex is calculated by BFS or SPFA. If there is such a path, the length of the path is the minimum number of button presses required; else the target configuration is not reachable.

Bibliography

1. Wu Yonghui, Wang Jiande. Data Structure Practice: for Collegiate Programming Contest and Education (Second Edition). (Traditional Chinese Version). GOTOP Information Inc. 2017.
2. Wu Yonghui, Wang Jiande. Data Structure Practice: for Collegiate Programming Contest and Education. (English Version). CRC Press. 2016.
3. Wu Yonghui, Wang Jiande. Data Structure Practice: for Collegiate Programming Contest and Education (Second Edition). (Simplified Chinese Version). China Machine Press. 2016.
4. Wu Yonghui, Wang Jiande. Programming Strategies Solving Problems: for Collegiate Programming Contest and Education. (Simplified Chinese Version). China Machine Press. 2015.
5. Wu Yonghui, Wang Jiande. Programming Strategies Solving Problems: for Collegiate Programming Contest and Education. (Traditional Chinese Version). GOTOP Information Inc. 2015.
6. Wu Yonghui, Wang Jiande. Algorithm Design Experiment: for Collegiate Programming Contest and Education. (Simplified Chinese Version). China Machine Press. 2013.
7. Wu Yonghui, Wang Jiande. Solutions and Analyses to ACM-ICPC World Finals (2004–2011). (Simplified Chinese Version). China Machine Press. 2012.
8. Wu Yonghui, Wang Jiande. Data Structure Experiment: for Collegiate Programming Contest and Education. (Simplified Chinese Version). China Machine Press. 2012.
9. Wu Yonghui, Wang Jiande. Data Structure Experiment: for Collegiate Programming Contest and Education. (Traditional Chinese Version) GOTOP Information Inc. 2012.

Index